W9-BAV-435

Mathematics: Form and Function

Saunders Mac Lane

Mathematics
Form and Function

With 116 Illustrations

Springer-Verlag
New York Berlin Heidelberg Tokyo

Saunders Mac Lane
Department of Mathematics
University of Chicago
Chicago, Illinois 60637
U.S.A.

AMS Classifications: 00-01, 00A05, 00A06, 03A05

Library of Congress Cataloging in Publication Data
Mac Lane, Saunders
Mathematics, form and function.
Bibliography: p.
Includes index.
1. Mathematics—1961- . I. Title.
QA39.2.M29 1986 510 85-22160

Typeset by House of Equations Inc., Newton, New Jersey.
Printed and bound by R.R. Donnelley & Sons, Harrisonburg, Virginia.
Printed in the United States of America.

9 8 7 6 5 4 3 2 1

ISBN 0-387-96217-4 Springer-Verlag New York Berlin Heidelberg Tokyo
ISBN 3-540-96217-4 Springer-Verlag Berlin Heidelberg New York Tokyo

Preface

This book records my efforts over the past four years to capture in words a description of the form and function of Mathematics, as a background for the Philosophy of Mathematics. My efforts have been encouraged by lectures that I have given at Heidelberg under the auspices of the Alexander von Humboldt Stiftung, at the University of Chicago, and at the University of Minnesota, the latter under the auspices of the Institute for Mathematics and Its Applications. Jean Benabou has carefully read the entire manuscript and has offered incisive comments. George Glauberman, Carlos Kenig, Christopher Mulvey, R. Narasimhan, and Dieter Puppe have provided similar comments on chosen chapters. Fred Linton has pointed out places requiring a more exact choice of wording. Many conversations with George Mackey have given me important insights on the nature of Mathematics. I have had similar help from Alfred Aeppli, John Gray, Jay Goldman, Peter Johnstone, Bill Lawvere, and Roger Lyndon. Over the years, I have profited from discussions of general issues with my colleagues Felix Browder and Melvin Rothenberg. Ideas from Tammo Tom Dieck, Albrecht Dold, Richard Lashof, and Ib Madsen have assisted in my study of geometry. Jerry Bona and B. L. Foster have helped with my examination of mechanics. My observations about logic have been subject to constructive scrutiny by Gert Müller, Marian Boykan Pour-El, Ted Slaman, R. Voreadou, Volker Weispfennig, and Hugh Woodin. I have profited from discussions of philosophical issues with J. L. Corcoran, Philip Kitcher, Leonard Linsky, Penelope Maddy, W. V. Quine, Michael Resnik, and Howard Stein. Some of my earlier views on various issues have been constructively examined by Joel Fingerman, Marvin J. Greenberg, Nicholas Goodman, P. C. Kolaitis, J. R. Shoenfield, and David Stroh. I am grateful to all those people—and to a number of others—even in the numerous cases where I have not followed their advice.

My wife, Dorothy Jones Mac Lane, enthusiastically supported the whole project from its inception. Her encouragement was vital.

Springer-Verlag in general, and the editorial staff in particular, has provided steady support in the process of turning my manuscript into a book.

Dune Acres, Indiana
July 4, 1985

SAUNDERS MAC LANE

Contents

Introduction

This book is intended to describe the practical and conceptual origins of Mathematics and the character of its development–not in historical terms, but in intrinsic terms. Thus we ask: What is the function of Mathematics and what is its form? In order to deal effectively with this question, we must first observe what Mathematics *is*. Hence the book starts with a survey of the basic parts of Mathematics, so that the intended general questions can be answered against the background of a careful assembly of the relevant evidence. In brief, a philosophy of Mathematics is not convincing unless it is founded on an examination of Mathematics itself. Wittgenstein (and other philosophers) have failed in this regard.

The questions we endeavor to answer come in six groups, as follows.

First, what is the *Origin* of Mathematics? What are the external sources which lead to arithmetic and algebraic calculations and thence to mathematical theorems and theories? Or, are there internal sources, so that some of these theories develop just from imagination and introspection? This is close to the traditional question: Is Mathematics discovered or invented?

Second, what is the *Organization* of Mathematics? Clearly a subject so large and diverse as Mathematics requires a quite extensive and systematic organization. Traditionally, Mathematics is often split into four parts: Algebra, analysis, geometry, and applied Mathematics. This subdivision is handy at first, say for the arrangement of undergraduate courses, but it soon needs refinement. Thus number theory is to be included, perhaps as a part of algebra, but often using analysis as a tool. Finite (or "discrete") Mathematics is presently popular–but is it algebra, or logic, or applied Mathematics? Algebra soon splits into group theory, field theory, ring theory, and linear algebra (matrix theory). These split up again: number theory can be elementary, analytic, or algebraic; research in group theory is sharply divided between finite groups and infinite groups, while ring theory is split into commutative and non-commutative ring theory, with different uses and different theorems. Analysis can be real analysis, complex analysis, or functional analysis. In geometry, algebraic geometry is based on projective geometry, differential geometry is close to parts of analysis, and topology has branches labeled point-set topology, geometric topology, differential topology, and algebraic topology. The fourth part, "applied Mathematics", is even more varied, since it may refer primarily to classical applied topics such as dynamics, fluid mechanics, and elasticity, or primarily to more recent applied topics such as systems science, game theory, statistics, operations research, or cybernetics. Finally, the active study of partial differential equations is in part applied Mathematics (especially when numerical methods are involved), in part analysis, and in part differential geometry (especially when invariant methods using differential forms are involved). But this list of subdivisions is incomplete, for example, it omits logic and foundations and their applications in computer science.

In sum, these subdivisions of Mathematics are imprecise and necessarily involve overlaps and ambiguity. The use of even finer subdivisions (as in the sixty-odd fields used by *Mathematical Reviews* to organize current research papers) still presents corresponding difficulties. Should we conclude that the real organization of Mathematics cannot be accomplished simply by subdivision into special fields? Are there deeper methods of organization? What is the proper order of the parts of Mathematics, and which branches belong first? Are there even parts of Mathematics which are unimportant or mistaken?

Since Mathematical ideas often arise in prescribed order, one may also ask whether a foundation of Mathematics provides a good organization of the subject.

Our survey will indicate that each part of Mathematics inevitably has an aspect which is formal. Factual problems necessitate calculations, but the calculations then proceed by prescriptions or by rule, rather than by continued attention to the facts of the case–yet the result of a good formal calculation does agree with the facts. Proofs in geometry flow by logical argument from axioms, but the resulting theorems fit the world. Therefore we must inquire as to the relation of the formal to the factual. Thus we begin the first chapter by exhibiting a few of the basic formal structures of Mathematics.

This leads to our third question: Are the formalisms of Mathematics based on or derived from the facts; if not, how are they derived? Alternatively, if Mathematics is a purely formal game, why do the formal conclusions fit the facts?

The fourth question is this: How does Mathematics develop? Is it motivated by quantitative questions which arise in science and engineering, is it driven by the hard problems which have arisen in the Mathematical tradition, or is it pushed by the desire to understand the tradition better? For example, how much does number theory owe to the repeated attempts to prove Fermat's last theorem? Is the solution of a famous problem the pinnacle of Mathematical accomplishment–or should there be comparable credit to the more systematic work in the introduction of new ideas by comparison, by generalization, and by abstraction? For that matter, how does abstraction come about, and how do we know which abstraction is appropriate?

These questions about the dynamics of the development of Mathematics touch on a further–and difficult–topic: How does one evaluate the depth and importance of Mathematical research?

Careful methods and canons of proof developed first in geometry (Chapter III). Subsequently the calculus worked well, but without careful proof, using dubious notions of infinitely small quantities. This led to the problem of finding a rigorous foundation for the calculus (Chapter VI). These two cases present the fifth general question, that about rigor. Is there an absolute standard of rigor? And what are the correct foundations of Mathematics? Here there are at least six competing schools of thought, as follows.

Logicism: Bertrand Russell asserted that Mathematics is a branch of logic, and so can be founded by a development from a careful initial statement of the principles of logic. Moreover, Whitehead and Russell carried out such a development in their massive (but now neglected) book *Principia Mathematica*.

Set Theory: It is remarkable that (almost) all Mathematical objects can be constructed out of sets (and of course sets of sets). Hence arises the view that Mathematics deals just with properties of sets and that these properties can all be deduced from a suitable list of axioms for sets–either the Zermelo–Fraenkel axioms, or these axioms with supplements, some perhaps still to be discovered.

Platonism: This set-theoretic description of Mathematics is often coupled with a strong belief that these sets objectively exist in some ideal realm. Indeed some thinkers, such as Kurt Gödel, may consider that we have special means (not the usual five senses) for perceiving this ideal realm. There are other versions of platonism for Mathematics, for example one in which the ideal realms are comprised of numbers and spatial forms (the "ideal triangle").

Formalism: The Hilbert School holds that Mathematics can be regarded as a purely formal manipulation of symbols, as though in a game. This is the manipulation done when we write rigorous proofs of Mathematical theorems from axioms. This idea was part of the Hilbert program: To show that some adequate system of axioms for Mathematics is consistent, in the exact sense that proofs in the system could never lead to a contraction, such as the contradiction $0 = 1$. To this end, the proofs were to be viewed as purely formal manipulations and were to be studied objectively by strictly "finite" (and hence secure) methods. As yet, such a consistency proof has not been achieved, and Godel's famous incompleteness theorem (to be discussed in Chapter XI) makes it unlikely that it can be achieved.

Intuitionism: The Brouwer school holds that Mathematics is based on some fundamental intuitions–such as that of the sequence of natural numbers. It holds, moreover, that proofs of the existence of Mathematical objects must proceed by exhibiting these objects. For this reason intuitionism objects to some of the classical principles of logic, more explicitly the tertium non datur (either p or not p). There are a number of variants of intuitionism, some emphasizing the importance of finding proofs which are constructive.

Empiricists claim that Mathematics is a branch of empirical science, and so should have a strictly empirical foundation, say as the science of space and number.

In recent years, these (and other) standard views as to the nature and foundation of Mathematics have not been very fruitful of new insights or understanding. For this reason, we do not wish in this book to assume any one such position at the start. Instead, we intend to examine what is actually present in the practice and in the formalism of Mathematics. Only then, with the evidence before us, will we turn to the question of what is and what ought to be a foundation of Mathematics.

Our last and most fundamental question concerns the Philosophy of Mathematics. This is actually a whole bundle of questions. There are ontological questions: What are the objects of Mathematics and where do they exist (if indeed they do exist)? There are metaphysical problems: What is the nature of Mathematical truth? This is a favorite question, given that the philosophers' search for truth often will use the truths of Mathematics as the prime example of "absolute" truth. There are epistemological problems: How is it that we can have knowledge of Mathematical truth or of Mathematical objects? Here the answers may well depend on what is meant by such truth or by such an object.

There are also more immediate or more practical questions. If Mathematics is just formal or just logical deductions from axioms, how can Mathematics be so unreasonably effective in science (E. Wigner)? Put differently, why is Mathematics of such major use in understanding the world?

The various schools on the foundations have correspondingly various attempts to answer these questions, none of them generally convincing. Often–especially in work by philosophers–they are anchored almost exclusively in the most elementary parts of Mathematics–numbers and continuity. Much more substantive material is at hand. This is why we begin with a fresh view of the variety of Mathematics.

To this end, Chapter I starts with the traditional idea that Mathematics is the science of numbers and space–but shows that this starting point can lead directly to some basic formal notions (transformation group, continuity, and metric space) in defiance of the usual historical order. The next chapter describes the natural numbers as a structure, with both surface and deep aspects. The traditional foundations of geometry are summarized in the third chapter, with emphasis on the ubiquitous role of groups of motions and on the remarkable observation that almost all the basic geometrical ideas can be developed in just two dimensions. The familiar (but extraordinary) fact that very many measures of magnitude (in time, space or quantity) can all be consigned to one structure–that of real numbers–is the subject of Chapter IV. The next chapter discusses the origins of the idea of "function" and the troubles in defining it. This leads through transformations to groups again and to the question: Why do the very simple group axioms lead to such deep structural results? The analysis of "effect proportional to cause" is the starting point of linear algebra (Chapter VII), but its ramifications (such as the notion of an eigenvalue) extend beyond algebra. The next chapter deals with some of the aspects of higher geometry: What is a manifold? Some of these ideas are closely tied to classical mechanics, which illustrates (Chapter IX) the intricate connection between applied and pure Mathematics. Chapter X in complex analysis returns to the study of functions–this time holomorphic functions; they are closely tied to the manifolds of Chapter VIII and to the origins of topology. At the end, the book returns to questions of foundations (Chapter XI) and then to the six philosophical questions raised above. With this sample of the extensive substance of Mathematics at hand, these questions take on a different and more illuminating form.

Our discussions of the scope of elementary Mathematics do assume some acquaintance with Mathematics; however, we endeavor to motivate and define explicitly all the Mathematical concepts which play a role in our discussion. Each defined word is italicized. A reference to §VII.6 is to the sixth section of chapter seven, while (VII.6.5) is to the fifth numbered equation of that section; references within a chapter omit the chapter number.

Since our survey touches upon many parts of classical elementary Mathematics, we assume that the reader has at hand some of his own familiar texts for possible reference. We add only occasional supplementary references to the Bibliography at the end, in the form Bourbaki [1940]. There are a number of references to *Survey* of *Modern Algebra* and to *Algebra*, both books written in some combination by Birkhoff and Mac Lane. *Homology* and *Categories Work* (short for "Categories for the Working Mathematician" refer to books by Mac Lane alone. We do note here a few other overviews of Mathematics. That magnificent multivolume monster by Bourbaki (for example, [1940]) is a splendid formal organization of many advanced topics, formulated in blissful disregard of the origins and applications which are important to our present purpose. On a more elementary level the 1977 essay by Gärding covers, with different emphasis, many of the topics on which we touch. Davis and Hersh [1981] has a more popular scope.

CHAPTER I

Origins of Formal Structure

Mathematics, at the beginning, is sometimes described as the science of Number and Space—better, of Number, Time, Space, and Motion. The need for such a science arises with the most primitive human activities. These activities presently involve counting, timing, measuring, and moving, using numbers, intervals, distances, and shapes. Facts about these operations and ideas are gradually assembled, calculations are made, until finally there develops an extensive body of knowledge, based on a few central ideas and providing formal rules for calculation. Eventually this body of knowledge is organized by a formal system of concepts, axioms, definitions, and proofs. Thus Euclid provided an axiomatization of geometry, with careful demonstrations of the theorems from the axioms; this axiomatization was perfected by Hilbert about 1900, as we will indicate in Chapter III. Similarly the natural numbers arise from counting, with notation which provides to every number the next one—its successor, and with formal rules for calculating sums and products of numbers. It then turns out that all these formal rules can be deduced from a short list of axioms (Peano–Dedekind) on the successor function (Chapter II). Finally, the measurements of time and space eventually are codified in the axioms (Chapter IV) for the real numbers. In sum, these three chapters II–IV present the standard formal axiomatization of the science of number, space, and time.

This development of the formal from the factual is a long historical process in which the leading concepts might very well have come in a different order. Our concern is not the historical order, but the very possibility of a development of form from fact. To illustrate this, we start again from number, time, space, and motion and build up directly some of the general concepts of modern Mathematics. Thus counting leads to cardinal and ordinal numbers and to infinite sets and transformations. The analysis of time leads to the notion of an ordered set and a complete ordered set; these concepts fit also with geometrical measurement. The study of motion (in space) and of the composition of two motions suggests the notion of a transformation group. Comparison of this notion of composition with the arithmetic operations of addition and multiplication leads by further abstraction to the concept of a group. On the other hand, motion involves continuity, and the formal analysis of continuity gives rise to a simple axiomatic description of space as a metric space or, more intrinsically, as a topological space. Thus this chapter introduces the idea of the formal in terms of certain basic structures: Set, transformation, group, order, and topology. With Bourbaki, we hold that Mathematics deals with such "mother structures". Against the historical order, we hold that they arise directly from the basic stuff of Mathematics.

1. The Natural Numbers

In order to list, label, count, enumerate, or compare it is convenient to use the single system of *natural numbers*, written in our conventional decimal notation as

$$0,1,2,3\,,\,\ldots\,,9,10,11,\ldots\,. \tag{1}$$

The *same* natural numbers could be written in other notations—with the base 2 instead of 10, or as Roman numerals, or simply as marks

$$\mathrm{I},\mathrm{II},\mathrm{III},\ldots\,. \tag{2}$$

These numbers are used to list in order the objects of some collection of things, or simply to label these objects, or to count the collection, or to (thereby) compare two collections. From these activities, several Mathematical concepts arise together

set·number·label·list.

At this point the word "set" simply means a collection of things: A grouping or assemblage S of objects (say, of physical objects or of symbols) such as the collection of two turtle doves, three french hens, four colley birds, or five gold rings—or the two collections

$$S = \{A,B,C\}, \qquad T = \{U,V,W\} \tag{3}$$

of three letters each, written with the conventional bracket notation for a set or collection. At this stage, the word "collection" is appropriate, because all that matters about a set (or collection) is that it is determined by specifying its elements; one does not yet need more sophisticated notions, such as sets whose elements are themselves sets, or sets of sets of sets, or sets of subsets.

In these terms, one can give semi-final descriptions of the (at first) highly informal operations of listing, labeling, counting, and comparing. To "list" a collection such as $\{A,B,C\}$ means to attach in regular order a numeral to each object in the collection; one usually begins with the numeral 1 and proceeds in order, say, as $\{A_1,B_2,C_3\}$. Note that the numerals will be adequate for this process in all cases only if there is always a next numeral; this is one origin of the idea that every natural number n has an immediate successor $s(n) = n + 1$. To "label" means to attach the same numerals to the objects of the collection, but irrespective of their order, as in $\{A_2,B_3,C_1\}$. To "count" a collection means to determine how many numerals (or which numerals) are needed to label all the objects in the collection. In this connection, note that the count, done properly, always comes out to the same answer. In particular, the numerals needed do not depend on the order in which the objects of the collections are counted: Whether it is $\{A_1,B_2,C_3\}$, $\{B_1,A_2,C_3\}$ or $\{C_1,B_2,A_3\}$, it always ends at the same 3. Comparing two collections, such as $\{A,B,C\}$ and $\{U,V,W\}$ means matching each object of the first collection with some object of the second, until both are exhausted, as in $\{A/W,\ B/V,\ C/U\}$. Of course, it might happen that one collection is exhausted before the other; the first is then "smaller" in the comparison. The result of this comparison does not depend on the order in which objects are matched: $\{A,B\}$ in any order is smaller than $\{U,V,W\}$. There are many pairs of collections to be compared, but it again turns out that it is not necessary to compare each pair; it is enough to compare finite collections with the standard initial segments of the positive natural numbers:

$$\{1,2,3\}, \qquad \{1,2,3,4,\}, \qquad \{1,2,3,4,5\}, \quad \text{etc.}$$

In this context, one says that the collection $\{A,B,C\}$ has the *cardinal* number 3, in symbols

$$\#\{A,B,C\} = 3. \tag{4}$$

As noted, this means that there is a *one-to-one correspondence* f

$$f{:}1 \mapsto A, \qquad 2 \mapsto B, \qquad 3 \mapsto C \tag{5}$$

which matches the standard collection $\{1,2,3\}$ to the collection $\{A,B,C\}$. The collection $\{U,V,W\}$ has the same cardinal number, by the correspondence

$$g{:}1 \mapsto U, \qquad 2 \mapsto V, \qquad 3 \mapsto W. \tag{6}$$

The formal definition of this matching process states that a *bijection* b (a one-to-one correspondence) from a collection S to a collection T is a rule

b which assigns to each element s in S an element $b(s)$ in T, in such a way that every element t of T occurs for exactly one s. This means that the *inverse* of b (b read backwards) is a bijection from T to S; thus the inverse of the bijection f of (5) above is

$$f^{-1}: A \mapsto 1, \qquad B \mapsto 2, \qquad c \mapsto 3. \tag{7}$$

"Composed" with the bijection g of (6) this gives a bijection, f^{-1} followed by g, directly from $\{A,B,C\}$ to $\{U,V,W\{$ as

$$g \cdot f^{-1}: A \mapsto U, \qquad B \mapsto V, \qquad C \mapsto W \tag{8}$$

Thus the elementary observation that the two collections $\{A,B,C\}$ and $\{U,V,W\}$ have the same cardinal number,

$$\#\{A,B,C\} = \#\{U,V,W\},$$

suggests the more general process of "composing" bijections, one followed by another. Indeed, these ideas about bijections can be used to provide a formal definition of the (cardinal) natural numbers (§II.8).

But, whatever the natural numbers are (or however they may be defined) their primary function is to serve in calculations of sums, products, or powers.

The *sum* of two numbers is the cardinal number one gets by combining two sets with the two given numbers, provided these sets are *disjoint*—that is, have no common elements. Thus if A, B, C, U, V above are all different, the sum $3 + 2 = 5$ is

$$3 + 2 = \#\{A,B,C,U,V\},$$

and similarly for other sums. The *product* $2 \cdot 3$ can be described "geometrically" as the cardinal number of a 2×3 square array

$$2 \cdot 3 = \# \begin{Bmatrix} (A,U)(B,U)(C,U) \\ (A,V)(B,V)(C,V) \end{Bmatrix}.$$

Here the three columns are three disjoint sets, so the product can also be described as an iterated sum

$$2 \cdot 3 = 2 + 2 + 2.$$

Similarly, the *exponential* 2^3 can be described as an iterated product

$$2^3 = 2 \cdot 2 \cdot 2\,;$$

it can also be described as the cardinal number of the set of all functions from a 3-element set $\{1,2,3\}$ to a 2-element set $\{0,1\}$.

These three arithmetic operations were invented (or discovered?) because they have all manner of practical uses in financial or scientific calculations. But, to make such calculations we never bother to reduce each operation to its original meaning, as this meaning has just been described. Instead, for the usual decimal notation, one may use a computer or employ the familiar rules: The addition and multiplication tables for the digits from 0 to 9, plus the rules for carry-over of tens. These rules are "formal" in the basic sense of the word: They do not refer to the meanings of the decimals or of the arithmetic operations (though they can be rigorously deduced from these meanings). Instead they simply specify what to do, and specify that correctly. Thus if one counts two disjoint collections as having 5 and 17 members, respectively, and then adds the decimals 5 and 17 according to the rules, the sum is always the count for the combined collection–and similarly for the product. To be sure, items can get lost from collections and calculators can make errors, but then there are further rules to make checks, like the rule of "casting out 9's" (replace each decimal by the sum of its digits, then add or multiply, according to the case). For numbers written in bases other than tens, there are corresponding rules for calculations and for checks (what does one cast out?).

This example gives a clear indication of what we intend to mean by *formal*: A list of rules or of axioms or of methods of proof which can be applied without attention to the "meaning" but which give results which do have the correct interpretation.

2. Infinite Sets

The collection of all the natural numbers,

$$\mathbf{N} = \{0,1,2,3,...\}, \tag{1}$$

starts with 0 and has to each number a successor; hence it is infinite. Historically, one started with 1 and not 0, but we need 0 as the cardinal number of the empty set.

The infinite set **N** of all natural numbers includes many finite subsets

$$\{0,1,2\}, \qquad \{1,3,5,7\}, \qquad \{2,4,16\},$$

as well as infinite sets, such as the set P of all positive natural numbers

$$P = \{1,2,3,4,...\},$$

the set E of all even positive numbers, and the set S of all positive multiples of 6. These various infinite sets may be compared as follows:

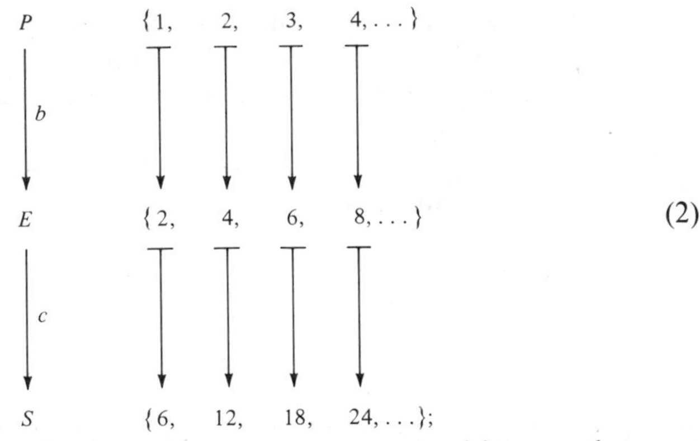

the result shows that there are just as many even positives as there are positives all told; $b(n) = 2n$ defines a bijection $b\colon P \to E$. Similarly $c(2m) = 6m$ is a bijection $c\colon E \to S$. In the comparisons (2), $c(b(n)) = 6n$ gives a "composite" bijection $c \cdot b\colon P \to S$.

A set X is called *denumerable* when there is a bijection $f\colon \mathbf{N} \to X$. Thus the comparisons (2) indicate that P, E, and S are all denumerable; as a matter of fact, *any* subset of **N** is either finite or denumerable.

Two sets X and Y have the same *cardinal number* when there is a bijection $f\colon X \to Y$. This definition includes the finite cardinals 0,1,... already discussed in §1, and the cardinal number called $\aleph_0$ (aleph-naught) of **N**, E, P, and all other denumerable sets. In this way, the elementary activity of counting leads to infinite cardinal numbers—of which $\aleph_0$ is only the first. We will later see that the set of all points on a line is infinite but not denumerable.

One can also formally describe when a set is infinite: When its cardinal number is not finite, or, equivalently, when it has a proper subset S for which there is a bijection $S \to X$.

Finitists hold that infinite sets (and geometrical infinities) are just convenient fictions, while only the finite is "real". This we must later consider. For that matter, is a finite set real? On the fourth day of Christmas, did my true love send me four colley birds or a set of four colley birds? Where is the set?

3. Permutations

A finite set, counted in any order, leads to the same (finite) cardinal number. The count is not changed by "permuting" the things counted. But one may also count how many permutations there are. Thus the set $\{1,2,3\}$ has six permutations

$$(123), \quad (231), \quad (312), \quad (213), \quad (321), \quad (132).$$

Such counts are useful in gambling or speculating. Choose three cards in succession from an (ordered) deck of thirteen; what is the chance that they come out in a direct or reverse order? It is the ratio of favorable cases [(123) or (321)] to the total number 6 of cases (of permutations). This is the root of probability, though in the end the definition of a probability must be more sophisticated than the simple ratio of favorable cases to total cases.

A permutation can be viewed "dynamically"—say, as an operation moving the original order (123) to the order (312) by the bijection

$$1 \mapsto 3, \qquad 2 \mapsto 1, \qquad 3 \mapsto 2.$$

This is usually written as a *cycle* (132), standing for $1 \mapsto 3 \mapsto 2 \mapsto 1$. Any permutation of $\{1,2,3\}$ can be viewed as a bijection

$$b\colon \{1,2,3\} \to \{1,2,3\}$$

As a bijection, it has an inverse, and any two permutations of $\{1,2,3\}$ have a permutation as their composite.

Permutations also arise in algebra. Thus, given the polynomial

$$(x_1 + x_2)(x_3 + x_4), \tag{1}$$

what permutations of the subscripts will leave the polynomial unchanged? To begin with, one may interchange 1 and 2, or interchange 3 and 4, or do both interchanges, or do neither. These we may list as the permutations

$$(12), \quad (34), \quad (12)(34), \qquad 1; \tag{2}$$

here (12)(34) is $1 \mapsto 2$, $2 \mapsto 1$, $3 \mapsto 4$, $4 \mapsto 3$; it is the composite of the two cycles (12) and (34). Also I (do nothing) is the "identity" bijection $1 \mapsto 1$, $2 \mapsto 2$, $3 \mapsto 3$ $4 \mapsto 4$. But the polynomial (1) is also left unchanged by the following four permutations which interchange the two factors:

$$(13)(24), \quad (14)(23), \quad (1324), \quad (1423). \tag{3}$$

This completes the list. Of the 24 possible permutations of the set $\{1,2,3,4,\}$ exactly eight leave this polynomial unchanged; of these eight, four leave the factors unchanged. One may also wonder at the sequence 24, 8, 4. One may also experiment with other polynomials. Thus the polynomial

$$(x_1 - x_2)(x_1 - x_3)(x_1 - x_4)(x_2 - x_3)(x_2 - x_4)(x_3 - x_4)$$

has more symmetries (12 permutations!) while the polynomial

$$(x_1 - x_2)(x_3 - x_4) \tag{4}$$

allows only four permutations (the *four group*)

$$(12)(34), \quad (13)(24), \quad (14)(23), \qquad I. \tag{5}$$

In this list, the composite of any two permutations still leaves the polynomial (4) unchanged, so the composite is also in the list. Such a list of permutations is called a *permutation group*. The combined list (2) and (3) is also such a group.

4. Time and Order

The passage of time suggests the ideas "before" and "after"; when the instant t of time comes before the instant t' we write $t < t'$. Moreover, if in turn t' is before t'', then it is apparent that t is also before t''. This can be stated formally in the *transitive law*

$$t < t' \quad \text{and} \quad t' < t'' \quad \text{imply} \quad t < t'' \tag{1}$$

for the "binary relation" $<$. Moreover, for any two distinct instants of time, one must come before. In different language, for all t and t'' exactly one of

$$t < t' \quad \text{or} \quad t = t' \quad \text{or} \quad t' < t \tag{2}$$

must hold. This statement is the law of *trichotomy*.

But the "before" and "after" of time is not the only example of these two laws. There is a "discrete" example. For natural numbers, $m < n$ means that n comes after m in the list of numbers succeeding m; here both laws (1) and (2) hold:

$$0 < 1 < 2 < 3 < \cdots . \tag{3}$$

The usual order of the positive and negative integers provides another instance of these laws:

$$\cdots \; -3 < -2 < -1 < 0 < 1 < 2 < 3 < \; \cdots \tag{4}$$

as does the usual ordering of the rational numbers, suggested by the display

$$\begin{aligned} \mathbf{Q}\colon -1/5 < \; \cdots \; < 0 \; \cdots \; < 1/5 \; \cdots \; < 1/4 \; \cdots \\ < 1/3 < \; \cdots \; < 2/3 \; \cdots \; < 1 \; \cdots . \end{aligned} \tag{5}$$

There are numerous other examples of these two formal laws. Hence it is handy to have a name for this combined situation, as it might apply to any set X (of instants of time *or* of integers or of rationals ...).

A *binary relation* $<$ on a set X specifies that $x < y$ is true or false for any two elements x,y in X; one might also say that the relation amounts to specifying a set: The set of all those ordered pairs (x,y) with $x < y$. A *linearly ordered set* is then a set X with a binary relation $<$ for which the laws (1) and (2) hold; in other words it is a set equipped with a transitive and trichotomous relation $<$. One can then invent (or discover?) many other examples of linearly ordered sets: Finite ones such as $1 < 2 < 3 < 4$ or long infinite ones such as

$$0 < 1 < 2 < 3 < \cdots < \omega < \omega+1 < \omega+2 < \cdots, \tag{6}$$

where ω is the first thing beyond all the finite natural numbers. (This linearly ordered set is actually the start of the infinite ordinal numbers.)

This definition is an easy first (of many) cases of a list of axioms describing a common situation with many different examples. As in other cases, the choice of axioms can vary. Thus, rather than using "before" and "after", the passage of time can be described by the notion "not later than", usually written $t \leqslant t'$. This alternative can be formalized for any linearly ordered set X. Define $x \leqslant y$ to mean $x < y$ or $x = y$. This binary relation on X is then

Transitive:	$x \leqslant y$ and $y \leqslant z$ imply $x \leqslant z$,
Reflexive:	$x \leqslant x$ for all x,
Antisymmetric:	$x \leqslant y$ and $y \leqslant x$ imply $x = y$.

Finally, corresponding to trichotomy, it has the property:

For all x and y in X, either $x \leqslant y$ or $y \leqslant x$.

Conversely, let any set X have a binary relation $\leqslant$ with these four properties, and *define* $x < y$ to mean that $x \leqslant y$ but $x \neq y$. Then X is indeed a linearly ordered set and the originally given relation $\leqslant$ is related to $<$ as before. In brief, the same notion of linear order can be defined in two formally different ways, via $<$ or via $\leqslant$. In general, the same situation may often be defined in two or more formally different ways.

One also asks when two "models" of the axioms are "essentially" the same–in the sense that the linearly ordered set of natural numbers has the same "order type" as the ordered set of even positive natural numbers:

$$2 < 4 < 6 < 8 < 10 < \cdots$$

So for linearly ordered sets X and Y an *order isomorphism* $f: X \to Y$ is defined to be a bijection of the set X on the set Y such that order is preserved: For all x_1 and x_2 in X,

$$x_1 < x_2 \qquad \text{implies } fx_1 < fx_2. \tag{7}$$

When there is such an isomorphism f, X and Y are said to have the same *order type*. (This is like the definition of "same cardinal number" except that now one also keeps in mind the order of the elements being compared.) One can then readily prove (say) that any linearly ordered set of 4 elements is order isomorphic to the standard such set: $1 < 2 < 3 < 4$.

A general question is then at hand: Can one describe a particular model of the axioms by giving enough additional axioms to determine the model uniquely (i.e. uniquely up to an order isomorphism?) In the present case, can one give properties of an ordered set X which imply the existence of an order isomorphism $X \to \mathbf{N}$ (or $X \to \mathbf{Q}$, the ordered set of rationals, or $X \to \mathbf{R}$, the ordered set of reals?)

The answers are "yes". To get at the case of the reals $\mathbf{R}$, one must formulate the sense in which a real number (an instant of time) can be approximated by rational numbers. For example, the real number π is determined by the usual sequence of decimal approximations

$$3.14, 3.141, 3.1415, 3.14159, 3.141592, \ldots.$$

Indeed, π is the "least upper bound" of this set of rational numbers. Formally, in a linearly ordered set X an element b is an *upper bound* for a subset S of X if $s \leqslant b$ for every s in S. Also, b is a *least upper bound* for S if no b' with $b' < b$ is an upper bound for S. This implies that if S has a least upper bound, that least upper bound is unique. (This is the sense in which π, for example, is determined uniquely by its decimal expansion). Also, the set X is *unbounded* if there is in X no upper bound and no lower bound. (For example, the ordered set $\mathbf{N}$ has a lower bound 0, hence is not unbounded).

The crucial property of the ordered set of real numbers is *completeness*: Every non-empty subset S with an upper bound has a least upper bound. The additional fact that every real number can be approximated by rationals can be made formal by stating that the set $\mathbf{Q}$ of rational numbers is "dense" in $\mathbf{R}$. Here a subset D of a linearly ordered set X is said to be *dense* in X if, for all $x < y$ in X there is always a d in D between x and y, so that $x < d < y$. It is then clear that the ordered set $\mathbf{R}$ is complete, unbounded, and has a denumerable dense subset. Also one can prove that any linearly ordered set X with these three properties is order isomorphic to $\mathbf{R}$ (see Hausdorff). In the proof one uses a characterization of the order type of $\mathbf{Q}$: It is denumerable, unbounded, and dense (as a subset of itself).

This result does provide a description of the *order* of the real numbers. In Chapter IV we will combine this with a description of their algebraic properties. These properties also arise from experience with the passage of time. Once intervals of time are measured by a clock (or an hourglass) one can *add* one interval to another, and regard each instant of time t as the end of an interval (from some starting time). This addition is then an operation which produces to each pair t, t' of instants their sum, $t + t'$, with properties such as $t + t' = t' + t$ and

$$(t + t') + t'' = t + (t' + t'')$$

–just like those for the addition of natural numbers. Again, different examples lead to the same formal law.

5. Space and Motion

Space can be regarded as something extended or as a receptable for objects or as a background for ideal "figures". These aspects are all closely tied to the notion of motion through space, while motion provides the notion of measuring distance in space. Space and motion crop up together everywhere, from physics to physical exercise.

Idealization of the notion of space suggests that chunks of space are made up of figures which are filled up with "points". A point is in space, but without extent. In the extreme analysis, the space consists just of points–but to make this work the points must have added structure, say that described by giving the distance $\rho(p,q)$ from the point p to the point q. This distance is to be measured along straight lines and is a number–at first, just some rational number. But some lines must be vertical (for balance) and others horizontal. Thence comes the idea of perpendicular lines (the word suggest the vertical, as in the perpendicular version of gothic architecture). This leads to right triangles. These lead in turn to the pythagorean theorem and the discovery that the hypotheses of an isosceles right triangle with both legs of length 1 is measured by $\sqrt{2}$–which cannot be a rational number (because $\sqrt{2} = m/n$ in lowest terms would give $m^2 = 2n^2$, forcing m and then n to be even). Thus it is that space, measured with distances, requires not rational numbers but real numbers.

Thus, given the real numbers, one is led to describe space–or a chunk of space–as a collection of points $p, q, \ldots$ together with a non-negative real number $\rho(p,q)$ which is the measure of the *distance* from p to q. It is the same as the distance from q to p:

$$\rho(p,q) = \rho(q,p) \qquad \text{for all } p,q\,; \tag{1}$$

it is zero only when the points coincide:

$$\rho(p,q) \geqslant 0; \qquad \rho(p,q) = 0 \qquad \text{if and only if } p = q. \tag{2}$$

Moreover the intent is that this distance is the shortest from p to q. (The straight line is the shortest distance between two points.) In particular, this means that the distance from p to q is not lessened when it is measured along two straight lines going through a third intermediate point r. This amounts to the (Figure 1) *triangle* axiom: For all p, q and r in X,

$$\rho(p,q) \leqslant \rho(p,r) + \rho(r,q). \tag{3}$$

Thus arises the concept of a *metric space*: A collection of points p, q together with the real number distances $\rho(p,q)$ which satisfy the axioms (1), (2), and (3). The evident chunks of space–a square, a cube, a cylinder, a blob, a dumbbell, each with the usual measure of distance–are all metric spaces in the sense of this definition, as is the whole of our ("ordinary") three dimensional space. Non-Euclidean geometry (Chapter III) provides natural examples of such spaces as do the curved spaces to be considered in Chapter VIII; there are also bizarre examples–such as "a space" with infinitely many different points, with distance 1 between any two different points (try to fit *that* into the plane). Despite such bizarre examples, many elementary properties of space can be formalized and studied for a general metric space. In other words, given numbers, the Mathematical study of space need not start with the conventional ideas of Euclidean geometry, but instead with an axiom system–that of metric space–which applies to many different examples of "space".

Motion can be described in any metric space–push the points around, keeping fixed their distances apart. More formally, if F is a *figure* (a collection of points) in a metric space X a motion of F will at each time t take each point p of F to a new position (a new point) $M_t p$ in X. This passage must be "continuous" (an idea to which we will soon return). Moreover, the motion must be *rigid*–the distance apart of any two points must stay the same during the motion; in other words, for all times t and all points p and q of F, the distance ρ must satisfy

$$\rho(M_t p, M_t q) = \rho(p,q). \tag{4}$$

We speak of such a motion $(p,t) \mapsto M_t p$ as a *parametrized motion* of the figure F, with t as the parameter.

It is perhaps easier to consider just a "completed" motion–the passage from the initial position p to the final position $M_{t_1} p$ at some chosen time

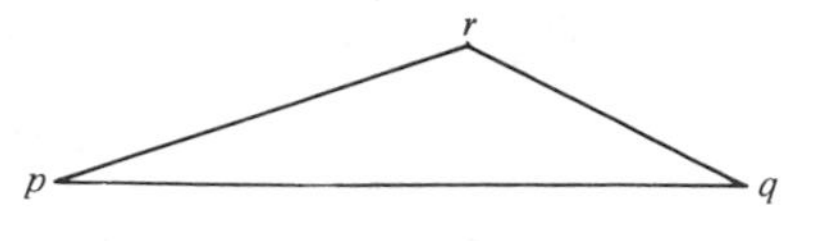

Figure 1

t_1. This is called a *rigid motion* M; it assigns to every point p of the figure concerned a new point Mp such that, for all p and q,

$$\rho(Mp,Mq) = \rho(p,q); \tag{5}$$

put briefly, a rigid motion is a bijection of space which preserves distances between points. For example, a rigid motion of an equilateral triangle into itself could be a rotation (by 120°, 240°) or a reflection of the triangle in one of the three altitudes or the identity motion (every point stays put). There are thus six such motions (symmetries) of the triangle. For motions of the plane as a whole, we will see in Chapter III the use of three typical motions: A *translation* (every line stays parallel to its original position), a *rotation* (one point is fixed) and a *reflection* (all the points on a line stay fixed. These are not all: Moving a triangle ABC into a congruent triangle $A'B'C'$ (Figure 2) may require a translation (A to A') followed by a rotation about A'; in other words, a composite motion.

From such examples arises the idea of the composition of two motions M and N–first move by M and then move the result by N, to give the *composite* motion C with

$$C(p) = N(Mp). \tag{6}$$

We write $C = N \cdot M$ for the composite and observe at once that if M and N are rigid motions, so is C. For parametrized motions the addition of time intervals usually corresponds to composites, in that

$$M_{s+t}(p) = (M_s \cdot M_t)(p). \tag{7}$$

The axioms for a metric space show that any rigid motion M keeps distinct points distinct. Indeed, $p \neq q$ implies by axiom (2) that $\rho(p,q) \neq 0$ and hence by the definition (5) of a motion that $\rho(Mp,Mq) \neq 0$), hence $Mp \neq Mq$ by axiom (2) again.

In studying the symmetry of a figure F, we usually consider a motion M of F "into" itself; that is, a motion M such that p in F moves to some $M(p)$ in F and such that every point q of F comes from some p in F, so that $q = M(p)$. By the above, the motion M is therefore a bijection of F to F, and so has an inverse M^{-1} which is also a rigid motion of F to F.

However, the reader might wish to construct an infinite figure F (say one in the plane) and a rigid motion M of F into F which is *not* onto F.

Figure 2

6. Symmetry

Symmetrical objects are all about us. There are many (man-made) symmetrical figures (Figure 1). Each of the figures has vertical symmetry, horizontal symmetry, and rotational symmetry. The vertical symmetry V can be construed as a reflection of the figure in its vertical axis, and similarly for the horizontal axis, H. The rotational symmetry can likewise be regarded as a 180° rotation R of the figure about its center. If we think of the figure as a metric space X, each of these symmetries is a rigid motion M of X onto itself, and these four motions are the only such. This suggests a definition of a *symmetry* of a figure F: A rigid motion of F onto itself. In particular the different figures of (1) have by this definition the *same* symmetry (later called the *four-group*).

By this definition, the composite of two symmetries of F is again a symmetry. Thus vertical reflection followed by another vertical reflection is the identity (which thus must count as a symmetry). Again, vertical reflection followed by horizontal reflection is the 180° rotation. This one may check by actual experiments with a rectangular card–or one may label the vertices of the rectangle by numbers 1, 2, 3, 4 so that V amounts to the permutation (12)(34), H is (14)(23), and the composite $H \cdot V$ (first apply V then H) is

$$1 \mapsto 2 \mapsto 3, \quad 2 \mapsto 1 \mapsto 4, \quad 3 \mapsto 4 \mapsto 1, \quad 4 \mapsto 3 \mapsto 2; \tag{1}$$

this is the permutation (13)(24) given by the 180° rotation. Thus the total list of symmetries for the Figure 1 is

$$(12)(34), \quad (14)(23), \quad (13)(24), \qquad I. \tag{2}$$

This is identical to the list (3.5) of permutations allowed by the polynomial $(x_1 - x_2)(x_3 - x_4)$ of (3.4). Thus the *same* symmetry turns up in both geometric and algebraic circumstances. This suggests that *the* underlying symmetry here–in this case the "four group"–must itself be something "abstract"; neither geometric nor algebraic; or perhaps both. It need not depend on numbers–the dumbell of Figure 1 has no convenient corners to be numbered!

There are many different types of such symmetries. In three dimensions, one has the symmetry of the regular tetrahedron, or of the cube, or of the icosahedron, or of the octahedron. In the plane there are symmetrical figures such as those of Figure 2. For the equilateral triangle there are six

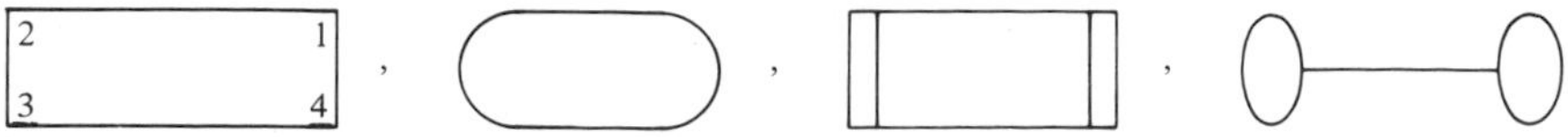

Figure 1

symmetries, accounting for all six permutations of the three vertices—or just as well, all six permutations of the three sides. For the square and also for the decorated square there are eight symmetries all told—four reflections (vertical, horizontal, and two reflections in the diagonals) and four rotations (counting the identity as a rotation through 360°!) If one labels the four vertices as in Figure 2, the eight symmetries turn out to be exactly the eight symmetries (3.2) and (3.3) listed in §3 for the polynomial $(x_1 + x_2)(x_3 + x_4)$. This again indicates that algebra and geometry have in common some underlying, more abstract, form.

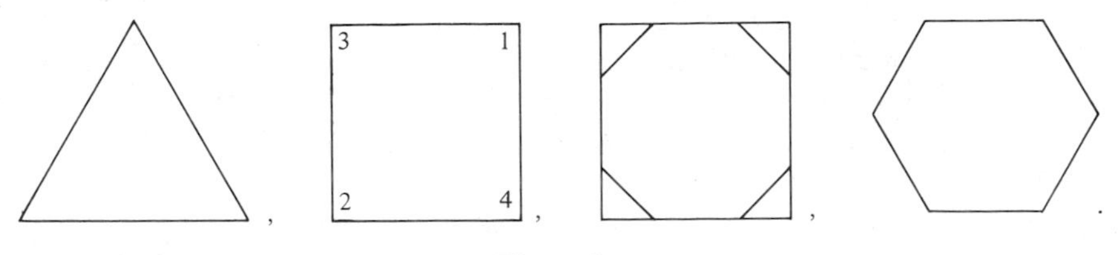

Figure 2

The frieze of a Greek temple, such as that suggested by the scheme (Figure 3) has "more" symmetry. One considers it as a "linear ornament", extending to infinity in both directions: one may picture it more schematically as in Figure 4, with nodes labeled by numbers. There are then infinitely many symmetries: Vertical reflection (n to $-n$), translation T to the right by two units and repeated such translation T^n, n times, as well as the inverse translation T^{-1} (two units left) and its iterates T^{-n}. There is also a different rigid motion S—translate one unit right *and* reflect in the horizontal axis. Then the composite $S \cdot S$ is just T, so that all the symmetries of this figure are "generated" by V and the "slide reflection" S and its inverse. If we erase the lower spikes in Figure 4 we get fewer symmetries (no S, but V and T). The reader may try to find linear ornaments with still different symmetries. (There are just seven sorts).

Three dimensional infinite symmetry comes in much greater variety. There the origin is not just from architecture, since the classification of three dimensional symmetries is the first step in the classification of crystals by the "crystallographic groups".

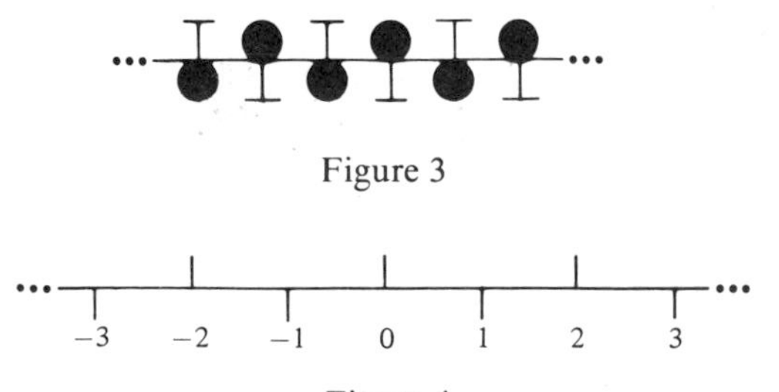

Figure 3

Figure 4

7. Transformation Groups

A permutation of a set, a symmetry of a figure, and a motion of Euclidean space are all examples of "transformations". A *transformation* T of a set X is a bijection $T\colon X \to X$; that is, a one-to-one correspondence $x \mapsto Tx$ on the elements x of X. Thus each transformation T has an inverse $T^{-1}\colon X \to X$; any two transformations S and T have a composite $S \cdot T$—first apply T and then S.

A *transformation group* G on a set X is a non-empty set G of transformations T on X which contains with each T its inverse and with any two transformations S, T in G their composite. This implies that G always contains the identity transformation I on X:

$$I = T \cdot T^{-1} = T^{-1} \cdot T. \tag{1}$$

A transformation group on a finite set (and especially on the typical finite set $\{1, 2, \ldots, n\}$) is usually called a *permutation group*. The *symmetric group* of *degree* n is the group of all $n!$ permutations of $\{1, \ldots, n\}$.

Symmetry groups of figures or formulas are the leading examples of transformation groups, and the source of the "abstract" concept. This is a typical example of Mathematical experience leading to a formal definition. But we are also led to explicate when two transformation groups are "essentially" the same. To do this, one may examine a case such as the representation in §6 of each symmetry of the square X by a permutation of the vertices of that square. This takes place by labeling the vertices by numbers, say by a function $f\colon \{1,2,3,4\} \to X$ which puts each number on the corresponding vertex. The labeled vertices are all different; that is, $fk = fm$ implies $k = m$; one says that the function f is *injective* (an *injection*, or *one-one into*). With these labels, each motion $T\colon X \to X$ of the square sends each vertex to a vertex, so determines a permutation ${}^{\#}T\colon Y \to Y$ of the set Y of vertices. Thus ${}^{\#}T$ does to k what T does to fk; in other words,

$$f({}^{\#}T)k = T(fk), \tag{2}$$

for $k = 1,2,3$, or 4. This equation can be written in terms of composites of functions as

$$f \cdot {}^{\#}T = T \cdot f \tag{3}$$

or displayed in a diagram of the corresponding functions as

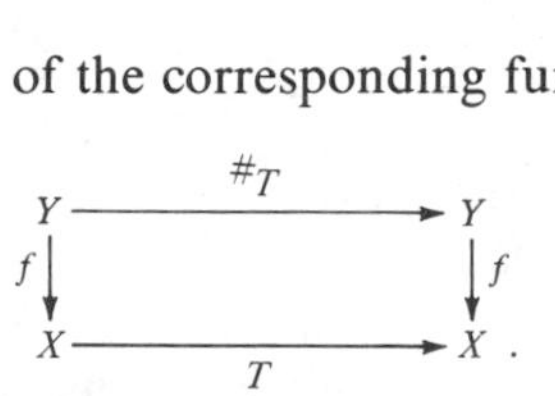

$$\begin{array}{ccc} Y & \xrightarrow{\;{}^{\#}T\;} & Y \\ {\scriptstyle f}\downarrow & & \downarrow{\scriptstyle f} \\ X & \xrightarrow[\;T\;]{} & X \end{array} \tag{4}$$

This exhibits f as comparing the action of ${}^{\#}T$ on the vertices Y with the action of T on X. This diagram is called *commutative* because (3) holds: Both paths from upper left to lower right have the same result. This example (and many others like it) suggests a general formalization of the idea of comparing a transformation group H on a set Y with G on X: A *map* of (H, Y) to (G, X) is a function $f\colon Y \to X$ and a function $\#\colon G \to H$ such that (4) commutes for every transformation T in G. In case this f is an injection (as in the case above), the equation (3) shows that giving f (giving the labels of the vertices) completely determines $\#$. If moreover f is a bijection, it has an inverse f^{-1} so that $\#$ can be described directly by

$$ {}^{\#}T = f^{-1} \cdot T \cdot f \,; \tag{5} $$

to find the permutation, label each vertex by f, look to see where the vertex goes, and read off its label (by f^{-1}).

This result does formalize the evident fact that the permutations of a typical set $\{1,2,3,4\}$ of 4 things represent also the permutations of any set of four things. Generally, if sets Y and X have the same cardinal number, by a bijection $f\colon Y \to X$, then the correspondence $\#$ of (5) is a bijection from the permutation group of X to that of Y. Note incidentally that $\#$ goes in the direction opposite to f.

However, this notion of a map is a bit complicated. Moreover, it doesn't directly handle all the desired comparisons. Thus in (6.1) the dumbell Y and the perimeter X of the rectangle clearly have the "same" symmetries, but there is no evident way to get a map $f\colon Y \to X$ to make such a comparison. Indeed, there is no such f—because the dumbell Y has a center point left fixed by all the motions and there is no such point on the perimeter of the rectangle. The two transformation groups in this case can at least be compared through some intermediary—mapping each (say) into a common (containing) such rectangle.

To summarize: symmetry forces us to consider transformation groups, and even forces thoughts as to more abstractions from this notion.

8. Groups

For any three transformations R, S, and T of a set X the iterated composite, by its definition, satisfies

$$ ((R \cdot S) \cdot T)x = R(S(Tx)) = (R \cdot (S \cdot T))x \,, $$

so composition of transformations is associative. Now, in a transformation group G, forget the fact that the elements T of G transform things, and use only the properties of composition. It is then a group in the sense of the following definition of an "abstract" group:

A *group* is a set G equipped with three rules, as follows:

(i) A rule assigning to any two elements s, t of G on element st, called their product, such that the product is *associative*,

$$r(st) = (rs)t, \tag{1}$$

for all r, s, t in G.

(ii) A rule determining an element e (the *unit*, often written as $e = 1$) of G such that, for all t in G,

$$te = t. \tag{2}$$

(iii) A rule assigning to each t in G an element t^{-1} in G such that

$$tt^{-1} = e. \tag{3}$$

In every transformation group, composition has these properties, so every transformation group is a group. Moreover (and vice versa) Cayley's theorem asserts that every group G arises in this way from a transformation group; just take the set X of points to be transformed to be the set G itself, while each t in G is the transformation sending x in G to the product tx in G. But transformations are not the only sources of groups. With multiplication taken to be the product, the positive rational numbers or the positive real numbers or the non-zero complex numbers constitute groups. If addition is taken to be a "product", the real numbers (the instants of time) form a group, as do the ordinary clock hours ($12 = 0$). Other groups, as we will see, arise in number theory. Groups such as these, where the product is commutative,

$$st = ts \tag{4}$$

for all s and t are called *abelian groups*.

There are many consequences of the simple axioms (i), (ii), and (iii) for a group. They include easy consequences such as the cancellation law ($st = s't$ implies $s = s'$) or the rules

$$te = t = et, \qquad tt^{-1} = e = t^{-1}t \tag{5}$$

which might as well (for the sake of symmetry) be used as axioms in place of (2) and (3). A group G may have *subgroups* S (a subset which is itself a group under the same multiplication (and inverse)). If G is finite, its cardinal number is called its *order*. One proves that the order of a subgroup is always a divisor of the order of the group; this serves to understand and explain some of the observations made above about the orders 8 and 4 of subgroups of the symmetric group of four things. There are all manner of constructions of particular groups. Thus to each positive n the *cyclic group*

Z_n of order n consists of all rotational symmetries of a regular n-gon; it is an abelian group. To any two groups G and H one may construct a group $G \times H$, their "product". It is abelian when G and H are. The "structure theorem" for finite abelian groups G states that every such group is a iterated product

$$G = Z_{m_1} \times Z_{m_2} \times \cdots \times Z_{m_k} \tag{6}$$

of cyclic groups. Moreover, the orders $m_1, \ldots, m_k$ of these factors can be chosen so that each is a multiple of the next (and their product is the order of G). We will be concerned with the origins of this theorem in number theory (the multiplicative group of residues prime to m, modulo m), in topology (the homology group of a finite complex described in terms of Betti numbers and torsion coefficients). We are also concerned with the question of the proper generality of such a theorem (is it really a theorem about finitely generated modules over a principal ideal ring (*Algebra*, p. 384)?). We are concerned with the additional concepts which such a theorem brings to attention–for example, the notion of direct product of groups and its eventual conceptual generalization to products of other types of objects (rings, spaces) and finally, to products of objects in a category (*Categories Work*, p. 68 or Chapter XI below).

For non-abelian groups G there is no structure theorem as simple as (6). For example, the symmetric group on 3 letters $\{1,2,3\}$ has order 6 but it is not cyclic nor is it a product of cyclic groups (though it does have subgroups of orders 2 and 3). For such non-abelian groups there are instead much deeper structural results (Chapter V). One may then ask why the very simple group axioms lead to such deep structure.

Return to the idea of comparing two groups. For the case of a bijection $f\colon Y \to X$ which gives a map of one transformation group (H, Y) to (G, X) we used $\#\colon G \to H$ with

$$^{\#}T = f^{-1} \cdot T \cdot f$$

for all T in G, as in (7.5). Then for any composite $S \cdot T$ in G one gets

$$^{\#}(S \cdot T) = f^{-1}(S \cdot T)f = (f^{-1}Sf) \cdot (f^{-1}Tf) = (^{\#}S) \cdot (^{\#}T)$$

Thus arises the definition: for any two groups G and H a *homomorphism* $\#$ or $b\colon G \to H$ is a function assigning to each s in G an element bs in H in such a way that

$$b(st) = (bs)(bt). \tag{7}$$

(It follows that $be = e$ and $b(t^{-1}) = (bt)^{-1}$). If b is a bijection, it is called an *isomorphism*; hence the simple formulation of the answer to a question from §7: Two figures have the same symmetry if their symmetry groups are isomorphic.

But geometry is not the only source of the idea of isomorphism. The familiar property of the logarithm (say to base 10),

$$\log(xy) = \log x + \log y$$

states that the logarithm is an isomorphism of the multiplicative group of positive real numbers to the additive group of (positive, 0, and negative) real numbers. Note also that $\log 1 = 0$ and $\log(x^{-1}) = -\log x$.

There are many other familiar examples of homomorphisms: absolute values of numbers, determinants of matrices, or the way in which each symmetry of a square produces a permutation of the diagonals of that square.

This discussion has summarized, in an especially striking case, the way in which some underlying informal idea arises and then is formalized by generalization and abstraction. First the study of motion, of symmetries and of permutations suggests the "idea" of composition. One formalization of this idea is that of a transformation group (we will later meet other formalizations of composition). The notion "transformation group" generalizes a variety of examples in a way which assists the understanding of the common properties and the comparisons between examples. But it turns out that some of the most useful properties involve not the composition of transformations, but just composition; moreover this behavior of composition is like that of addition or multiplication of numbers. By an effort of abstraction this leads to the more abstract notion "group" and to the study of the extensive properties of such groups.

Our presentation implicitly argues that the notion of a group comes early (and prominently) in the order of mathematical ideas. Historically, this was not the case. Geometry was treated by axioms and not via groups (as it might have been–see Chapter III). Classification of crystals by their symmetry groups apparently was not developed till the 19th century. The first groups explicitly recognized as such were permutation groups, and the first explicit use of group notions came with Galois (1832), who used homomorphisms (i.e., normal subgroups) to prove theorems about the solution of algebraic equations (Galois theory: §V.7). For the rest of the 19th century groups (usually described with confused definitions) were chiefly permutation groups. When Cayley defined an abstract group in 1852, nobody paid any heed. When he repeated his definition in 1871 he found *three* different groups of order 6 (not two), because he failed to recognize an isomorphism $Z_6 \cong Z_2 \times Z_3$. In 1905 Burnside published a definitive monograph "Theory of Finite Groups". It dealt in fact with abstract groups, but called them "groups of substitutions". That doubtless expresses the intuitive base for group theory. It would seem that only the 20th century saw the full utility of the notion that a variety of examples could be understood well by an axiomatic description of the concept.

The axioms for a group provide a pattern for other axioms of algebraic structure. Instead of stating an axiom "there exists in G an element e such that always $te = t$", our axiom (ii) has specified that the element e is "given". Indeed it can be "given" as a function $e: \{*\} \to G$ mapping the one point set $\{*\}$ into the element e of the set G. Such a function is a *nullary operation* (on the set G). Thus the group axioms provide three operations

$$c: G \times G \to G, \qquad e: \{*\} \to G, \qquad -1: G \to G \tag{8}$$

a *binary operation* (multiplication), a nullary operation (unit), and a *unary operation* (inverse). These operations are required to satisfy certain identities (1), (2), and (3) which can be regarded as identities between "composites" of the initial operations (8).

Much the same pattern applies to operations of addition and multiplication (the axioms (§IV.3) for a ring or a commutative ring) and for the axioms on the algebraic operations for lattices, vector spaces, and the like.

Groups have been variously generalized. There are, for example, generalizations made by deletion of axioms. Drop the unary operation of inverse (and the axiom (iii) pertaining thereto) and one has the axioms for a *monoid.* Drop also the axiom (ii) for the unit e to get the axioms for a *semi-group*, and observe that there are various motivations for these deletions; semi-groups arise in the operation of finite state machines (the sequences of states form a semi-group) and in the composition of operations in functional analysis–but semi-groups do not have as rich a structure as do groups (How does one account for such varying richness of structures?) We will repeatedly examine generalizations by deletion.

These and many other cases illustrate the general notion of an algebraic structure: A set X with nullary, unary, binary, ternary . . . operations satisfying as axioms a variety of identities between composite operations. "Universal algebra" is concerned with the general properties of such structure. There is also a "many-sorted" universal algebra for those structures involving more than one set. A first example (two sorts) is a transformation group: A set X together with a group G of transformations on X. An even more decisive example is that of a ring R and a left module (§VII.11) over that ring. More recently, many-sorted universal algebra has proved useful in the computer science of data types.

9. Boolean Algebra

Another example of an algebra is provided by the operations such as the *intersection* and the *union* of subsets S and T of a given set X. If we write $x \in S$ for "x is an element of S" and $\Longleftrightarrow$ for "if and only if", these operations are specified by giving the elements of the resulting subset of X as follows:

$$\text{Intersection} \quad x \in S \cap T \Leftrightarrow x \in S \quad \text{and} \quad x \in T, \tag{1}$$

$$\text{Union} \quad x \in S \cup T \Leftrightarrow x \in S \quad \text{or} \quad x \in T, \tag{2}$$

$$\Rightarrow \quad x \in S \Rightarrow T \Leftrightarrow \text{if} \quad x \in S, \text{ then} \quad x \in T, \tag{3}$$

$$\Leftrightarrow x \in T, \quad \text{or not} \quad (x \in S).$$

They correspond exactly to the three propositional connectives "and", "or", and "if then". They may also be pictured by Venn diagrams; if the set X is taken to be all the points in a rectangle while S and T are respectively the points inside the ovals S and T, then two of these operations may be indicated by shaded areas as in Figure 1. There is also a unary operation, the *complement* $\neg S$ of S:

$$x \in \neg S \quad \Leftrightarrow \quad \text{not } (x \in S) \tag{4}$$

These various operations $\cap$, $\cup$, $\Rightarrow$, $\neg$ satisfy certain algebraic identities which can all be deduced from a suitable list of axioms, the axioms for *Boolean Algebra*. Thus the set $P(X)$ of all subsets of X is a Boolean algebra.

There also are operations on infinite families of sets. Thus if S_i is a subset of X for each i in some "index" set I, the (infinite) Union and intersection are defined by

$$x \in \bigcup_i S_i \quad \Leftrightarrow \quad \text{For some } i \text{ in } I, \qquad x \in S_i, \tag{5}$$

$$x \in \bigcap_i S_i \quad \Leftrightarrow \quad \text{For every } i \text{ in } I, \qquad x \in S_i \tag{6}$$

These operations correspond to the logical quantifiers "There exists an i" and "For all i", respectively. These connections with logic will be explored further in Chapter XI.

Boolean algebra provides a Mathematical way of representing properties, in that each property H of elements of a set X determines a subset of X; namely, the subset S consisting of all those elements which have the property

$$S = \{x \mid x \in X \text{ and } x \text{ has } H\}. \tag{7}$$

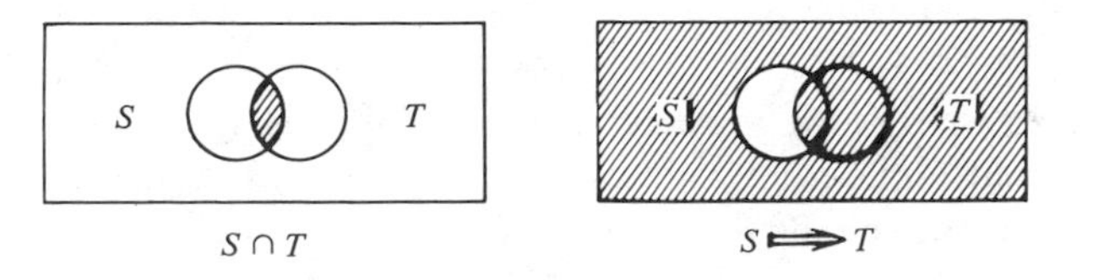

Figure 1. Boolean operations.

This subset is sometimes called the *extension* of the property H, to emphasize the notion that differently formulated properties may have the same extension—and that Mathematics has to do with extensions rather than with meanings. This in turn involves the "extensionality" axiom for sets—that a set is completely determined just by specifying its elements. This means that the equality of two subsets of X is described by the statement

$$S = T \iff (\text{For all } x \text{ in } X,\ x \in S \iff x \in T), \tag{8}$$

while the inclusion of one subset S in another is described by

$$S \subset T \iff (\text{For all } x \text{ in } X,\ x \in S \Rightarrow x \in T); \tag{9}$$

here the arrow $\Rightarrow$ stands for "implies".

This inclusion relation is transitive, reflexive, and antisymmetric, as these properties were defined in §4 above. In general, an *ordered set* W is a set W (such as $P(X)$) with a binary relation (such as $S \subset T$ for $S, T \in W$) which is transitive, reflexive, and antisymmetric. An ordered set is often said to be *partially* ordered (a *poset*) because it need not satisfy the "trichotomy" property which holds for a linear order.

It is important to recognize that many orders are just partial orders and not total orders (i.e., *not* linear). However, in many domains of the application of Mathematics to social phenomena, there is a strong tendency to order ideas, people, and institutions in a *linear* way—for example, according to rank on some imagined numerical measure. The more relevant notion of partial order seems little known and less used.

Diagrammatic presentation of an inclusion relation is suggestive. Thus the various inclusions of the subsets of a three-element set can be pictured by the rising lines in Figure 2, where the bottom symbol $\varnothing$ denotes the *empty* subset. The Boolean operations on subsets may be visualized in this figure. For example, the union $\{1,2\}$ of the subsets $\{1\}$ and $\{2\}$ is the smallest subset which lies "above" both the subsets $\{1\}$ and $\{2\}$; in this way it is the least upper bound, as defined in §4, of $\{1\}$ and $\{2\}$. Generally, the union $S \cup T$ of two subsets S and T of a set X has the properties

$$S \subset S \cup T, \qquad T \subset S \cup T, \tag{10}$$

$$S \subset R \text{ and } T \subset R \Rightarrow S \cup T \subset R, \tag{11}$$

which state that it is the least upper bound of S and T in the partial order given by inclusion. In an exactly dual way, the intersection $S \cap T$ is the greatest lower bound of the subsets S and T. In other words, both these Boolean operations can be described directly in terms of inclusion,

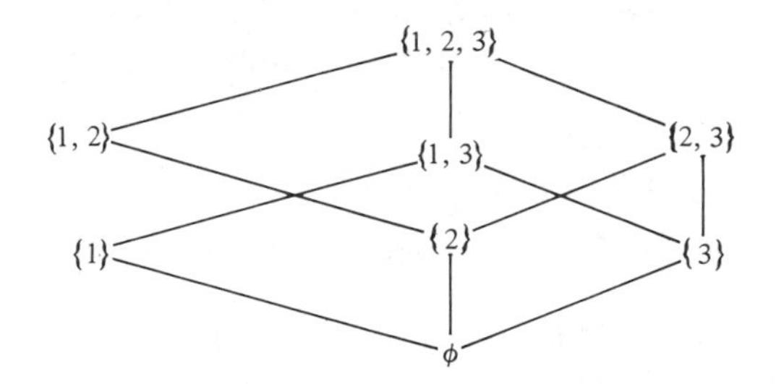

Figure 2. Lattice of subsets.

without any use of membership. In Chapter XI we will see further examples of sets treated without the use of elements.

There are corresponding definitions for other inclusion relations. In general (and in view of diagrams like that above) a poset is said to be a *lattice* when it has a top element 1, a bottom element 0 and when each pair of elements have a least upper bound (called their *join*) and a greatest lower bound (called their *meet*). The lattice of subobjects of an algebraic object is a way of describing some of the structure of that object.

10. Calculus, Continuity, and Topology

Many notions besides those of transformation groups arise from the mathematical analysis of motion. The complex motions of the planets and the varying velocities of falling bodies suggest the idea of "rate of change": Velocity as rate of change of distance or acceleration as rate of change of velocity. These ideas were codified in the notion of the derivative, subsequently formalized (Chapter VI) in the rigorous foundation of the calculus, as based on the axioms for the real numbers. This uses the definition of the derivative by means of limits and thus the consideration of a class of "good" functions–those which are differentiable. As a first example of this circle of ideas, we examine here another good class–the functions which are continuous.

A rigid motion $M: F \to F$ of a figure is continuous because (by rigidity) the distance from Mp to Mq must equal that from p to q. For a function $f: \mathbf{R} \to \mathbf{R}$ on the real numbers $\mathbf{R}$ continuity means considerably less: Just that fx and fy will be close if the originals x and y are sufficiently close. This formulation is still pretty vague; it should mean that one can make fx and fy "as close as you please" by requiring x to be "suitably close" to y. This is still vague. "As close as you please" should mean "within a specified measure δ (a positive real number) of closeness; "suitably close" should mean that one can specify a measure of closeness (again a positive real number ϵ) which will do the job. All this (and we have telescoped a long and painful historical development) comes down to make the familiar (but meticulous) $\epsilon - \delta$ definition of continuity: A function f: $\mathbf{R} \to \mathbf{R}$ is *continuous* at a point $a \in \mathbf{R}$ if

$$\text{For all real } \epsilon > 0 \text{ there is a real } \delta > 0 \text{ such that, for all } x \text{ in } \mathbf{R}, \quad (1)$$

$$\textit{If } |x - a| < \delta, \text{ then } |f(x) - f(a)| < \epsilon. \quad (2)$$

If this statement holds for *all* points $a \in \mathbf{R}$, the function f is called continuous; the class of all such continuous functions is called C.

Note that the statement involves both propositional connectives ("if . . . then") and the so called "bounded" quantifiers (For all real numbers, there exists a real number). Thus it is that careful formulations lead to the use of concepts of formal logic.

Topological and metric spaces arise from analysis of this definition of continuity. The inequalities used in the definition arise from ideas of approximation (approximations of the value $b = f(a)$ to within the accuracy ϵ) and so implicitly involve the open interval $I_\epsilon(b) = \{y \mid |y-b| < \epsilon\}$ of center b and "radius" ϵ. In the familiar representation of the function f by its *graph* (the set of points $(x, f(x))$ in the plane), this open interval appears as an open horizontal strip of width 2ϵ around $y = f(a)$ (Figure 1). The definition is concerned with those points $x \in \mathbf{R}$ for which $f(x)$ lands in this interval $I = I_\epsilon(b)$—this set of points is usually called the *inverse image* of I under the function f, in symbols:

$$f^{-1}I = \{x \mid x \in \mathbf{R} \text{ and } f(x) \in I\}.$$

Indeed, if $x_0 \in f^{-1}I$ (that is, if $f(x_0) \in I$), then one can prove from the definition of continuity that there is an open interval (on the x axis) of center x_0 wholly contained in $f^{-1}I$. This amounts to the

Theorem. *The function f: $\mathbf{R} \to \mathbf{R}$ is continuous for all $a \in \mathbf{R}$ if and only if the inverse image $f^{-1}I$ of every open interval of $\mathbf{R}$ is a union of open intervals.*

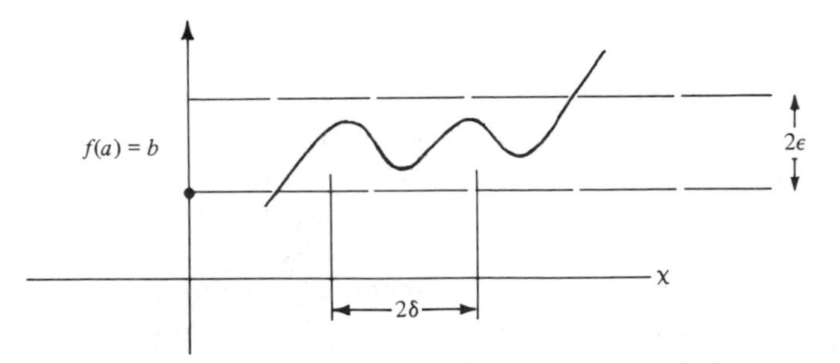

Figure 1

In other words, continuity can be described wholly in terms of open intervals.

Continuity is also needed for functions f defined on only a part of the real line, for functions of two or more variables, or for functions defined on a surface, etc. But this new definition requires no new ideas. The absolute value $|x - a|$ which appears in the definition (1) and (2) is just the distance $\rho(x,a)$ on the line $\mathbf{R}$ from the point x to the point a. Hence the same definition will work with any suitable notion of distance–that is, for a metric space.

Definition. If X and Y are metric spaces, a function $f\colon X \to Y$ is continuous at a point $a \in X$ if for all real $\epsilon > 0$ there is a real $\delta > 0$ such that for all x in X, if $\rho(x,a) < \delta$ then $\rho(f(x),f(a)) < \epsilon$.

In particular, this defines continuity for functions of two real numbers x,y. Just regard (x,y) as the (coordinates of) a point in the plane $\mathbf{R} \times \mathbf{R}$ with the usual (pythagorean) metric $\rho_1 = \rho$,

$$\rho((x,y),(a,b)) = [(x - a)^2 + (y - b)^2]^{1/2},$$

Then in the definition of continuity the open interval consisting of all x with $|x - a| < \delta$ is replaced by the *open disc* with center (a,b) and radius δ. This disc consists of all points (x,y) in the plane with

$$(x - a)^2 + (y - b)^2 < \delta^2. \tag{3}$$

As in the theorem above, one can prove that a function $f\colon \mathbf{R} \times \mathbf{R} \to \mathbf{R}$ of two real variables is continuous if and only if the inverse image of each open interval in $\mathbf{R}$ is a union of open discs in $\mathbf{R} \times \mathbf{R}$.

But the plane is also a metric space with a distance ρ_2 described as the "shortest path along the rectangular grid", so that

$$\rho_2((x,y), (a,b)) = |x - a| + |y - b|,$$

or with the distance given by

$$\rho_3((x,y), (a,b)) = \operatorname{Max}(|x - a|, |y - b|).$$

For a function $f(x,y)$ of two real variables, any one of these distance formulas may be used to define the continuity of f in the usual $\epsilon - \delta$ style. These different distances all yield the *same* continuous functions, so that continuity, viewed invariantly, depends not on distance but on something else more intrinsic. What is it? The answer is well-known. Each distance ρ_1, ρ_2, or ρ_3 will determine a series of "open circular discs"–circular in that metric–of the respective forms (interior of) circle, diamond, or square (Figure 2). Within each disc we can readily draw a smaller disc of the

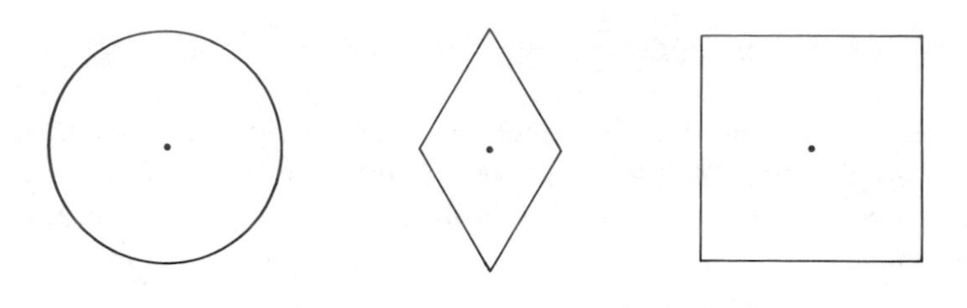

Figure 2

other forms with the same center—so we get the same $\epsilon - \delta$ continuous function by just choosing different δ's to a given ϵ.

What is the intrinsic formulation? In the alternative description of continuity stated in the theorem above there appeared unions of open intervals and unions of discs. In any metric space X the open *disc* with center a and radius δ may be defined to be the set of all points x in X with $\rho(x,a) < \delta$. Now define an *open set* U in X to be any union (finite or infinite) of open discs. Equivalently, a subset U is *open* in X if to each point $a \in U$ there is a $\delta > 0$ such that $\rho(x,a) < \delta$ implies $x \in U$ – every point a in U is contained in an open disc centered at a and within U. Now the continuity of $f: X \to Y$ is expressed in terms of open sets: f is continuous if and only if the inverse image of any open set of Y is an open set of X. This is the desired intrinsic formulation, independent of the perhaps accidental choice of a metric. Specifically, the three different metrics $\rho = \rho_1$, ρ_2, and ρ_3 described above for the plane all yield the same open sets, because any union of circular open discs is also a union of open squares or of open diamonds, and conversely. In this way, the notion of "open set" is more intrinsic than that of distance.

This suggests that a space can be defined directly in terms of its open subsets. A *topological space* is a set X in which certain of the subsets U are distinguished and called the *open* sets, and in which these open sets are required to satisfy the following three axioms:

1. The intersection of two open sets is open;
2. The union of any collection of open sets is open;
3. X itself and the empty subset $\varnothing \subset X$ are open.

A *topology* on a set X is then the specification of open subsets which satisfy these axioms. Thus any metric X determines a topology, in which the open sets are the unions of discs open in the metric. There are also topologies not defined by way of a metric. For example, there is a topology on the set $\mathbf{N}$ of natural numbers for which the open subsets are the empty subset and all those subsets S with finite complement (in $\mathbf{N}$). There are many other examples of topologies.

The definition of continuity (the inverse image of any open set is open) now applies to any function $f: X \to Y$ between topological spaces

X and Y. The three axioms on open sets are enough to prove most of the basic facts about continuous functions–for example, the fact that the composite $x \mapsto g(f(x))$ of two continuous functions g and f is again continuous.

To describe continuity at a single point of a space, one may use the notion of "neighborhood"" A *neighborhood* of a point a in a topological space X is any open set of X which contains a. One then says that a function $f: X \to Y$ between topological spaces in continuous at one point $a \in X$ if to each neighborhood V of $f(a)$ there is a neighborhood U of a for which $f(U) \subset V$. This definition agrees with the previous notion of continuity at a point for a metric and expresses the intuitive idea that "nearby" points in U go into nearby points in V. Moreover, f is continuous if and only if it is continuous in this sense at each point $a \in X$.

Extensive experience has shown that this description of a "topology" in terms of open sets and neighborhoods is extraordinarily effective in formulating all sorts of Mathematical facts in a geometric form. The concept of "topology" has been appropriately abstracted from the many examples of "continuity".

The notion of a topological space was first presented by F. Hausdorff in a famous (and beautiful) book *Mengenlehre*. His definition was formulated differently, in terms of selected neighborhoods, and included an added axiom (the Hausdorff separation axiom): Two distinct points have disjoint neighborhoods. A topological space with this property is called a *Hausdorff space*.

We have now seen a number of Mathematical concepts which are described as *sets-with-structure*. Thus a linearly ordered set is a set equipped with a binary relation $<$ having certain specified properties. A group is a set equipped with a binary, a unary, and a nullary operation, which together satisfy certain identities. A Boolean algebras is similarly a set with appropriate operations. A topological space X is a set-with-structure, where in this case the "structure" consists of a specified collection of the subsets of X, namely the collection of all open sets. This kind of structure is quite different in style from the algebraic structures. There are also structures of a mixed kind. For example, there are cases of motions (e.g., translations or rotations) which deal with a set of motions which is both a group and a space. This leads to the notion of a *topological group*. Such a group is a set G which is both a group and a topological space and in which the group operations–both the product $G \times G \to G$ and the inverse $G \to G$–are continuous. It is this last condition which ties the two structures together (to make the definition complete, one must know how the topology on G induces, in a natural way, a topology on $G \times G$). As in this case, most composite axiomatic structures (combinations of two kinds of structure on the same set) involve one or more axioms expressing the formal connection between the two structures–here between the group structure and the topology.

Here is another example of a mixed structure: A *linearly ordered group* G is a set which is a group and also has a linear order, with the added axiom that $a \leqslant b$ in G and $1 \leqslant c$ implies both $ac \leqslant bc$ and $ca \leqslant cb$. This added axiom is the one which ties together the two structures of order and of multiplication. There are many examples of such linearly ordered groups–positive rational numbers or real numbers under multiplication, or integers with multiplication replaced by addition.

We will see that many Mathematical notions can be described as set-with-structure.

11. Human Activity and Ideas

This chapter, starting from the study of number, space, time, and motion, has led to the description of various formal notions–especially cardinal number, permutation, linear order, group, continuity, and topology. Each notion represents a type of formalization in Mathematics. The formalization may take the guise of a rule (e.g., a multiplication table), a simple definition (the same cardinal number), a more subtle definition (that of continuity), a list of axioms describing the common properties of several systems (linear order), a less evident such list (a group), or a list of axioms deemed sufficient to describe exactly one object (the real numbers as an ordered set). In some cases, like that of topological space, the axioms serve to help understand the common features of a wide variety of situations.

These formal notions arise largely from premathematical concerns which can best be described as "human cultural activities". For this reason, our analysis of the genesis of Mathematics will note a number of such activities. Often it is illuminating to say that the activity leads first to a somewhat nebulous "idea", which is finally formalized, perhaps formalized in several different ways. For example, the process of counting suggests the idea of "next"–the next item to be counted or the next number to be used in the count or the next thing in some ordered list. This general idea "next" may then be formalized by a rule for adding one to each decimal or by the axioms on the operation which provide to each natural number its successor. The idea "next" appears in other forms: The next (infinite) ordinal beyond a given set of ordinals or the next step (after choice of alternative) in some computer program. Or the frequent observation of steady changes may suggest the (nebulous) idea of steady change, formalized (say) by what we called a parametrized motion.

This type of source for Mathematical form, in the cases we have noted so far, may be summarized in a table, where each activity suggests an idea and its subsequent formalizations (Table 1).

This tabulation is intended to be suggestive but not dogmatic. Each "idea" is intended to have some intuitive content; it may serve as the car-

Table 1.1

Activity	Idea	Formulation
Collecting	Collection	Set (of elements)
Counting	Next	Sucessor; order Ordinal number
Comparing	Enumeration	Bijection Cardinal number
Computing	Combination (of nos)	Rules for addition Rules for multiplication Abelian group
Rearranging	Permutation	Bijection Permutation group
Timing	Before and after	Linear order
Observing	Symmetry	Transformation group
Building, shaping	Figure; symmetry	Collection of points
Measuring	Distance; extent	Metric space
Moving	Change	Rigid motion Transformation group Rate of change
Estimating	Approximation	Continuity Limit
	Nearby	Topological space
Selecting	Part	Subset Boolean algebra
Arguing	Proof	Logical connectives
Choosing	Chance	Probability (favorable/total)
Successive actions	Followed by	Composition Transformation group

rier for the well known phenomenon of "Mathematical Intuition". The same idea may arise from different activities, and may well be the background for several different formalizations. We have tried to use familiar words to describe each idea but this does not represent any established consensus or precise definition. On the other hand, each notion, as conventionally formalized, has a rigorous definition (within some context).

The table is by no means complete; as the reader keeps it in mind he may find new examples in subsequent chapters.

Even after the basic Mathematical notions have been developed out of these activities and ideas, there continue to be inputs from outside Mathematics. These inputs often take the form of Mathematical questions arising in other sciences and requiring application of Mathematics. Thus the primitive sort of study of motion noted above becomes later the subject of dynamics (in physics) or that of celestial dynamics in astronomy. The study of social changes in part becomes the study of marginal costs or econometrics. In general, under the genesis of Mathematics we intend to include all sorts of inputs from scientific and other cultural activities.

Some formal Mathematical notions have a more complex origin. Such is the case for the notion of a "set". The idea of a collection is surely there when we count, but on this level it is hardly a useful candidate for formalization. Infinite collections also arise, perhaps at first in observations and in Euclid's proof that there are infinitely many prime numbers–but then one soon has other infinite collections. They are often subsets of (say) the set of all natural numbers, but the notion of a subset is not really forced on our attention until we try to describe the completeness of the ordered set of reals (Every bounded subset has a least upper bound) or the principle of Mathematical induction (Every set of natural numbers containing zero and the successor of each of its elements contains all the numbers). Even here we might dispense with subsets: Completeness can be described by convergent sequences and induction can be described by properties. But Boolean algebra is unthinkable without subsets. The more sophisticated notion of a set whose elements are themselves sets does arise later. The set of integers modulo 6 will be described as the set whose elements are the congruence classes such as $\{1,7,13,19, \ldots\}$, right now a topological space is most clearly defined as a set with a specified set of its subsets (namely, the open subsets). However, in both of these cases the use of sets of sets can be avoided by using relations: the relation of congruence module 6 (Gauss) or the relation stating that the subset U is a neighborhood of the point p. The real motivation for the full use of set theory lies much deeper, and will be explored in Chapter XI, where we will note the curious fact that abstract set theory arose from the study of trigonometric series!

12. Mathematical Activities

The genesis of the more complex mathematical structures tends to take place within Mathematics itself. Here there are a variety of processes which may generate new ideas and new notions. We list a few of these processes in tentative form for further refinement after our more detailed studies.

(a) *Conundrums.* Finding the solution of hard problems is one of the driving forces of Mathematical development. Fermat asserted without proof that the equations $x^n + y^n = z^n$ for $n > 2$ have no solutions in integers. As we will see in Chapter XII, this apparently innocent diophantine equation was one historical source of the whole development of algebraic number theory in the 19th century–and so was the principal origin of such algebraic notions as that of "ideal"—although the arithmetic theory of quadratic forms also played a role.

The problem of solving polynomial equations by formulas involving radicals was a historically important conundrum. For quadratic polynomials the solution is easy, by the familiar "quadratic formula". Early algebraists found no such formulas for solutions of the general equation of 5th degree. Using permutations of the roots, it was eventually showed (by Lagrange) that such a solution was impossible—but the first real insight into the reasons for the impossibility came with Galois in 1832, (see Chapter V); this was the point where the notion of a group first explicitly arose.

Our presentation has in effect argued that the notion of a group could have arisen otherwise—but in historical perspective the solution of different Mathematical problems is a vital element in the progress of the science (and is often viewed as *the* characteristic aspect of that science).

(b) *Completion*. The whole list of natural numbers arises by starting with the first few 0, 1, 2, 3, . . . , 9 and asking that there always be a successor. Then subtraction, alas, is not always possible—until one creates all the integers. To insure the possibility of division, one must then have all the rational numbers, and so on to the real numbers and then to the complex numbers. In many other cases, the need to complete a structure under some partially defined operation brings out a new structure.

(c) *Invariance*. A non-trivial homogeneous equation

$$ax + by + cz = 0$$

has infinitely many non-zero solutions, but all can be expressed as sums of multiples of some two solutions—because, as we know, the set of all solutions (x,y,z) is a plane through the origin in 3-space and any vector lying in that plane is the sum of multiples of two suitable such. Again the solutions of the homogeneous linear second order differential equation $d^2x/dt^2 = -k^2x$ all have the form

$$x = A \cos kt + B \sin kt\,;$$

they are expressed here as linear combinations of two particular solutions $\cos kt$ and $\sin kt$. These two parallel situations serve to suggest the structure of a vector space (Chapter VII), the idea of a basis for such a space, and the need to describe its properties independently of any one choice of basis.

(d) *Common Structure (Analogy)*. This example exhibits also the motive of finding a common structure (here, that of a vector space) underlying different but similar phenomena (here, geometry, linear equations, and linear differential equations). Another such instance is given by a description (§4) of linear order. The symmetry group as the commonality of two superficially different symmetrical figures is another example. Again, the definition of continuity for a function on a metric space exhibits the common features of the definition of continuity for functions of one or several real variables.

(e) *Intrinsic Structure*. Sometimes a (hidden) formal structure will seem to explain the facts. Thus a permutation group on n letters always has order dividing $n!$; the explanation is provided by the well known theorem on the possible orders of a subgroup of finite order. The observation of §9 above that different metrics in the plane yield the same continuous functions $f(x,y)$ of two variables is explained by shifting attention from the (various) metrics to the (common) topology which they define in the plane. There will be many such cases in which the appropriate abstract notion serves to provide an understanding of Mathematical phenomena. Perhaps the progress of Mathematics depends on a counterpoint between solving conundrums and searching for the concepts which provide better understanding.

(f) *Generalization*. The process of generalization in Mathematics takes a number of different forms. There are generalizations from many concrete instances to a "general" law; thus $2 + 3 = 3 + 2$ and $4 + 7 = 7 + 4$ lead to the commutative law $x + y = y + x$ for addition. Or some case, already general, is extended to be more so: Finite cardinals suggest infinite cardinals, but the finite laws do not all carry over. Similarly, finite groups generalize to infinite groups–but theorems on the orders of subgroups drop out.

The generalization from analytic geometry in 2 and 3 dimensions to that in n dimensions is straightforward (but it took a long time historically); the generalization was not really relevant or interesting until there was considerations of "events" described by more than three numbers (coordinates) *and* a recognition that a geometrical language helped to visualize and describe such events. Axiomatic structure (like the structure of a group) may be generalized by deletion of axioms. One of our aims is to examine enough cases to obtain a more complete typology of generalization in Mathematics.

(g) *Abstraction*. This process typically consists in getting (some of) the same results under weaker or more "abstract" hypotheses. A standard example of abstraction is the shift from groups of transformations to "abstract" groups. In the concrete case, the elements of a transformation group are actual transformations acting on a specified set, and the group multiplication is the composition of transformations; this multiplication is automatically associative. In the abstract case, the elements of a group and their multiplication can be arbitrary (i.e., "abstract")–but the multiplication must satisfy the necessary group axioms, in particular the associative law. Under this process of abstraction, two transformation groups on different sets may turn out to be "the same" (i.e., isomorphic). However, no new groups turn up in this process, in view of the famous theorem of

Cayley (§8) which asserts that every (abstract) group is isomorphic to a group of transformations. Hence all those theorems about transformation groups which do not involve the elements being transformed (e.g., which are not theorems about transitive or primitive permutation groups) remain valid for abstract groups. Much the same happens when the "algebra of sets" is abstracted to Boolean algebra.

Other examples of abstraction can involve real generalization. Thus Hilbert and others considered number rings—subsets of the complex numbers which are closed under addition, subtraction, and multiplication. The abstract notion of a ring (due to Emmy Noether) is that of a set of elements with operations of addition, subtraction, and multiplication subject to suitable axioms (§IV.3)—and such an abstract ring need not be isomorphic to any ring of numbers.

In these cases, abstraction amounts to considering a structure by neglecting the provenance of its elements, but keeping "all" the operations on these elements, with (one hopes) all the formal properties of these operations formulated as axioms and their consequences.

(h) *Axiomatization.* This process typically asks: Given a long list of "all" the theorems on a given topic, can one deduce them all from a suitable shorter list—a list which will then constitute the axioms for that topic. The right choice of axioms can lead to great insight and better understanding. Thus Euclid produced congruence axioms for triangles sufficient to prove all the known theorems about congruence of such figures. In more recent times, one first observed that much of the geometry of three-dimensional Euclidean space could be described in terms of vectors $v = (x,y,z)$ with vector addition

$$(x,y,z) + (x',y',z') = (x + x', y + y', z + z')$$

and the multiplication $(x,y,z) \mapsto (ax,ay,az)$ of a vector by a scalar. Axiomatization then asks: what short list of properties of addition and multiplication by a real scalar will give all these theorems? The answer is the usual list (Chapter VII) of axioms for a vector space over the reals, plus the assumption that there is a basis of three vectors. Many other successful examples of axiomatization spring to mind. Processes of abstraction, as recounted above, also will usually involve axiomatization. The task of axiomatization can be hard, especially for such quasi-mathematical subjects as mechanics, thermodynamics or (in economics) utility theory.

(i) *The Analysis of Proof.* One way to find axioms is to ask for a minimum list of properties needed to carry out a given proof or proofs. For example, the axioms for a commutative ring are essentially the minimal list needed for the standard algebraic manipulations of addition, subtraction, and multiplication. Again, new Mathematical ideas can sometimes be disentangled by analyzing what makes a given proof tick. A striking example is the proof that a function $f\colon I \to \mathbf{R}$, continuous on a closed interval I of the reals, is uniformly continuous there. A straightforward direct proof can be given, using the basic properties of the real numbers. This proof, originally given by the German Mathematician Heine, and further developed by the French Mathematician Emil Borel, leads to the Heine–Borel theorem: If the closed interval I is the union of an (infinite) collection of opens sets U_i, so that $I = \cup U_i$, then it is a union of a finite number of these open sets,

$$I = U_{i_1} \cup \cdots \cup U_{i_h},$$

for some finite list of indices $i_1, \ldots, i_h$. In current terminology, this property states that I is a compact subset (of $\mathbf{R}$) and so leads to the idea of compactness for topological spaces.

At the end of our study of structure, we will return to a more detailed examination of these processes, internal to Mathematics, for the generation of new notions. They play a role counterpuntal to the input of problems from the sciences outside Mathematics. Both are accompanied by the continued search for deeper properties of the notions already at hand.

13. Axiomatic Structure

In the next three chapters we will indicate how number, space, and time can be described by axioms; that is, by axioms for the natural numbers, the Euclidean plane, and the real line which describe these structures uniquely. In classical terminology, these axiom systems are *categorical*, in the sense that any two "models" of the axioms, taken within an inclusive set theory, are isomorphic–as in the case described in §4 for the reals (we will also note another "first order" version of these axioms where there can be non-isomorphic non-standard models). Thus these structures are closely attached to the traditional view that (say) the axioms of Euclidean geometry describe one specific object–physical space.

In this first chapter, we have deliberately followed a different order of axiomatics, emphasizing those systems of axioms (linear order, group, metric space) which have many essentially different models. This use of axioms is historically more recent than the categorical axiomatization of geometry. In particular, it allows for the view that the formal systems studied in Mathematics come in a great variety and are intended primarily to help organize and understand selected aspects of the "real world" without being necessarily exact descriptions of a part of that unique world. For example, our presentation allows that the first step in the formalization of space could be the description of figures and chunks of space as models of metric space and not as subsets of Euclidean space. This is by no means the conventional view.

Nevertheless, this chapter has started from the conventional idea of Mathematics as the science of number, time, space, and motion, to go beyond these topics to related more general formal notions of cardinal number, permutation, order, transformation, group, and topological space. Mathematical experience, as suggested in our subsequent chapters, shows that each of these notions plays a basic role in Mathematics. We have deliberately put them first to let the reader judge their importance. This does not mean that they need be prior to the classical description of number and space, but simply that they appear in parallel to these classical notions.

The linear order of a book does not allow the actual presentation to be in parallel.

CHAPTER II

From Whole Numbers to Rational Numbers

1. Properties of Natural Numbers

Various human activities such as listing, counting, and comparing lead, as we saw, to the natural numbers

$$\mathbf{N} = \{0,1,2,3,4,\ldots\}$$

and to the operations of addition, multiplication, and exponentiation on these numbers. These operations have a variety of general properties. For example, addition for all natural numbers k, m, and n satisfies the equations

$$m + 0 = m, \qquad m + n = n + m, \tag{1}$$

$$k + (m + n) = (k + m) + n. \tag{2}$$

These rules can be proved from the definitions of the operations. For example, the *commutative* law (1) holds because, when two disjoint finite sets are combined, the cardinal number of the combined set does not depend on which of the two sets is taken first. On the other hand, the rules are *formal* in the sense that they can be used directly without attention to their "meaning". For example, the *associative* law (2) tells me that if I add a long column of figures in three successive groups, subsequently combined, the final result will be the same, irrespective of the order in which the three are combined. A similar rule will work for more than three groups. Moreover, these (long-established) rules are inviolate: If it doesn't turn out as they specify, I know that I have made a mistake somewhere. This is the merit of a formal rule: Once firmly established, it can be applied mechanically and is an infallible guide.

Multiplication has corresponding formal properties:

$$m \cdot 1 = m, \qquad mn = nm, \tag{3}$$

$$k(mn) = (km)n. \tag{4}$$

Together, addition and multiplication satisfy the *distributive* law

$$k(m + n) = km + kn. \tag{5}$$

Again, this law can be used formally, without attention to its origin in the definitions of addition and multiplication, as suggested in the following display:

k [m | n] = k [m] [n] (6)

There are many other properties of these operations. For example, every square, on division by 4, leaves a remainder 0 or 1 (never 2 or 3). If $b > 1$, then every natural number n can be written in terms of b as

$$n = a_k b^k + a_{k-1} b^{k-1} + \cdots + a_1 b + a_0 \tag{7}$$

for some natural k and with coefficients a_i all satisfying $0 \leqslant a_i < b$. In particular, if $b = 10$, this is the decimal expansion of n, and its properties lead to the familiar formal rules for manipulating decimals.

2. The Peano Postulates

Each of these properties, and many more, of the addition and multiplication of natural numbers could be demonstrated directly from the definitions of these operations on finite cardinal numbers. Such proofs would be cumbersome. However, a remarkable fact emerges: Both addition and multiplication can be described just in terms of the number zero and the single operation "add 1", and their properties can be derived from a short list of axioms on the single operation. These axioms are the Peano postulates. The idea is that the natural numbers can be listed, starting with zero, so that to each number n there is always a "next" number, its *successor* $n + 1$, and so that this process exhausts all the natural numbers. Thus we can state formally:

The (natural) numbers $\mathbf{N}$ with zero and "successors" s form a collection with the following five properties (the Peano postulates):

(i) 0 is a number;
(ii) If n is a number, so is its successor sn;
(iii) 0 is not a successor (i.e., sn is never 0);
(iv) Two numbers n, m with the same successor are equal (i.e., if $sn = sm$, then $n = m$);
(v) Let P be a property of natural numbers. If 0 has P, and if sn has P whenever n does, then P holds for all natural numbers.

This is a typical description of a structure by axioms. There are certain primitive (or undefined) terms: here the terms "number", "zero", and "successor". The statements of the axioms use only these terms and the standard logical connectives: "if . . . then", "not", "and", "equality", "for all", "there exists". Such a statement is called a formula (or a formal statement) in the language of Peano arithmetic (for more detail, see Chapter XI). In particular, a "property" of the number n, as used in postulate (v), should be one which is described by such a formula, involving n.

The induction axiom (v) is vital; it expresses the intuitive idea that taking successors exhausts all the natural numbers. It is very useful practically, in proving all sorts of formulas involving general n (for example, the formula for the sum $1 + 2^2 + 3^2 + \cdots + n^2$ of the first n squares) and for proving such results as the binomial theorem.

Sometimes the induction axiom is formulated in terms of sets rather than properties, as follows:

(v′) If S is a set of numbers containing 0 and if every n in S has its successor in S, then S contains all (natural) numbers.

This axiom implicitly refers to "all" subsets of **N**, so it is sometimes called a "second order" axiom, because the quantifier "all" is applied not just to elements of **N**, but also to subsets. More specifically, this form of the axiom means that we are considering the natural numbers in a context of sets, and that proofs of theorems about natural numbers may use not just the Peano axioms, but properties of sets, as these might be formulated in axioms for set theory. In this respect, it is like the completeness property of the real numbers (§I.4).

The set-theoretic induction axiom (v′) does include the property-theoretic version (v), because the usual axioms for set theory do specify that every (formal) property of elements of a set **N** does determine a subset of **N**. This transition from properties of numbers to sets of numbers is a familiar one. The use of properties may be called "intensional", because a property is described by a formula. Thus the properties "n is odd" and "n leaves the remainder 1 on division by 2" are verbally different, but describe the same set $\{1,3,5, \ldots\}$. On the other hand, as in §1.9, the use of sets is extensional: As soon as two sets include the same elements, they are equal. The "extent" of the set is all that matters.

However, the induction axiom (v) for properties is weaker than that for subsets. Since a property, as explained, can be expressed in a finite list of words in a fixed language, the number of properties of natural numbers is denumerable. However, for the usual notions of sets, a "diagonal" argument (see Chapter XI) shows that the number of subsets of **N** is not denumerable but larger. This observation has consequences. One can formulate theorems about natural numbers which are true within set theory but which cannot be proved from the Peano axioms with induction in the form (v). Using sets, one can also construct a model of these axioms which is "non-standard" in the sense that the numbers are not exhausted by taking successors (Chapter XI). However, this is not so for the set-theoretic version with axiom (v′). As we will soon see, one can prove that these Peano postulates do determine the natural numbers up to an isomorphism.

From the Peano postulates one can define all the familiar arithmetic operations by *recursion*—specifying in succession the results of the operation. Thus addition (explicitly, the operation "add k") is defined by the two recursion equations

$$k + 0 = k, \qquad k + sn = s\,(k + n); \tag{1}$$

multiplication (multiply by k) is defined by the recursion

$$k \cdot 0 = 0, \qquad k(sn) = k + kn; \tag{2}$$

while exponentiation to the base k is given by the recursion

$$k^0 = 1 \qquad k^{sn} = k \cdot k^n. \tag{3}$$

The other properties (commutative, distributive, etc.) of these operations may then be proved by induction on n. Each of these pairs of formulas defines a function $f(n) = k + n$, kn, or k^n by first giving the value $f(0)$ and then the value $f(sn)$ in terms of some known function g of $f(n)$; in (1) g is s, in (2) it is $k + -$, and in (3) it is $k \cdot -$. The principle involved may be stated as a

Recursion Theorem. *If X is a set, a an object of X and $g\colon X \to X$ a function, then there is a unique function $f\colon N \to X$ with*

$$f(0) = a, \qquad f(sm) = g(fm), \tag{4}$$

the latter for all natural numbers m.

A proof of this theorem uses axiom (v′) and must depend upon the set-theoretic definition of "function" as a table of values (see Chapter V), but the idea behind any such proof is a straightforward use of induction, as follows. For each n *let* **n** denote the finite set $\mathbf{n} = \{0,1,2,3, \ldots, n\}$ and consider the following property P of n: There is a unique function $f_n\colon \mathbf{n} \to X$ which satisfies (4) for $m = 0,1,2, \ldots, n-1$. Clearly, P holds for $n = 0$. If n has the property P, we have the (unique) function f_n whose values may be listed as

0	1	2	$\cdots$	n
$f_0 0$	$f_1 1$	$f_2 2$	$\cdots$	$f_n n$

The desired value of f at $n+1$ can then be adjoined to this table; it is $g(f_n n)$; from the (formal) definition of function it follows that this gives a (unique) function f_{n+1}: $\mathbf{n} + 1 \to X$. Thus $n + 1$ has the property P. Hence by induction all numbers n have P. This gives a list of functions $f_0, f_1, \ldots, f_n, f_{n+1}, \cdots$ which "fit together" and so (again by the definition of a function) produce the desired function f on all of $\mathbf{N}$.

The recursion theorem can be stated conveniently in diagrammatic form, if we observe that an object a of the set X is the same thing as a function a: $1 \to X$ from the one-point set 1 into X (the function whose only value is a). Thus we are given the functions represented by solid arrows in the diagram

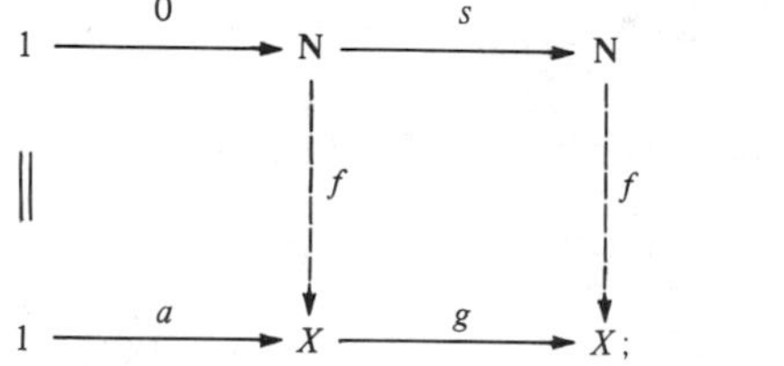

(5)

the recursion theorem then asserts that there is a unique function f (shown by the two dotted arrows) such that the diagram "commutes"—that is, such that $f0 = a$ (left-hand square commutes) and $g \cdot f = f \cdot s$ (right-hand square commutes).

In this theorem, f is said to be defined by "primitive recursion" from g. More elaborate versions of the theorem are true. For example, one might assume instead of g a function $h = h(x,n)$ of two variables (a function h: $X \times \mathbf{N} \to X$) and replace the second equation of (4) by the condition $f(sm) = h(fm,m)$. Recursions with a "parameter"—such as the parameter k in (1), (2), and (3)—are also valid.

By recursion one may also prove the

Uniqueness Theorem. *The set-theoretic Peano postulates determine the collection* $\mathbf{N}$ *of natural numbers uniquely, up to an isomorphism of the structure given by* 0 *and successor.*

If $\mathbf{N}'$ is another set with a distinguished object $0'$ and a "successor" function s': $\mathbf{N}' \to \mathbf{N}'$, a *homomorphism* of the $0,s$ structure is a function f: $\mathbf{N} \to \mathbf{N}'$ with

$$f(0) = 0', \qquad fsn = s'fn, \qquad \text{for all } n. \tag{6}$$

Such an f is an *isomorphism* of $\mathbf{N}$ to $\mathbf{N}'$ if it is a bijection (a one-to-one map of $\mathbf{N}$ onto $\mathbf{N}'$). By the recursion diagram (5), with X replaced by $\mathbf{N}'$, there is such a homomorphism. Since $\mathbf{N}'$ with s' also satisfies the Peano postulates, the same recursion diagram produces a unique homomorphism g: $\mathbf{N}' \to \mathbf{N}$ in the opposite direction, as in the bottom row of the commutative diagram below:

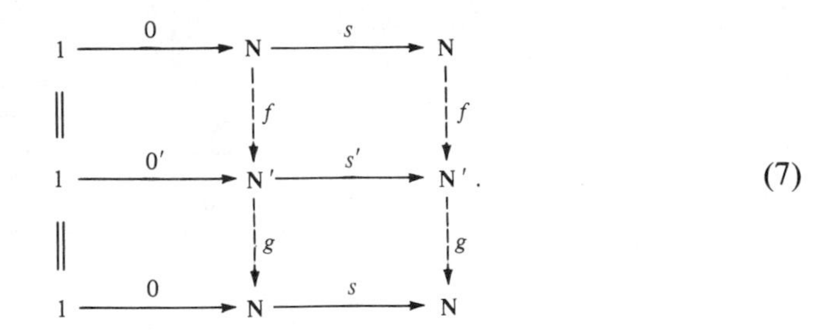

(7)

Now compare the composite function $g \cdot f$: $\mathbf{N} \to \mathbf{N}$ with the identity function I: $\mathbf{N} \to \mathbf{N}$. They both make the diagram

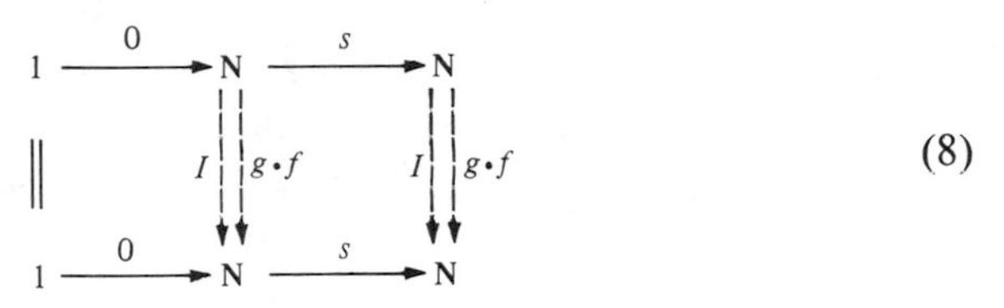

(8)

commute. Hence, by the uniqueness assertion of (5), $g \cdot f = I$. Similarly, $f \cdot g$ is the identity function. Thus f has g as a two-sided inverse under composition, and so is a bijection.

This result is typical of the axiomatic description of sets with structure. At best, such a description can determine the model only "up to isomorphism". As in this case, an isomorphism means a bijection from one model to another which "preserves" all the primitive terms involved in the axioms–as in (6) above. In this case, there are in fact many different but isomorphic models. For instance, if 100 is viewed as the zero, then the even natural numbers starting with 100 form a model for the Peano postulates when the assignment $n \mapsto n+2$ is taken to be the successor function.

3. Natural Numbers Described by Recursion

The Peano postulates are not the only possible axiomatic description of the natural numbers. Instead, one can take the recursion theorem as the (sole) axiom. In detail, this axiom assumes that the natural numbers are a set $\mathbf{N}$ with a distinguished object 0 and a function s: $\mathbf{N} \to \mathbf{N}$ which together satisfy (for all $a \in X$ and all g: $X \to X$) the recursion theorem, as pictured in the diagram (2.5). This approach to the natural numbers was first made explicit by Lawvere; it is described in some detail in (the first edition of) Mac Lane–Birkhoff.

The logical equivalence of the two approaches is readily verified. Thus, we have already seen that the Peano postulates imply the recursion axiom. Conversely, one may prove that the recursion axiom implies all the Peano postulates. The most interesting part of this demonstration is that for the axiom of mathematical induction, for a subset S of $\mathbf{N}$, as summarized in the following diagram:

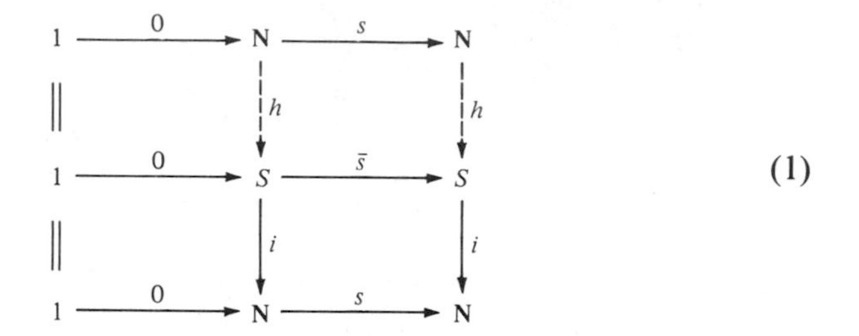

Since S is a subset of $\mathbf{N}$ each of its objects x is in $\mathbf{N}$, so the assignment $x \mapsto x$ is a function i: $S \to \mathbf{N}$ (the inclusion function, as displayed in the lower part of (1)). The induction assumptions on S state that 0 is in S, giving the function 0: $1 \to S$, and also that each n in S has its successor sn in S, so that the assignment $n \mapsto sn$ for n in S is a function $\bar{s}$: $S \to S$ as shown. By the recursion axiom, there is a function h: $\mathbf{N} \to S$ with $h0 = 0$, $\bar{s}h = hs$, as displayed in (1). The composite function $f = i \cdot h$: $\mathbf{N} \to \mathbf{N}$ then satisfies the same recursion conditions $f0 = 0$, $sf = fs$ as the identity function $\mathbf{N} \to \mathbf{N}$. Since our axiom asserts that the conditions determine the function uniquely, f must be the identity. Thus, each number n is $n = fn = i(hn)$, which states that n is the element hn of S and hence that the elements in S include all the elements n of $\mathbf{N}$.

This case illustrates a general point: The axioms needed to describe a Mathematical structure (here to describe the structure of $\mathbf{N}$, unique up to isomorphism) are themselves by no means unique. The recursion theorem of (2.5) is an especially convenient form of axiom; it states that the diagram $1 \to \mathbf{N} \to \mathbf{N}$ is "universal" (that is, maps uniquely into every other such diagram $1 \to X \to X$).

4. Number Theory

Once the Peano postulates are at hand, they yield all manner of specific results. Division is sometimes but not always possible, but if one tries to divide m by n one obtains a quotient q and a remainder r, which may be 0 but in any event less than n, as in the equation $m = nq + r$ with $0 \leqslant r < n$. This result is known as the *division algorithm*. Those natural numbers which have no divisors (except, of course, for themselves and 1) are the *primes*; they appear in a curious irregular order:

$$2, 3, 5, 7, 11, 13, 17, \cdots .$$

Every number n can be factored into a product of primes (some of which may be repeated). No matter how this factorization is obtained, the resulting prime factors are unique, except, of course, for their order. The proof of this unique factorization theorem rests on the division algorithm. From the prime factorization of two numbers one may read off their greatest common divisor; however, this also could be found directly from the numbers by the Euclidean algorithm, which is just an iteration of the division algorithm.

The curiously irregular sequence of primes noted above is infinite, by a proof which goes back to Euclid. One is soon led to try to estimate how thick the primes are. If $\pi(n)$ denotes the number of primes less than or equal to n, the prime number theorem (proved with more sophisticated means) will tell how fast $\pi(n)$ grows as n approaches infinity. Again, if we arrange all the numbers according to their remainder on division by 3, we get the following three arithmetic sequences

$$\begin{array}{ccccl} 0 & 3 & 6 & 9 & 12, \ldots, \\ 1 & 4 & 7 & 10 & 13, \ldots, \\ 2 & 5 & 8 & 11 & 14, \ldots . \end{array}$$

Except for the prime 3 in the first sequence, all the primes must fall in the last two sequences. It turns out that there are an infinite number of primes in each of these two arithmetic sequences, and that they are, in a sense, equally distributed between those two sequences. More generally, Dirichlet's theorem asserts that any arithmetic sequence $nd + r$, for fixed d and r and increasing n, will have an infinite number of primes, provided only that d and r have no common factors except 1.

Every number can be written as a sum of at most four squares or of at most nine cubes. These results have relatively elementary proofs; by much deeper analysis for Waring's problem, similar results hold for higher powers. By trial, one can verify that each small even number can be written as a sum of two primes. Goldbach (in 1742) conjectured that this was always true. To date, no one has proved this to be so. The best results to date are Vinogradoff's: Every sufficiently large odd number r is a sum of three primes, and Chen's: Every sufficiently large even number is a sum $p + b$, where p is a prime and b is either a prime or a product of two primes.

Problems in Diophantine equations ask for solutions in integers and in natural numbers. The equation $x^2 + y^2 = z^2$ has infinitely many (well-known) solutions in non-zero integers x, y, and z, but the equation $x^4 + y^4 = z^4$ has none. Fermat stated, and no one has yet proved, that $x^n + y^n = z^n$ for $n > 2$ has none. That the numbers of such solutions is finite has just recently been proved (the Mordell conjecture). Pell's equation $x^2 - Dy^2 = 1$ has an infinite number of integer solutions, of relevance to algebraic number theory.

This is but a small sample of the wealth of questions arising for the natural numbers. All these results are ultimate consequences of the structure specified with such simplicity by the five Peano postulates.

5. Integers

To keep accounts of gains and losses, subtraction is needed. Within the natural numbers, subtraction is not always possible, but it becomes possible when the set **N** of all natural numbers is expanded to the set **Z** of integers. One can formally define the integers (and arithmetic operations upon them) in several ways. Perhaps the simplest is that of adjoining to **N** a new copy of the positive numbers, each prefixed by $-$, as $-1, -2, -3, -4, \cdots$. Then addition of the old and new integers is defined for natural numbers n and m in **N** in cases:

$$\begin{aligned} n + (-m) &= n - m, && \text{if } n \geqslant m, \\ &= -(m-n), && \text{if } n < m, \\ (-n) + (-m) &= -(n+m). \end{aligned}$$

With this definition, subtraction is always possible in **Z**; moreover, similar definitions describe the appropriate multiplication and order in **Z**.

Another approach to **Z** observes that subtraction amounts to solving for x an equation $n + x = m$; the ordered pair (m,n) is then introduced formally so as to denote "the" solution to this equation. The familiar rules for manipulating differences $m - n$ translate to give definitions of sum and product of such pairs by the formulas

$$(m,n)+(m',n')=(m+m', n+n'),\ (m,n)(m',n')=(mm'+nn', mn'+m').$$

But beware: the pairs (m,n) and $(m+h,n+h)$ should count as the same, hence one defines $(m,n) = (r,s)$ if and only if $m + s = n + r$, and verifies that this artificial equality satisfies the expected rules; in particular, that sums and products of equals are equal. The integers, defined to be these pairs with this equality, do not literally contain the natural numbers from which we started, but the meaning of subtraction suggests that each n in **N** be identified with the pair $(n,0)$; this identification preserves addition and multiplication. Stated more formally, this says that the function $\mathbf{N} \to \mathbf{Z}$ given by $n \mapsto (n,0)$ carries sums to sums, products to products, inequalities to inequalities, and distinct numbers to distinct integers; it is thus a monomorphism of the structure described by $+$, $\times$, and $\leqslant$.

These two constructions of the integers give essentially the same result. Specifically, the map $n \mapsto (n,0)$, $-m \mapsto (0,m)$ is an isomorphism (for the structure $+$, $\times$, and $<$) of the integers as first constructed to those described by pairs and equality. In either construction, the integers $\mathbf{Z}$ can be described as the minimal way, unique up to isomorphism, of embedding $\mathbf{N}$ in a larger structure in which subtraction is always possible, preserving all the algebraic properties of $+$ and $\times$. Here as elsewhere, what matters is not an exact description of what an integer *is*, but a description of the structure of all the integers, up to isomorphism.

6. Rational Numbers

To keep accounts, one often needs to divide numbers evenly into parts—and this often cannot be accomplished with whole numbers. Fractions provide the answer. They are introduced individually, as 1/2, 2/3, 1/5, 4/5, etc., and then manipulated in the evident way:

$$\frac{m}{n} + \frac{m'}{n'} = \frac{mn' + m'n}{nn'}, \qquad \frac{m}{n} \cdot \frac{m'}{n'} = \frac{mm'}{nn'}. \tag{1}$$

This practical process suggests a corresponding formalization. The initial observation is that division is not always possible within the set $\mathbf{N}^+$ of all positive natural numbers. Hence one is led to introduce the set $\mathbf{Q}^+$ of all ordered pairs (m,n) of *positive* natural numbers, defining addition and multiplication by the evident translations

$$(m,n) + (m',n') = (mn' + nm', nn'), \qquad (m,n)(m',n') = (mm',nn') \tag{2}$$

of the practical rules (1) and taking care to define the equality of (m,n) and (r,s) by $ms = nr$. When m in $\mathbf{N}^+$ is identified with the pair $(m,1)$ this again defines a minimal expansion of the set $\mathbf{N}^+$ to a larger set in which division is always possible and in which all the rules of arithmetic still hold. As before, there is nothing unique about the formulation of this construction. Instead, one might have used only those pairs (m,n) in which m and n have no common factor (except 1); in that case, one must modify addition and multiplication in (1) to reduce each answer to lowest terms. With this inconvenience, the "artificial" definition of equality of pairs is avoided. Again, what matters is only the resulting structure up to isomorphism.

The system $\mathbf{Q}$ of *all* rationals may then be obtained from $\mathbf{Q}^+$, the positive rationals, by simply adjoining zero and negative rationals. Alternatively, one may construct $\mathbf{Q}$ directly from $\mathbf{Z}$ by using all pairs (a,b) of integers a,b in $\mathbf{Z}$, with the same addition and multiplication as in (2)—and the important proviso that the "denominator" b is never 0.

As in previous cases, what matters is not the explicit definition of a rational, but the resulting structure.

7. Congruence

A typical clock runs up to the figure of 12 hours and then repeats, but one can still do arithmetic on the limited list of hours: Seven hours after nine o'clock is four o'clock. Similarly, in the decimal system there are only ten digits $0,1,2,\ldots,9$; the usual rules for addition and multiplication, ignoring the carryover to the tens' place, work perfectly well for the manipulation of these digits by themselves:

$$6 + 7 = 3, \qquad 8 + 7 = 5, \qquad 8\cdot 3 = 4, \qquad 3\cdot 9 = 7.$$

These rules ignore all the multiples of 10; in a sexagesimal system there are similar rules which ignore all the multiples of sixty. "Casting out nines" is a rule for checking arithmetic calculations. This rule for checking a multiplication says: Add up the digits of each factor, multiply the resulting sums, and check this against the digit sum of the original purported answer. Thus 32 times 27 calculates to 864. To check, 32 becomes 5, 27 becomes 9, 5 times 9 is 45, with digit sum 9. This checks with the digit sum in the purported answer, which is $8 + 6 + 4 = 18$ with digit sum 9. What happens here is that 32 is replaced by 5, casting out the difference which is 27, or three nines. The reason it works is that factors differing by a multiple of nine will have a product differing (at most) by a multiple of 9. In brief, arithmetic operations are valid "casting out 9's" or "modulo 9".

These examples each involve the use of a modulus: 12, 10, 60, or 9, as the case may be. The general procedure is similar. For integers a, b, and any natural number $m \neq 0$ as modulus, one writes $a \equiv b \pmod{m}$, or says that a is *congruent* to b for the modulus m, when the difference $a - b$ is a multiple of m. Then one readily proves the arithmetic rules: If $a \equiv b$ and $c \equiv d$, both mod m, then

$$a + c \equiv b + d \pmod{m}, \qquad ac \equiv bd \pmod{m}. \tag{1}$$

This congruence modulo m behaves like equality; also it is reflexive, symmetric and transitive. ("Transitive" is defined in §1.5; a relation such as $\equiv$ is *symmetric* when $a \equiv b$ implies $b \equiv a$ for all a and b.)

Two integers a and b are congruent modulo m if and only if they leave the same remainder r, with $0 \leqslant r < m$, upon division by m. As a result, calculations modulo m amount to calculations with a finite list of objects (to wit, with the remainders $0,1,2,\ldots,m-1$). All the rules for addition and multiplication–commutative, associative, and distributive laws–still hold for these finite calculations. Thus the remainders modulo m form an (abelian) group under addition. Under multiplication, the non-zero remainders modulo a prime p also form a group of $p - 1$ elements. This is not the case for a composite modulus m, such as $2\cdot 3$, because there $2\cdot 3 \equiv 0 \pmod 6$ so that neither 2 nor 3 can have a multiplicative inverse modulo 6. To get a multiplicative group for such a composite modulus m, on must use only those remainders r which have no factor in common with the modulus m. The number of such remainders is denoted by $\phi(m)$, while ϕ is called Euler's ϕ-function. For a prime p or for integers m,n with greatest common divisor 1 one readily calculates that

$$\phi(p) = p-1, \phi(p^k) = (p-1)p^{k-1}, \phi(mn) = \phi(m)\phi(n). \qquad (2)$$

These formulas provide for a computation of any $\phi(m)$ from the prime decomposition of m. We cite them to emphasize that the formulation of congruence arises both from practice (multiplying hours or digits) and from number theory.

To say that calculations with congruences *are* calculations with the remainders is a bit artificial. Thus modulo 5 one could replace the five remainders 0, 1, 2, 3, 4 by the remainders -2, -1, 0, 1, 2 or by -4, -3, -2, -1, 0. Here, as always, mathematicians strive for an invariant formulation. Each remainder r stands for (and may be replaced by) the "congruence class" $C_m r$ of *all* integers a with $a \equiv r \pmod{m}$. To add the class $C_m r$ to the class $C_m s$ one may then take any representative a in $C_m r$, any b in $C_m s$, add a and b, and take the class of this sum $a + b$ as the sum $C_m r + C_m s$. One must then prove that the resulting sum of classes doesn't depend on the representatives a and b chosen—but this fact is just a restatement of the rule (1) for adding two congruences. With this fact established, we see that the collection $\mathbf{Z}_m$ of all these congruence classes C_m forms a system with binary operations of addition and multiplication, and that the function C_m from $\mathbf{Z}$ to $\mathbf{Z}_m$, as in

$$C_m\colon \mathbf{Z} \to \mathbf{Z}_m, \qquad a \mapsto C_m a,$$

carries the addition and multiplication of integers to that of congruence classes. (It is thus a first example of a homomorphism of $+$ and $\times$.) This gives an "invariant" formulation of the calculation with remainders.

Thus we have (at least) three descriptions of the algebra of integers modulo m: As the ordinary integers taken with a new equality, congruence modulo m; as the algebra of remainders modulo m; or as the algebra of congruence classes, modulo m. The last description is the more invariant–and the more sophisticated, since it involves a set whose elements are sets (a collection of congruence classes). However, all three constructions yield isomorphic results, and the results are useful (and practically indispensable) for the statement of simple number theoretic facts. For example, for any integer x, one has always

$$x^2 \equiv 0 \quad \text{or} \quad 1 \pmod{4}, \qquad x^2 \equiv 0,\ 1, \quad \text{or} \quad 4 \pmod{8}.$$

Another problem is that of finding a common solution x for two (simultaneous) congruences:

$$x \equiv b \pmod{m}, \qquad x \equiv c \pmod{n}. \tag{3}$$

In this situation, the "Chinese remainder theorem" states that if m and n have no common factors (except 1), there always is a solution, unique modulo the product mn.

8. Cardinal Numbers

The question arises: What, after all, *is* a natural number? One explanation says that it is a *cardinal number*. For this purpose, as in §1.2, define two sets S and S' to be *equinumerous* (or, *cardinally equivalent*), in symbols $S \equiv S'$, when there is a bijection b: $S \to S'$; that is, a one-to-one correspondence between S and S'. This relation between sets is reflexive (because the identity function is a bijection), symmetric (because the inverse of a bijection is again such), and transitive (because the composite of two bijections is again such).

Arithmetic operations work appropriately under this equivalence relation. To add two collections S and T, first make sure that they are disjoint (have no objects in common), then take as sum the collection $S + T$ of all objects in either S or T. This *disjoint union* is preserved by bijections: if b: $S \to S'$ and c: $T \to T'$ are bijections, they determine together a bijection $S + T \to S' + T'$. Therefore

$$S \equiv S' \quad \text{and} \quad T \equiv T' \qquad \text{imply} \qquad S + T \equiv S' + T'. \tag{1}$$

To get a product of S by T one needs to add T copies of S; that is, to count the points in a rectangle $\square$ S units wide and T units high. Call such a rectangular set $S \times T$ the *product* of the sets S and T; it can be described as the set of all the *ordered pairs* $\langle s,t \rangle$ where $s \in S$ and $t \in T$. The use of bijections shows, much as in (1), that

$$S \equiv S' \quad \text{and} \quad T \equiv T' \qquad \text{imply} \qquad S \times T \equiv S' \times T'. \tag{2}$$

Exponentiation is repeated multiplication; thus $S^3 = S \times S \times S$ is the set of all ordered triples $\langle s_1,s_2,s_3 \rangle$ of elements s_i in S (with possible repetitions). This is equivalently the collection of all functions $i \mapsto s_i$ on a three-element set $\{1,2,3\}$ to the set S. Generally, the exponential set S^T denotes the set of all functions f: $T \to S$. Here one again has

$$S \equiv S' \quad \text{and} \quad T \equiv T' \qquad \text{imply} \qquad S^T \equiv S'^{T'} \tag{3}$$

Explicitly, a bijection b: $S \to S'$ yields by $f \to b \cdot f$ a bijection $S^T \to S'^T$, while a bijection c: $T \to T'$ yields by $f' \to f' \cdot c$ a bijection $S^{T'} \to S^T$, going in the opposite direction.

On this basis, a finite cardinal *number* is just a finite set taken "modulo" cardinal equivalence; that is, with cardinal equivalence taken as the equality. In other words, a number is "represented" by a finite set, and two sets count as the same if they are in bijection. Alternatively, if we want not a "representative" of the number, but a single object, we may define *the* cardinal number of S to be the set of all sets S' equinumerous with S:

$$\text{card } S = \{S' \mid S \equiv S'\} \tag{4}$$

The arithmetic rules above for congruence then justify definitions of the sum and the product of cardinal numbers by

$$(\text{card } S) + (\text{card } T) = \text{card}(S + T), (\text{card } S)(\text{card } T) = \text{card}(S \times T) \tag{5}$$

This definition of cardinal number is strongly analogous to the invariant definition of a congruence class module m, as in §7. Historically, congruence classes understood well before the more general definition of cardinal number. There is a difference: The congruence class $C_m r$ of r modulo m consists of all integers a in a *given* set of integers with $a \equiv r \pmod{m}$, but the cardinal of S, as defined in (4), consists of all sets S' in the whole world with $S' \equiv S$.

Some discomfort with this wide scope of things is in order. If one is using here the "naive" idea that a set S is *any* collection whatever of identifiable things (and that two sets are equal when they contain exactly the same things), then there can be paradoxes. To formulate one paradox, note that there can be sets (such as the set of all infinite sets) which are members of themselves–and other sets not members of themselves. Thus Bertrand Russell, noting that we can define a set by specifying its elements, considered the set R consisting of those sets which are not members of themselves:

$$R = \{S \mid \text{not } (S \in S)\}. \tag{6}$$

In other words

$$S \in R \qquad \text{if and only if not} \qquad (S \in S). \tag{7}$$

He then asked: is R a member of itself, or not? If so, then $R \in R$, so by (7), not ($R \in R$). If not so, then not ($R \in R$), so by (7) read backwards, with $S = R$, one has $R \in R$- a contradiction either way.

This *Russell Paradox* exhibits a difficulty with the naive notion of set, when coupled with the idea that one can specify a set by collecting all the objects with a given property P to form a set T:

$$T = \{X \mid X \text{ has the property } P\}. \tag{8}$$

(Call this the naive *comprehension* axiom.) One (standard) way to avoid this trouble is to apply this comprehension axiom *only* to construct subsets of some already given set W—as, for example, subsets of the set $\mathbf{Z}$. This "bounded" comprehension axiom then reads: Given a set W and a property P of sets, one may form the set

$$T = \{X \mid X \in W \text{ and } X \text{ has the property } P)\}. \tag{9}$$

Later (in Chapter XI) we will examine this axiom in the context of a systematic axiomatization of "set" and "element of a set". For this purpose, we also assign to each set U its *power set*: All subsets of U:

$$PU = \{S \mid S \subset U\}. \tag{10}$$

For the present, to explicate cardinal numbers, start with some (infinite) initial set V_0, called a *type* or a "universe"; for example, V_0 might be the set of natural numbers or the set of all real numbers. The "next" type V_1 is to consist of all the subsets S of V_0, so can be described as $V_1 = PV_0$. Now we can describe the cardinal number of such a set S as the set of all equinumerous S' which are also subsets of V_0

$$\text{card } S = \{S' \mid S' \in V_1 \text{ and } S \equiv S'\}. \tag{11}$$

This cardinal number is then a set in the next higher type $V_2 = PV_1$. This approach makes the cardinal number depend on a choice (or hierarchy) of types, but the definition (11), in contrast to (4), uses only the bounded comprehension axiom so that Russell's paradox no longer arises. The use of successive "universes" V_0, V_1, V_2 is a highly simplified version of the "type theory" invented by Russell to avoid his paradox. (More care is needed in details; for example to arrange that a type is closed under products.)

Any such approach defines the finite cardinal numbers in terms of sets and bijections and derives their arithmetic properties (addition, multiplication, exponentiation) from properties of sets (and bijections). It serves also to define the cardinal number of an infinite set, and arithmetic operations upon such infinite sets.

In this way the natural question: What is infinity? receives a numerical answer. There are different sizes of infinity, and they can be measured by suitable "cardinal" numbers which are subject to arithmetic operations.

9. Ordinal Numbers

The cardinal number of a finite set S does not depend on the order in which the elements of S are presented or counted. However, for other purposes, we often list the elements of a set in a given order, as we count first, second, third,..., and not just one, two, three,.... This leads to the notion of an ordinal number. Such a number is attached to a suitable linearly ordered set $(P,<)$. If P' is a second such set, an *order-isomorphism*, as in §I.4, is a bijection $f\colon P \to P$ which preserves the order, in the sense that $p_1 < p_2$ in P implies $fp_1 < fp_2$ in P'; when there is such an isomorphism, the ordered sets are said to be *ordinally equivalent* (in symbols, $P \sim P'$). Then a finite ordinal number can be defined to be a finite linearly ordered set P taken modulo ordinal equivalence–that is, with ordinal equivalence as equality. Alternatively, as with congruence classes of integers modulo m or as with cardinal numbers, we can define finite ordinal numbers as equivalence classes in a type V_1

$$\text{ord } P = \{P' \mid P' \in V_1 \text{ and } P \sim P'\} \tag{1}$$

Arithmetic operations apply to ordered sets. Thus, if P and Q are disjoint ordered sets, we can order their (disjoint) union $P + Q$ by taking the elements of P, in order, before those of Q, in order. The product $P \times Q$ of two linearly ordered sets has a *lexicographic* order defined by

$$(p_1,q_1) < (p_2,q_2) \quad \text{iff} \quad p_1 < p_2 \text{ or } p_1 = p_2 \quad \text{and} \quad q_1 < q_2 . \tag{2}$$

If Q is finite, the exponential set P^Q of all functions $f\colon Q \to P$ has a similar lexicographic order: Given functions f and g, find the first q in the given order of Q for which $fq \neq gq$, and then specify that $f < g$ if and only if $fq < gq$. For each of these three operations we have the rules for equivalence, such as:

$$P \sim P' \quad \text{and} \quad Q \sim Q' \qquad \text{imply } P \times Q \sim P' \times Q'. \tag{3}$$

The product of two ordinal numbers is then defined by setting

$$(\text{ord } Q)(\text{ord } P) = \text{ord}(P \times Q)\,; \tag{4}$$

here the reversal of the factors follows tradition. The result is independent of the choices of P and Q within the classes ord P and ord Q because of the rule (3). The definitions of sum and exponents for ordinals is analogous. For finite ordinals, these operations agree with the corresponding arithmetic operations on finite cardinals, because the cardinal number of a finite ordered set does not depend on the order.

This is not so for infinite sets. If P is an infinite ordered set, the set ord P defined in (3) is usually called the *order type* of P. For example, the denumerably infinite set $\mathbf{N}^+$ of positive integers with its usual order has an order type quite different from that of the set $\mathbf{N}^-$ of negative integers ($\mathbf{N}^+$ has a first element, but $\mathbf{N}^-$ has none). The order type of $\mathbf{N}^-$ is not regarded as an infinite ordinal number.

For ordinal numbers, the leading idea is that there is always a "next" one—the first ordinal "beyond" a given set of ordinals. More generally, this should mean that any set of ordinal numbers has a first element; this leads to the definition: A (linearly) ordered set P is *well-ordered* when every non-empty subset $S \subset P$ has a first element in the given order. For the natural numbers $\mathbf{N}$, the axiom of induction is equivalent to the assertion that every non-void subset T of $\mathbf{N}$ has a first element—that is, to the requirement that $\mathbf{N}$ in its given order is well-ordered.

Generally, an *ordinal number* is defined to be the order type in the sense (3) of some well-ordered set. This definition provides, for example, for many infinite but denumerable ordinal numbers. The ordinal number of the (well-ordered) set $\mathbf{N}$ is usually written as ω; then the arithmetic operations apply to ω as well as to finite ordinals. For example, $\omega + \omega = \omega 2$ is the order type of two copies of $\mathbf{N}$, one following the other. On the other hand, 2ω is the order type of ω copies of 2, one following the other; hence $2\omega = \omega$ and $2\omega \neq \omega 2$! The first few infinite ordinals come in order as

$$0, 1, 2, \ldots, \omega, \omega+1, \omega+2 \ \ldots, \omega 2, \omega 2+1, \omega 2+2, \ldots, \ \ldots, \tag{5}$$

where each ... stands for the whole string of natural numbers from 3 on.

This display suggests that the ordinal numbers are themselves well-ordered, where ord $P <$ ord Q means that P is order isomorphic to an "initial segment" of Q. Moreover, each ordinal number is the order type of the ordered set of all preceding ordinals. This motivates an alternative description of an ordinal number as the ordered set of all the preceding ordinals:

$$0 = \varnothing,\ 1 = \{0\},\ 2 = \{0,1\}, \ldots, \ \omega = \{0,1,2, \ldots\}. \tag{6}$$

In other words, each ordinal nun ,er is a member of each succeeding ordinal. This suggests a formal definition of an ordinal number as a set whose elements are linearly ordered by the membership relation. We will return to this idea in §XI.2 when we analyze more closely the notion of a set.

10. What Are Numbers?

This chapter has proposed several alternative answers to this question, as follows:

(a) The natural numbers are just a sequence of marks

$$|,\ ||,\ |||,\ ||||,\ |||||, \ldots$$

for use in counting or labeling. Any other sequence of marks would do as well, provided the sequence always provides a way to write down the next (the successor) mark. This is the case with decimal notation

$$0, 1, 2, \ldots, 10, 11, 12, \ldots, 100, \ldots, 200, \ldots.$$

(b) Natural numbers are finite sets, with equinumerous sets regarded as equal numbers, while successor means "adjoin one more element".
(c) Natural numbers are cardinal-equivalence classes of finite sets.
(d) Natural numbers are ordinal-equivalence classes of finite ordered sets, while successor means "adjoin one new element, to come after all the others".
(e) Natural numbers are finite sets of sets, linearly ordered by the membership relation, as in (6).

In each of these cases, the natural numbers as described do satisfy the Peano postulates. From this multiplicity of answers to our question, we must conclude that there is no answer to the question: What is a natural number? There are various alternative concrete descriptions, depending on the sort of counting intended or on the prior assumptions. In each case, the description provides "numbers" which do satisfy the Peano postulates. Hence we conclude that one does not define what a natural number "is", by itself. Instead, one defines the system of *all* natural numbers, with successor operation. Then **N** is *any* such system which satisfies the Peano postulates. This means that there are many such systems within set theory—but that they are all isomorphic, just as in the case above of the decimal notations.

Note that the postulates themselves are by no means unique; for example, the Peano postulates may be replaced by the recursion theorem as an axiom. Here, as in other axiomatic descriptions of mathematical objects, there are a variety of choices for lists of axioms. Number theory, like other subjects in Mathematics, is not the study of a unique model nor yet the examination of a unique axiomatic system—it is rather a study of the form exemplified by the various models and specified in the axioms.

To summarize: The natural numbers start out from elementary operations of counting, listing, and comparing; they then develop into effective tools for calculation. The rules for calculation are formal and can be organized as the consequences of simple systems of postulates. The consequences of these postulates include the remarkably varied and rich properties studied in number theory—properties by no means apparent in the original processes of counting and listing. They are none the less implicit in these elementary human activities. Number theory is inevitable.

But number theory is not self-contained. First, calculations with natural numbers do not allow all subtractions or all divisions, and so require the constructions of other systems of numbers such as **Z** and Q. Secondly, number theory suggests and requires calculations with congruences and then with integers modulo m—Mathematical objects which are really not simply numbers, but classes of numbers. Finally numbers also lead inexorably to the study of sets and functions. Thus bijections are required to explain when two finite sets have the same cardinal number, binary relations are required to understand the ordinal numbers implicit in the listing process, while subsets are required to formulate the induction axiom. Then the explication of congruence modulo m and of cardinal and ordinal numbers requires sets of sets, namely, sets of congruence classes (see also §4).

Thus understanding simple and elementary Mathematical ideas leads inevitably to general and abstract notions.

References. There are many sources for number theory; one good one is Hardy and Wright [1983]. For a truth about natural numbers not demonstrable from the Peano postulates, with induction on properties, see the last article in Barwise: Handbook of Logic.

CHAPTER III

Geometry

This chapter is concerned with the origins of elementary plane geometry, its close relation to groups of transformations, and the notions of non-Euclidean geometry. Subsequently, n-dimensional geometry will appear with linear algebra in Chapter VII, manifolds and spaces with curvature in Chapter VIII, and topology, in its connections with complex analysis, in Chapter X. These and other aspects of geometry pervade Mathematics.

1. Spatial Activities

Geometrical figures and facts arise from a wide array of activities and observations. Some involve motion; thus in watching motion we see objects falling in vertical lines, water waves expanding in circles from a dropped stone, duneland grasses moved by the wind to describe semicircles on the sand, tree branches oscillating back and forth in the wind, long straight lines of ocean waves approaching a beach, and the like. There are also motions which we initiate and then watch—the thrown ball falling, the log rolling downhill, the circle turned into a wheel for a cart, and so on. Some activity involves construction. Thus a post or a column will stay upright in balance if it is set perpendicular to the floor as a vertical; a three-legged stool is more likely to be steady than one constructed with four legs; rods joined to form a triangle stay rigid, but rods joined in a square may wobble. Also, pieces of a board cut apart will fit together again to make up the same size board. Mapping a labyrinth or painting a scene calls for reproducing a variety of shapes, each on a smaller scale. Fitting an object inside another highlights differences in shapes. To check ahead of time that a fit will be possible may require a measurement of length or of circumference. Estimating separation at a distance requires lines of sight, angles, and triangles. These are but a few of the many activities that go into the formation of geometrical ideas.

From these activities one may disengage various figures: Circles and ovals; lines, horizontal, vertical, and perpendicular; triangles, equilateral or otherwise; squares and rectangles; cubes, cylinders, as with logs or rods, and spheres, as with oranges or balloons. One also discovers a variety of facts about these figures: A triangle is rigid because three sides (or two sides and the included angle) determine the triangle up to rigid motion. A right triangle can come in various shapes, but in every case the square constructed on the longest side is the sum of the squares on the two shorter legs (a fact said to have been useful in re-establishing property lines after the Nile had been in flood). Two different perpendiculars to the same line would never meet. Two triangles with corresponding angles equal have corresponding sides proportional–and so on.

After a considerable array of facts about geometrical figures are at hand, some connections between these facts come to light. Once the area of a triangle can be computed from the lengths of its sides, it is possible to fit four congruent triangles together so as to prove the Pythagorean theorem, by adding up the five areas in Figure 1. Or, the Pythagorean theorem can be deduced from facts about ratios of sides in similar triangles.

Explicitly (Figure 2), the perpendicular from the right angle divides the hypotenuse c in Figure 2 into pieces h and k with $h/a = a/c$ and $k/b = b/c$. Clearing of fractions, $hc = a^2$ and $kc = b^2$, so that $(h + k)c = c^2 = a^2 + b^2$, q.e.d. Many more such logical relations between geometrical facts gradually appear.

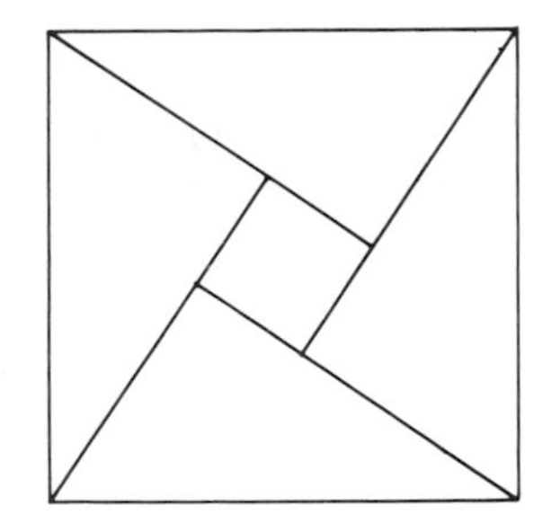

Figure 1. The Pythagorean theorem.

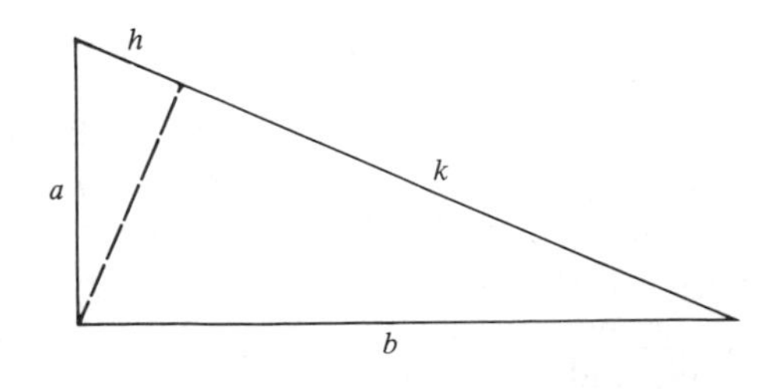

Figure 2

Nevertheless, it is truly remarkable that our experiences of spatial extent and of motion through space can be organized so completely. By taking certain figures–line, plane, segment, angle, and circle–as basic ones, and by assuming certain simple facts as axioms, other geometric figures can be constructed and other geometric facts deduced from the assumed axioms. Some of the resulting facts are surprising; for example, the three medians of a triangle meet in a point, as do the three altitudes. The resulting deductive structure of Euclidean geometry is a model of Mathematical method.

2. Proofs Without Figures

In elementary Euclidean geometry, proofs of various facts about figures are developed from the axioms, with occasional implicit use of intuitively evident properties of the figures concerned. For instance, to find the perpendicular bisector of a segment AB one constructs two circles with centers A and B and both with radius AB; these circles obviously intersect on opposite sides of the line AB, ... But what is meant by a "side" of AB and why must the circles intersect? (See Figure 1.) If *all* facts are to be deduced from axioms, then these plausible observations must also be proved from the axioms, rather than from inspection of the figure. Once this austere requirement was recognized, it turned out to be possible to add suitable new axioms to those of Euclid so that all arguments could be free of any intuitive or pictorial content. This meant that geometry was not really "about" the figures themselves, but was about certain corresponding notions defined exclusively in terms of basic notions to be taken as primitive or undefined. This austere program was systematically carried out by David Hilbert in a book, *Grundlagen der Geometrie* (B. G. Teubner, first edition 1899, twelfth edition 1977). His axioms were for both plane and solid geometry; we will examine only those for the plane. He presented the axioms in the five following groups:

Group I (Incidence Axioms). These axioms involve only three primitive notions: "point", "line", and "the point P lies on the line l". They require:

(1) Two distinct points P and Q lie on one and only one line. (Except for the meticulous logical formulation, this is the familiar requirement that "two points determine a line".)
(2) There are at least two points on every line.
(3) There are three points P, Q, R not all on a line.

With these axioms, one can define a *triangle* (the figure formed by three non-collinear points). However, one cannot yet define "right angle".

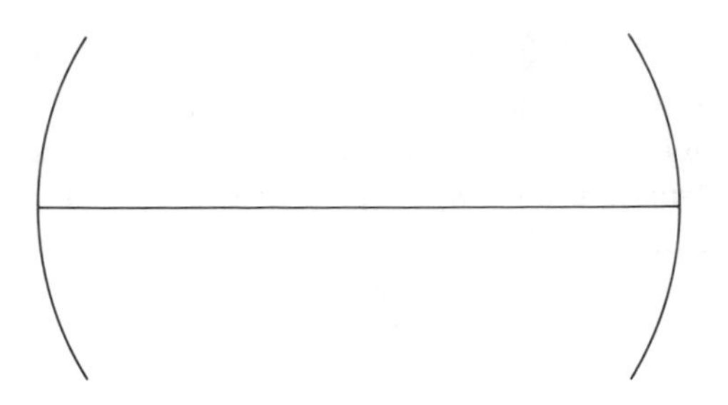

Figure 1. Do the circles meet?

Group II (Axioms of Order). In Euclidean geometry a line does not have a selected direction, so we cannot describe the order of points P, Q on a line in terms of binary relations such as "P lies to the left of Q" or "P comes before Q" or "$P < Q$". Instead, the description is given means of a *ternary* relation "B lies between A and C", available when A, B, and C are all known to lie on a line. The axioms then read:

(1) If B is between A and C, then B is between C and A.
(2) If A and B are distinct points on a line l, there exists on l a point C between A and B and a point D such that B is between A and D.

(In elementary geometry, one often speaks of "prolonging the line segment AB beyond B. This axiom provides the existence of at least one point D in this prolongation.)

(3) If A, B, and C are three distinct points on a line l, then exactly one of these three is between the other two.

With these axioms, one can define the *segment* AB for $A \neq B$, to consist of all the points C between A and B on the line determined by A and B. One is now in a position to formulate such pictorially obvious facts as "A line divides the plane into two parts". Specifically, this means that given a line l, all points in the plane but not on l can be put in one of two nonempty disjoint sets U and V (the two "sides" of l) such that any segment AB joining a point A in U to a point B in V meets l in one point, while any segment joining two points of U (or, two points of V) does not meet l. To establish this "intuitively evident" separation property, it is necessary to assume an additional axiom:

(4) Pasch's Axiom. If a line l meets the segment AB of a triangle ABC and does not contain C, then l meets either AC or BC (Figure 2).

In pictures, this states that a line which crosses into the triangle over side AB must again come out, either through AC or through CB (or through C). From the austere, no-pictures, point of view, it was this axiom which was most sharply missing in the Euclidean formulation. This axiom will show that each triangle divides the plane into two parts, an "inside" and

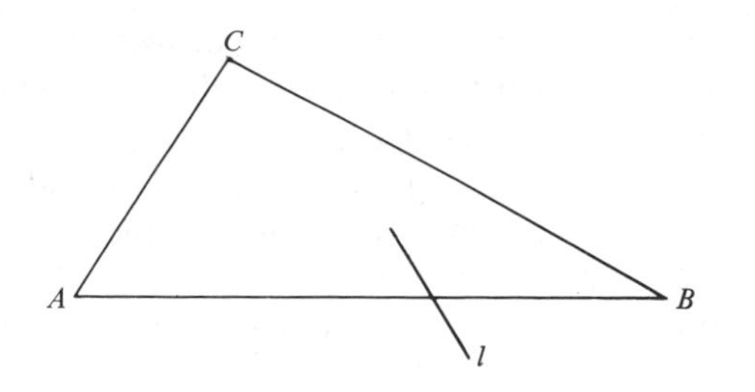

Figure 2. Pasch's axiom.

an "outside". One can also prove that any closed polygon (which does not cross itself) divides the plane into two parts, and inside and an outside—but the proof is quite involved because the polygon may be convoluted, with many reentrant angles.

With these axioms, one can also define that is meant by a *ray* or "half line" starting at a point A. If B is a second point and l the line determined by A and B, then the ray r from A containing B consists of all the points between A and B, the point B, and all points D with B between A and D. An *angle* $\angle\, rs$ between two rays r and s can then be defined as the figure formed by two (distinct) rays r and s from the same point. In particular, this defines the (three) angles of a triangle ABC. One also defines straight angle (formed by two rays on the same line).

Group III (Congruence). Next one introduces two new undefined terms, "Segment AB is congruent to segment $A'B'$", in symbols $AB \equiv A'B'$, and "Angle rs is congruent to angle $r's'$", in symbols $\angle rs \equiv \angle r's'$. The corresponding axioms require:

(1) Given a segment AB and a ray r from A', there is a unique point B' on r with $AB \equiv A'B'$.

In simpler language, this states that we can "lay off" the length AB on a given line from A' in a given direction.

(2) Congruence of segments is reflexive, symmetric, and transitive.
(3) $AB \equiv A'B'$ and $BC \equiv B'C'$ imply $AC \equiv A'C'$, provided B is between A and C and B' between A' and C'.

This amounts to describing the addition of segments.

(4) Given an angle $\angle rs$ and a ray r' from A' on a line l, there is a unique ray s' from A' on a given side of l so that $\angle rs \equiv \angle r's'$.

This axiom specifies that one can "lay off" the angle $\angle rs$ from a given ray r' at A' and on a given side of the line of r'.

(5) If two triangles ABC and $A'B'C'$ have $AB \equiv A'B'$, $BC \equiv B'C'$ and $\angle B \equiv \angle B'$, then also $\angle A \equiv \angle A'$ and $\angle C \equiv \angle C'$.

Given this much about the two triangles, one proves that $AC \equiv A'C'$ and hence that the triangles are congruent. This is the familiar first congruence theorem (side-angle-side, or SAS) of Euclidean geometry. It is conventionally "proved" by moving one triangle till its parts coincide with the other. Though we regard such "motion" as lying in the practical origins of geometry, it use in a formal axiomatic proof is not acceptable; hence this axiom. From the congruence axioms, one can define right angles.

Two lines are defined to be *parallel* when they do not meet (have no point in common). One then requires the following famous axiom:

Group IV (The Parallel Axiom). Given a point A not on the line m, there is at most one line through A parallel to m.

Group V (Continuity Axioms). Examples show that the axioms stated so far do not yet insure that the Euclidean plane contains all the points that should be there, and no more. To achieve this, two more axioms are needed. To state the first, notice that a segment AB on a ray from A can be laid off repeatedly to give n points $B = B_1, B_2, \ldots, B_n$ on the ray with $AB_1 \equiv B_iB_{i+1}$ and each B_i between A and B_{i+1}, for $i = 1, \ldots, n-1$ (Figure 3).

(1) *Archimedian axiom*. Given C on that ray from A, there is a natural number n with C between A and B_n.

That is, multiples of AB_1 will eventually exceed the given segment AC; in other words, C, as measured by the unit AB, cannot be infinitely far away.

(2) *Completeness*. The system of points and lines satisfying the given relations of incidence and congruence cannot be a part of a larger system of points and lines with the same relations satisfying all the same axioms.

These are the two axioms of continuity as formulated in Hilbert's book. However, this pair of axioms can be replaced by a single axiom which concerns the division of a line l into two rays. Each point 0 on l divides l into two rays; if S is the set of all points A on one ray and T that on the other (and 0 is put in neither ray), then 0 lies between any point A of S and any point B of T. The axiom desired is essentially a converse to this:

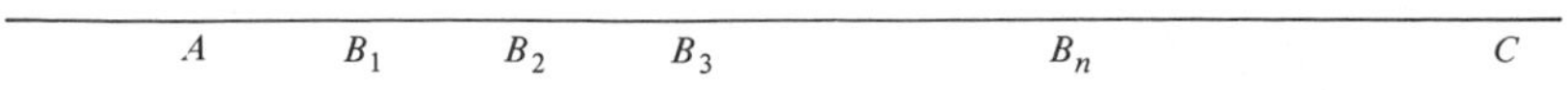

$A \quad B_1 \quad B_2 \quad B_3 \quad B_n \quad C$

Figure 3. The Archimedean axiom.

Dedekind's Axiom. Let all the points on a line l be divided into two non-empty disjoint subsets S and T in such a way that no point of S is between two points of T and no point of T is between two points of S. Then there is a unique point 0 on l with the following property: For points A, B not 0 on l, 0 is between A and B if and only if A lies in S and B in T, or A in T and B in S.

In other words, the point 0 divides the line into S and T. We will see in Chapter IV that this axiom is modeled on a similar axiom for the real numbers.

From this list of axioms all of plane geometry can be developed in austere and logical completeness, even though the brevity of our presentation does not make clear the austere beauty of the resulting development.

3. The Parallel Axiom

The parallel axiom (Group IV above) may be intuitively less convincing than the other axioms, and so has been the source of extensive discussions. The axiom as we have formulated it above requires only that there be *at most* one parallel to a given line m through a given point not on m. This formulation is chosen because we can and will prove (from the previous axioms) that there is at least one such parallel.

For the proof, consider a line k which meets two other lines l and m in points A and B, respectively; in classical terminology, k is *transversal* to l and m, as in the Figure 1. In this figure, angles (like α and β) between k and l or m which have one ray along AB are called *interior* angles. For this figure, we prove

Lemma 1. *If the interior angles α and β on the same side of the transversal k to l and m have sum $\alpha + \beta = 180°$, then the lines l and m are parallel.*

(Note incidentally that the statement, carefully formulated, makes use of the "sides" of k.)

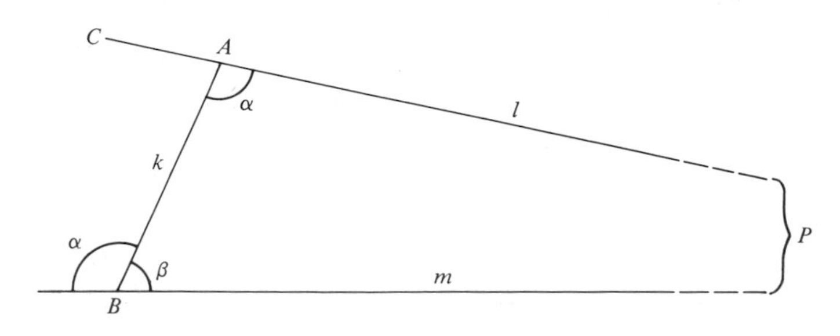

Figure 1

Proof. If l and m are not parallel, then by the definition of parallel these lines must meet in some point P. By symmetry, we may take P to be on the same side of k as the angles α and β. Then (and with covert references to Figure 2, which cannot "really" exist) we lay off on m, on the ray opposite BP, a segment $BP' \equiv AP$. Since $\angle P'BA + \beta$ is a straight angle, so equal to 180°, it follows that the angle $P'BA$ must equal α. Therefore the triangles $P'BA$ and PAB have corresponding sides ($P'B \equiv PA$), angles α, and sides $BA \equiv AB$ congruent, so by the basic congruence axiom (side-angle-side, or SAS), they are congruent triangles. Hence angle $P'AB$ of the first triangle is congruent to angle PBA (or β) of the second. Since $\beta + \alpha = 180°$, this in turn means that the angle $P'AB + BAP = 180°$ is a straight angle, so that $P'AP$ is a straight line. It must then be the given line l, which is thus revealed to be a line meeting m in two *different* points P' and P on opposite sides of k (note the essential use of "sides" of a line, which can be defined by using Pasch's axiom). From this contradiction, we deduce that l must have been parallel to m.

Theorem 2. *Through a point A not on a line m there exists at least one line l parallel to m.*

Proof. On the line m there is at least one point B. Join this point to A. The segment AB makes with m two angles γ and β which together form a straight angle, so $\gamma + \beta = 180°$. Lay off the angle γ at A, with one side along the segment AB and the other on a ray AP along the (new) line l'; then BAP, on the same side as β, is equal to γ (Figure 3). By the lemma, l' must then be parallel to m, as desired.

Up to this point, we have not invoked the parallel axiom.

Theorem 3. *When a line k is transversal to two parallel lines l and m, the alternate interior angles formed are equal.*

In Figure 1, with l parallel to m, this means that β equals $\angle BAC$.

Proof. In Figure 1 we first show that the angles α and β on the same side of k have sum 180°. If not, take γ with $\beta + \gamma = 180°$, l' as in Figure 3,

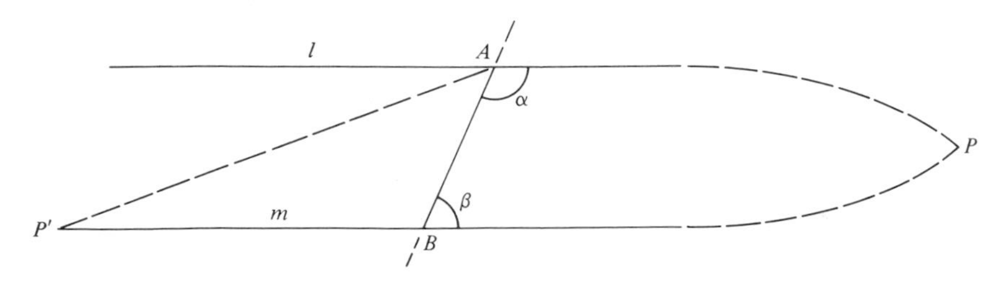

Figure 2

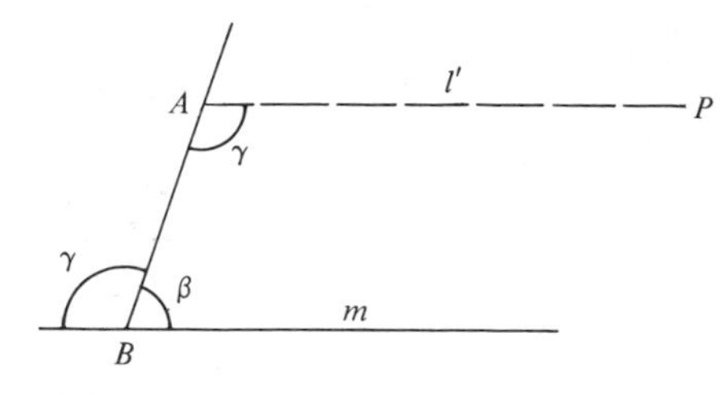

Figure 3

so that l' is parallel to m by Lemma 1. But the parallel axiom asserts there is only *one* parallel to m through A. Therefore $l = l'$, so that $\alpha = \gamma$ and $\alpha + \beta = 180°$. The requisite equality of alternate interior angles follows.

(In describing the proof, we have really used the figure. The reader should convince himself that this prop is not necessary.)

Corollary. *The sum of the three interior angles of a triangle is* 180°.

PROOF. Through a vertex A of the triangle construct a line l parallel to the base of the triangle. Then three angles at A on one side of l add up to 180°; by alternate interior angles, these equal the three angles of the triangle (Figure 4).

This result is characteristic for Euclidean geometry; in fact, the parallel axiom itself can be proved from the assumption that the sum of the angles in a triangle is always 180°. From it one can also prove that the sum of the interior angles of any quadrilateral is 360°. Similarly, the sum of the interior angles of an n-sided polygon in the plane must be $(n-2)180°$—though the proof takes some care where the polygon is not convex. But the basic fact, illustrated by the various corollaries, is that about the sum of the interior angles of a triangle.

There are other (equivalent) formulations of the parallel axiom. Thus for Euclid the axiom stated that for a transversal k as in Figure 1, if the sum $\alpha + \beta$ of the interior angles on one side of k is less than 180°, then the lines l and m meet on that side of k.

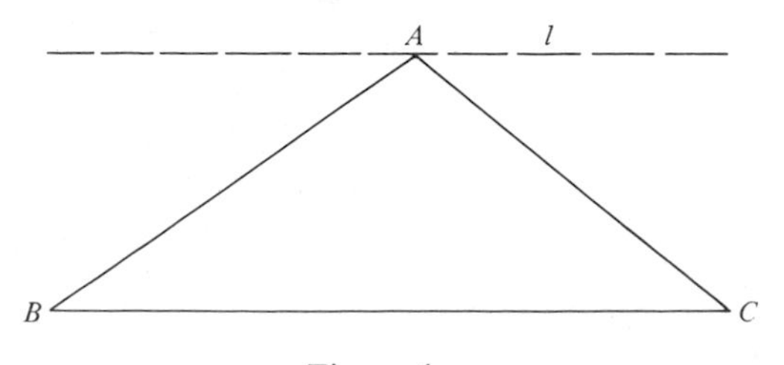

Figure 4

4. Hyperbolic Geometry

A set of axioms is said to be an *independent* set if no one of these axioms can be deduced from the others. It is desirable and appropriate (though not necessary) that the axioms for a basic structure, such as that of the Euclidean plane, be independent. In particular, there is the question: Is the parallel axiom independent, or can it be deduced from the others? This question has had considerable historical importance. For example, one might try to prove the parallel axiom by assuming the contrary (more than one parallel to m through a point A) and deducing a contradiction. There were several attempts to do this, most notably one in which Saccheri in 1733 deduced a large number of consequences, some of them perhaps bizarre–but none a contradiction. Nevertheless, he concluded that Euclid's parallel postulate was "vindicated". Then in the 19th century Bolyai, Lobachevsky, and Gauss took the opposite view, preparing to develop systematically a non-Euclidean geometry (specifically, a *hyperbolic* geometry) on the assumption that there is *more* than one parallel, and hence that the angle sum in a triangle is not 180°. When this is done systematically, it turns out that the angle sum is always less than 180° and that the difference between 180° and the sum is proportional to the area of the triangle.

This striking development raised (at least) two questions: "Is the resulting geometry consistent?", and "Does it fit the real world?" To answer the latter question, one must propose a specific "real world" interpretation of the primitive concepts of the geometry–say, by taking a straight line to be the path (in vacuum) of a ray of light, while an angle is the thing measured by a surveyor "turning off" with a transit the angle between two rays of light. With this interpretation, it appears that Gauss (who was also active as an astronomer) measured the angle sum for the triangle formed by chosen "points" on the peaks of three convenient mountains in Germany; the resulting angle sum was 180°, within the accuracy of the measurements then made. While the result indicates that there is not a flagrant deviation from Euclidean geometry on this interpretation, it does not provide any clear decision between the reality of Euclidean and hyperbolic geometry. It even suggests that there might never be such a decision, in view of the inevitable margin of error in the measurements made in any such interpretation. The terms involved in the interpretation are also open to question; for example, in general relativity theory the path of a light ray may not be "straight" in the intended sense. This ultimately brings up another and more profitable thought: Any geometrical axioms, Euclidean or non-Euclidean, offer a mathematical structure which may be open to a variety of different interpretations to suit a variety of geometrical (or even non-geometrical) circumstances.

There remains the question of the consistency of the assumptions of hyperbolic geometry. By definition, these assumptions are consistent if they lead to no contradiction—but it is clearly not enough to just observe that they haven't yet led to a contradiction. One way to approach this consistency would be to provide within Euclidean geometry an *interpretation* of hyperbolic geometry, say, by proposing certain Euclidean objects to serve as the "pseudo-points", the "pseudo-lines", and the "pseudo-distance" of a hyperbolic geometry, and proving from the Euclidean axioms that these "pseudo" objects do satisfy the axioms for point, line, and distance in hyperbolic geometry.

This can be done. Let C be a circle in the Euclidean plane. Let all Euclidean points *within* C count as pseudo-points, while diameters of C and the arcs of Euclidean circles which are orthogonal to C count as pseudo-lines, and a pseudo-point is on a pseudo-line in the evident (usual) sense of a point on a circle. It is then true that two distinct pseudo-points A and B do determine a unique pseudo-line. For either A and B lie on a diameter, or among the circles passing through A and B there is (by a continuity argument) exactly one which is orthogonal to the circle C. The axioms of incidence in Group I then hold, while with the usual Euclidean relation of betweeness the axioms of Group II also hold.

The notion of pseudo-congruence can then be defined in a standard way by first introducing a "pseudo-distance". Given pseudo-points A, B on a pseudo-line as in Figure 1, let S and T be the Euclidean points in which the (orthogonal) circle meets the base circle C, and let AT, BT, etc. denote the Euclidean lengths of corresponding arcs on this circle. Now define a (pseudo) distance by

$$\operatorname{dis}(AB) = \log_e \left[\frac{AT}{AS} \bigg/ \frac{BT}{BS} \right], \qquad (1)$$

and observe that this means that as B approaches T along the pseudo-line, with A fixed, the pseudo-distance AB will approach infinity. Then with this "metric" (i.e. in this metric space), the length of the whole pseudo-line will be infinite, as one might wish. All the congruence axioms for segments thus hold. The angle between two pseudo-lines at a point of inter-

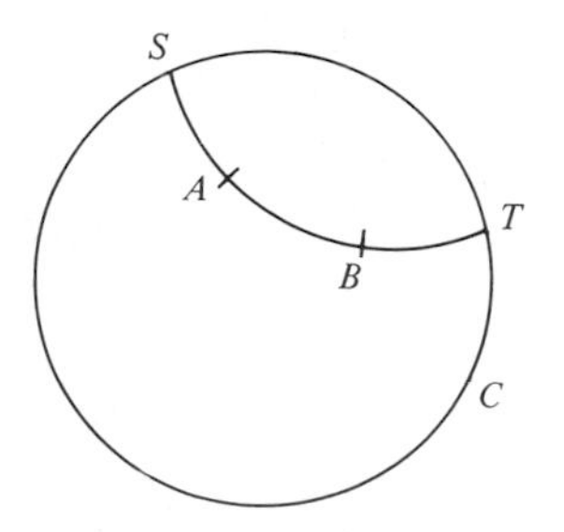

Figure 1

section is then taken to be the Euclidean angle between the Euclidean circles. One can then define congruent angles and lay off such angles as required in the axiom. Moreover, drawing a triangle formed by three pseudo-lines will readily show that the angle sum in such a (pseudo) triangle will turn out to be less than 180°. All the axioms for hyperbolic geometry hold in the interpretation. In particular, Figure 2 suggests that through a pseudo-point A not on a pseudo-line m there will be many pseudo-lines not meeting m, and hence parallel to m.

This Euclidean interpretation of hyperbolic geometry proves that hypergeometry are multiple. On the one hand, this geometry arises from measurements: Do the angles in a triangle of light rays really add up to 180°, This contradiction could then hold for the pseudo-points and the pseudolines of this model, and so would apply to those points and circles of Euclidean geometry–it would then be a contradiction in Euclidean geometry. In this way, the interpretation provides a proof of relative consistency. We will return later (Chapter XI) to the deeper question of absolute consistency.

The intellectual origins of the structure of a hyperbolic non-Euclidean geometry are multiple. On the one hand, this geometry arises from measurements: Do the angles in a triangle of light rays really add up to 180°, or to something else? On the other hand, this geometry also arises from a formal study of the axioms for geometry. Is the parallel axiom necessary, or can it be deduced from the other axioms? If not, what does this say about geometries? The result is clearly the conclusion that there can be different forms of geometrical theories. In the actual historical development, this 19th century conclusion was a tremendous shock. The result emphasizes the subtle nature of the primitive terms of a formal axiomatic system: The straight line, assumed as a primitive term in the Hilbert axiomatics, can indeed be variously interpreted–in particular, by objects, such as our pseudo-lines, which are intuitively by no means straight.

The development of non-Euclidean geometry represents a major change in the nature of Mathematics, from a science (of number and space) to a study of form.

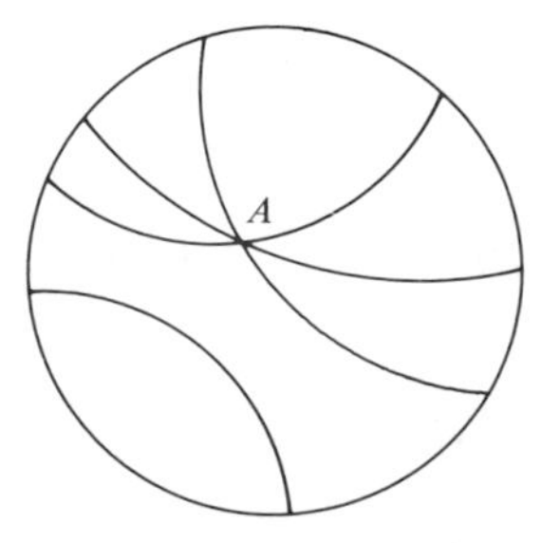

Figure 2

5. Elliptic Geometry

There is another non-Euclidean plane geometry, called elliptic geometry, in which there are *no* parallel lines. The axioms needed to describe such a structure must deviate from the parallel axiom and from other axioms of Euclidean geometry, since the latter axioms by themselves (as in §3) suppose the existence of some parallel lines. Instead of examining these axioms, we will describe elliptic geometry by Euclidean models.

The proof of Lemma 3.1 already suggests what must happen in an elliptic plane: Two perpendiculars at different points to a line k are likely to meet on *both* sides of k (so something must break down in the Euclidean description of the two "sides" of a line). More explicitly, if the perpendiculars at two points A and B on k meet at some point P, then the triangle APB has its base angles at A and B equal, hence is isosceles. Then take another point C on k so that $AB = BC$ (and B is between A and C); then $PB \equiv PB$, $AB \equiv BC$, so the triangle PBA is congruent to PBC, and then the corresponding sides PA and PC are equal (Figure 1). Therefore PC is a third perpendicular to k, and all three of these perpendiculars to k (at points A, B, and C) meet at the single point P—and there will be many more such perpendiculars from P to various points along k.

This configuration seems implausible in our usual plane, but it suggests a different interpretation. Let a "point" be a Euclidean point on some fixed sphere, while "line" is a great circle on that sphere. Then the equator is a "line", and so is the Greenwich meridian—and the various meridians from the north pole N do realize on the sphere the curious behavior suggested in Figure 2. In this interpretation one may define congruence in terms of distance (the length of arc along a great circle) and angle measure (the usual angle between two great circles). The result is a geometry in which the appropriate axioms hold. In particular, it is clear from examples that the sum of the three interior angles in a triangle for this geometry is always greater than 180°—a result opposite to that in hyperbolic geometry. This model is sometimes called double elliptic plane geometry—"double" because any two lines (two great circles) meet in two points (diametrically opposite on the sphere).

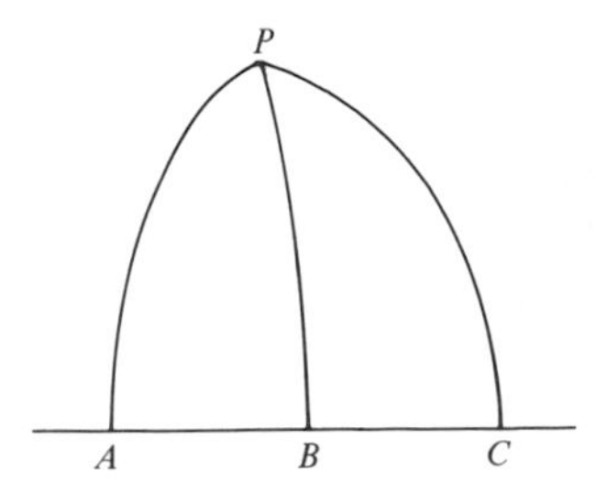

Figure 1

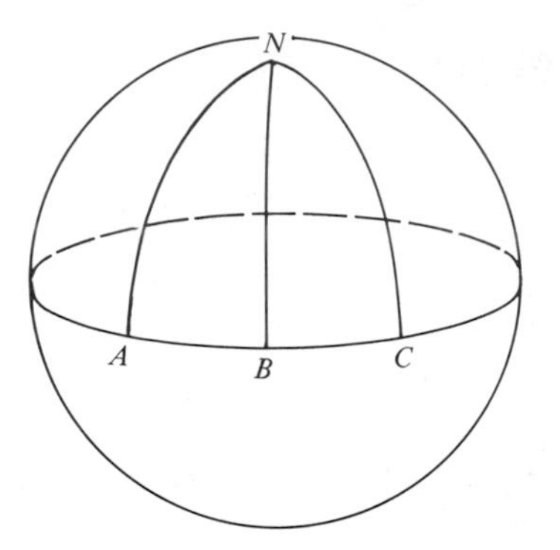

Figure 2. Elliptic geometry.

The axiom that two "lines" meet in a "point": can be rescued by constructing a single elliptic interpretation: Identify diametrically opposite points in the previous model. Thus a "point" of this new model is a pair of diametrically opposite points on the fixed sphere, a "line" is a great circle on the sphere, the "distance" from A to B is the shorter of the great-circular arcs from A to B or its diametrical opposite, and so on. Put differently, the geometry of great circles on the sphere *is* a non-Euclidean geometry.

This is a special case of the geometry of geodesics on a curved surface S. A curve γ on S is said to be a *geodesic* if, given two nearby points A and B on γ, the length of γ from A to B is less than the length of any other curve from A to B on the surface. (It is necessary to speak here of "nearby" points; on the sphere, a great circle from the north pole N is no longer a shortest distance when one travels along it beyond the south pole.) Hyperbolic plane geometry can also be represented by such a model. Take the tractrix, a curve in the xy plane given by the equations in a parameter θ,

$$x = \sin\theta, \qquad y = \cos\theta + \log(\tan\theta/2), \tag{1}$$

as in Figure 3. Rotate this curve about the y axis to form a surface. The

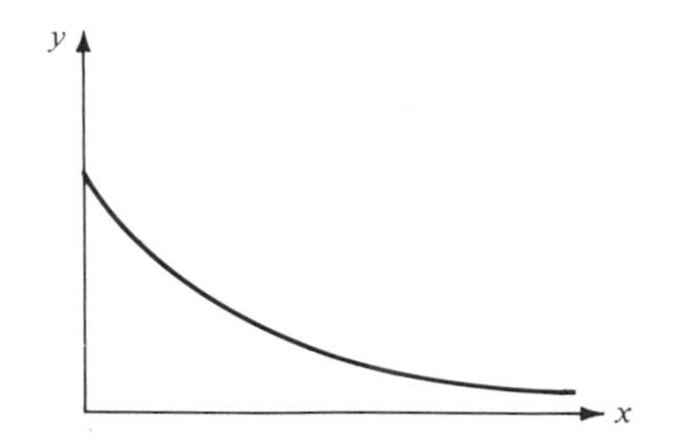

Figure 3

geodesic curves on this surface form a model of hyperbolic geometry. The surface itself turns out to have constant but negative curvature, in the same sense that the sphere has the same (constant but positive) curvature at all of its points. In this and related ways, plane geometry evolves to geometry on a curved surface (see Chapter VIII).

6. Geometric Magnitude

Hilbert's (and Euclid's) axioms for Euclidean geometry have been expressed in terms of congruence, and not in terms of distance, as it is usually measured by a real number. In other words, the geometric approach has been formulated independently of any use of real numbers (but does employ natural numbers, to state the Archimedean axiom). In fact, the geometric approach can be used to give a geometric description of magnitudes (i.e., of real numbers). We now sketch briefly how this might be done.

Fix on a line k and choose on it a point 0 (as origin) and another point U, calling the segment $0U$ the *unit* magnitude. Then by a *magnitude* one means any segment $0D$ from the origin on this line k; one of the two rays from 0 on k is chosen to be positive, and the segments $0D$ on that ray are called *positive* magnitudes (Figure 1). These magnitudes are ordered; specifically, any negative magnitude is by definition less than any positive magnitude; again, given two positive magnitudes $0D'$ and OD, the first is less than the second when D' lies between 0 and D on k. Integral magnitudes $(0U_2, 0U_3, \cdots)$ may be obtained by simply laying off the same segment $0U$ repeatedly along the ray $0U$.

A segment AB anywhere in the plane is then measured by some (positive) magnitude: Simply lay off AB as a segment $0D$ along the positive ray on k from 0, and use the corresponding magnitude $0D$ as the measure. These magnitudes can be added (in the evident way) and multplied, by using similar triangles (see §IV.3). A magnitude $0D$ is *rational* if there are integers m and n with $m(0D) = n(0U)$. There are *irrational* magnitudes, such as the hypotenuse of an isosceles right triangle with legs of unit length. Any irrational magnitude can be approximated by rational ones.

Such a geometric theory of magnitude was present in Euclid's geometry. The theory there avoided the choice of any unit of measurement, and so operated wholly with *ratios*. Thus what we have called the magnitude $0D$ would appear in Euclid as a ratio $0D/0U$. Euclid then used proportions, which are simply equalities of ratios; our approach would

0 U U_2 U_3 D

Figure 1

also need such proportions to compare magnitudes for different units. A crucial point in the Euclidean approach was a definition of such a proportion. This definition, due to Eudoxus, uses integral multiples as follows: An equality $0D/0U = AE/AV$ holds if and only if, for all whole numbers m and n,

$$m\,0D > n\,0U \quad \text{implies} \quad m\,AE > n\,AV,$$

and

$$m\,0D < n\,0U \quad \text{implies} \quad m\,AE < n\,AV.$$

In effect, this describes the ratio $0D/0U$ in terms of the set of those fractions n/m with $n/m < 0D/0U$; similar ideas will appear in our description in Chapter IV of real numbers in terms of rationals. The essential observation is that the axioms for plane geometry suffice to give a geometric theory of real numbers.

Angular magnitudes may also be compared and measured. Here an *angle*, as in the axioms, is an angle rs between two rays r and s emanating from a common origin 0. Once a unit angle has been chosen, it can be bisected repeatedly to yield binary fractions of this unit and then, by approximations, real number measures of all angles. Once the perimeter 2π of the unit circle has been determined, one may take as unit an angle of one radian–that is, the angle at the center of a circle of radius 1 which is subtended by a circular arc of length 1. This assigns to a right angle the familiar measure $\pi/2$, and gives measures ranging from 0 to π for all the angles between two rays r and s. Angle measures greater than π apply only to angles *from* a ray r *to* a ray s in (say) the counterclockwise direction–and this uses the notion of an orientation, to be discussed below in §8.

For Greek Mathematics, as this discussion indicates, magnitudes were geometric rather than arithmetic; western Mathematics, as we will see in Chapter IV, has reversed this emphasis. Oswald Spengler, in his book *The Decline of the West,* has argued that this means that there are two wholly different "Mathematics", as parts of two different cultures. Our position is rather that congruence and geometric ratios on the one hand and Dedekind cuts on the other are just two different careful formalizations of the *same* underlying idea of magnitude–and that this point exemplifies the unity of idea behind the inevitably varied form.

7. Geometry by Motion

Geometry need not be static. Intuitively, it is concerned with the ways in which "objects" can be moved around in an ambient "space". From this point of view space is really there just as a receptacle for motion. For example, the congruence theorems for triangles in plane geometry can be viewed as descriptions of conditions when one triangle can be moved so as to coincide with another. The motions involved include translations, rotations, and reflection. All are familiar from early practical activities; all are examples of "rigid motions".

A *rigid motion* of the plane is a bijection $A \mapsto A'$, $B \mapsto B'$ on the points of the plane such that each segment AB is congruent to its image $A'B'$; in other words, it is a transformation preserving distance (§I.5). From this it follows that every angle is congruent to its image, for every angle $\angle ABC$ is part of a triangle ABC which is moved to an image triangle $A'B'C'$ with corresponding sides congruent–and hence, by a congruence theorem, with corresponding angles congruent. Since a straight line is the shortest distance between any two of its points, it also follows that a rigid motion must take the points on a line l into points on some line l'; in other words, it moves l to l'. Moreover, it preserves betweenness on l: If C is between A and B on l, then the image point C' is between A' and B' on l'. A rigid motion also takes each side of l into one of the sides of l'; indeed, D and E lie on the same side of l when the segment CD does not meet l; consequently the images D' and E' lie on the same side of the image line l'. All told, a rigid motion of the plane is an "automorphism": An isomorphism $A \mapsto A'$, $l \mapsto l'$ of the whole structure of the plane.

From the definition it follows at once that the composite of two rigid motions is again rigid, and also that the inverse of any rigid motion is itself a rigid motion. Hence the rigid motions of the plane form a group.

Theorem 1. *Any rigid motion which leaves each of three non-collinear points fixed is the identity.*

Indeed, if the rigid motion M leaves two distinct points 0 and A fixed, it must leave the whole line l through 0 and A fixed. Since a third point not on the line is fixed, the motion carries each side of the line l into itself. Since any other point B on one side of the line is determined by the distances $0B$ and AB in the triangle $0AB$, this point B must also be fixed.

The several familiar kinds of rigid motions can be described and analysed directly from the axioms, although they are often described by equations in cartesian coordinates.

Intuitively, a *translation* is a motion which moves every point in the same "direction" by the same amount. Formally, a translation is a motion such that, if $A \mapsto A'$, $B \mapsto B'$, then AA' is parallel to BB' and $A'B'$ is parallel to AB. Given $A \mapsto A'$ and B, these conditions fix B' as the fourth vertex of the parallelogram with sides AA' and AB, as in Figure 1. It follows that, given points A and A', there is a unique translation T which moves A to A'. If C is a third point in the plane which is similarly translated to a point C', consideration of the congruent triangles

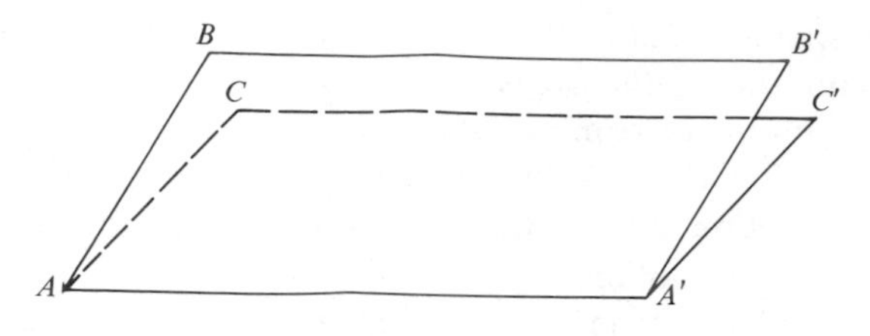

Figure 1

$ACB \equiv A'C'B'$ yields the congruence $BC \equiv B'C'$. This last congruence means that the translation T sending B to B' and C to C' does preserve distance. Hence T is a rigid motion, according to our definition of such motions.

The identity is a translation, the inverse of a translation is a translation, and the composite of two translations is again a translation. Hence all the translations of the plane form a group under composition, called the *translation group* H of the plane. It is an abelian group. If each translation T is represented by a vector (from a chosen point 0, the origin, to the image $T0 = 0'$ of that point), the translation group H is just the group of these vectors under the operation called vector addition. Once 0 is chosen, the possible translations of the plane are determined by the points $0'$ of the plane; that is, by the vectors $00'$ in the plane.

The intuitive idea of a rotation is direct: Take a circular disc, fix the center 0 and spin the disc. This idea leads to a formal definition: A *rotation* R of the plane is a rigid motion which leaves exactly one point fixed; it is called a rotation "about" that point. For completeness, the identity transformation is also counted as a rotation (about every point in the plane). From this definition it follows that the inverse of a rotation is a rotation–about the same point 0. All the rotations about a point 0 do form a group, but to show this one needs

Theorem 2. *If the distinct segments* $0A$ *and* $0A'$ *are congruent, with* $0 \neq A$, *there is exactly one rotation* R *about* 0 *taking* A *to* A'.

To construct such an R, first assume that 0, A and A' are not collinear. We need to construct the image X' of each point $X \neq 0, A$. Since the desired rotation must preserve angles, the angle $X'0A'$ must be chosen equal to the given angle $X0A$; this is surely possible, since the congruence axiom III.4 of §2 specifies that a given angle between rays can be "laid off" on a given side of another ray. It remains to choose which side of $0A'$. We know in general that all the points on one side of $0A$ must go into points on one side of $0A'$; after inspecting Figure 2 we propose the rule (which could be formulated without "looking" at a figure) that the side of $0A$ containing A' rotate to the side of $0A'$ *not* containing A. In more detail:

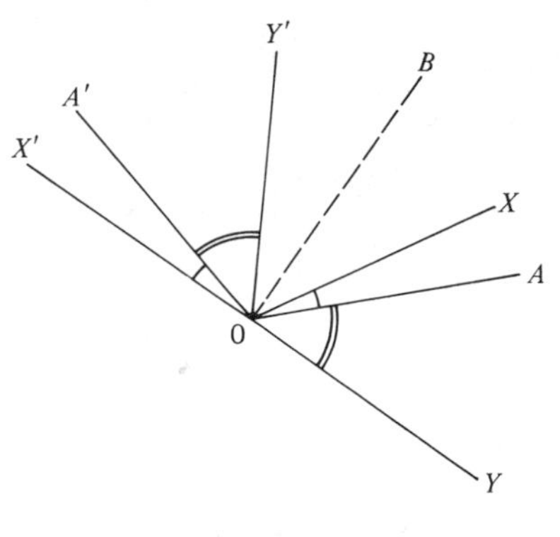

Figure 2

(i) If X and A' are on the *same* side of $0A$, put the image point X', with $0X \equiv 0X'$, on the side of $0A'$ *opposite* A;
(ii) If Y and A' are on *opposite* sides of $0A$, put the image point Y', with $0Y \equiv 0Y'$, on the side of $0A'$ *containing* A.

This prescription covers all cases. From the figure, it then appears that $\angle X0Y \equiv \angle X'0Y'$; from the axioms of congruence one can prove that this is always the case, so that the rotation $X \mapsto X'$, $Y \mapsto Y'$ does carry each angle at 0 into a congruent angle and each segment from 0 into a congruent segment. From this it then follows by the congruence axioms that this rotation R carries every segment XY into a congruent segment $X'Y'$. Therefore R is a rigid motion leaving only 0 fixed, as desired.

It remains to prove that this transformation R is the *only* rotation about 0 taking A to A'. But the only alternative would be to make the opposite choices in rules (i) and (ii) above; in this case one sees that the ray $0B$ bisecting the angle $A0A'$ (see Figure 2) must go into itself, so that $B \equiv B'$ would be an added fixed point, contrary to the definition of a rotation. (Incidentally, the opposite choice of the rules (i) and (ii) would construct a different rigid motion; namely, the motion reflecting the whole plane in the angle bisector $0B$.)

In the excluded case when A and A' are on the same line through 0, with $A \neq A'$, the image of each point X may be constructed by prolonging $X0$ over 0 to the point X' opposite X, with $0X \equiv 0X'$. This motion $X \mapsto X'$ is then a rotation about 0 by the angle π, sometimes called a "half-turn". In each case, the rotation R constructed is the only rotation with $A \mapsto A'$. From this uniqueness it follows that the composite of two rotations about 0, if not the identity, leaves only 0 fixed, hence is a rotation. Therefore all these rotations about 0 form a group.

It is remarkable that the simple idea "Hold 0 fixed and swing A to A' leads to such a subtle proof, using sides of lines. This subtlety reflects the difference between the Euclidean study of figures and the more "dynamic" idea of moving all the points in the plane.

Note however that rotation R, as constructed, simply specifies where each point eventually ends up (A to A', X to X') and not how the point "moves" there. In other words, this rotation R is not a motion R_t parametrized over time t, in the sense described in §I.5. Thus R is neither clockwise nor counterclockwise from A to A'! On the contrary, we will use the properties of rotations given in Theorem 2 to explain, in §8, the choice of a "counterclockwise" sense.

Next we combine rotations and translations. Since a rigid motion which leaves fixed three non-collinear points must be the identity (Theorem 1), two rigid motions which have the same effect on three non-collinear points must be equal. This observation makes it possible to compute several useful composites. Thus if R is a rotation about the point 0, while T is a translation taking 0 to some point $0'$, then the composite

$$RTR^{-1} = T_1 \tag{1}$$

is the translation T_1 taking 0 to the point $R0'$. This can be proved formally from the axioms, or checked visually by showing that the composites RT and T_1R have the same effect on the points of the right angle $A0B$ in Figure 3. Similarly (Figure 3 again), if T translates 0 to $0'$ and A to A', while R rotates $0A$ to $0A''$, then the composite

$$TRT^{-1} = R_1 \tag{2}$$

is the rotation about $0'$ taking $0'A'$ to $0'(TA'')$. Recall here that TRT^{-1} is called a *conjugate* of R, so this equation states that translation T conjugates a rotation about 0 into a rotation (through the same angle) about $0'$. Similarly, equation (1) states that the conjugate of a translations by a rotation is another translation.

These equations enable us to prove a familiar fact:

Theorem 3. *Composites of rotations and translations of the plane form a group. If* 0 *is any chosen point, an element of the group can be written uniquely as a product $T \cdot R$ with T a translation and R a rotation about* 0. *This group is called the group E_0 of* proper rigid motions.

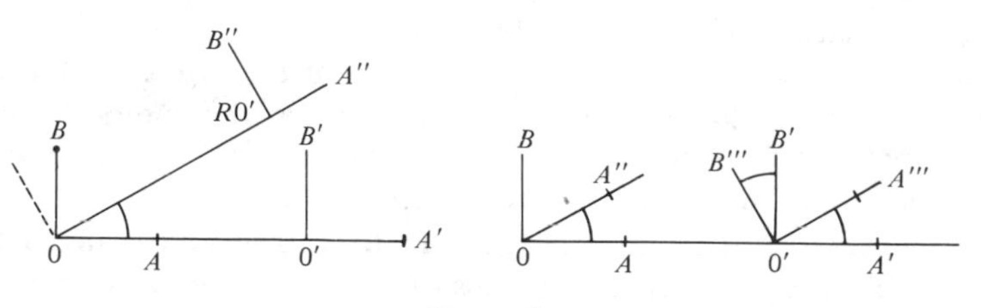

Figure 3

Proof. Consider the collection of all composites of a finite number of rotations and translations; it is a group, since the inverse of a translation or a rotation is again such. By equation (2), any rotation R_1 about a point $0'$ may be replaced by translations and a rotation R about 0. By equation (1), any composite $R \cdot T$ may be replaced by a composite $T_1 \cdot R$ for some translation T_1. Hence all the composites can be rewritten in the desired form $T \cdot R$. This form is unique, for take any point $A \neq 0$. Then $T \cdot R$ takes the segment $0A$ to a congruent segment $0'A'$, so (Figure 4), T must be the translation taking 0 to $0'$ and R the rotation taking $0A$ to $0A_1$ where $A_1 \equiv T^{-1}(A')$.

This last argument also proves the following familiar fact:

Corollary. *If* $0A$ *and* $0'A'$ *are congruent segments, with* $0 \neq A$, *there is a unique proper rigid motion taking* 0 *to* $0'$ *and* A *to* A'.

In other words, a proper rigid motion is determined by its effect on any two distinct point 0 and A.

The *reflection* L in a line l is also a rigid motion. Under this reflection, each point A is sent to the point A' with l a perpendicular bisector of AA'. Thus (Figure 5) A' is on the opposite side of l (and at the same distance from l). Again, AB is congruent to $A'B'$, so reflection is a rigid motion by our definition. It has an evident physical meaning (reflect, as in a mirror at l). It differs from translation and rotation in that it "turns the plane over".

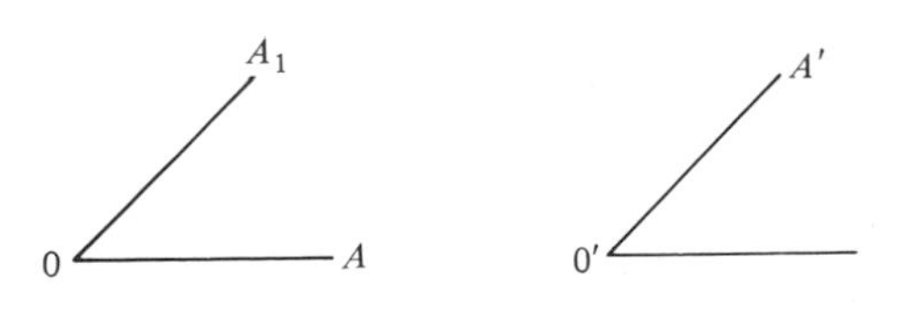

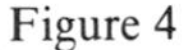
Figure 4

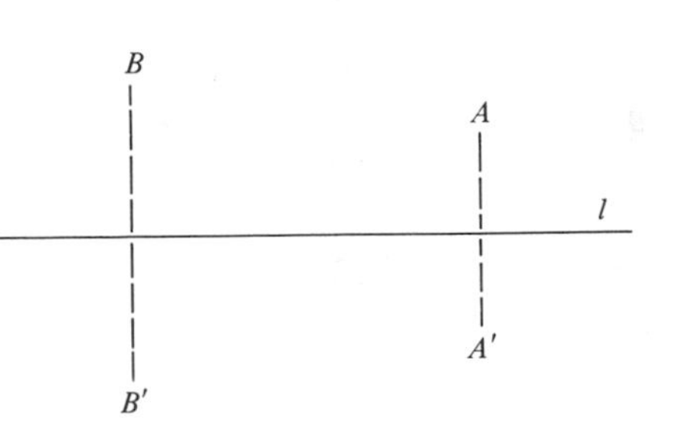

Figure 5

Theorem 4. *Let l be any line in the plane. Any rigid motion of the plane is either a proper rigid motion P or a composite $P \cdot L$ where L is the reflection in the line l.*

PROOF. The motion is determined by what happens to each of three non-collinear points; that is, by what happens to a triangle. (Hence the importance of triangles in Euclidean geometry.) So consider the triangle $0AB$ with side $0A$ along the given line l and its image $0'A'B'$ under the motion (Figure 6). Since $0A \equiv 0'A'$, there is a proper motion (rotation R followed by translation T) which takes $0A$ to $0'A'$; one can then place the image B' on the desired side of l, by using first a reflection L in l, if necessary. Hence the motion has a unique representation as either $T \cdot R$ or $T \cdot R \cdot L$, with R a rotation about 0.

These theorems serve to show that groups (transformations and their composites) are firmly embedded (though not explicit) in classical Euclidean geometry.

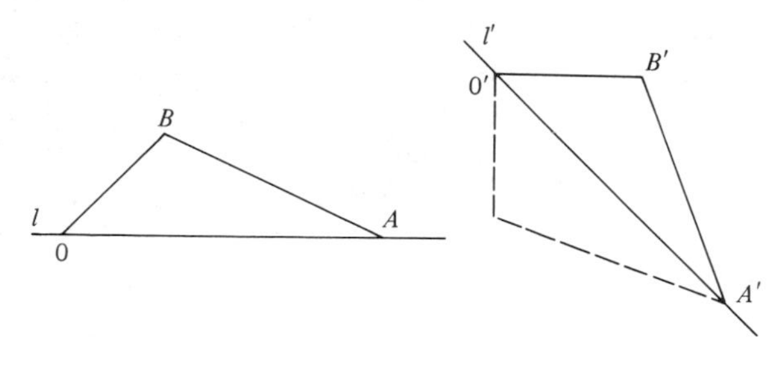

Figure 6

8. Orientation

Hitherto "angle" has meant the angle *between* two rays r and s from a point 0. For trigonometry and elsewhere we also need to use the angle *from* the ray r *to* the ray s. But this is ambiguous (Figure 1) unless we specify that we mean the "counterclockwise" angle from r to s. But this is

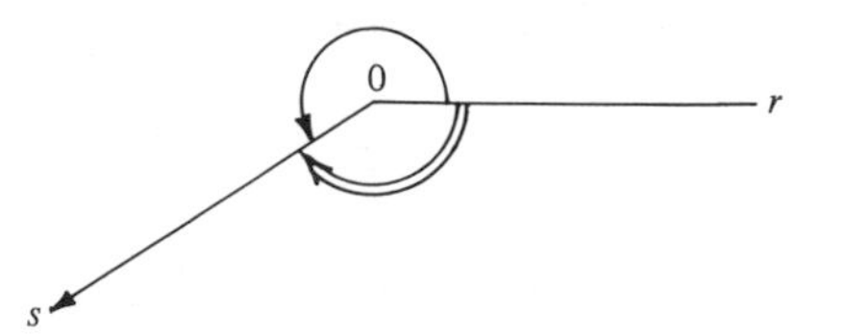

Figure 1. Clockwise and counterclockwise.

meaningless for the plane as it has been axiomatized, because the counterclockwise sense, viewed from in front of the plane, will be clockwise when viewed from behind the plane. Thus we must make a deliberate choice of one of the two possible "senses" for angles. Such a choice is called on *orientation* of the plane, and the choice of an orientation is an additional structure on the Euclidean plane.

Practically speaking, someone (the first clockmaker?) chooses a direction of rotation for his clock; then one can carry this clock (or copies, such as wrist watches) around so as to determine the same clockwise sense for all other clocks. Here to "carry around" means a proper rigid motion; this idea we can now formalize (in the plane).

To fix the ideas, consider, instead of clock-hands, ordered right angles from $A0$ to $B0$, with legs say of unit length, as in Figure 2. Here $\angle A0B$ and $\angle A'0'B'$ have the same sense, while $\angle A''0''B''$ has the opposite sense. Formally, two such ordered right angles have the same *sense* if and only if the first can be carried into the second by a proper rigid motion (a rotation followed by a translation). Since there is exactly one such proper motion carrying $0A$ into the congruent segment $0'A'$, there are exactly two possible senses, as given by the opposite perpendiculars $0'B'$ and $0'B'''$ in the middle of Figure 2. Put differently, the ordered right angles with unit legs fall into two distinct classes (or "orbits") under the action of the group of all proper rigid motions. Choosing one of these classes is the choice of a "sense" or an *orientation* of the plane; for instance, the chosen sense may be called the *counterclockwise* one. The choice is usually made by picking some one right angle, from $0A$ to $0B$, to be counterclockwise.

Once the orientation is made by this choice of $\angle A0B$, one can introduce the usual four "quadrants" I–IV for this right angle and then cartesian coordinates, since we now have a "positive" direction $0A$ or $0B$ on each axis of Figure 3. We can also consider and measure ordered angles (say, the angle from the ray $0A$ to the ray $0t$ in Figure 3) on the familiar radian scale from 0 to 2π, so that these angles are real numbers taken modulo 2π. We can also speak of the "left" side of the ray $0A$; it comprises all the

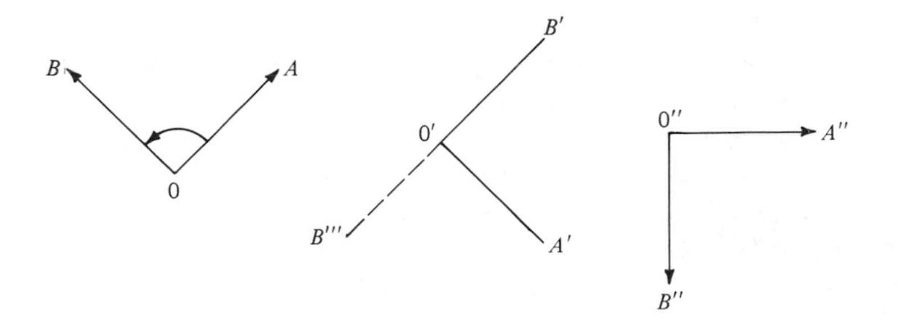

Figure 2

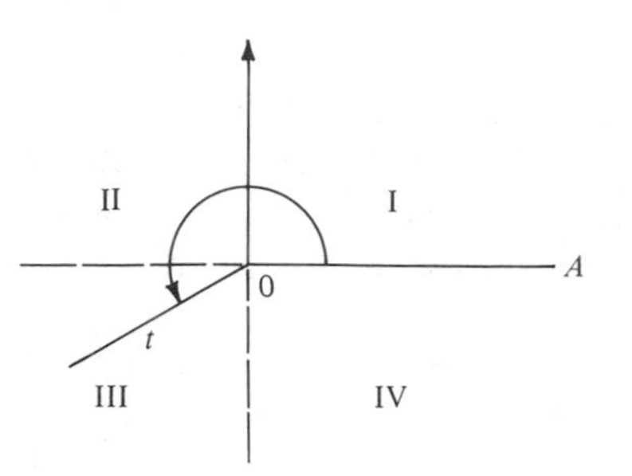

Figure 3

rays from 0 in quadrants I and II. We will write sLr for rays r and s from 0 to say that s is on the *left side* of r.

A plane with orientation is really not the same object as one without. The plane with an orientation has more structure–namely, the choice of the orientation. At the same time, it has less symmetry; the automorphism group of the oriented plane is the group of all proper rigid motions (i.e., no reflections), while that of the unoriented plane is the group of all rigid motions, including the reflections. This is a first example of a striking observation: A geometrical "thing" (the plane) can be formalized in different ways to give different mathematical objects. In the oriented plane, one has angles up to 2π and coordinates as well–but these are not present in the unoriented plane described by the Euclidean axioms.

The added structure of the oriented plane has been described as a "choice of an orientation". It can be formulated in other ways, so as to resemble an "axiomatic" structure. For example, consider the relation sLr for "s is on the left side of r", where r and s are rays from a common point 0. If $-r$ denotes the ray opposite r, but along the same line through 0, this relation has the following properties:

(i) sLr if and only if $(-r)Ls$,
(ii) sLr if and only if not $(-s)Lr$,
(iii) If a translation carries the rays r and s at a point 0 to rays r' and s' at a point $0'$, then rLs implies $r'Ls'$.

These three statements can be regarded as the added axioms describing an oriented plane. From them one can prove directly, without using rotation, that a Euclidean plane has exactly two orientations. On this basis one can then define a rotation about the point 0 to be a rigid motion leaving 0 fixed and mapping rays r, s from 0 to rays r', s' in such a way that rLs implies $r'Ls'$. One can then easily prove that there is exactly one such rotation carrying a segment $0A$ into a specified congruent segment $0A'$ (as in Theorem 2 of §7).

These related notions "rotation" and "orientation" are known to be subtle–and are often left to the intuition. Our discussion, both in

Theorem 2 of §7 and in the formulation of the relation L, does not follow any standard source in the literature. It is intended to show how the very natural "idea" of rotation requires a subtle formulation. Still, the idea is a familiar one: In three dimensions, one cannot rotate a left-hand glove into a right-hand one!

9. Groups in Geometry

Rigid motions not only carry points into points, but they also carry segments into (congruent) segments, angles into (congruent) angles, and so on; one says that the group E of all rigid motions "acts" on the set of segments or on the set of angles, and that the "orbit" of an angle is the set of all congruent angles. This example motivates a general definition. A group G is said to *act* on a set X when for each g in G and each x in X there is given an element gx in X (the result of g acting on x) in such a way that

$$g_1(g_2x) = (g_1 \cdot g_2)x, \qquad 1x = x \tag{1}$$

for all g_1 and g_2 in G, for the identity 1 of G, and for all x in X. The *orbit* of an element x in X under this action is the set Gx of all gx for g in G, and the action is *transitive* when $Gx = X$ for all x; that is, when to each x and y in X there is at least one g with $gx = y$. In any case, the elements f in G which leave a given point x fixed form a subgroup F_x of the group G, called the subgroup *fixing* x or the *isotropy* group of x.

For example, when the group E_0 of proper rigid motions acts on the points of the plane, the isotropy group F_x of each point x is the group of rotations about x. Any two such groups F_x and F_y are isomorphic. All the proper motions carrying x to y can be written as products TR for R in F_x and T any motion (say a translation) carrying x to y. All these motions constitute a so-called "coset"

$$TF_x = \{\text{all } TR \mid R \in F_x\} \tag{2}$$

of F_x. The whole group E_0 consists of the various disjoint cosets TF_x, $T'F_x$ corresponding to the various points $Tx = y$, $T'x = z, \ldots$ of the plane.

These observations hold in general. For any group G acting on a set X, $hx = kx$ for h and k in G if and only if $k^{-1}h$ is in the isotropy group F_x. Thus all the elements k in the coset

$$hF_x = \{\text{all } hf \mid f \in F_x\} \tag{3}$$

map x to the same point on the orbit, so that the set of cosets corresponds to the set of points in the orbit, as in the bijection

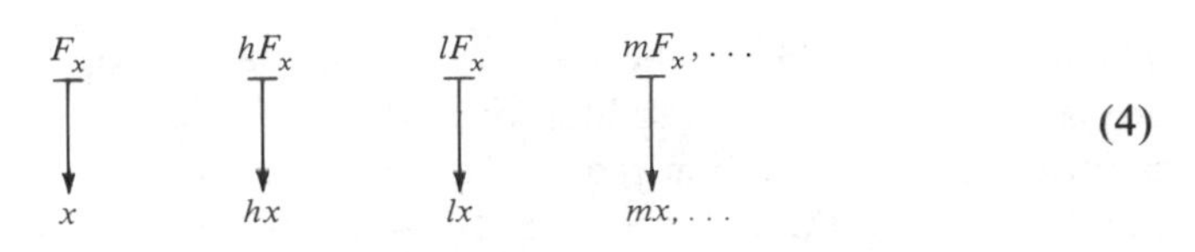

This is the (geometric) origin of the notion of a "coset". If points x and y are in the same orbit, so that $hx = y$ for some h, then $F_y = hF_xh^{-1}$; this states that the isotropy subgroups F_x and F_y are *conjugate* subgroups of G, isomorphic under the bijection $g \mapsto hgh^{-1}$. This is the geometric origin of the notion of "conjugation".

The group E of all rigid motions acts on the set X of all segments AB. Here the orbit of each segment is the set of all congruent segments, while the subgroup fixing a segment AB consist of four motions, $\{I, L, R, L \cdot R\}$, where L is the reflection in the line of AB and R is the half turn (rotation by 180°) about the midpoint of AB.

Similarly, the reader may examine the associated action of the group E on the set of triangles, on the set of lines, on the set of rays, and on the set of circles in the plane.

Consider the subgroups of E:

$$E \supset E_0 \supset H \supset \{1\}, \tag{5}$$

where E_0 is the group of proper rigid motions and H the group of translations. The plane has just two orientations, call them $\uparrow$ and $\downarrow$. The whole group E acts on the set of these orientations, in that each rigid motion M induce a permutation $\sigma(M)$ of the set $\{\uparrow, \downarrow\}$; thus when $M \in E_0$ is a proper rigid motion, the orientation is unchanged, so $\sigma(M)$ is the identity permutation, while a reflection L interchanges the two orientations. For the composite $M \cdot M'$ of two motions the effect is

$$\sigma(M \cdot M') = \sigma(M)\sigma(M'); \tag{6}$$

this equation states that σ preserves composites, hence is a *homomorphism* of groups. This homomorphism carries the subgroup E_0 to the identity permutation and its coset E_0L (all products ML, for M in E_0) to the permutation interchanging $\uparrow$ and $\downarrow$.

A *direction* (e.g., the "direction" of a ray r_0) in the plane may be described formally as the orbit of the ray r_0 under the translation group. Every rigid motion M carries directions to directions, so defines an action of E on the set of directions. Let $\sigma(M)$ denote the permutation of the set of directions induced by the rigid motion M; thus σ is a homomorphism, as in (6), while the set of all motions M with $\sigma(M)$ the identity is the subgroup H of translations.

In the terminology of group theory, this subgroup H is a normal subgroup of E because every conjugate MTM^{-1} of a translation is a translation. We showed that every proper motion in E_0 had the form $T \cdot R$, with

T a translation and R a rotation about a fixed point 0, so that E_0 is the union of the cosets of H

$$HR = \{\text{all } TR \mid T \in H\}.$$

The cosets correspond to the rotations R, so that the cosets may be said to form a group E_0/H isomorphic to the group of rotations about 0.

These observations, and many like them, indicate the very close relation between Euclidean geometry and group theory—so close that one might say that groups were implicit (though never explicit) in traditional geometry. For these reasons, it is clear that the basic ideas of group theory belong early in the conceptual order of mathematical structures. Historically, groups did not appear until the 19th century, implicitly with Gauss and others and explicitly with Galois. When they were fully recognized, they were applied promptly to geometry, by Klein, Lie, and others. In particular, hyperbolic and elliptic geometries also involve appropriate groups of motions—as does solid geometry.

10. Geometry by Groups

Space is not just a static array of geometric figures; it is primarily a field for motion. This can be stated more emphatically in the observation that the axioms for the Euclidean plane can be formulated wholly in terms of the group E of rigid motions.

We will sketch briefly how this can be done. In any group, an element h of order 2 (an element $h \neq 1$ with $h^2 = 1$) is called an *involution*. In the group E of all rigid motions the reflection L_k in the line k and the rotation (half-turn) R_A by 180° about the point A are both involutions.

Theorem 1. *A rigid motion M of the plane which is an involution is either a reflection L_k in some line k or a half-turn R_A about some point A.*

Proof. Since $M^2 = 1$ but M is not the identity, there is a point B with $MB = C \neq B$. Then $MC = M^2B = B$, so the motion M interchanges the two points B and C. Let l be the line which is the perpendicular bisector of the segment BC, as in Figure 1. The points on l are the points equidistant from B and C; hence the given involution M must carry points on l to points on l (it thus leaves line l as a whole fixed). Moreover, M must leave fixed the midpoint 0 of the segment BC; 0 lies on l. Take a point P on l but different from 0. Then M must send P to another point on l at the same distance from B and C—thus either to P itself or to the point P' on the opposite side of 0 with $P0 = 0P'$. The reflection L_l sent P to P; the half-turn R_0 sends P to P'. Since a rigid motion is determined by what it does to three non-collinear points, this proves that the given involution M is either L_l or R_0.

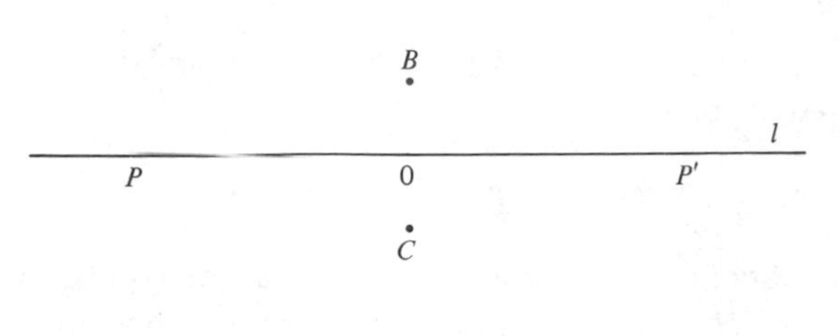

Figure 1

Theorem 2. *Every rigid motion is a composite of involutions.*

Proof. It will suffice to show that every rotation and every translation is a composite of involutions. To this end, consider the composite of two reflections $L_k \cdot L_l$ in distinct lines k and l. If k and l are not parallel, they meet in a point 0, say at an angle ϕ. The composite motion $L_k \cdot L_l$ leaves (Figure 2) the point 0 fixed and preserves orientation, hence is a rotation about 0. Now L_l leaves l fixed, while L_k takes l to the line l' through 0 making with k the same angle ϕ as does l, but on the opposite side of k. Therefore $L_k \cdot L_l$ is the rotation about 0 through the angle 2ϕ. Hence every rotation is a composite of two reflections.

In the remaining case k and l are parallel, so have a common perpendicular m. The composite $L_k \cdot L_l$ then turns out to be translation along m by twice the distance from k to l.

The objects of plane geometry can now be identified with elements in the group E of rigid motions. Since each rotation R_A by 180° is the composite of two reflections, the points A may be identified with those involutions which can be written as the composite of two involutions, while the lines k are identified with those involutions (to wit, the L_k) which are not such a composite. The point A lies on the line k if and only if $L_k \cdot R_A = R_A \cdot L_k$. If the motion M takes the point A to the point B, then the half-turn R_B is clearly MR_AM^{-1}, so the motion M conjugates the involution R_A to the involution R_B. One may similarly describe the effects of any motion on a line, represented as an involution.

This start can be continued to give a complete list of axioms on the group E sufficient to describe E as (generated by the involutions

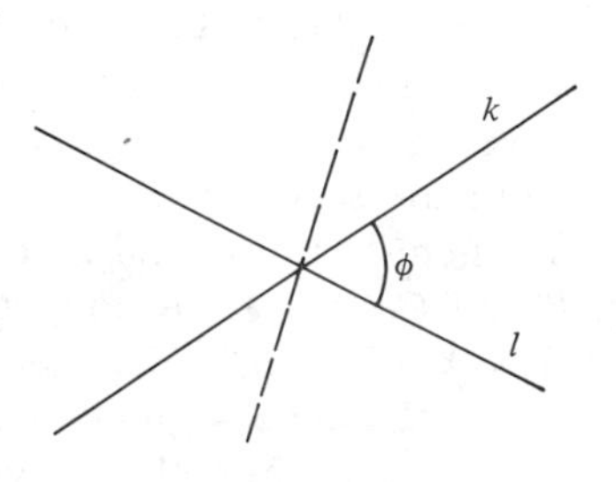

Figure 2

representing) the points and lines of the Euclidean plane. Details are given in Bachmann (1973) and in Guggenheimer (1967). The reader may wish to formulate for himself some geometric facts in group-theoretic form. For instance, $AB = BC$ states that B is the midpoint of AC!

11. Solid Geometry

The geometry of "our" space is three dimensional, so is harder to visualize on paper. As for the plane, experience reveals many facts about solid figures. It again turns out that these facts are logically related to each other and so can be derived from suitably chosen axioms. These axioms are stated in terms of basic figures: points, lines, planes, spheres, and the like. As formulated in Euclid or as presented in modern form in Hilbert they involve little more than the axioms for plane geometry. The surprising fact is thus the observation that so much of geometry is already contained in the geometry of the plane (but not yet in the geometry of the line). For example, the Pythagorean theorem for plane right triangles readily produces the distance formulas of analytic geometry, both in the plane and in space. Again, angles necessary for plane geometry (and not present in one line alone) are vital to trigonometry, periodic functions, navigation by spherical trigonometry, and the group of the circle—but the corresponding notion of "solid angle" is very little used. It was surely a remarkable discovery (was it made by the Greeks?) that concentration on the plane would yield so much of the ideas of geometry. This discovery rested in turn on the prior formulation of the "idealized" notion of a plane (or of a line), but this latter discovery is not so remarkable (it is surely suggested by diagrams in the sand or on paper, by maps, etc).

The group of rigid motions of Euclidean solid geometry is a much bigger and richer group than that of the plane—but it is still true that any rigid motion of space is a composite of rotations, translations, and reflections. For example, a rotation R is a rigid motion which leaves fixed all the points on some line k; then k is called the *axis* of rotation. It is a rotation through an angle θ, because each plane perpendicular to the axis is rotated by R through the same angle θ (watch the orientations!). In this sense a solid rotation is just a combination of plane rotations in parallel planes.

All the rotations about a point 0 (i.e., all with axis through 0) form a group. This is a consequence of one of the fascinating facts about such solid rotations, the rule for the composition of two rotations about the same point 0. Take the sphere of radius 1 and center 0, and consider a rotation R about the axis $A0$ through an angle θ followed by a rotation S about a different axis $0B$ through an angle ϕ. To find the composite $S \cdot R$, draw the arc AB of a great circle joining A to B and then construct great circular arcs AC and BC making angles $\theta/2$ and $\phi/2$, respectively, with

AB (Figure 1). This gives a "spherical triangle" ABC. One may show that $S \cdot R$ leaves C fixed and hence is a rotation (through what angle?) about $C0$. This curious fact about rotations is one of the results of solid geometry much used in the study of rotations in mechanics. However, even this apparently "solid" geometrical result has its predecessor for the plane: There *is* a similar description of the composite of two rotations (about different points) in the plane.

The orientation of 3-dimensional space also involves no essentially new ideas. To describe it, consider three perpendicular rays x, y, z from a point 0 (they are, in effect, coordinate axes). Then the ordered triple (x,y,z) is defined to have the same orientation as a similar triple (x',y',z') if and only if there is a rotation leaving 0 fixed and carrying x to x', y to y', and z to z'. This definition provides for exactly two orientations at 0 (just as in the plane). For, one can always rotate x to x' (about an axis perpendicular to the plane determined by x and x'). The new position of y can then be rotated (about the axis x') to the position y' (Figure 2). The resulting position of z must then be perpendicular to x' and y' and hence must coincide either with z' or with its opposite–therefore there are just two orientations at 0. Note that the triple (x,y,z) has the same orientation as its cyclic permutation (y,z,x)–just rotate x to y (about z) and then the new y (about the old y) to z. In other words, the orientation of (x,y,z) is

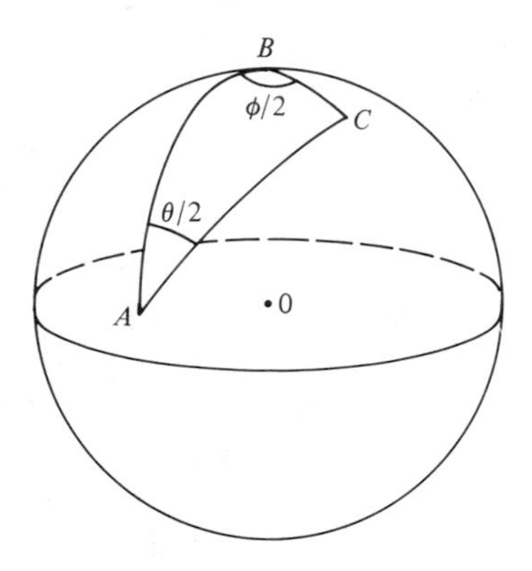

Figure 1

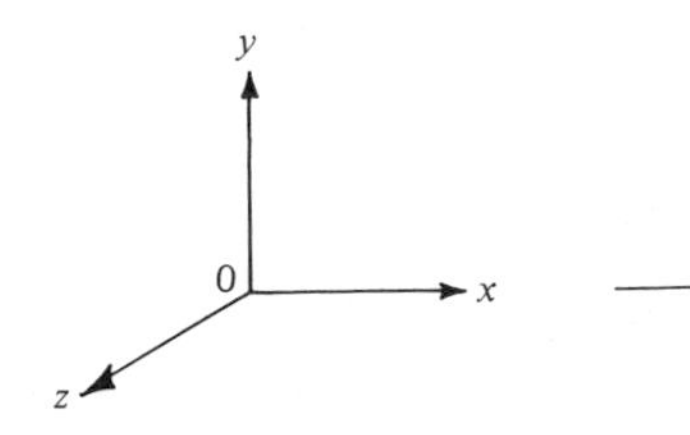

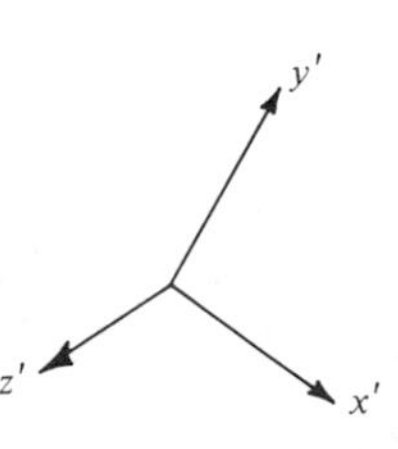

Figure 2

unchanged by a cyclic permutation (an *even* permutation). However, it is changed by interchanging x and y to give (y,x,z), which is an *odd* permutation. The orientation determined by the first three fingers, in order, of a left hand is called the "left-handed" orientation.

This description of just two orientations extends readily to dimensions higher than three. In general, an axiomatic treatment of n-dimensional Euclidean geometry in terms of the appropriate "flat" subspaces (points, lines, planes, hyperplanes . . .) is readily possible–and a bit dull.

In all these respects, three dimensions is much like two dimensions–but there are important points of difference, such as the much greater variety of figures in 3-space: Spheres, cylinders, pyramids, regular solids, irregular solids, and the like. Decisive new notions, such as the Gaussian curvature of a surface at a point, arise from the study of these figures, as we will see in a later chapter. There we will also examine a wide variety of other geometries.

12. Is Geometry a Science?

This chapter has summarized the elementary aspects of the remarkable way in which primitive experiences of space and motion can be arranged and refined in a precise deductive system organizing and relating geometric facts and figures. At first blush, the resulting system of Euclidean geometry can appear to be *the* science of space, just as mechanics is *the* science of motion or biology is *the* science of life-forms. But if it were to be *the* science of space, it should be the unique such; this is not the case. We will not attempt here a precise description of what is meant by a *science*. Geometry was "extracted" from experience, but it seems clear that we do not test the theorems of geometry by experiments, as we would the propositions of physics. Even if measurements were to show the Pythagorean theorem false for some proposed right triangle, we would not doubt the proofs of that theorem. We would be more likely to alter the definitions used in those measurements of distance or of right angles, or we would perhaps decide that those measurements of distance and of an angle belonged to some different geometric theory. In the language of Karl Popper, statements of a science should be falsifiable; those of geometry are not.

We are more concerned with the positive aspects of the question: *What*, then, is geometry? It is a sophisticated intellectual structure, rooted in questions about the experience of motion, of construction, of shaping. It leads to propositions and insights which form the necessary backdrop for any science of motion or of engineering practices of construction. But at the same time, the structure of geometry is not unique. The axioms can be formulated as in the Euclidean tradition, as it was refined by Hilbert–or they can be put in group-theoretic form. Moreover, non-Euclidean geometries of plane or of space are indeed possible and useful, and they can be reduced to axiomatic form. Hence geometry is a variety of intellectual structures, closely related to each other *and* to the original experiences of space and motion.

There arise from this study other structures which are less geometric—distance and angle as geometrical magnitudes, algebraic manipulations of these magnitudes, and thus real numbers, developed geometrically. There also arise structures which are not geometric at all—groups are implicitly present in the transformations of geometry. Logic is (historically) first fully exemplified in the deductive structure of Euclidean geometry. Continuity and topology are hidden there as well.

The development of geometry also turns up a number of general ideas. The very formulation of axiomatic geometry requires ideas of line, plane, angle, and triangle. Later developments turn up more subtle ideas, such as that of orientation or of composition of motions. Geometry is indeed an elaborate web of perceptions, deductions, figures, and ideas.

CHAPTER IV

Real Numbers

1. Measures of Magnitude

This chapter will explore the origins and development of the system of real numbers. These numbers form the most central structure in all of Mathematics. Like other structures it arises from more primitive human activities–in this case, from the measurement and comparison of magnitudes of various kinds.

Comparison of magnitudes, in its simplest qualitative form, may just assert that "A is bigger than B" without specifying by how much A is bigger. The idea of such qualitative comparisons leads to the formal notion of a linear order, as already discussed in §I.4. Now we are concerned with the associated quantitative question "A is how much bigger than B?". Such questions arise in a number of different regards "A is how much further away" or "how much heavier", or "how much longer", or "how much later". Such comparisons are not limited to just two objects A and B, but may well intercompare the sizes of three, four, or many objects. This makes it effective not just to compare two objects, but to locate all the relevant objects on some on *scale* of sizes. Once a unit of size is chosen, the scale becomes a scale of numbers. It is a familiar but nonetheless remarkable fact that one single scale of numbers will be applicable to each of many types of quantitative comparisons: To distance, to weight, to length, to width, to temperature, to time, to height, and so on. Once a unit is chosen, each of these magnitudes exemplifies one and the same scale: That of real numbers, considered as a scale laid out as the points of a line with chosen origin and unit point; and so emphasizing the interpretation of the scale by distances:

−2 −1 0 1 2 3

The ubiquity of this scale may account in part for the prominence of the real numbers in Mathematics. This ubiquity can also be read as a statement about the nature of the world: All sorts of physical magnitudes can be reduced to one scale—a situation which has sometimes misled social scientists to use fake magnitudes.

The importance of this scale depends also on the fact that the scale is complete. It is not limited just to the whole number points and not even just to the rational points. All the points at irrational distances from 0 are to be included, as, for example, $\sqrt{2}$. By the Pythagorean theorem, this number, considered as a length, is the diagonal of a unit square; it cannot be expressed as a rational distance as $\sqrt{2} = m/n$. (Proof: reduce to lowest terms, then one of m or n is odd, but $(\sqrt{2})^2 = 2 = m^2/n^2$ gives $2n^2 = m^2$, so m and then also n must be even.) There are many other such irrationals which are *algebraic* (roots of a polynomial equation with integer coefficients, as in $x^2 - 2 = 0$). In addition, there are numbers on the scale which are *transcendental* (i.e., *not* algebraic). The first example is π, the ratio of circumference to diameter in any circle, but the proof that π is transcendental is not easy. Another transcendental number is the base e of natural logarithms.

There are many more transcendentals. Completeness (to be formulated below) is the assurance that they are all there. They are not all in use at any one time—but their presence for potential use is what makes the scale effective.

2. Magnitudes as a Geometric Measure

The basic scale of magnitude can be regarded as either an arithmetic or as a geometric structure. The arithmetic approach starts just with the scale of whole numbers (positive, zero, and negative), then adjoins the magnitudes represented by the rational numbers, and finally those represented by irrational (algebraic and transcendental) numbers. This strictly arithmetic view of magnitude will be developed below, in §6.

The scale of magnitudes may also be regarded as primarily geometric. Then the basic model of a magnitude is a line segment AB (or, if you will, the length of that line segment). Then, as discussed in Chapter III on geometry, two line segments AB and CD represent the *same* magnitude when they are congruent, so that CD can be "moved" to exactly cover AB. Thus a magnitude is not really a single segment AB, but all the segments CD which are congruent to AB—in set-theoretic terms, the magnitude AB is the (congruence) class of all segments CD congruent to AB. Also the magnitude AB can be compared with any other magnitude RS by "laying off" the segment AB along the ray of the segment RS, with A at the position of R; the congruence axioms of plane geometry provide precisely for such a comparison. It follows that the points on a single line represent all the possible magnitudes of segments so that the line (with zero and unit point chosen) *is* the desired scale of magnitudes.

Angles present another type of geometric magnitude. As in Chapter III, the axioms of plane geometry provide for the congruence of angles and the comparison of angular magnitudes. The reduction of these *angular* magnitudes to *linear* magnitudes (i.e., to the real number scale) involves a subtle development. First one measures the circumference of a circle, say a circle S^1 of radius 1. This can be done by inscribing regular polygons in the circle, starting say with an equilateral triangle and using successive bisections (by ruler and compass) to get inscribed regular polygons of 6, 12, 24, . . . sides, as suggested in Figure 1. The (rectilinear) perimeter of each such polygon can then be calculated; as the number of sides approaches infinity, these perimeters approach a definite limit, the real number 2π.

Now the radian measure of an angle θ at the center of the circle is to be the length of the arc which θ subtends on the circumference. Also, bisecting the angle halves the arc length, so this gives the usual radian measures π for a straight angle and $\pi/2$ for a right angle. Successive bisections (again by straightedge and compass) then provide angles with radian measures $\pi/2^n$. Any angle can be approximated, as closely as desired, by multiples of such angles: Any angle $\theta = \angle ACP$ at the center C of a circle of radius 1 has measure t_0 (Figure 2) described as follows:

$$\text{measure}(\theta) = t_0, \qquad 0 \leqslant t_0 < 2\pi, \qquad t_0 = \text{length of arc } AP. \quad (1)$$

Any real number t can be written modulo 2π as $t = t_0 + 2\pi k$ for some integer k, and so determines the angle $\theta = \theta_t$ with measure $\theta_t = t_0$. The function $t \mapsto \theta_t$ then "wraps" the whole real line around the circle S^1, sending t to the point P; i.e., to the angle $\theta_t = \angle ACP$. Moreover this *wrapping function* is periodic:

$$\theta_{t+2\pi} = \theta_t . \quad (2)$$

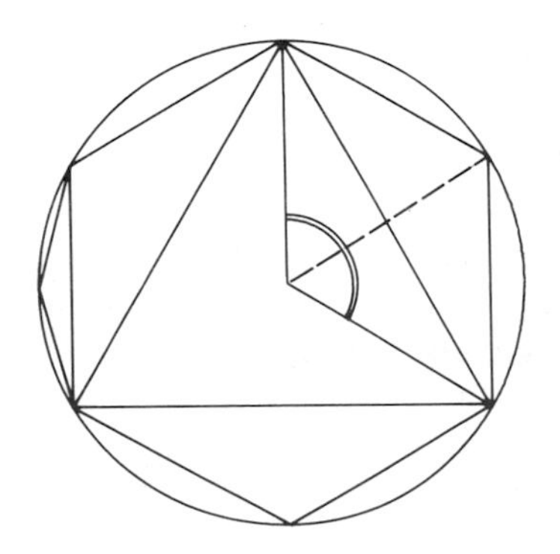

Figure 1

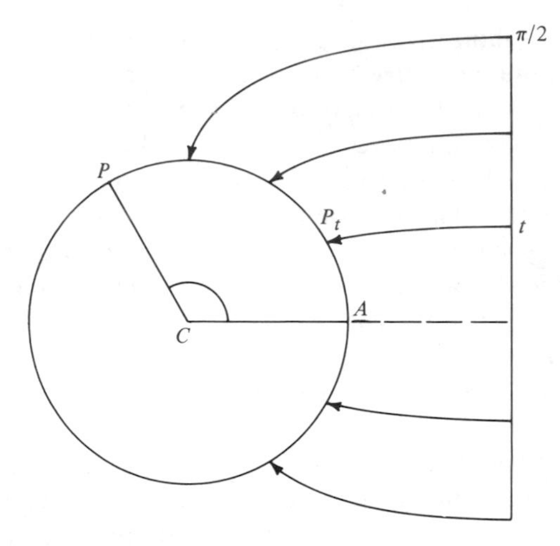

Figure 2

Alternatively, the methods of calculus could be used to determine circular arc lengths and hence radian measure and the wrapping function θ_t. In either way, this function plays a central role in analytic trigonometry (§9). The development of Euclidean geometry from its foundations in the axioms also provides a treatment of the magnitudes of areas (of triangles, rectangles, circles, etc.) and a comparison of these magnitudes with those measured along the real line. This is carried out in Hilbert's *Foundations of Geometry*. Thus one might say that the axiomatic foundations of geometry must provide for the comparison and reduction to a single scale of the various types of geometric magnitudes–including also volumes. The same must be done in any completed system of non-Euclidean geometry.

Measurements can also reduce other types of physical magnitudes to a geometric form. Thus weights on a balance are compared with numbers (the number of weights needed to achieve balance), while the weights measured on other instruments give the magnitude of a weight by a pointer position along some geometric scale. A thermometer does the same for the magnitudes of temperature, while a clock hand reduces time to angular magnitudes.

To summarize: Many comparisons or measurements of various magnitudes can be reduced to the single scale provided by the real numbers. And in those cases where a single real number does not suffice to measure a magnitude, it is often fitting to use several such numbers–as when the size of a plane figure is given by its width and its height, or of a solid figure by width, height, and depth. In other words, that scale of real numbers has many different realizations.

3. Manipulations of Magnitudes

When two objects are put together, side by side or end to end, the magnitude of this combination is just the sum of the two separate magnitudes. This operation, called *addition*, arises for all sorts of magnitudes–for distance, weight, time, height, area, and the like. Geometrically, the addition of two segments consists in laying off one segment after the other along the line. This geometric operation corresponds exactly to the arithmetic operation of adding numbers.

A second operation on magnitudes, that of multiplication, is suggested both by the multiplication of numbers and by geometric formulas; thus the area of a rectangle is obtained by multiplying its base by its height. A complete geometric description of the multiplication of segments requires more than one line in the plane. Thus to multiply two positive linear magnitudes x and y one may represent them on two intersecting linear scales: Segment $0A$ and then AB on the first line with measures 1 and x, respectively, and then $0A'$ with measure y on a second ray from 0. Then drawing (Figure 1) BB' parallel to AA' constructs similar triangles $0AA'$ and $0BB'$, while the proportionality theorem for similar triangles makes $0A'/0A = A'B'/AB$ and hence shows that $A'B'$ represents the magnitude xy. This may be regarded as the geometric *definition* of multiplication of magnitudes.

Other types of magnitudes such as weight or time may more easily be multiplied first by whole numbers–thus to multiply the weight of a given item by three, take the combined weight of three such items. By division, this extends first to multiplication by rational numbers and then by continuity to the multiplication of a weight by any number, rational or irrational. It is again remarkable that one gets the same operation of multiplication for *all* these types of magnitudes.

To summarize: The "practical" operations of addition and multiplication on various types of magnitudes lead to the algebraic operations of sums and product for the real numbers on the linear scale. The various rules for these manipulations of numbers were well known before they

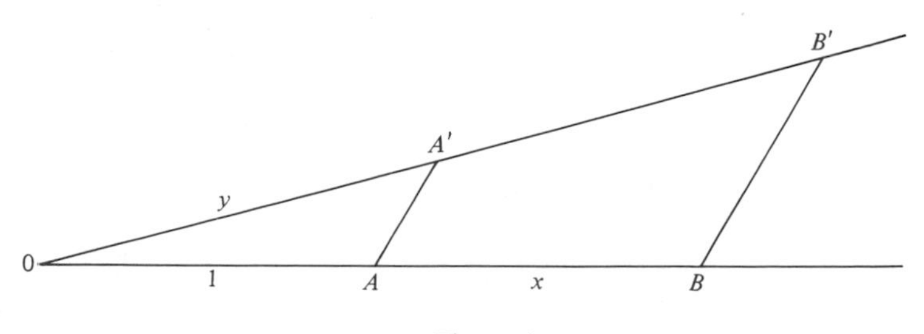

Figure 1

were codified by axioms. We record the codification: The real numbers form an abelian group under addition, under addition and multiplication they form a commutative ring, moreover, one which is a field. Here a *commutative ring* R is a set of elements (numbers) which is an abelian group under an operation of addition, which has an associative and commutative binary operation of multiplication with a unit (a number 1 such that $r \cdot 1 = r$ for all r in R) and in which both distributative laws hold:

$$a(b + c) = ab + ac \qquad (b + c)a = ba + ca. \tag{1}$$

(Since multiplication is commutative, either one of these two laws would suffice; it is a curious observation, repeated in other axiom systems, that here a single axiom, the distributive law, suffices to tie together the separately axiomatized additive and multiplicative structures.) Finally, a *field* is a commutative ring in which every equation $xa = 1$ with $a \neq 0$ has a solution x, necessarily unique. The remarkable fact is that *all* the *algebraic* rules for the manipulation of sums and products of real numbers follow from this simple list of axioms—a list far more perspicuous than the axioms of plane geometry, reflecting in part the fact that the line (the scale) is geometrically simpler than the plane, and that the axioms for the line more strictly separate the algebraic structure from the order structure (§4).

For the immediate purposes of this chapter, we could have formulated these axioms just as properties of the real numbers. With the more general terms (ring and field) defined above, we can note that there are already at hand other examples of fields (the rational numbers $\mathbf{Q}$ and $\mathbf{Z}/(p)$, the integers modulo a prime). There are many more examples of commutative rings, including the system $\mathbf{Z}$ of integers and $\mathbf{Z}_n$, the integers modulo any n. More generally, a *ring* is defined by dropping the commutative law $ab = ba$ in our definition of commutative ring, but retaining *both* distributive laws (1). A *division ring* is a ring in which every equation $xa = 1$ and $ay = 1$ for $a \neq 0$ has a solution x or y. These general notions, involving a possibly non-commutative multiplication, are recorded here for convenience. They do not belong here in the order of ideas, which now should concern only the scale of reals and its remarkable properties.

4. Comparison of Magnitudes

Many practical observations about magnitudes amount to the determination of which of two magnitudes is the greater. In geometry, the notion of betweenness provides for such a comparison of the magnitudes of two segments. On a line, the directed segment AB is less than the directed segment AC if and only if B lies between A and C. If the real numbers b and

c represent these segments AB and AC, then this defines the relation $b < c$ (b less than c) for real numbers b and c. The axioms appropriate to this relation "less than" can be derived (though we will not carry this derivation out) from the geometrical axioms for betweenness. They state that the real numbers form an "*ordered field*". Here a field F is said to be *ordered* if it is linearly ordered by a binary relation $a < b$ with the following two properties, for all field elements a, b, and c:

$$a < b \quad \text{implies that} \quad a + c < b + c, \tag{1}$$

$$a < b \quad \text{and} \quad 0 < c \quad \text{imply that} \quad ac < bc. \tag{2}$$

(Note that here too there is just one axiom connecting order and addition, and just one axiom connecting order and multiplication.) Observe also that the rational numbers (as well as the reals) form an ordered field in this sense.

These axioms providc for all the usual manipulations of inequalities. In particular, a real number c is *positive* if and only if $c > 0$, while the *absolute value* $|b|$ of a number b is b or $-b$ according as $b \geqslant 0$ or $b < 0$. Then one proves rules such as

$$|ab| = |a|\,|b|, \qquad |a + b| \leqslant |a| + |b|.$$

These absolute values formulate thc notion of magnitude without direction, and are convenient for expressing ideas of "nearness". Thus a is near to b when $|a - b|$ is small.

Inequalities and absolute values are the customary formal tools for expressing ideas of approximation. These ideas arise in many ways, perhaps most vividly in the observation that a real number, considered as an infinite decimal, can be successively approximated by finite decimals; that is, by rational numbers with denominators powers of 2 or of 10. To formalize this idea, one may start by recalling the Archimedean axiom from geometry (§III.2): that multiples of any segment $0A$ with $0 \neq A$ suffice to bound *all* larger segments–any segment $0B$ is exceeded by some multiple of the first segment $0A$. In arithmetic terms, this amounts to observing that the order of the real numbers satisfies the following "Archimedean" law:

Archimedean Law. If a and b are positive, then there is a natural number n such that $na > b$:

In other words, no matter how small the positive number a may be, every other positive number b will be exceeded by some multiple of a. This amounts to asserting that there are no infinitesimal real numbers a–though for other purposes, as we will see, it is convenient to think about infinitesimals.

The Archimedean law also implies the important property that each real number can be approximated–to any desired degree of accuracy–by a rational numbers. This amounts to saying that there is a rational between any two reals.

Theorem. *If* $0 < b < c$, *there is a rational* m/n *with*

$$b < m/n < c. \tag{3}$$

The idea of the proof is that $c - b > 0$, so must exceed the reciprocal $1/n$ of some integer n. This can be formalized as follows. Since $c - b > 0$, the Archimedean law provides a natural number n with $n(c-b) > 1$; that is, with $c > b + 1/n$. By the same law and mathematical induction, there is also a smallest natural number m with $m \cdot 1 > nb$. This means that $b < m/n$. Since m is the smallest such natural number, we must also have $m - 1 \leqslant nb$; that is, $m/n \leqslant b + 1/n$ and hence $m/n < c$. Thus m/n is between b and c, just as required.

A similar argument will show that the interval from b to c contains a rational with denominator a power of 2. This suggests again that every such number, such as b, can be approximated by rationals with denominators powers of 2.

Now that the notions of "order" and "absolute value" are formalized, one may also give a formal definition of the idea that a sequence converges to a limit. To say that a sequence $\{a_n\} = a_1, a_2, \cdots$ of real numbers a_n converges to a real number b as limit should mean that the successive terms a_n get close to b–ultimately closer than any preassigned measure. Here "ultimately" is to mean "after some index n", while the preassigned measure is to be a positive (but "small") real number, usually written $\epsilon > 0$. The idea "ultimately closer than this ϵ" is better expressed in the opposite order, that given such an ϵ one can find an index n beyond which a_n is indeed that close to b. In formal language, this becomes the

Definition. The sequence a_n of real numbers has the real number b as a limit if and only if to each real $\epsilon > 0$ there exists a natural number k with the property: if $n > k$, then $| a_n - b | < \epsilon$.

In this formal definition, the anthropometric ideas (given an ϵ, we can find an index n) have been replaced by the use of logical quantifiers (For *all* $\epsilon > 0$ there *exists* a k . . .). The particular choice of k is not relevant, for once k works for an ϵ, so will any larger natural number k'. The informal idea that a_n gets close to b seems to have disappeared, but it is covered in the phrase "for *all positive* ϵ"–and hence in particular for a small ϵ. Indeed, once the statement is true for one ϵ, it automatically holds for any larger ϵ_2, and with the same k. For instance, by the Archimedean law there is to each positive ϵ_2 a natural number $m > 0$ with $1/m < \epsilon_2$. Hence for convergence it is enough to require that to each natural number m there exists k so that if $n > k$, then $| a_n - b | < 1/m$. In this way *different* formal statements deseribe the *same* intuitive idea of convergence.

This sophisticated definition of convergence to a limit is a remarkable achievement, since it both codifies the intuitive intentions about convergence and makes possible proofs of the various expected formal consequences. For example, if a sequence $\{a_n\}$ converges to limits b and b', then necessarily $b = b'$ (because they differ by less than any $\epsilon > 0$). Also, if $\{a_n\}$ converges to b and $\{c_n\}$ to d, then $\{a_n + c_n\}$ converges to $b + d$ and $\{a_n c_n\}$ to bd. Moreover, in a convergent sequence the successive terms must not only get close to the limit, but also "ultimately" close to each other. This can be expressed formally by the theorem that any convergent sequence $\{a_n\}$ is Cauchy—where a sequence is said to be *Cauchy* if to every real $\epsilon > 0$ there is a natural number k such that $n > k$ and $m > k$ imply $| a_n - a_m | < \epsilon$.

The ideas of approximation represented in the definition of convergence are essentially the same as those involved in the definition (§I.4) of continuity. Indeed, using the axiom of choice (!), one can prove that a function f: $\mathbf{R} \to \mathbf{R}$ on the reals is continuous at a real number b if and only if f maps every sequence a_n converging to b into a sequence $f(a_n)$ converging to $f(b)$. In brief, f continuous means exactly that f preserves convergence.

Infinite sums also come up and must be explained in terms of convergence. Thus every unending decimal fraction is really an infinite sum; for example, 1/6 is the unending decimal

$$.166666\ldots = 1/10 + 6/10^2 + 6/10^3 + 6/10^4 + \cdots .$$

For m a positive integer, the binomial theorem gives

$$(1 + x)^m = 1 + mx + \frac{m(m-1)}{1\cdot 2} x^2 + \cdots + x^m,$$

but when m is a fraction this becomes an infinite series in powers of x

$$(1 + x)^m = 1 + mx + \frac{m(m-1)}{1\cdot 2} x^2 + \cdots$$

$$+ \frac{m(m-1)\ldots(m-n-1)}{1\cdot 2\ldots n} x^n + \cdots .$$

The formula suggest an infinite number of additions. This infinite operation is not literally possible; instead one approximates by finite sums.

The study of convergent series is essentially equivalent to that of convergent sequences. An *infinite series* is a formal infinite sum

$$c_1 + c_2 + c_2 + \cdots$$

of real numbers c_i, and is said to *converge* to the limit b if and only if the sequence $s_n = c_1 + c_2 + \cdots c_n$ of "partial sums" of the series converges to b. One could, vice versa, define convergent sequences in terms of convergent series, and the understanding of either notion is improved at the hand of examples of series which do not converge—as for instance with the harmonic series $1 + \frac{1}{2} + \frac{1}{3} + \frac{1}{4} + \cdots$. Infinite series also occur essentially in complex analysis (§X.7), to expand known functions and to define new ones.

Comment. Here is an example of an ordered field in which the Archimedean law fails. It is suggested by the manipulation of the power series which occurs in analysis. Introduce a formal symbol t and consider all possible infinite power series which occurs in analysis. Introduce a formal symbol t and consider all possible infinite power series of the form

$$s = a_{-n}t^{-n} + a_{-n+1}t^{-n+1} + \cdots + a_{-1}t^{-1} + a_0 + a_1 t + a_2 t^2 + \cdots ,$$

with real number coefficients and with only a finite number of negative powers of t—but with no convergence required. Add two such "formal" series by adding the corresponding coefficients, and multiply two such series in a purely formal way; for example

$$(a_0 + a_1 t + a_2 t^2 + \cdots)(b_0 + b_1 t + b_2 t^2 + \cdots)$$
$$= a_0 b_0 + (a_0 b_1 + a_1 b_0)t + (a_0 b_2 + a_1 b_1 + a_2 b_0)t^2 + \cdots .$$

By suitable calculations, one may verify that these elements form a field, called the field of formal power series $\mathbf{R}((t))$. One may order this field by specifying that $s > 0$ if and only if its first non-vanishing coefficient a_{-n} is positive. Then also $s > 0$ if and only if $s - t$ is positive. This means, in particular, that $1 > 0$ and also that $t > 0$; however, no integral multiple of t will be as large as 1—so that t is a sort of infinitesimal, while t^{-1} is "infinitely large" compared to 1. Moreover, the infinite series s of elements in this field is convergent in the sense of our definition of convergence—using among the $\epsilon > 0$ all the powers $\epsilon = t^n$ for $n > 0$.

5. Axioms for the Reals

The practical understanding of real magnitudes and their uses in approximations leads eventually to the idea of characterizing the field $\mathbf{R}$ of real numbers by a suitable list of axioms. It is not enough to require just that $\mathbf{R}$ be an ordered field, for there are many such fields, including the field $\mathbf{Q}$ of rational numbers and the formal power series field $\mathbf{R}((t))$. The crucial feature is a completeness axiom, stated in §I.4: Every non-empty set of reals with an upper bound has a least upper bound. An ordered field with this property is called a complete ordered field. The real numbers form such a field. The formal power series do not.

This completeness axiom implies the Archimedean law. For suppose to the contrary that there were positive reals $a > 0$ and $b > 0$ such that no multiple na exceeded b. Then the set S of all multiples na, for n a natural number, has an upper bound, namely b. By the completeness axiom, S then has some least upper bound, call it b^*; thus $b^* \geqslant na$ for all n. This also means that $b^* \geqslant (n+1)a$ for all n, and so that $b^*-a \geqslant na$ for all n. Thus $b^* - a$ is less than b^* and is also an upper bound for S, a contradiction to the choice of b^* as a least upper bound for S.

The force of the completeness axiom is to insure that all the real numbers that ought to be there are there. For example, the irrational $\sqrt{2}$ must be there, as the least upper bound of the set 1, 1.4, 1.41, 1.414,... of rationals (those approximating $\sqrt{2}$). Similarly π must be there, as the least upper bound of the set of decimals 3, 3.1, 3.14, 3.141, 3.1415, 3.14159,.... Indeed, since there is a rational between any two reals, any real can be expressed as the least upper bound of a set of rationals. The completeness axiom is also used in more sophisticated ways, for example in the proof of Rolle's theorem and of the mean value theorem of the calculus (Chapter V).

There are other equivalent forms of the completeness axiom. Instead of requiring that every bounded set of reals have a least upper bound, one may require any one of the following:

Dedekind Cut Axiom. If the set $\mathbf{R}$ of reals is the union of two disjoint non-empty subsets L and U such that $x \in L$ and $y \in U$ imply $x < y$, then there is a real number r such that $x \leqslant r$ for all $x \in L$ and $r \leqslant y$ for all $y \in U$.

Cauchy Condition. Every Cauchy sequence of real numbers has a limit.

Weierstrass Condition. If a series $c_1 + c_2 + \cdots + c_n + \cdots$ of positive real numbers $c_n > 0$ is such that there is an upper bound b for all partial sums $s_n = c_1 + \cdots + c_n$, then the series converges.

It is illuminating to establish the equivalence of the different completeness axioms. The Dedekind cut axiom is perhaps the more geometric (and has already appeared (§III.2) in the completeness axioms for the geometry of the plane): Any cut of the real line into a lower part L and an upper part U must be a cut *at* some real number r. Indeed, each real number determines just two such cuts, one with L consisting of all $x \leqslant r$, the other with L consisting of all $x < r$. In this way, real numbers can be completely described by cuts.

The completeness axioms also serves to determine the real numbers uniquely, up to an isomorphism of ordered fields. We sketch the proof.

Suppose that $\mathbf{R}'$ is any complete ordered field, with unit element $1'$ for multiplication. Then $0 < 1'$, for otherwise $1' < 0$, which would give $0 < -1'$ and so $1' = (-1')(-1') > 0$, a contradiction. For each natural number $n > 0$ the multiple $n1'$ must then be positive, and all these elements $n1'$ are different elements in our field $\mathbf{R}'$. Therefore $\phi(n) = n1'$ defines an injective map $\phi: \mathbf{N} \to \mathbf{R}'$. Because $\mathbf{R}'$ is a field, this ϕ can be extended to a map $\phi: \mathbf{Q} \to \mathbf{R}'$ by setting $\phi(n/m) = \phi(n)/\phi(m)$ whenever $m \neq 0$; in other words, $\mathbf{R}'$ contains a copy $\phi(\mathbf{Q})$ of the ordinary rational numbers $\mathbf{Q}$. Also, ϕ preserves the order. By the Archimedean law, each element r' in $\mathbf{R}'$ must then be a least upper bound of a set L' of these rationals, indeed it is the least upper bound of the image $\phi(L)$ of the set

$$L = \{n/m \mid m \neq 0, \phi(n/m) < r'\}.$$

This set L has the special property that $x < y$ and $y \in L$ imply $x \in L$, so it is the lower half of a "cut" in the rationals. Since it is bounded, it has an (ordinary) real number r as least upper bound, and r in turn determines L. The one-to-one correspondence $r \mapsto r'$ then maps the ordinary real numbers on the given complete ordered field $\mathbf{R}'$. One can show that it preserves sums, products and order, hence it is the desired order isomorphism $\mathbf{R} \cong \mathbf{R}'$.

Note that this argument makes essential use of sets, such as the sets L of rationals. In this it is like the use of sets in the induction axioms to prove that the Peano postulates uniquely determine the natural numbers.

We now have two different axiomatic descriptions of the reals: Here, the real *field*, described as a complete ordered field, in §I.4 the real *continuum*, described as an unbounded ordered set with a denumerable dense subset. Each description determines the set of reals uniquely, up to an isomorphism of the structure concerned. However the structures are drastically different. The real continuum has only the order structure, and there are many automorphisms of this structure. The real field has both order and algebraic structure, and its only automorphism (by a proof like that just above) is the identity automorphism. These differing structures on the same "thing" (here the reals) are much like the differing structures on the plane (without and with orientations, as in §III.8). These differing structures furthermore reflect practical differences. Thus the ordered continuum handles comparisons of many items where one knows only which of two items is the larger, with no measure of "how much" larger.

In either structure, the real numbers form a "continuous" scale. Physicists (and others) sometimes suggest that an "atomic" or "discrete" scale would be more "real"; finitists propose finite scales of magnitude. These proposals turn out to be hard to execute–the continuous scale of reals works smoothly!

6. Arithmetic Construction of the Reals

The axiomatic approach assumes that the real numbers are already there, as geometrical or other magnitudes, and describes them–uniquely up to an isomorphism–by axioms. There is a wholly different and deliberately arithmetic approach, in which one starts from the natural numbers and constructs the reals as sets of natural numbers. Since we already have constructed the rationals, it will be sufficient (and more appropriate) to construct the real numbers directly from the rationals and sets of rationals in such a way that the resulting reals do satisfy the axioms.

This construction can be done in several ways, in parallel to the different forms of the completeness axiom. The construction by Dedekind cuts in the rationals is perhaps the most direct. Define a Dedekind cut in the field **Q** of rational numbers to be a pair of disjoint non-empty subsets (L, U) with union **Q** such that $x \in L$ and $y \in U$ imply $x < y$; to get uniqueness, require that the "lower set" L have no maximal element. Any such cut is determined by the set L alone, because it is a bounded non-empty set L of rationals, with no maximal element, which does not contain all rationals but has $x' \in L$ whenever $x' < x \in L$. Call such a set L a real number. To add two such real numbers L and L', take the sum L'' to be the set of all sums $x + x'$ of rationals $x \in L$ and $x' \in L'$. Multiplication is more delicate, because the product of two negative rational numbers is positive. If both the real numbers L and L' contain positive rationals, take their product LL' to be the set of all products xx' of rationals $x \in L$ and $x' \in L'$, with at least one of x, x' positive. The remaining cases of multiplication can then be handled by replacing the set L by its negative, defined as the set of all rational y with $x + y < 0$ for every $x \in L$. Finally, define the order by specifying that $L \leqslant L'$ if and only if $L \subset L'$. A systematic proof (cf. Landau [1951]) then shows that the real numbers L so constructed do form a complete ordered field **R**. Moreover, the rationals **Q** can be embedded in **R** by the monomorphism (of ordered fields) which sends each rational x into the set L of all rationals y with $y < x$. (This may recall the uniqueness proof of §5.)

There is an alternative construction of the reals as Cauchy sequences of rationals. Since different such Cauchy sequences may have "the same" limit, one must for this construction consider two Cauchy sequences $\{a_n\}$ and $\{b_n\}$ as equivalent when the sequence $a_n - b_n$ converges to 0. Under suitable operations of addition and multiplication, and with a suitable linear order, these equivalence classes of Cauchy sequences form a complete ordered field. By the uniqueness of such fields, this field is isomorphic to the field of reals constructed by Dedekind cuts. However, in other foundations of Mathematics in terms of elementary topoi the Dedekind and Cauchy sequence constructions may yield differing results (see Johnstone [1977]).

It was primarily George Cantor who developed this construction of real numbers by Cauchy sequences. Weierstrass has a quite similar construction related to his version of the completeness axiom. He considered sequences $c = \{c_n\}$ of positive rational numbers c_n with bounded partial sums; that is, such that there is a rational b with $c_1 + \cdots + c_n \leqslant b$ for every n. If c' is a second such sequence, then $c \leqslant c'$ is defined to mean that for each n there is a natural number m with

$$c_1 + \cdots + c_n \leqslant c'_1 + \cdots + c'_m .$$

If both $c \leqslant c'$ and $c' \leqslant c$, then c and c' are said to be equivalent. The equivalence classes of such sequences then constitute the desired positive real numbers, with operations of addition and multiplication defined by

$$c + d = \{c_n + d_n\}$$

$$c \cdot d = \{c_1 d_1, c_1 d_2 + c_2 d_1, c_1 d_3 + c_2 d_2 + c_3 d_1, \ldots\}.$$

(This latter is essentially the formal product of the two intended infinite series.)

This construction really amounts to the use of the sequence $s_n = c_1 + \cdots + c_n$ which is both monotone and bounded:

$$s_1 < s_2 < \cdots < s_n < \cdots \leqslant b.$$

In particular, it must therefore be a Cauchy sequence. The choice of a construction by series or by sequences then depends on one's preference–and Weierstrass, in all of his extensive work, was inclined to use series. However, one cannot say that a real number *is* a Dedekind cut or *is* an (equivalence class of) Cauchy sequences; the cut or the sequence is only a means to an end, the construction of the reals, defined only up to an isomorphism. The one idea of a real number has different formalizations.

All of these constructions are *arithmetic*, in that they build up the real numbers from the arithmetic of the rational numbers. They are, however, not purely arithmetic, since each of the constructions also makes intrinsic use of set theory–the set of all Dedekind cuts, each cut itself a set, or the set of all equivalences classes (= sets) of Cauchy sequences. Hence the appearance, in the constructions, that the reals are "purely" arithmetic is deceptive. It would be more accurate to say that the scale of real magnitudes requires both sets and arithmetic, as well as geometric understanding.

Our axioms for the reals are the standard ones, but there are other models of the reals–as indeed there are multiple models for many mathematical ideas. The "non-standard" reals, built by a sophisticated construction, contain both the standard reals and actual infinitesimals; they can be used to prove theorems (for the ordinary reals) (Lightstone and Robinson [1975]).

Other approaches restrict attention to those real numbers which are explicitly "constructible" in some specific sense, perhaps by means of recursive functions of natural numbers. An efficient version of constructive analysis is due to Bishop [1967]; the corresponding real numbers are described in Myhill [1973]. Earlier, L.E.J. Brouwer developed intuitionism, with emphasis on real numbers given by choice sequences (Troelstra [1977]; see also Dummett [1977]). In my view, such requirements of constructivity tend to emphasize arithmetic at the expense of geometry. Since the real numbers should subsume *both* arithmetic and geometric aspects of mensuration, these practical uses seem better formalized by the standard axioms for the field of real numbers.

7. Vector Geometry

The real numbers provide a one-dimensional geometry—a line with a chosen origin and unit. The higher dimensional analog is vector geometry. Thus in the Euclidean plane, choose an origin 0 and a segment $0U$ representing the unit of distance. Then each point P in the plane is represented by the directed segment $0P$, called a *vector* v. Vectors $0P$ and $0Q$ can be added by the usual parallelogram law: Form the parallelogram with edges $0P$ and $0Q$, and take the diagonal vector $0R$ as the sum (Figure 1). Under this addition, the vectors form an abelian group; it is just the group of all translations of the plane (§III.7). Using the chosen unit segment $0U$, each real number r can be laid off along the line $0U$. The geometric construction of multiplication by r (Figure 3.1) then defines an operation of multiplying the vector $v = 0R$ by the real number (scalar) r to give the vector rv along the line of $0R$. If $r > 0$ it is in the direction of v; if $r < 0$, it is in the opposite direction, in both cases with $|r|$ times

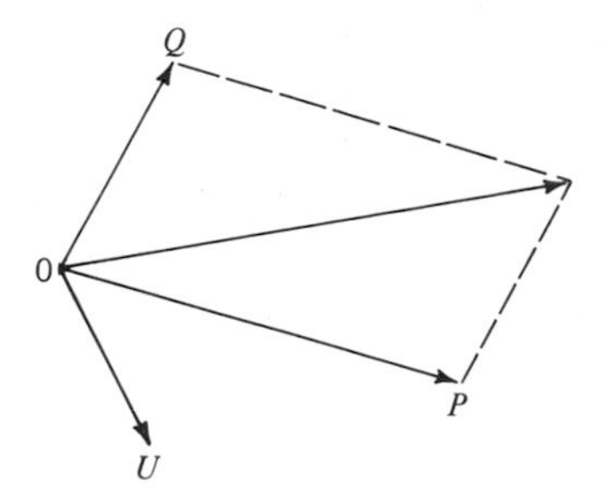

Figure 1

the length. This operation of multiplication of a vector by a scalar has the following formal properties

$$r(v + w) = rv + rw, \qquad 1v = v, \tag{1}$$

$$(r + s)v = rv + sv; \qquad (rs)v = r(sv). \tag{2}$$

for all vectors v and w and all scalars $r,s \in \mathbf{R}$. In view of these properties of addition and of scalar multiplication, we say that the vectors form a *vector space* over the real numbers.

Many geometric facts can be established readily by the use of vector algebra. For example, this algebra gives an easy proof of the theorem that the three medians of a triangle meet in a point.

Vectors may also be introduced in 3-dimensional spaces (and in higher dimensional spaces); moreover, exactly the same formulas (1) and (2) apply, whatever the dimension. More generally, a *vector space* V over a field F is an abelian group (under addition) with another operation $F \times V \to V$, written $(r,v) \mapsto rv$ for elements $r \in F$, $v \in V$, which satisfies with addition the four laws (1) and (2) above. One can also describe formally the "dimension" of such a space. In the plane, if u_1 and u_2 are two non-collinear vectors, every vector v in the plane can be written as a "linear combination" $v = x_1u_1 + x_2u_2$ with unique choices for the scalars x_1 and x_2. In three dimensions, we need three such vectors u_1, u_2 and u_3, along three "axes". More generally, vectors $u_1, \ldots, u_n$ in V are said to form a *basis* for the vector space V if every vector v in V has a unique expression

$$v = x_1u_1 + \cdots + x_nu_n;$$

then the x_i are the *coordinates* of v relative to the basis u_i. A first theorem of vector geometry (= linear algebra) states that any two (finite) bases for a given vector space must have the same number of elements. This number n is the *dimension* of the space.

Vectors also provide a convenient description for certain functions transforming the plane into itself (and leaving the origin fixed): Transformations such as expansions from the origin ($v \mapsto rv$) for a fixed r, compressions on a line through the origin, and shears. All of these transformations preserve the vector operations. More generally, a *linear transformation* on V is a function $t\colon V \to V$ which preserves both addition of vectors and multiplication of vectors by scalars. This means that the equations

$$t(v + w) = tv + tw, \qquad t(rv) = r(tv) \tag{3}$$

hold for all vectors v and w and all scalars r. For example, a rotation about the chosen origin 0 is a linear transformation (it maps the parallelogram of addition into a parallelogram). The same notion of linear transformation applies to higher dimensional vector spaces, and is the cornerstone of the study of linear algebra and matrix theory (Chapter VII).

The general idea that some geometrical facts can be handled well by appropriate algebraic operations has now appeared in the use of real numbers for one-dimensional geometry and of vectors for higher dimensions. The addition of vectors was possible only with the choice of an origin 0. Without a choice of origin one still has some algebraic operations such as the construction of the midpoint, written $(1/2)P + (1/2)Q$, of two points P and Q or of the point $(2/3)P + (1/3)Q$ which is one third of the way from P to Q. The general such operation is the formation of the *weighted average*

$$P = w_0 P_0 + w_1 P_1 + \cdots + w_n P_n \tag{4}$$

of $n + 1$ points with weights real numbers w_i with sum $w_0 + \cdots + w_n = 1$. A transformation preserving such averages is called an *affine* transformation; the resulting geometry (in any number of dimensions) is *affine geometry*. It is possible to write a complete set of axioms for the weighted average operations (4), but they are cumbersome (*Algebra*, first edition). Even the simplest ternary operation $P_0 - P_1 + P_2$ of this type has not proven to be very useful. It seems more efficient to reduce to vector algebra by choosing an origin. Addition is a very convenient operation! Binary operations are much handier than ternary ones.

8. Analytic Geometry

Another and earlier reduction of plane geometry to algebra is provided by the familiar method of cartesian coordinates. Given an orientation, a choice of origin and unit, and two perpendicular coordinate axes, each point P is represented by its coordinates, a pair (x,y) of real numbers. Each line in the plane may be described as the set of those points whose coordinates satisfy a linear equation, while the distance between two points is given by the familiar formula in coordinates, derived from the Pythagorean theorem. In this way, all sort of geometric facts about the plane are handled by algebraic machinery; in effect, the plane is reduced to the cartesian product $\mathbf{R} \times \mathbf{R} = \mathbf{R}^2$ of two copies of the real line $\mathbf{R}$. However, this reduction does depend on the choice of origin *and* of axes, and one must betimes verify that truly geometric facts are independent of the choice of coordinates.

Geometry and coordinates arise first in dimensions 2 and 3. The need for higher dimensional geometry is motivated by phenomena which need specification of more than three coordinates: Events in space-time need four (position and time); in dynamics, the initial conditions for a particle need six, three for position and three for velocity. The use of such a six-dimensional *phase space* is a first example of the importance of mechanics as a motivation for mathematical developments (Chapter IX). To be sure, this example could be (but usually is not) described without explicit use of coordinates, as the product of a three-dimensional position space and another three-dimensional velocity space. A general n-dimensional Euclidean space may be constructed as the n-fold product $\mathbf{R}^n$ of real lines, with points the n-tuples of real numbers $(x_1, \ldots, x_n)$. This is a vector space over $\mathbf{R}$, with basis the n "unit vectors" $(0, \ldots, 0,1,0, \ldots, 0)$. Such use of coordinates allows for the efficient treatment of higher-dimensional phenomena in dimensions where the corresponding geometric intuitions are weak or non-existent. It serves to extend geometric ideas beyond the ordinary three dimensions.

Much of mathematical physics deals with phenomena in three dimensions and so often requires a formulation in triads of equations, one for each coordinate in $\mathbf{R}^3$. When $\mathbf{R}^3$ is regarded as a vector space, many of these equations can be written as single vector equations. This accounts for the popularity of vector analysis in Physics. It includes the use of inner products, to which we now turn.

9. Trigonometry

Trigonometry is essentially a procedure for turning angular measures into linear measures. This appear directly in the definition of the two basic trigonometric functions $\sin\theta$ and $\cos\theta$ of the angle θ. In the oriented plane take a point P on the unit circle with center at the origin 0 so that the segment $0P$ makes the given angle θ with the x coordinate axis. Then $\cos\theta$ and $\sin\theta$ are defined to be the x and y coordinates of P; since the circle has radius 1, this immediately gives the identity

$$\cos^2\theta + \sin^2\theta = 1. \tag{1}$$

This defines $\sin\theta$ and $\cos\theta$ for angles θ of all sizes (not just for an angle θ in the first quadrant, as displayed in Figure 1).

When angles are measured (in radians) by numbers t we can also think of the sine and the cosine as functions of the *number* t, so that

$$\operatorname{Sin} t = \sin(\text{angle of } t \text{ radians}). \tag{2}$$

Thus there are really two legally different functions: The Sine of a *number*, here with capital S, and the sine of an angle, with lower case s. This pedantic (but real!) difference is usually ignored. It implicitly

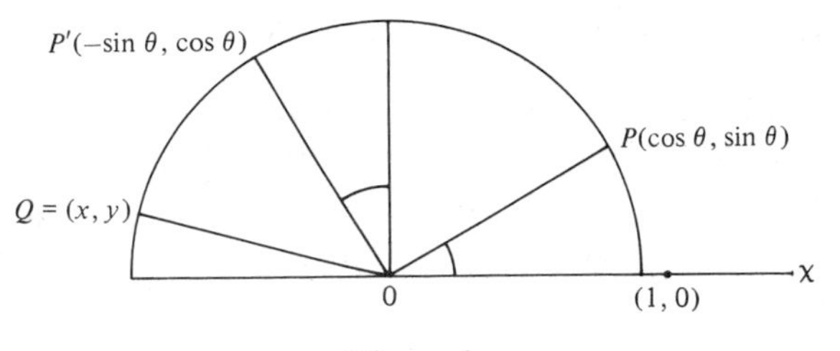

Figure 1

involves the wrapping function θ_t of §2 above. Recall that this function sends each real t to the point $P = (\cos\theta_t, \sin\theta_t)$ on the circle such that the length of the counterclockwise circular arc from (1,0) to P is congruent to t, modulo 2π. Then the definition (2) reads

$$\operatorname{Sin} t = \sin(\theta_t), \qquad \operatorname{Cos} t = \cos(\theta_t): \tag{3}$$

it really amounts to a composite function $t \to \theta_t \to \sin(\theta_t)$. Since the wrapping function has $\theta_{t+2\pi} = \theta_t$ we get

$$\operatorname{Sin}(t + 2\pi) = \operatorname{Sin}(t), \qquad \operatorname{Cos}(t + 2\pi) = \operatorname{Cos}(t). \tag{4}$$

In other words, Sin and Cos are periodic functions, of period 2π, as in the familiar graph of Figure 2. Now that this has been stated, we drop the S in Sin and the function θ_t; they would just get in the way of trigonometric manipulations, but we emphasize that the wrapping function accounts for radian measure and the periodicity (4) of the trigonometric functions. Indeed, the whole study of periodic functions, as carried on in Fourier analysis, concerns the expression of more or less arbitrary functions $f(t)$ of period 2π in terms of the periodic functions $\sin nt$ and $\cos nt$ for the natural numbers n.

A rotation of the circle of Figure 1 about the origin through the angle θ will carry the point (1,0) to our point P and the point (0,1) to a corresponding point P'; thus this rotation has the effect

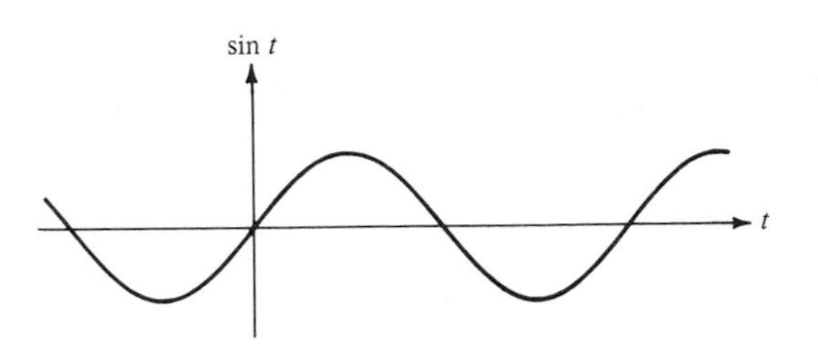

Figure 2

$$(1,0) \mapsto (\cos \theta, \sin \theta), \qquad (0,1) \mapsto (-\sin \theta, \cos \theta) \tag{5}$$

Since (1,0) and (0,1) form a basis for the 2-dimensional vector space, any point Q with coordinates (x,y) can be written as the linear combination $(x,y) = x(1,0) + y(0,1)$ of the two basis vectors. Now rotation is, as already noted, a linear transformation, so preserves linear combinations and thus by (5) has the effect

$$(x,y) \mapsto (x \cos \theta - y \sin \theta,\ x \sin \theta + y \cos \theta).$$

When one writes (x',y') for the coordinates of the rotated point, this yields the customary equations in coordinates

$$\begin{aligned} x' &= x \cos \theta - y \sin \theta, \\ y' &= x \sin \theta + y \cos \theta \end{aligned} \tag{6}$$

for a rotation $(x,y) \mapsto (x',y')$ of the plane about the origin. Also, if one takes $Q = (x,y)$ to be that point Q on the unit circle of Figure 1 for which $0Q$ makes the angle ϕ with the positive x-axis, then $x = \cos \phi$, $y = \sin \phi$. Now the rotation by θ clearly carries Q to a point Q' where $0Q'$ makes the angle $\phi + \theta$ with the positive x-axis. The coordinates (x',y') of Q' are then $\cos(\phi + \theta)$, $\sin(\phi + \theta)$ and the equations (6) for rotation become

$$\begin{aligned} \cos(\phi + \theta) &= \cos \phi \cos \theta - \sin \phi \sin \theta, \\ \sin(\phi + \theta) &= \cos \phi \sin \theta + \sin \phi \cos \theta. \end{aligned} \tag{7}$$

These "addition formulas" for the trigonometric functions are thus another expression of the relation of angular to linear measure.

This relation also accounts for the use of trigonometric functions in calculating the connections between the sides and the angles of a triangle, as suggested by the Euclidean congruence theorems. In a triangle $0AB$ with sides a, b and c, as illustrated in Figure 3, drop a perpendicular from A to the opposite side at M. If θ and ϕ are the angles of the triangle at 0 and

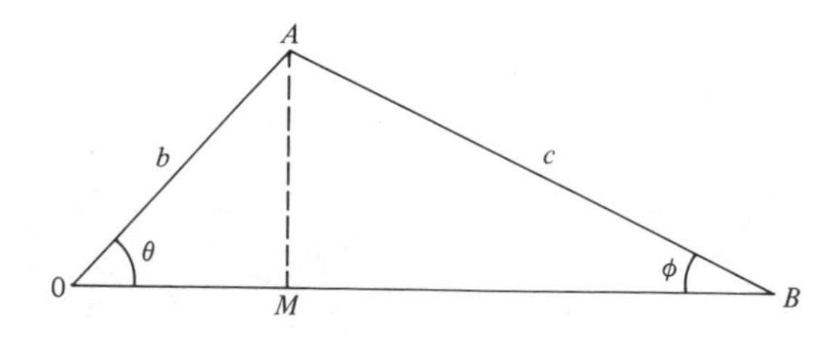

Figure 3

at B, the definition of the sine function shows that the length AM can be expressed either as $b \sin \theta$ or as $c \sin \phi$; this gives the *law of sines*: $b/c = \sin \phi / \sin \theta$. On the other hand, one may compute the length c of the side AB from the Pythagorean theorem for the right triangle AMB to be

$$c^2 = (b \sin \theta)^2 + (a - b \cos \theta)^2, \tag{8}$$

$$c^2 = a^2 + b^2 - 2ab \cos \theta.$$

This is the *law of cosines*. When $u = 0B$ and $v = 0A$ are regarded as vectors of the respective lengths $|0B| = a$, $|0A| = b$, the expression $ab \cos \theta$ which appears in (8) is called the *inner product* $u \cdot v$ of the two vectors; it is a scalar (a real number). If A has coordinates (x_1, y_1) and B coordinates (x_2, y_2), the lengths a, b, and c in (8) can be written in terms of these coordinates by the Pythagorean theorem; the formula (8) then becomes a formula for the inner product $u \cdot v$:

$$(x_1, y_1) \cdot (x_2, y_2) = x_1 x_2 + y_1 y_2. \tag{9}$$

This again is a linearization of angular phenomena.

With these ideas and formulas we have presented all of the essential concepts of trigonometry in less than 6 pages. It is curious (and troublesome) that the standard elementary presentations of trigonometry have been inflated to inordinate lengths—an inflation hardly justified by the inevitable subtlety wrapped up in the comparison of linear and angular magnitude.

In (9), the inner product is expressed in terms of a choice of coordinates. However, as in other cases, the inner product can be described in a more invariant way by suitable axioms. It has three characteristic properties: *Linearity*, *symmetry*, and *positive definiteness*, as expressed by the three equations

$$u \cdot (t_1 v_1 + t_2 v_2) = t_1 (u \cdot v_1) + t_2 (u \cdot v_2), \tag{10}$$

$$u \cdot v = v \cdot u, \tag{11}$$

$$u \cdot u \geqslant 0, \qquad u \cdot u = 0 \quad \text{only if} \quad u = 0, \tag{12}$$

valid for all vectors u, v, v_1, and v_2 and all scalars t_1 and t_2. All the properties of the inner product follow from these equations. Indeed, given these equations—for a product $u \cdot v$ of vectors in a two dimensional vector space over $\mathbf{R}$—one can prove that there is a choice of basis and corresponding coordinates for which the given inner product is expressed by the formula (9). Such a basis is called a *normal orthogonal basis* (for the given inner product).

We will extend these ideas to higher dimensions in §VII.5.

10. Complex Numbers

Each extension of the number system to a larger system is driven by the need to solve questions which the smaller system cannot always answer. Subtraction is not always possible in the natural numbers; hence construct the system **Z** of integers. Division (by non-zero integers) is not always possible in **Z**; hence construct the rationals **Q** where division is possible. Cauchy sequences in **Q** need not converge to a limit in **Q**; hence construct the reals **R** to include such limits. Finally, there are plausible polynomial equations such as $x^2 + 1 = 0$ and $x^2 + x + 1 = 0$ with real coefficients which do not have real solutions; hence construct the complex numbers **C** to provide solutions for such equations.

Were there a solution i for $x^2 + 1 = 0$, there would also be combinations $a + bi$ for any real numbers a and b. They could be manipulated by using the algebraic rules for the reals and the fact that $i^2 = -1$. This would yield in particular the following rules for addition and multiplication:

$$(a + bi) + (c + di) = (a + c) + (b + d)i, \tag{1}$$

$$(a + bi)\cdot(c + di) = (ac - bd) + (ad + bc)i. \tag{2}$$

However, the symbol i imagined here cannot possibly be a real number, because the square of any non-zero real is necessarily positive. Therefore this i was originally (in the 18th century) thought to be an "imaginary" number–not in any sense real. Early in the 19th century, these practical rules were turned into a formal definition. One does not need the actual symbol i; a complex number $a + bi$ could be *defined* arithmetically to be just an ordered pair (a,b) of real numbers. When the addition and multiplication of such pairs is defined, following (1) and (2), by the rules

$$(a,b) + (c,d) = (a+c,b+d), \qquad (a,b)(c,d) = (ac-bd,ad+bc) \tag{3}$$

it follows readily that these pairs do form a field, call it **C**. The given real numbers are embedded monomorphically in **C** by the map $a \rightarrow (a,0)$; in particular 1 becomes (1,0). Moreover, by the multiplication rule (3) the pair $i = (0,1)$ has $i^2 = (-1,0)$. Then any pair (a,b) can be rewritten as a linear combination

$$(a,b) = (a,0) + (0,b) = a + bi.$$

In other words, this strictly arithmetic construction has produced things which act exactly like the desired complex numbers, and i, defined as a pair (0,1), need no longer be regarded as imaginary, although this word is still used for i and its real multiples bi.

This arithmetic construction may also be viewed geometrically. The complex numbers, rewritten as $z = x + iy$ for real x and y, correspond exactly to the points (x,y) in the cartesian plane (Figure 1) with the real numbers $x + i0$ placed along the x axis, as usual, while the purely imaginary numbers iy lie along the y axis. This provides a geometric construction of the complex numbers with each complex number z a vector $0Z$ from the origin, while the addition of complex numbers is just addition of the corresponding vectors. The complex number z may also be described with polar coordinates: The counterclockwise angle θ from the positive x-axis to the vector $0Z$ and the length $r \geqslant 0$ of this vector; one calls θ the *argument* of z and r its *absolute value*, and writes

$$z = r(\cos\theta + i\sin\theta), \qquad r = |z|, \qquad \theta = \arg z. \tag{4}$$

This polar form is especially useful for the multiplication of complex numbers. Thus if $w = s(\cos\phi + i\sin\phi)$ is a second complex number, the product zw, as defined by (3), can be rewritten, using the addition formulas for sin and cosine, as

$$zw = rs(\cos(\theta + \phi) + i\sin(\theta + \phi)). \tag{5}$$

In other words

$$|zw| = |z|\,|w|, \qquad \arg(zw) = \arg z + \arg w; \tag{6}$$

thus multplication of complex numbers multiplies their absolute values and adds their arguments. This may be stated geometrically: Multiplication by the complex number w is a transformation $z \mapsto zw$ of the z-plane into itself which rotates the plane counterclockwise about the origin by the angle $\arg w$ and at the same time stretches the plane away from the origin uniformly by the factor $|w|$. This geometric description of complex multiplication illustrates the tight connection between the algebra of complex numbers under addition and multiplication and the geometry of the plane under translation, rotation, and expansion. Thus i, once imaginary, now means rotate by 90°! The subsequent development of calculus for complex numbers, as we will see in Chapter X, illustrates many more such relations between algebra and geometry.

The complex numbers thus are constructed by adjoining to the field of real numbers one solution of the polynomial equation $x^2 + 1 = 0$. It is a

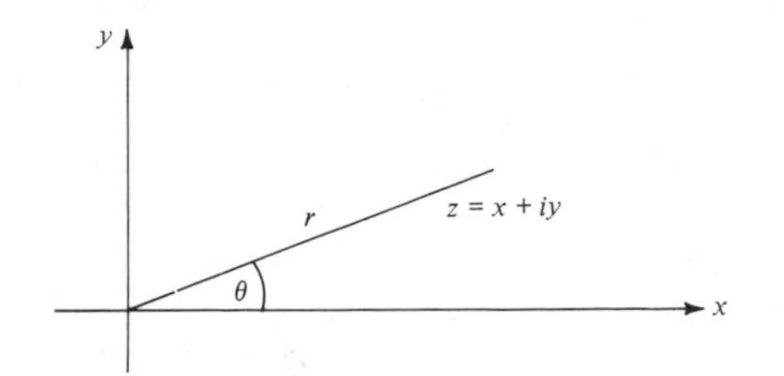

Figure 1

remarkable fact that this one adjunction suffices to provide solutions for *all* such polynomial equations. Indeed, the fundamental theorem of algebra first proved by C.F. Gauss asserts that any equation $a_nx^n + a_{n-1}x^{n-1} + \cdots + a_1x + a_0 = 0$ with complex coefficients a_i and with $a_n \neq 0$ has at least one complex number as root. The proof is not easy. At least one form of the proof (presented in Birkhoff–Mac Lane, Chapter V) uses heavily the geometric realization of the complex numbers in the complex plane. Another proof will be indicated in §X.9.

This construction of the complex numbers does not depend on any special virtues of the equation $x^2 + 1 = 0$; other polynomial equations can serve as well. For example, the equation $x^2 + x + 1 = 0$ also has no real root, because the minimum of $x^2 + x + 1$ for x real lies at $x = -1/2$ and is $3/4 > 0$. One might then introduce a new symbol ω for a non-real root of this equation and manipulate the combinations $a + b\omega$ for a and b real, using the rule $\omega^2 = -\omega - 1$, derived from this equation, in multiplying two such symbols. The resulting symbols again form a field; however, this field is isomophic to $\mathbf{C}$ under the map sending ω to $-1/2 + (\sqrt{3}/2)i$. Indeed $\omega^3 = -\omega^2 - \omega = 1$, so this symbol ω is in fact a (complex) cube root of 1. By the rule for products of complex numbers, the argument of any cube root of 1 must be 0 or $\pm 2\pi/3$; we have taken $\omega = \cos(2\pi/3) + i\sin(2\pi/3)$.

There could be a different choice as

$$\omega' = \cos(4\pi/3) + i\sin(4\pi/3) = \cos(2\pi/3) - i\sin(2\pi/3).$$

The change from ω to ω' amounts to replacing i by $-i$. (Switching the two roots of $x^2 + 1 = 0$.) For any complex numbers, this replacement $a + bi \mapsto a - bi$ carries sums to sums and products to products—because the definition of sum and product depend only on the rule that $i^2 = -1$, a rule equally valid for i and for $-i$. This replacement, $a + bi \mapsto a - bi$, called *complex conjugation*, is then an automorphism (an isomorphism with itself) of the field of complex numbers. It is a first example of the automorphisms studied in the Galois theory of polynomial equations (see §V.7).

11. Stereographic Projection and Infinity

Division by zero does not work; if attempted, it leads inevitably to paradoxes. Nevertheless, it is suggestive to think that the reciprocal $w = 1/z$ of a complex number z should approach infinity as z approaches zero. This does not mean that there is a *number* ∞ in the field $\mathbf{C}$ of complex numbers, but it does mean that there is a geometric model which adds a "point" at "infinity" to the complex plane.

Place a sphere of radius 1 so that it is tangent to the complex plane at its south pole, at the origin of the plane. Then project each point P of the sphere onto the plane. This means that one prolongs the line NP from the north pole N of the sphere to P till it meets the plane at a point P' (Figure 1). This provides a transformation $P \mapsto P'$ which is a bijection from the sphere (omitting the north pole) to the plane. It is called *stereographic projection*. It carries the south pole of the sphere to the origin in the plane and the equator to the circle of radius 2 with center at the origin. Each line L in the plane is the projection of a circle passing through the north pole of the sphere–it is the circle cut out from the sphere by the plane passing through the line L and the pole N. Also, if two lines L and L' in the plane meet at an angle θ, the corresponding circles on the sphere meet at the same angle θ on the sphere. Thus stereographic projection is a *conformal* transformation, in the sense that it preserves angles–though it manifestly does not preserve distances.

Under this stereographic projection all the points P on the sphere reappear as points P' on the plane–except for the north pole N on the sphere. But one can "extend" the plane by adding a point ∞, called the point "at infinity" and then specify that stereographic projection sends the north pole N on the sphere to ∞ on the extended plane; it is then a bijection from the whole sphere to the whole (extended) plane. As a point P' on the plane moves along a straight line L away from the origin, the corresponding point P on the sphere moves toward the north pole N. Hence the point ∞ at infinity in the extended plane lies on all lines in the plane, moreover, as a complex number z "approaches ∞" in absolute value, the corresponding point in the extended plane approaches this point ∞. This can be formalized by defining "neighborhoods" of ∞ so as to make the extended plane a topological space.

The virtue of the extended plane (or the corresponding *Riemann Sphere*) is that functions such as $w = 1/z$ are now defined for *all* z. Specifically, for $z = re^{i\theta} \neq 0$, the corresponding $w = 1/z$ is $(1/r)e^{-i\theta}$ (draw a picture), while for $z = 0$, $w = \infty$ and for $z = \infty$, $w = 0$. The

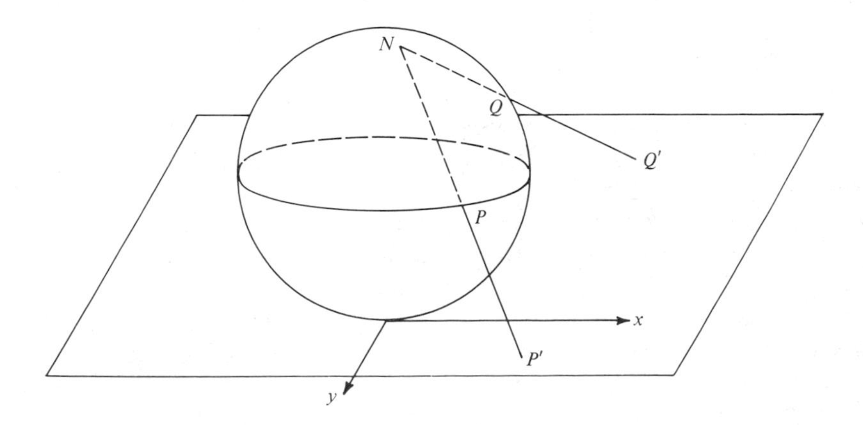

Figure 1. Stereographic projection.

resulting transformation $w = 1/z$ of the extended plane to itself can be shown to be conformal (with the inevitable definition of angles at the point ∞). The same holds for the more general *fractional linear transformations*

$$w = (az + b)/(cz + d), \qquad ad - bc \neq 0,$$

where the point $z = -d/c$ which makes the denominator zero is sent to the point at ∞ in the w-plane. These transformations form a group, with many fascinating properties. But the extended complex plane is useful in many other ways, for the effective geometric understanding of more general functions $w = f(z)$ of a complex variable z (Chapter X).

12. Are Imaginary Numbers Real?

As the name "imaginary" for i recalls, the introduction of the complex numbers in Mathematics met with doubts and resistance, and it was thought that numbers such as $4i$ were not really there, hence were "imaginary". This doubt can be put as a matter of principle as follows: If Mathematics is the science of number, space, time, and motion, and if number covers just whole, rational and real numbers, then there is no place in this science for these imaginary numbers. Therefore, the gradual general acceptance (after 1800) of the complex numbers as reasonable and useful does mark a departure from the "number, space, time, and motion" concept of Mathematics. In other words, once the construction of the complex numbers is accepted as "real", it becomes more natural to accept many further "constructions"–of higher-dimensional spaces, quaternions, groups, and the like. The complex numbers represent a major first step in developing the present scope of Mathematics.

The acceptance of complex numbers depends only in small part on the standard "explanation" (§10) that a complex number is a point in the plane or an ordered pair of real numbers. The ordered pair approach does represent the fact that a complex number has a real part and an imaginary part, but the formal definition of ordered pair is not really relevant. The points in the plane can be interpreted as complex numbers only after a choice of origin and axes. It is not just that geometry justifies the complex numbers; it is also that the algebra of complex numbers helps explain geometry. Addition gives translation, multiplication rotation.

The complete acceptance of complex numbers came primarily in their many uses in helping to understand other parts of Mathematics. Thus, they produce two square roots for every non-zero real number; this is just a small part of the benefit from the fundamental theorem of algebra. Particularly striking is the so called "irreducible" case of the cubic: After the formula (with cube roots) for the general solution of a cubic equation was found, it developed that there can be real cubics with all real roots for which the formula inevitably involves the use of complex numbers (For details, see for example Birkhoff–Mac Lane, Chapter XV, Theorem 22.) In other words, there are phenomena with real numbers which can be properly explained only with complex numbers. In electricity and magnetism, complex numbers provide a convenient formalism. Finally, many ordinary functions (e^x, $\sin x$) of a real variable are better understood when they are extended to be functions (e^z, $\sin z$) of a complex variable z. As we will see in Chapter X, the use of complex numbers yield a much deeper understanding of the nature of a "well behaved" function. In short, the real "foundation" for complex numbers lies in their manifold uses in better understanding of Mathematics. This is typical of the conceptual expansion of Mathematics.

The complex numbers are just one of the many new aspects which expanded the scope of Mathematics in the 19th century. Others include the use of integers modulo m and congruence classes, non-Euclidean geometry, n-dimensional geometry, decompositions of algebraic numbers into "ideal" factors (§XII.3), non-commutative multiplication in groups and quaternions, infinite cardinal and ordinal numbers, and the various necessary uses of sets and of logic, especially in the foundation of calculus (Chapter VI). These 19th century developments made obsolete the simple view that Mathematics is the science of number, space, time, and motion.

13. Abstract Algebra Revealed

The Peano axioms describe the natural number uniquely, up to isomorphism. The axioms for the real numbers describe these numbers uniquely, again up to an isomorphism. Such is not the case with the axioms for a commutative ring or a field–there are non-isomorphic models. Our successive constructions of the number systems

$$\mathbf{N}, \mathbf{Z}, \mathbf{Q}, \mathbf{R}, \mathbf{C}$$

have aimed to preserve the appropriate algebraic properties. All of these systems, save the natural numbers, are commutative rings and the final three are fields. Thus properties deduced from the axioms for a field hold in all three cases–as well as in other fields such as the finite fields $\mathbf{Z}/(p)$ for prime p. The familiar formula for finding roots for a quadratic equation applies to any one of these fields, as do the less familiar solutions of cubic and quartic equations. The same applies to the solution of simultaneous linear equations and to determinant formulas for their solution–although the determinant can also be given a geometric interpretation as an area (for 2×2 determinants) or as a volume in the 3×3 case (Birkhoff–Mac Lane, §10.3). Matrices and linear algebra work for every field. In this way the rules for the manipulation of ordinary numbers lead to codified rules for the manipulation of all sorts of numbers, and the various newer sorts of numbers were constructed so as to obey these rules. But it turns out that addition and multiplication apply not just to "numbers" but also to other mathematical objects: To polynomials, to congruence classes of integers, to formal power series (§4) and the like. Algebra replaces numbers by symbols; it is then possible that these symbols do not stand for numbers or "quantities". Hence it develops that the actual subject matter of algebra is not the manipulations of numbers under addition and multiplication, but the manipulation of *any* objects satisfying the rules (the axioms for a ring or for a field) for such manipulation. Algebra is thus inevitably abstract, even though this was not fully recognized until the 1920's when Emmy Noether and her disciples saw clearly that many algebraic phenomena could be grasped more effectively when stated and proved abstractly.

14. The Quaternions–and Beyond

The real numbers provide an algebraic formulation of one-dimensional geometry in algebraic terms, by means of a field. The complex numbers provide a corresponding field for geometry in dimension 2. Why not more such fields for geometry in higher dimension? The answer is that there cannot be such fields.

But there is something more–in four dimensions; that is, in $\mathbf{R}^4$. Pick a basis $1, i, j, k$ of the vector space $\mathbf{R}^4$, so that any q in $\mathbf{R}^4$ has the form of a "quaternion"

$$q = t + xi + yj + zk \tag{1}$$

with real coefficients t, x, y, z. For this basis, introduce the multiplication table

$$i^2 = j^2 = k^2 = -1, \qquad ij = -ji = k. \tag{2}$$

These equations (and the associative law) imply that $jk = -kj = i$ and $ki = -ik = j$. Hence by distributivity we can find the product of two "quaternions" q, each of the form (1), while the sum of two quaternions is found by term-by-term addition. With these operations the quaternions (1) do form a division ring (as this was defined in §3). Division is possible, because for any quaternion $q \neq 0$ the multiplicative inverse is $(t - xi - yj + zk)/(t^2 + x^2 + y^2 + z^2)^{1/2}$.

The striking fact is that the multiplication (2) of the quaternions is *not* commutative. There is no alternative multiplication table (2) which would make it commutative; that is, there is no way to define a multiplication on the vector space $\mathbf{R}^4$ extending the multiplication on $\mathbf{R}$ so that this multiplication, with vector addition, makes $\mathbf{R}^4$ a field; i.e. a system with commutative multiplication. In a field, the equation $x^2 + 1 = 0$ can have at most two solutions; in the division ring of quaternion there are many: $\pm i$, $\pm j$, $\pm k$, for example. Moreover, the multiplication table (2) is not just "pulled out of the air"; it can be used to describe rotations in three space much as the complex numbers of absolute value 1 describe rotations in the plane. Also the multiplication rule for two "pure" quaternions ($t = 0$) reads

$$(xi + yj + zk)(x'i + y'j + z'k) = -(xx' + yy' + zz')$$

$$+ (yz' - y'z)i + (zx' - z'x)j + (xy' - x'y)k.$$

The first term on the right is the "inner product" of the vectors (x,y,z) and (x',y',z'); the remaining terms form the "vector product". It is no wonder that texts in Physics still write vectors with the three basis units i, j, and k.

All this suggests that the quaternions are inevitable (though perhaps shocking to the commutativity-minded). They are. Consider any division ring D which contains the real numbers $\mathbf{R}$ as a subring; then D is a vector space over $\mathbf{R}$. A famous theorem then asserts: If D is finite-dimensional (over $\mathbf{R}$), then it must be isomorphic (as a ring, preserving $\mathbf{R}$) to the quaternions, or to the complex numbers, or to the reals themselves. There are no such finite dimensional division rings beyond dimension 4, and in dimension 4 the quaternions (possibly with a different choice of basis) is the only possibility!

15. Summary

Measures of many different sorts of magnitudes are all reduced to and codified by the real numbers. The simple formal axioms for the reals involves both algebraic aspects (the field axioms on addition and multiplication), ordinal aspects (the linear order), and aspects of continuity (the completeness requirement). Of these, the completeness requirement is the most geometric and the deepest–and has the most varied expression. For the algebraic axioms it is remarkable that multiplication appears to have no natural geometric representation on the line, but only one in the plane. One dimensional geometry is evidently quite weak.

The arithmetic constructions of the reals from the rationals are varied, straight-forward, and not of great weight. More significant is the use of the reals to reduce problems of geometry to questions in algebra, by way of vectors or of coordinates. The reduction of angular measure to linear measure is the source of trigonometry and the starting point for the algebraic treatment of vector spaces with inner products. It is also vital to the introduction of complex numbers. The real numbers (in dimension one) and the complex numbers (dimension two) both satisfy the axioms for a field–but in dimensions higher than 2 there is no corresponding field of "numbers".

Originally, it was the manipulation of magnitudes which led to the algebraic operations of addition and multiplication–but the same operations, with corresponding properties, hold for many systems which are not "magnitudes", hence the development of algebra as abstract. This abstraction is a natural sequel to the recognition of the real existence of imaginary numbers.

References. The systematic construction of number systems can be found in many texts, for example in Gleason [1966]. There is also a famous classical presentation by Landau [1951]. He was an expert on austere and rigorous formulation. The historical aspects are well suggested in Sondheimer–Rogerson [1981].

CHAPTER V

Functions, Transformations, and Groups

Firm formal definitions come first in any systematic foundation of Mathematics. Hitherto, however, we have not formally defined the notion "function", although we have used the notion extensively and consider it central to the organization of Mathematics. This delay was deliberate, since it has allowed the assembly of part of that wide variety of examples from which arises the general and abstract notion of function. Though some may hold that "abstract notions are difficult to understand" we hold with G. Kreisel that these notions "in fact, are usually introduced to make concrete situations intelligible" (*Math Reviews* **37** (1969) # 1224).

1. Types of Functions

Functions probably first appear with the practical experience that the size of some one magnitude depends on the size of another–the weight of a block of ice depends on the size of the block, or the distance travelled depends on the speed, or the area of a rectangle depends on its dimensions, or the angle subtended by a circular arc depends on the length of that arc. Thus practical problems, physical problems, geometric facts, and algebraic manipulations all indicate that one "quantity" may depend upon others. This leads to an idea of functional dependence, often described in suggestive but imprecise ways. A modern version of such an informal description would say that a function acts like a machine which, given any number as input, will produce as output some other number, "depending" on the input.

Algebraic operations provide many examples of functions of numbers. For instance, the operations "add 2 to the given number" or "multiply the given number by 3" or "take the square of the given number" are functions which we indicate by the "assignment" notation (with a barred arrow $\mapsto$) to designate the destination of the given number x, y, or z:

$$x \mapsto 2 + x, \qquad y \mapsto 3y, \qquad z \mapsto z^2. \tag{1}$$

One such function, followed by a second one, will give a *composite* function. In elementary parlance, "substituting" $2 + x$ for y and then $3y$ for z gives the composite of the three functions (1) as the quadratic function

$$x \mapsto 36 + 36x + 9x^2. \tag{2}$$

If the input x is a real number, so is the output; the result is a function on the reals to the reals, in symbols $\mathbf{R} \to \mathbf{R}$. Or, one may restrict the input x to be rational (or allow x to be a complex number), obtaining thereby from the same formula different functions $\mathbf{Q} \to \mathbf{Q}$ or $\mathbf{C} \to \mathbf{C}$, respectively.

Many algebraic operations produce functions which are not defined for *all* real numbers. This is usually the case for a rational function (the quotient of two polynomials). Thus the assignment

$$x \mapsto (3x^2 - 1)/(x^2 - 3x + 2)$$

produces from x a real number if and only if the denominator is not zero; that is, if and only if $x \neq -1$ and $x \neq -2$. It is then a function to the set $\mathbf{R}$ of reals from that set with -1 and -2 deleted: $\mathbf{R} - \{-1, -2\} \to \mathbf{R}$. The assignment $x \mapsto {}^{+}\sqrt{x}$ (take the positive square root of x) acts only on non-negative reals (or on non-negative rationals). (The two-valued expression $\pm\sqrt{x}$ does not count for us as a function.) A function may be determined by one of several different operations, with the choice depending in the location of the input number. Thus the absolute value of a real number may be described by the assignments

$$x \mapsto x \text{ if } x \geqslant 0, \qquad x \mapsto -x \text{ if } x < 0.$$

There can be all manner of such combinations, often represented by graphs. In some cases, the function constructed may not be readily visualized by a graph, as for instance for the assignments $x \mapsto 0$ if x is rational and $x \mapsto 1$ if x is irrational.

Geometric definitions also produce numerical functions such as the trigonometric functions $\theta \mapsto \sin\theta$ and $\theta \mapsto \cos\theta$ with their familiar periodic graphs. These functions are *not* polynomial or rational, but can be expressed or even defined by formulas, specifically by infinite power series (see Chapter X) such as

$$\sin\theta = \theta - \theta^3/3! + \theta^5/5! - \cdots.$$

These series, or others technically more convenient, may then be used to calculate the familiar tables of the trigonometric functions. There is also an inverse function "arc sin", intended to send each number x into an angle θ with $\sin\theta = x$. For real angles, only the values between

-1 and $+1$ are possible for $x = \sin \theta$, and there are then many choices for the angle. To get a function, one may choose the angle always between $-\pi/2$ and $\pi/2$ inclusive; this gives the function "arc sin" from the interval consisting of all real x with $-1 \leqslant x \leqslant 1$ to the interval of all real θ with $-\pi/2 \leqslant \theta \leqslant \pi/2$. Recall that such inverse functions are needed both in solving triangles and in performing certain integrations, such as $\int (1-x^2)^{-1/2}dx$. Another basic function which is not algebraic is the exponential e^x and its inverse, the logarithm $\log x$ – defined only for positive real x.

"Variable quantities" in physics and other sciences are essentially magnitudes measured (according to established scientific rules) by real numbers. Thus weight, length, volume, temperature, velocity, and the like are such quantities. Various elementary laws of physics assert that one such quantity is a function of another. Thus, for materials of fixed density ρ, weight w is $w = \rho V$, where V is the volume. The distance transversed by a falling body is a function of the time t of fall, according to the familiar formula $s = gt^2/2$. For an ideal gas, pressure P, volume V and temperature T are related by the law $PV = kT$, for a suitable constant k. Thus each of these three quantities is a function of the other two. There are many such relations between variable quantities, and the study of the manner of their variation is at the origin of much of Mathematics.

2. Maps

Functions of points arise in geometry. The problem of representing the globe or a part of the globe on a piece of paper is the problem of mapping a portion of the sphere S^2 on the Euclidean plane $\mathbf{R}^2$, by some function $S^2 \to \mathbf{R}^2$. The path of a projectile in space is specified by giving its coordinates x, y, and z as they depend upon time t by three functions $x(t)$, $y(t)$, $z(t)$; that is, by a single function from (part of) the time axis into $\mathbf{R}^3$. These and many similar examples provide "maps"–that is, functions–from one set X of objects, points, or numbers into some other set Y. In other words, functions need not be numerical. The operation of rotating a plane about a given point and through a given angle maps each point of the plane into another point, hence is a function $\mathbf{R}^2 \to \mathbf{R}^2$, expressed in coordinates via trigonometry as in the linear equations (IV.9.5). More generally, any linear transformation T of the plane is a function $\mathbf{R}^2 \to \mathbf{R}^2$ which takes linear combinations to linear combinations. Hence, writing each vector of $\mathbf{R}^2$ as a linear combination $(x,y) = x(1,0) + y(0,1)$ of the standard basis vectors, the whole transformation T is determined by the images $T(1,0) = (a,c)$ and $T(0,1) = (b,d)$ of the two basis vectors; indeed one may use these numbers to write T as

$$T(x,y) = (ax + by, cx + dy). \tag{1}$$

We recall the convenient shorthand in which such a linear transformation T, expressed thus in terms of coordinates, is represented by a 2×2 *matrix* of real numbers,

$$\begin{bmatrix} a & b \\ c & d \end{bmatrix}; \tag{2}$$

the composition of two such transformations then corresponds to the familiar operation of the multiplication of two 2×2 matrices. There are corresponding rules for $n \times n$ matrices in the case of n-dimensional space. In any event, a matrix (2) is essentially a shorthand record of a function on vectors to vectors, represented in coordinates as in (1); a matrix is a model example of a Mathematical object which is neither a number nor a point.

There are many other sorts of non-numerical functions. Thus in Chapter I we already met correspondences (§I.1), bijections (§I.1), rigid motions (§I.5), symmetries (§I.6), permutations (§I.3), transformations (§I.7), and unary operations (§I.8). For these we have already used the modern notation $f\colon X \to Y$ for a function on the domain X to a codomain Y, with $x \mapsto fx \in Y$ for each x in X. There are all sorts of such. On a group G, the inverse is a function $G \to G$, while group multiplication is a function of two variables in G. In a metric space, the distance is a real-valued function of pairs of points. In Boolean algebra, "intersection" and "union" are functions of pairs of sets. In geometry, "length" is real-valued function of curves. Historically, functions were perhaps first explicitly recognized as functions $\mathbf{R} \to \mathbf{R}$ in the calculus, as with Newtons "fluents" (functions of time). Now the more general notion of an arbitrary function $X \to Y$, in parallel to the notion of a set, plays a powerful unifying role in Mathematics.

3. What Is a Function?

The various intuitive ideas about functions and functional dependence are helpful but vague. Here are some of them.

Formula. A function is a formula in a letter x. When x is replaced by a number, the formula produces a number, the value of the function for the given argument x.

This description is not very helpful for notions of functional dependence of variable quantities in physics, but it does describe the elementary Mathematical functions well: Polynomial, rational, algebraic, trigonometric, and exponential functions. This ideas of "formula" does need to be specified. It should include algebraic or "analytic" formulas, perhaps also the formulas given by infinite series, but it doesn't seem to encompass functions defined by several different formulas in different portions of the domain. The essential problem remains: What sorts of formulas are envisaged? Are all functions given by formulas? "Formulas" depend on the symbolism, but functions depend upon the facts.

Rule. The variable y is a function of the variable x when there is given a rule which to each value of x produces the corresponding value of y.

This description, and its variants, has for generations puzzled the students of calculus. It has a pleasant generality: Any "rule" will do. Also, the use of rules clearly takes care of the case of functions defined by different formulas in different portions of the domain; any such collection of alternative formulas is clearly a *rule*. Nevertheless this is not a formal definition, since it uses the undefined words "correspond" and "rule". Even if one defines a rule as something expressed in a specified formal language, there are troubles. (In the usual formal languages, the set of all formal expressions is denumerable, while the set of possible functions on **R** to **R** is not denumerable.)

Graph. A function is a curve in the (x,y) plane, such that each vertical line $x = a$ meets the curve in at most one point with coordinates (a,b). When it does so meet, the number b is the value of the function at the argument a. For other arguments a, the function is undefined.

This description emphasizes the geometric aspect. It is persuasive for functions of real numbers which are smooth or at least continuous, but it doesn't fit well with a function which jumps from 0 to 1 as the variable changes from rational to irrational. It involves also the (undefined) notion of a curve, and so makes arithmetic depend upon geometry.

Dependence. The variable quantity y is a function of the quantity x if and only if a determination of the value of x also fixes the value y, so that y depends on x.

This, a physicist's definition, is also not formal.

Table of Values. A function is determined by a table of values, which opposite each entry for the first quantity x lists the corresponding numerical value for the second quantity y.

This is a hard-nosed, no nonsense definition; evidently inspired by tables of trigonometric or logarithmic functions. Trouble is, the actual tables are finite while most of the intended functions have infinitely many different values. It is not clear what would be meant by an infinite table.

Syntax. A function f on the set X to the set Y is a symbol f such that whenever the term x stands for an element of X, then the string of symbols fx stands for an element of Y, the value of f at the argument x.

This doesn't really describe functions, but just the use of symbols for functions. It is a mute protest against the confusion of standard notations, in which $f(x)$ ambiguously denotes a function of x and a value of that function. (Thus, strictly speaking, "sin" is the trigonometric function, while "sinθ" denotes its value at the angle θ.)

Enough. The variety of these descriptions of "function" illustrate well one of our theses as to the nature of Mathematics: Human activities and facts about phenomena together indicate many examples of dependence of one item upon another. This leads to useful but informal ideas about dependence and functions–and poses the problem of producing a formal definition, necessary to make unambiguous Mathematical statements about functions.

4. Functions as Sets of Pairs

A formal definition of "function" must be stated in the context of some axiomatic system. There are at present two such definitions. One of these directly axiomatizes the notion of function in terms of the composition of functions (categories, in Chapter XI). The other operates in terms of axioms for sets, to be given in full in Chapter XI. These axioms assume that everything (in Mathematics) is a set, and are formulated in terms of one primitive notion, *membership*, written $x \in A$ for "x is a member of (i.e., an element of) the set A". The *equality* of sets is then defined by (cf. (I.9.8)

$$A = B \Longleftrightarrow \text{ for all } x, \quad (x \in A \Longleftrightarrow x \in B), \tag{1}$$

where the double headed double arrow $\Longleftrightarrow$ is short for "if and only if". An axiom (the axiom of extensionality) then requires that equals can be substituted for equals:

$$A = B \Rightarrow \text{ for all } C, \quad (A \in C \Rightarrow B \in C); \tag{2}$$

here the double arrow $\Rightarrow$ is short for "implies". Another axiom asserts that for any two sets a, b there is a set $\{a,b\}$ such that

$$x \in \{a,b\} \Longleftrightarrow (x = a \text{ or } x = b); \tag{3}$$

in other words, $\{a,b\}$ is the set whose only elements are a and b. In particular, $\{a,a\}$ is the set whose only element is the set a; it is usually written $\{a\}$. On this basis we can *define* the *ordered pair* of two sets a, b to be the set

$$<a,b> = \{\{a\}, \{a,b\}\}. \tag{4}$$

This is a set whose elements are sets of sets. One can then prove that

$$<a,b> = <a',b'> \Longleftrightarrow (a = a' \text{ and } b = b'), \tag{5}$$

which states that this set $<a,b>$ has the property one would expect of an ordered pair of any two things. Other axioms prove that, given two sets X and Y, there exists a set

$$X \times Y = \{<x,y> \mid x \in X, y \in Y\} \tag{6}$$

whose elements are exactly all the ordered pairs of elements $x \in X$ and $y \in Y$. This *cartesian product* of sets has already been used, for example in the description of the coordinate plane as the cartesian product $\mathbf{R} \times \mathbf{R}$. We also use the formal definition of a *subset* $S \subset X$: A set such that $x \in S$ implies $s \in X$.

Definition. A *function* f on the set X to the set Y is a set $S \subset X \times Y$ of ordered pairs which to each $x \in X$ contains exactly one ordered pair $<x,y>$ with first component x. The second component of this pair is the value of the function f at the argument x, written $f(x)$. We call X the *domain* and Y the *codomain* of the function f.

This provides a formal definition which, in plausible ways, does match the intent of the various preformal descriptions of a function. Specifically, it does provide y, depending on x. If one imagines a list of all the ordered pairs $<x_1,y_1>$, $<x_2,y_2>$, $<x_3,y_3>, \ldots$ in the set S, then this list is just the (usually infinite) table of values of the function. If one visualizes the sets X and Y as "spaces" of some sort, then the cartesian product $X \times Y$ is a "space" (in Figure 1 a cylinder) and the function S is a subset of the product which meets each "vertical" subspace $\{x\} \times Y$ in exactly one point. Thus S is a curve in this product space. In view of such examples, the set S of ordered pairs is often called the *graph* of the function—though in our definition, the function *is* its graph.

From this formal definition (and the axioms of set theory) one can derive all the formal properties of functions. For example the definition of equality for sets provides an equality condition for functions. Two functions f,g: $X \to Y$ with the same domain and the same codomain are equal

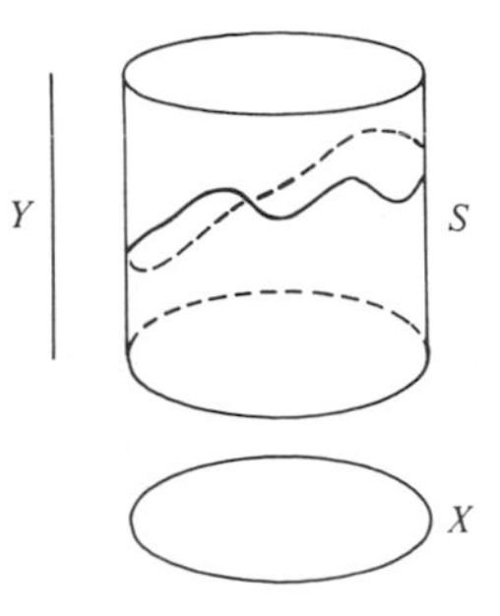

Figure 1

(as sets of ordered pairs and thus as functions) if and only if, for all $x \in X$, $f(x) = g(x)$ in Y.

Our examples of functions have involved many instances of functions constructed by composition, such as $(2x + 2)^2$ or $\cos 7\theta$. There is a corresponding formal definition. Given any two functions f and g, when the codomain of f is the domain of g, as in

$$X \overset{f}{\mapsto} Y \overset{g}{\mapsto} X,$$

one can define the *composite* function $h = g \cdot f$ to be the following set of pairs

$$\{<x,z> \mid x \in X, z \in Z, \exists y \in Y \text{ with } <x,y> \in f \text{ and } <y,z> \in g\}. \quad (7)$$

Here $\exists y$ is a (logical) quantifier meaning "there exists a y"; moreover we have identified (as we should) f with "its" set S of ordered pairs. One readily shows that this set (7) of ordered pairs is indeed a function $g \cdot f \colon X \to Z$ and that its value, for each argument $x \in X$, is

$$(g \cdot f)(x) = g(f(x)), \qquad x \in X. \quad (8)$$

Here the usual convention that $f(x)$ means "apply the function f to the argument x" has the consequence that the composite function $g \cdot f$ means "first apply f, then apply g". The definition of the composite function $g \cdot f$ is then equivalent to the statement that (8) holds, and that

$$\text{domain}(g \cdot f) = \text{domain } f, \qquad \text{codomain}(g \cdot f) = \text{codomain } g. \quad (9)$$

This description of composite functions does match the examples such as $\cos 7\theta$. It illustrates the fact that the formal definition of functions by ordered pairs is needed only at the start, and soon is replaced by equivalent (derived) statements like (8) and (9) in more conventional language, *not* explicitly using ordered pairs. For example, the formula (8) suffices to prove *associativity*, a central property of composition: Given three composable functions

$$X \xrightarrow{f} Y \xrightarrow{g} Z \xrightarrow{h} W,$$

one necessarily has $h \cdot (g \cdot f) = (h \cdot g) \cdot f$.

In defining the composite $g \cdot f$ we have required that the domain of g be exactly the codomain of f. In this view, a function f involves all of these things: A domain, all the pairs $<x, fx>$, *and* the codomain. The set of pairs determines the domain, but does not determine the codomain. Changing the codomain changes the function. Thus if $\mathbf{R}^+$ is the set of non-negative reals, the assignment $x \mapsto x^2$ defines a function $\mathbf{R} \to \mathbf{R}^+$ and a different function $\mathbf{R} \to \mathbf{R}$ with the *same* values but a larger codomain.

This convention (that a function carries with it a specified codomain) is usually not made in elementary Mathematics. However, it is useful in keeping composites in order and it is essential for some concepts in topology (see §XI.9 below). Also it is essential to make sense of statements such as "this function is a bijection". For example, the set of all pairs $<a,a+2>$ for a an integer determines a bijection $\mathbf{Z} \to \mathbf{Z}$ and also gives a function $\mathbf{Z} \to \mathbf{Q}$ which is not a bijection.

Formally, this definition of function is fine–but it does appear to rest upon a possibly quite artificial notion of an ordered pair. Also, why should an ordered pair be construed to be a set? However, with this construction the ordered pair (4) has exactly the formal property (5) one might require of such a pair. The crucial observation is that this formal definition does indeed provide for all the intended practical properties of functions. In particular, two functions are equal if and only if they have the same domain, the same codomain, and the same values.

The definition (7) of the composition of functions does include the composition of bijections, as this was defined in Chapter I at the hand of several examples. For economy of presentation, we should have defined "composition" just once. However, the repetition of the definition may serve to make a philosophical point: The definition of composition formalizes an idea ("carry out two operations in succession") which has arisen from a wide variety of activities. These activities include the successive comparison of sets for size (§I.2), the "operation" of composing two permutations, the essentially "static" compositions of symmetries (§I.6) and the more "dynamic" composition of motions (§I.5). Moreover, there is the composition of arithmetic operations by substitution, as in (1.2). Composites such as these, as presented by formulas (e.g. sin 8θ), are perhaps the most elementary, since they can be handled by rules for manipulating formulas in ways which require no formal definition of "composite". On the other hand, the composition of motions is more geometric and hence more "visible". The philosophical point is that all these varied human activities fall–quite naturally and easily–under the one general notion of "composition", and the formulation of this general notion helps to organize and understand the various cases. This, we hold, is typical of Mathematical abstraction.

Binary relations, such as linear order, inclusion, and congruence relations, provide another example of a set-theoretic formalization of a practical notion. Thus in ordinary language the phrases "brother of" or "larger than" provide two-place relations which appear in context in phrases such as "A is the brother of B" or "s is larger than t". There is also a notion of composite relation; for example, "uncle of" means "brother of a parent of".

The composition of functions is a special case of the composition of relations. A (binary) *relation* from the set X to the set Y is a subset $R \subset X \times Y$; the intention is that it consist of those ordered pairs

$<x,y>$ of elements $x \in X$ and $y \in Y$ which stand in the intended relation to each other. One may often write yRx for $<x,y> \in R$. If $S \subset Y \times Z$ is a second relation from Y to Z, the *composite* relation $S \cdot R$ is defined to consist of those ordered pairs $<x,z>$ for which there exists at least one element $y \in Y$ such that $<x,y> \in R$ and $<y,z> \in S$. This composition is associative. One may develop a consequential algebra of relations extending Boolean algebra. There are many Mathematical examples of binary relations, such as the inclusion relation $\subset$ for subsets of a set and the linear order relation for real numbers. There are also cogent philosophical arguments for the importance of relations. (One cannot reduce all relations between objects x and y to conjunctions of separate properties of x and properties of y.) Nevertheless, despite various attempts, the properties of relations do not seem to play a central role in Mathematics; instead, relations are often replaced by operations or functions. For example, congruence is replaced by congruence classes.

There are a few useful general facts about functions on arbitrary sets. They all have been suggested by specific examples. Thus a function $f\colon X \to Y$ may have as "inverse" some function g from Y to X–just as square root is inverse to square and log inverse to exponential. To define this formally, first introduce for each set X the identity function $1_X\colon X \to X$ which sends each element $x \in X$ to itself. A more vivid description states that the graph of the identity function is the diagonal subset Δ of $X \times X$, consisting of all the ordered pairs $<x,x>$ for $x \in X$. If two functions $f\colon X \to Y$ and $g\colon Y \to X$ have composite $g \cdot f = 1_X$ the identity, then the function g is said to be a *left inverse* of f, while f is called a *right inverse* of g. Thus the definition of a logarithm, $10^{\log_{10} x} = x$, rewritten

$$\exp_{10}(\log_{10} x) = x, \qquad x > 0,$$

states that *log*: $\{x \mid x > 0\} \to \mathbf{R}$ is a right inverse of exp (it is also a left inverse). The operation "take the positive square root" is a function $\{x \mid x \geqslant 0\} \to \mathbf{R}$. The equation $x = (\sqrt{x})^2$ means that "square" is its left inverse–but not its right inverse. In general again, $g \cdot f = 1_X$ and $f \cdot g = 1_Y$ together mean that g is a (two-sided) inverse of f. When $f\colon X \to Y$ has such an inverse g, that inverse is unique; moreover the sets X and Y have the same size (formally, have the same *cardinal number*).

When any $f\colon X \to Y$ is regarded as a comparison of the set X to the set Y one naturally may ask when f hits all elements of Y (is "surjective") or when f keeps distinct elements of X distinct (is "injective"). Formally, a function $f\colon X \to Y$ is called an *injection* when $x_1 \neq x_2$ in X implies $fx_1 \neq fx_2$ in Y; intuitively, an injection maps the set X in one-to-one fashion onto some subset of Y, called the *image* of f. Unless X is the *empty* set ϕ, a function f with domain X is an injection if and only it has a left inverse g; specifically, g takes each element of the image back to the necessarily unique element of X from which it came, and the remaining elements of Y (if any) back to any old element of the (non-empty!) set X.

A function $f: X \to Y$ is a *surjection* when to each $y \in Y$ there is at least one element $x \in X$ with $fx = y$. The axiom of choice, commonly assumed among the axioms of set theory, asserts that any surjection f with a non-empty domain X has a right inverse g. Specifically, such a right inverse "chooses" to each $y \in Y$ an element $x \in X$ with $fx = y$, and the axiom of choice is there to provide for such a potentially infinite collection of choices.

Finally, a function $f: X \to Y$ is a *bijection* if and only if it is both an injection and a surjection; that is, if and only if there exists to each $y \in Y$ exactly one $x \in X$ with $fx = y$. Thus a bijection establishes a one-to-one correspondence between the elements of X and those of Y. A function f is a bijection if and only if it has a two sided inverse.

One may note that these definitions are formal, but make very little use of the ordered pairs used in the formal definition of a function.

A function $f: X \to Y$ which is not injective "collapses" some of the elements of X, and the collapse may be formalized by a binary relation E_f "has the same f-image as" on the domain X. Specifically, for x and y in X, take xE_fy to mean $fx = fy$; the relation E_f may be called the "kernel" or the "kernel pair" of f. It is reflexive, symmetric, and transitive. A reflexive, symmetric, and transitive relation E on a set X is usually called an *equivalence* relation or a *congruence* relation; the prime example is congruence modulo m on the set $\mathbf{Z}$ of integers. In that case, congruence is the kernel of the projection $\mathbf{Z} \to \mathbf{Z}_m$ to the integers modulo m. By the same construction, any equivalence relation E on a set X is the kernel of a surjection $p: X \to Y$: one simply takes $p(x) = \{x' \mid x' \in X \text{ and } xEx'\}$, and Y to be the set of all these *equivalences classes*. Each element x' in X belongs to exactly one of these equivalence classes, while p turns the equivalence relation E in X into equality in Y. This important construction has already been used to construct cardinal numbers, ordinal numbers, the set of "directions" in the plane (§III.9) and the Cauchy real numbers. It is a handy set-theoretic way of specifying new Mathematical objects.

This section has indicated how the notion of a function, originally suggested by formulas and functional dependence, can be formalized and applied to sets quite different from sets of numbers, and so provide a setting for the comparison and mapping of arbitrary sets.

5. Transformation Groups

The analysis of the symmetry of figures, of formulas, of ornaments or of crystals leads inevitably, as in Chapter I, to the study of transformation groups. Such groups also arise in geometry, as groups of rigid motions, and can even provide a foundation for Euclidean geometry (§III.10). In this section we will note additional examples of transformation groups.

The *symmetric group* S_n on n letters consists of all the $n!$ permutations of the set $\{1,2,\ldots,n\}$ of the first n positive integers; any other set of n elements has the same symmetry group. Every permutation can be written as a composite of *transpositions*, where a transposition interchanges two integers and leaves all others unchanged. A permutation is *even* if and only if it can be expressed as a product of an even number of transpositions. Within S_n the even permutations form by themselves a transformation group A_n with $n!/2$ elements, called the *alternating* group, a *subgroup* of S_n. For example, all the permutations of four letters x_1,x_2,x_3,x_4 which do not change the sign of the product

$$(x_1 - x_2)(x_1 - x_3)(x_1 - x_4)(x_2 - x_3)(x_2 - x_4)(x_3 - x_4)$$

constitute the alternating group A_4. For an equilateral triangle, each symmetry is determined by the permutation induced on the vertices of the triangle, so the symmetry group of the triangle is isomorphic to S_3. For analogous reasons, the symmetry group of an equilateral tetrahedron is isomorphic to S_4 while the alternating group A_5 appears as the symmetry group of the icosadedeon. In general, finite symmetry groups can always be expressed as subgroups of some permutation group S_n.

Linear algebra is replete with transformation groups, beginning with the group of translations of the line, which is isomorphic to the additive group of real numbers. A linear transformation $t\colon V \to V$ of a vector space into itself is said to be *non-singular* if and only if it is a bijection. The set of all such non-singular transformations on V is clearly a transformation group, called the *general linear group* $GL(V)$. When V is the space F^n of n-tuples of elements of a field F, such a non-singular transformation is represented, as in §2 above, by an $n \times n$ matrix with entries in F and determinant $\neq 0$. The group of all these non-singular matrices is written $GL_n(F)$; especially prominent are the groups $GL_n(\mathbf{R})$ and $GL_n(\mathbf{C})$. The subgroup of those real matrices with determinant 1 is called $SL_n(\mathbf{R})$. In a vector space over $\mathbf{R}$ with an inner product, the linear transformations which preserve the inner product are called orthogonal, and constitute the *orthogonal* group $O_n = O_n(\mathbf{R})$. The matrices representing such an orthogonal transformation must have determinant ± 1; those with determinant $+1$ preserve orientation and constitute the *special orthogonal* group SO_n. There are corresponding constructions for complex numbers. In an affine space A, the bijections $t\colon A \to A$ which preserve weighted averages (that is, satisfy $t(\Sigma w_i P_i) = \Sigma w_i t P_i$ for weighted averages of points P_i, expressed as in (IV.7.4)), are called *affine* transformations. In this way each type of geometry can be summarized by the transformation group of all those bijections of the space which preserve the intended geometric structure. This was the central observation of Felix Klein's *Erlanger Program* for geometry (1872). Two figures count as "equivalent" in a geometry when there is a transformation of the group carrying the first figure into the second. Thus in the Euclidean plane two triangles are equivalent in this sense if and only if they are congruent, in the oriented Euclidean plane two triangles are equivalent if and only if they are congruent *and* have the same orientation, while in the affine plane any two non-degenerate triangles are equivalent.

This description of the Erlanger program fits all manner of geometries. Thus topology involves for each topological space X the group of all those bijections $t\colon X \to X$ which are continuous with a continuous inverse. More generally, a *homeomorphism* $t\colon X \to Y$ of topological spaces is a continuous bijection with a continuous inverse. Thus the surface of a sphere is not homeomorphic (why?) to the torus (the surface of a doughnut), but is homeomorphic to the surface of an ellipsoid or of a cube. However the latter homeomorphism cannot be "smooth" (at the corners of the cube); this suggests still another kind of geometry, that of C^∞ structure, to be discussed in Chapter VIII.

All told, the initial and quite intuitive observations of symmetry, of motion, and of transformations in geometry and of their composition leads to the formal set-theoretic notion of a transformation group. Its study involves algebra, geometry, and continuity.

6. Groups

The definition of a group G by axioms, as given in §I.8, arises by abstraction from the notion of a transformation group—one ignores the fact that the elements are transformations of something but retains the composition operation, the identity, and the inverse and requires the associative law (valid automatically in case the group elements are indeed transformations). No further axioms are needed to formalize the properties of composition, because these axioms suffice to prove the Cayley theorem that every (abstract) group G is isomorphic to a transformation group G'. Specifically, each element $g \in G$ can be regarded as a transformation $g'\colon x \mapsto gx$ of the elements x of the *set* G. By the associative law, $h(gx) = (hg)x$, so the composition hg of group elements matches exactly the composition $h'g'$ of the corresponding transformations; thus the bijection $g \mapsto g'$ is an isomorphism of the abstract group G to the transformation group G'. It is called the (left) *regular representation* of G, because each element g of G is "represented" by the operation "multiply on the left by g".

Other groups arise not as groups of transformations but as groups of numbers under multiplication (or addition; cf. §I.8). Most of these groups are abelian, but algebraic considerations also yield examples of non-abelian groups, such as GL_n and the multiplicative group of non-zero quaternions (§IV.14). From given groups one may construct new groups as products or as semidirect products (§8). One may also construct groups *ad hoc* by listing the elements and writing a suitable multiplication table. Indeed, the richness of group theory may be ascribed in part to its multiple origins.

The simplest (multiplicative) groups G are those "cyclic" groups generated by a single element a, and so consisting just of the identity and powers $a,a^{-1},a^2,a^{-2},\ldots$ of that element. The *order* of an element a is by definition the least positive integer m with $a^m = 1$, the identity element. This order may be infinite, in which case the cyclic group is infinite–and isomorphic to the additive group $\mathbf{Z}$. The order m may be finite, in which case the (cyclic) group is isomorphic to the group of all rotations of a regular m-gon into itself and to the additive group $\mathbf{Z}_m$ of integers modulo m. In any finite group, the order of each element is a divisor of the order (number of elements) of the whole group. Much of the structure of a particular group may be revealed by examining the lattice of its subgroups and their inclusion relations–where a subset S of a group G is a *subgroup* if and only if it is non-empty and closed under composition and inverse.

The comparison of two (multiplicative) groups G and H is made by a function or "morphism" preserving the group structure. Specifically, as in §I.8, a *homomorphism* $t\colon G \to H$ of groups is a function t such that $t(g_1 g_2) = (tg_1)(tg_2)$ for all pairs of elements $g_1,g_2 \in G$. It follows also that $t(g^{-1}) = (tg)^{-1}$ for all g and that $t(1) = 1$. When first studied, homomorphisms were taken to be always surjective (*onto* the codomain H) but it is more convenient to allow the codomain to be larger than the image. In particular, when S is a subgroup of G, the inclusion $S \subset G$ can be considered as a function from S to G which is a homomorphism. In general, a homomorphism t is called a *monomorphism* when t is an injection, an *epimorphism* when t is a surjection, and an *isomorphism* when a bijection. A homomorphism $t\colon G \to G$ of G to itself is an *endomorphism*, and an *automorphism* when it is also a bijection. In fact, the set Aut G of all automorphisms of G is itself a transformation group; it expresses the symmetry of G. The mono-, epi-, endo-, and auto-terminology applies also to "morphisms" of other types of algebraic systems such as rings, fields, or monoids–but any homomorphism of fields is necessarily a monomorphism.

Homomorphisms arise naturally for transformation groups. For example, the group Q of the square has 8 elements (four rotations, four reflections). Each symmetry $t \in Q$ induces a permutation t' of the diagonals; the permutations of the (two) diagonals constitute the symmetric group S_2, and the correspondence $t \mapsto t'$ is an epimorphism $Q \to S_2$. Again each symmetry $t \in Q$ must either leave the square right side up or turn it over; in other words, t determines a permutation t'' of the two faces of the square, and this correspondence $t \mapsto t''$ is an epimorphism $Q \to S_2$–different from the previous epimorphism! More elaborate examples can be constructed from the group of symmetries of a cube.

A homomorphism $t\colon G \to GL_n(F)$ of a group into a general linear group is called a *representation* of G. The systematic study of such representations reveals many deeper properties of groups. They include the representation of G by permutations, since any permutation of the symmetric group S_n can be regarded as a permutation of the n unit vectors of the vector space F^n, and so can be represented (isomorphically) as an $n \times n$ matrix with entries zero except for one entry 1 in each row and each column.

For each element k of a group G, the operation $g \mapsto kgk^{-1}$ is called *conjugation* by k. It is an automorphism k_*: $G \to G$, called an *inner automorphism*. Moreover, the correspondence $k \to k_*$ is itself a homomorphism $G \to \text{Aut } G$ –trivial if G is abelian. For a transformation group G, conjugation by k has "geometric" meaning; for instance (§III.9), it carries the subgroup fixing one point x isomorphically onto the subgroup fixing the point $k(x)$; cf. (III.9.3) and (III.9.4).

Any homomorphism maps subgroups to subgroups. A subgroup N of a group G is called a *normal* subgroup if and only if it is mapped onto itself by every inner automorphism k_* of G; that is, if and only if $n \in N$ and $k \in G$ imply $knk^{-1} \in N$. For any homomorphism t: $G \to H$, the *image* is the subset (subgroup) of all elements $t(g)$ for $g \in G$ and the *kernel* K is the set of all those $n \in G$ with $t(n) = 1$ in H. One proves readily that the kernel K is always a normal subgroup of G. Moreover every normal subgroup N is the kernel of an epimorphism. This basic result we state in the correct (but not customary) conceptual form, as in our book *Algebra*:

Theorem. *For any normal subgroup N of G there is a group H and an epimorphism t: $G \to H$ with kernel exactly N. If s: $G \to L$ is any group homomorphism with kernel containing N, there is a unique homomorphism s': $H \to L$ with composite $s' \cdot t = s$, as in the commutative diagram*

(1)

One says that s' is *induced* by s. The group H may be constructed explicitly, for example by taking as its elements $t(g)$ the sets gN of all multiples gn of g by an element $n \in N$. Such a set gN is called a *coset*, as in §III.9; observe that g and gm for m in N give the *same* coset. To make the set H of all cosets into a group, their product is defined by $(g_1 N)(g_2 N) = (g_1 g_2)N$. Then $g \mapsto gN$ is an epimorphism to the group of cosets, and it follows that any s as stated does factor uniquely through t, as in the commutative diagram (1). Because of this factorization, one says that t is *universal* among the homomorphisms from G with kernel containing N.

The explicit description of the elements of H as cosets is irrelevant, since the group H is uniquely determined (up to an isomorphism) by its universal property–for any other homomorphism s: $G \to H'$ with the same universal property, the diagram (1) produces an isomorphism s': $H \cong H'$. This group H, constructed by cosets or otherwise, is called the *factor group* (quotient group) G/N. A very simple example is the subgroup (n) of all multiples of an integer n in the additive group $\mathbf{Z}$; the factor group is the additive group $\mathbf{Z}/(n)$ of congruence classes modulo n, because these congruence classes are exactly the (additive) cosets of (n). Again, A_4 is a normal subgroup of the symmetric group S_4, and $S_4/A_4 \cong S_2$. In the oriented plane, the group of translations is a normal subgroup of the group of all proper rigid motions (cf. (III.7.1)) and the corresponding quotient group is isomorphic to the group of rotations about a point. Such examples, plus others from Galois theory (§7), led to the general construction–and manipulation–of factor groups. Contrary to common custom, they are not best understood in terms of cosets.

7. Galois Theory

Symmetry and its formalization by transformation groups arises not only in geometric situations, but also in purely algebraic cases. We have already noted a first example, in the operation $\mathbf{C} \to \mathbf{C}$ of conjugation $x + iy \mapsto x - iy$ for complex numbers. This operation interchanges the two complex cube roots ω and ω^2 of 1. Conjugation leaves only the real numbers fixed, and can be viewed geometrically as a reflection of the complex plane in the real axis. As a symmetry group, it consists of just two transformations: conjugation and the identity. It arises from the equation $x^2 + 1 = 0$, since it interchanges the two roots of this equation.

Now consider a polynomial equation of degree n such as

$$f(x) = a_n x^n + a_{n-1} x^{n-1} + \cdots + a_1 x + a_0 = 0, \qquad a_n \neq 0. \quad (1)$$

We will assume that the coefficients $a_n, a_{n-1}, \ldots, a_0$ lie in some subfield F of the field $\mathbf{C}$ of complex numbers, and that the polynomial f is *irreducible* over F–that is, cannot be factored in F as $f(x) = g(x)h(x)$ into two polynomials g and h of lower degree.

Galois theory is concerned with formulas for or properties of the roots of $f(x) = 0$. By the fundamental theory of algebra, there is at least one complex root α_1 and hence a corresponding factorization of $f(x)$ as $f(x) = (x - \alpha_1)g(x)$, where the second factor $g(x)$ is a polynomial (of degree $n - 1$) with complex coefficients. This polynomial in its turn has a complex root α_2 so continuation of this process ultimately gives n roots and a factorization

$$f(x) = a_n(x - \alpha_1)(x - \alpha_2)\ldots(x - \alpha_n). \quad (2)$$

From the fact that f is irreducible, one may prove that these n roots $\alpha_1, \ldots, \alpha_n$ are all different.

For $n = 2$, the familiar process of "completing" the square produces formulas involving a square root for the two roots α_1 and α_2. For polynomials of degrees 3 and 4 there are similar but more complicated (and less useful) such formulas, involving cube roots and fourth roots (Survey, §V.5). The Galois theory will reveal why there can be no corresponding formulas valid for the roots of all polynomials f of degree 5–or higher. To begin with, shift attention from the n roots $\alpha_1, \ldots, \alpha_n$ to the set of all complex numbers which can be obtained by rational operations (addition, subtraction, multiplication, and division) from these roots and the elements of the given field F. All these numbers constitute a subfield

$$N = F(\alpha_1, \ldots, \alpha_n) \supset F$$

of the field of complex numbers. It is called a *splitting field* for the polynomial f over the base field F, because it is a smallest field containing F in which the polynomial f "splits" into linear factors. In fact (forgetting the complex numbers) this property determines the field $F(\alpha_1, \ldots, \alpha_n)$ up to an isomorphism leaving the elements of F all fixed. In the case of the polynomial $x^2 + 1$ over $\mathbf{R}$, the splitting field is just the field $\mathbf{C}$ of complex numbers.

Now consider the symmetries of the splitting field N relative to F. A symmetry is by definition a transformation t: $N \to N$ which is an automorphism of fields (i.e., a bijection which preserves sums and products) and which leaves fixed all the elements of the base field F. All such symmetries constitute a transformation group G, the *Galois group* of the splitting field N–and of the polynomial f–over the base field F. Since each symmetry t leaves all the coefficients of f fixed, it must carry any root α of $f(x) = 0$ into another root of this polynomial. Hence such a symmetry induces a permutation of the roots $\alpha_1, \ldots, \alpha_n$ of f; since N is generated by these roots, the symmetry is determined by the permutation, so that G is isomorphic to a subgroup of the symmetric group S_n. This means in particular that the Galois group G is finite. It can be described either as a group of automorphisms of N of as a group of permutations of the roots.

Linear algebra provides a more specific measure of the size of the Galois group. If we neglect multiplication in the splitting field, then addition, plus multiplication by elements of F, still make N a vector space over the field F–and the dimension of this vector space is equal to the order of the Galois group. For algebraic reasons, this dimension is called the *degree* $[N:F]$ of N over F. (This is a striking simple case of the use of geometric dimensions in algebra.)

For example, the easy equation $x^3 - 5 = 0$ over the rationals has three roots: $\sqrt[3]{5}$, $\omega\sqrt[3]{5}$, and $\omega^2\sqrt[3]{5}$, where the first root is the real one, while ω is a complex cube root of unity. The whole splitting field over $\mathbf{Q}$ then can be built up in two steps from Q–first the real root, then the others–as

$$\mathbf{Q} \subset \mathbf{Q}(\sqrt[3]{5}) \subset \mathbf{Q}(\sqrt[3]{5}, \omega\sqrt[3]{5}, \omega^2\sqrt[3]{5}) = N = \mathbf{Q}(\sqrt[3]{5},\omega).$$

It has degree 6 over $\mathbf{Q}$ and has a subfield $\mathbf{Q}(\sqrt[3]{5})$ of degree 3 over $\mathbf{Q}$, as well as another subfield $\mathbf{Q}(\omega)$ of degree 2.

For any polynomial equation $f(x) = 0$ over F any solution of the equation by "radicals" (roots of equations $x^p - a$, say with p a prime) is found to involve a process of building up the splitting field N by stages, with intermediate fields K. Here enters the fundamental theorem of Galois theory, which establishes a bijection between the intermediate fields K, with $N \supset K \supset F$, and the subgroups S of the Galois group G. Each intermediate field K determines a subgroup, call it $K^{\#}$:

$$K^{\#} = \{\text{all } t \in G \mid \text{for all } b \text{ in } K,\ tb = b\}.$$

Each subgroup S of G determines an intermediate field, call it S^b

$$S^b = \{\text{all } b \in N \mid \text{for all } s \text{ in } S,\ sb = b\}.$$

A subtle argument (*Algebra* or *Survey*) then shows that these two assignments $K \mapsto K^{\#}$ and $S \mapsto S^b$ are mutually inverse functions and so provide the asserted bijection from intermediate fields to subgroups.

In this bijection, the whole field N is left fixed only by the identity automorphism, so $N^{\#} = 1$ is the one element subgroup of the Galois group G. We may then picture the correspondence briefly as follows

$$\begin{array}{ccc} 1^b = N & \Leftrightarrow & 1 = N^{\#} \\ \cup & & \cap \\ S^b = K & \Leftrightarrow & S = K^{\#} = \mathrm{Gal}(N{:}K) \\ \cup & & \cap \\ G^b = F & \Leftrightarrow & G = F^{\#} = \mathrm{Gal}(N{:}F). \end{array}$$

Note that a larger subgroup S will leave fewer elements b fixed, so corresponds to a smaller subfield, hence the subgroups appear "upside down" in the diagram. Also the subgroup $K^{\#} = S$ consists of all the automorphisms of N which leave K pointwise fixed; it is just the Galois group of N over K, as indicated in the figure. Also one can prove that $S = K^{\#}$ is a normal subgroup of G if and only if K is itself a splitting field over F; in this case the factor group G/S is exactly the Galois group of K over F. Factor groups here enter essentially.

The bijection between intermediate fields and subgroups has remarkable consequences. First the Galois group G, as a permutation group, is finite. Hence it has only a finite number of subgroups. Consequence: There are only a finite number of intermediate fields! (For $x^3 - 5$ there are four such.)

Next, building up the splitting field N in stages amounts to the consideration of the corresponding chains of subgroups

$$G = G_0 \supset G_1 \supset G_2 \supset \cdots \supset G_{m-1} \supset G_m = 1 \tag{3}$$

of a group G. If each successive group G_k in this chain is a normal subgroup of its predecessor G_{k-1} and if each quotient group G_{k-1}/G_k is simple (§9), for $k = 1, \ldots, m$, this chain is called a *composition series* for G: it exhibits G in pieces G_{k-1}/G_k. Clearly any finite group G has such a composition series: Just choose G_1 to be a maximal proper normal subgroup of G, and continue down. There can be many different such composition series, but the Jordan–Holder theorem (with a perspicuous proof, *Algebra*, p. 430) asserts that any two such series for a finite group G have the same length of m and, up to order and isomorphism, the same set of factor groups

$$G/G_1,\ G_1/G_2,\ \ldots,\ G_{m-1}/G_m$$

called the *composition factors* of G. A group is said to be *solvable* when these factors are all cyclic groups.

This name comes from the striking theorem which states that a polynomial f over F has roots given by formulas involving pth radicals (for various prime numbers p) if and only if the Galois group of f over F is solvable. The essential reason is this: When the field F already contains all the pth roots of unity, the Galois group over F of an irreducible equation $x^p - a$ is a cyclic group of order p. Since pth roots of unity also lead to cyclic Galois groups, this means that fields built up by radicals must have solvable Galois groups.

The most general equation of 5th degree will have as Galois group the whole symmetric group S_5. A composition series for this group reads

$$S_5 \supset A_5 \supset 1,$$

where A_5 is the alternating group of all even permutations; the quotient group S_5/A_5 is cyclic of order 2, and the alternating subgroup A_5 can be shown, by direct calculation, to be a simple group. This is the group-theoretic reason for the insolvability of the quintic equation by radicals.

In the hands of Evariste Galois, this Galois theory was the occasion for the explicit discovery of the uses of groups, normal subgroups, and factor groups. Note in particular that the use of factor groups is essential–just the axiomatic description of symmetries by a "group" is not enough!

This short sketch has omitted many interesting aspects of the theory. The splitting fields can be characterized directly, independent of any one polynomial which is split. The idea also works for splitting fields of infinite degree, provided however that the Galois group is then considered as a topological group, while its *closed* subgroups corresponding to intermediate fields. It applies to finite fields, and provide an easy proof that for each prime p and each natural numbers n there is up to isomorphism exactly one finite field with p^n elements–and this field is the splitting field of the polynomial $x^{p^n} - x$ over the field $\mathbf{Z}/(p)$ of the integers modulo the prime p. However, the Galois theory does not apply without restrictions to infinite extensions of such fields (called fields of *characteristic* p, because the p-fold sum $1 + 1 + \cdots + 1$ is zero in these fields).

The Galois theory is a decisive demonstration of the utility of group-theoretic and abstract methods: Practical formulas for solving equations are best understood by groups of symmetries, their subgroups and factor groups. The Galois theory is an example of a piece of Mathematics which is deep but not difficult. Originally, however, it was obscure; it was not understood by the contemporaries of Galois, and some expositions published as late as 1925 were wholly confusing. It became really perspicuous only when the advent of modern abstract methods in algebra, in the hands of Richard Dedekind, Emmy Noether, B. L. van der Warden, and Emil Artin, yielded the present elegant and conceptual proofs of the basic theorem of Galois theory (see for example, Artin [1949] or the treatments in *Survey* [1977] or in *Algebra*, second edition; [1979]). A text by Kaplansky [1972] provides a wealth of examples.

8. Constructions of Groups

The remarkable richness of group theory rests in part on the extensive variety of specific groups, especially finite groups. Initially, Mathematicians wanted to know all groups, and so were tempted to try to list all the possible finite groups of a given order, of course counting isomorphic groups as the same. Thus, there are two different groups of order 4, the cyclic group Z_4 and the group of 4 symmetries of a rectangle (the so-called *four group*). In a group of order 5, any non-identity element must have order 5, so any such group is (isomorphic to) the cyclic group $\mathbf{Z}_5$. There are two different (i.e., non-isomorphic) groups of order 6, the cyclic group $\mathbf{Z}_6$ and the symmetric group S_3, and only one group $\mathbf{Z}_7$ of order 7 because 7 is prime. The continuation of such a simplistic catalog, however, soon becomes cumbersome and is not very enlightening. Too much depends on the arithmetic properties of the order of the group.

Instead, long experience with examples has developed more useful constructions.

If a prime p divides the order n of a group G, as $n = p^e n'$ with n' relatively prime to p, then G has at least one subgroup P of order p^e, and any two such subgroups are conjugate. These subgroups P are called the *Sylow* subgroups of G; one shows that any subgroup S of G with order some power of the prime p is necessarily contained in a Sylow subgroup P (i.e., one of order p^e). The investigation of these subgroups (*Algebra*, chap. XII) is a first step in the examination of the arithmetic influence of the order of a group on its structure.

New groups may be constructed, in various ways, from given groups. If A and G are given groups, their (direct) *product* $A \times G$ consists of the ordered pairs (a,g) of elements $a \in A$ and $g \in G$ with termwise multiplication, as in

$$(a,g)(b,h) = (ab,gh). \tag{1}$$

The functions $(a,g) \mapsto a$ and $(a,g) \mapsto g$ provide two homomorphisms $\pi_1: A \times G \to A$ and $\pi_2: A \times G \to G$, called *projections*. Given any homomorphisms $t_1: H \to A$ and $t_2: H \to G$ from a group H to the factors A and G, there is a unique group homomorphism $s: H \to G$ which produces t_1 and t_2, by composition with the projections, as $t_1 = \pi_1 \cdot s$ and $t_2 = \pi_2 \cdot s$. This is expressed by the (commutative) diagram of group homomorphisms

$$\begin{array}{ccccc} & & H & & \\ & \swarrow^{t_1} & \downarrow s & \searrow^{t_2} & \\ A & \xleftarrow{\pi_1} & A \times G & \xrightarrow{\pi_2} & G \end{array} \tag{2}$$

This property of the projections characterizes the product $A \times G$, up to isomorphism; one says that π_1 and π_2 are a "universal" pair of morphisms to the pair (A,G) of groups.

The direct product $A \times G$ of abelian groups is clearly abelian. For abelian groups, the construction of direct products gives a satisfactory description of finite abelian groups A. Every such group is (isomorphic to) a direct product of a number of cyclic groups of prime power orders:

$$A \cong Z_{p_1^{e_1}} \times Z_{p_2^{e_2}} \times \cdots \times Z_{p_k^{e_k}}. \tag{3}$$

There can be various such representations of A, but the prime numbers $p_1, \ldots, p_k$ and their exponents $e_1, \ldots, e_k$ are the same in all; they are "invariants" of the group A.

No such simple direct product decomposition applies to infinite abelian groups. To be sure, the multiplicative group of positive rational numbers is a sort of product of infinitely many infinite cyclic groups (one generated by each prime number) but there is no such decomposition of the quotient group $\mathbf{Q}/\mathbf{Z}$ of the additive group of rational numbers. Some questions about the structure of "large" abelian groups remain quite unsolved!

The direct product $A \times G$ contains isomorphic copies $A \times 1$ and $1 \times G$ of the given factors A and G; each is a normal subgroup of $A \times G$ and each element $(a,1)$ of the first factor commutes with every element $(1,g)$ of the second. This is not the case with the *free product*

$A*G$. Its elements are all possible *words* $a_1g_1a_2g_1 \cdots a_ng_n$ which are formal products of elements $a_i \in A$ and $g_i \in G$, with each a_i (except possibly for a_1) and each g_j (except possibly for g_n) $\neq 1$. Two such words are multiplied by juxtaposition and subsequent cancellation (when possible); thus ag multiplied by $g^{-1}bg^1$ becomes $agg^{-1}bg^1 = (ab)g$, where ab is the given product in factor A. With some care, one may prove that this multiplication is indeed associative, and that $a \mapsto a \cdot 1$ and $g \mapsto 1 \cdot g$ are monomorphisms $k_1: A \to A*G$ and $k_2: G \to A*G$ which enjoy the following "universal" property: To any morphisms $t_1: A \to H$ and $t_2: G \to H$ into an arbitrary group H there is exactly one morphism $s: A*G \to H$ which yields t_1 and t_2 as the composites $t_1 = s \cdot k_1$ and $t_2 = s \cdot k_2$; that is, which makes the following diagram

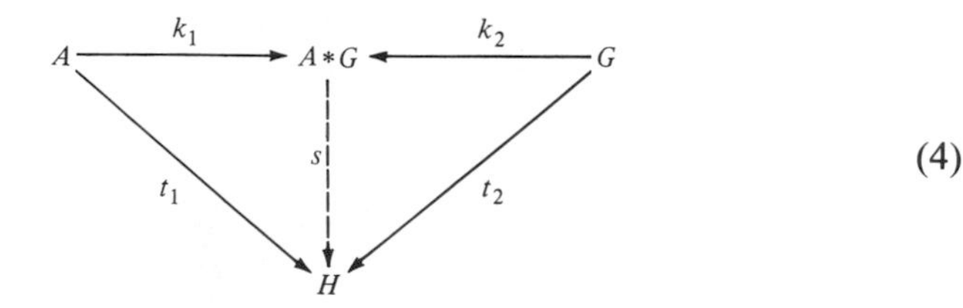

(4)

commutative. This diagram may be obtained from the direct product diagram (2) by simply reversing all the arrows; one says that the diagrams are *dual* to each other. This diagram (4) means that $A*G$ is the "most general" group generated by A and G.

In the complex plane the modular group provides a striking example. This group consists of all the fractional linear transformations

$$z \mapsto (az + b)/(cz + d) \tag{5}$$

of the plane (z a complex number) with integral coefficients and such that the determinant $ad - bc = 1$. This group (which plays a considerable role in number theoretic investigations) can also be described as the multiplicative group of all 2×2 matrices of integers

$$\begin{bmatrix} a & b \\ c & d \end{bmatrix}$$

which are unimodular (i.e., have $ad - bc = 1$), modulo the subgroup consisting of the two matrices $\pm I$, where I is the identity matrix. (The matrix $-I$ yield the identity transformation (5).) The group contains a subgroup of order 2, generated by the transformation $z \mapsto -1/z$ and a subgroup of order 3, generated by $z \mapsto (z - 1)/z$. It is exactly the free product of these two cyclic groups!

The free product $F = \mathbf{Z} * \mathbf{Z}$ of two infinite cyclic groups is called the *free group* on two generators (namely, the generators, call them a and b, of the two cyclic factors $\mathbf{Z}$). Given any two elements g and h of any group

H, the universal property (4) implies that three is a unique homomorphism $t: F \to H$ with $t(a) = g$ and $t(b) = h$. Therefore any group H which can be generated by a finite number of elements can be represented (usually in many ways) as a factor group of a free group $\mathbf{Z} * \cdots * \mathbf{Z}$ on a finite number of generators. A surprising theorem asserts that *Any subgroup of a free group is free*. There is a purely algebraic proof, but the reasons lie in geometric facts, depending on representation of a group as a group of paths in a space.

One may systematically construct groups from subgroups and factor groups. Thus given groups A and G, one may seek to construct all the groups E with A as normal subgroup and $G \cong E/A$ as the corresponding factor group. Then E is called an *extension* of A by G. If k is the inclusion of A in E and π the projection on G, the extension can be pictured as a sequence of group homomorphism

$$1 \to A \xrightarrow{k} E \xrightarrow{\pi} G \to 1, \tag{6}$$

where 1 designates the group with just one element. This sequence is said to be *exact* because at each node the image of the incoming homomorphism is "exactly" the kernel of the outgoing homomorphism. Thus exactness at A means that k is a monomorphism, and exactness at E means that the (normal) subgroup kA is the kernel of the projection π. The direct product $A \times G$ yields one such extension E, but there are many others. For instance, if E is the symmetric group S_n and A the alternating subgroup A_n, the group G in (6) is cyclic of order 2, but S_n is not the direct product $A_n \times \mathbf{Z}_2$.

In any such sequence, the requirement that A (or $k(A)$) be normal in G means that each element $e \in E$ induces an inner automorphism $a \mapsto eae^{-1}$ on A, call it ϕe. Hence ϕ is a homomorphism $E \to \operatorname{Aut} A$. If A is abelian (and in some other cases), A will be in the kernel of this map, so that ϕ "induces" a homomorphism $\theta: G \to \operatorname{Aut} A$. Conversely, given A, G, and such a homomorphism $\theta: G \to \operatorname{Aut} A$, sending each $g \in G$ to $a \mapsto {}^g a$, one may construct a group E on the set $A \times G$ by the multiplication

$$(a,g)(b,h) = (a\,{}^g b, gh). \tag{7}$$

Then $a \to (a, 1)$ and $(a,g) \to g$ does yield an exact sequence (6). This group E is called the *semidirect product* of A and G, with "operators" θ.

The dihedral group Δ_n is such a semidirect product. For each natural number n, the group consists of all the $2n$ symmetries of a regular n-gon Δ_n (thus, for $n = 3$, of an equilateral triangle). Call a transformation t of Δ_n even $(+1)$ if it leaves the n-gon right side up and odd (-1) if it turns the n-gon over; these labels constitute an epimorphism $\pi: \Delta_n \to \mathbf{Z}_2$ to the cyclic group (± 1) of order 2; its kernel is the normal subgroup consisting of all n possible rotations of the n-gon. If R is such a rotation though $2\pi/n$

and D any odd transformation (a reflection), the equations $R^n = 1$, $D^2 = 1$, and $DRD^{-1} = R^{n-1}$ show that Δ_n is a semidirect product. The group of orientation preserving rigid motions of the Euclidean plane is similarly a semidirect product, with A the normal subgroup of all translations and G the group of rotations about a point.

In the semidirect product (7), the function $g \mapsto (1,g)$ is a right inverse σ of the projection π; one says that the exact sequence (6) is *split* by σ. Many other extensions are not split. In particular, when A is abelian (and written additively) all extensions of A by G with operators θ can be described by giving a multiplication on the set $A \times G$ by the formula

$$(a,g)(b,h) = (a + {}^{g}b + f(g,h), gh)$$

where the function $f\colon G \times G \to A$ satisfies an identity sufficient to insure that this multiplication is associative (*Homology*, p. 111). This identity, in its turn, has repercussions in the study of division rings (where the multiplication must be associative) and in algebraic topology (where a great deal of information is codified in exact sequences).

Other aspects of groups enter vitally in number theory, in harmonic analysis, in quantum theory (the eight-fold way) and elsewhere. Mackey [1978] is a good reference.

9. Simple Groups

Any group G has itself and 1 as normal subgroups; it is *simple* when it has no other normal subgroups. The Jordan–Holder theorem (§7) for composition series indicates that any finite group may be constructed as an iterated extension of simple groups, though the order in which the simple factors may occur in the series can vary. For this and many other reasons, it would be useful to know *all* the finite simple groups. They include the cyclic groups of prime order, the alternating groups A_n for $n \geqslant 5$, and others. From an examination of other examples, W. Burnside had, about 1900, conjectured that all finite simple groups except the cyclic ones had even order. This conjecture was proved true, with a complicated argument, by Feit and Thompson in 1962. Their methods were novel and powerful and were then further extended, till there is now a complete listing of all possible finite simple groups. This list begins with certain familiar infinite families: The cyclic groups of prime order, the alternating groups A_n for $n \geqslant 5$ and 16 families suggested by geometric constructions; for example, one may modify the definition of the special orthogonal group by replacing the field of real numbers by a finite field, so as to get a finite group. There are then 26 simple groups which do not fit in any such systematic infinite family. They are called the *sporadic* groups. For example the largest one, called the "monster", has order the following product of prime powers:

$$2^{46}\cdot 3^{20}\cdot 5^{9}\cdot 7^{6}\cdot 11^{2}\cdot 13^{3}\cdot 17\cdot 19\cdot 23\cdot 29\cdot 31\cdot 41\cdot 47\cdot 59\cdot 71\,.$$

This number is approximately $8\cdot 10^{53}$. This large group has been constructed as a group of the transformations preserving certain geometric structure in a space of 196,883 dimensions! This dimension, in its turn, is closely connected with coefficients in the expansion of certain modular functions.

These remarkable results serve to illustrate the richness of group theory.

10. Summary: Ideas of Image and Composition

The practice of mapping one thing (a point, a number, or some other object) into some image thing arises in many connections: In the successor operation for the natural numbers, in addition and multiplication in arithmetic, in algebraic manipulation, in trigonometric formulas, in infinite series, and in all sorts of geometric transformations. These experiences lead to the notion of forming the image of a thing and the images of all the things in some collection. In its turn, this notion has various intuitively attractive but inexact formulations, all dealing with the way in which one thing or quantity may "depend" on another. This in turn finally leads to the precise but meticulous concept of a function as a suitable set of ordered pairs of things; from this formal definition the abstract properties of functions flow easily.

Repetition or iteration of such functions leads to the notion of composition–one map followed by another. This notion of composition turns out to be extraordinarily fruitful. It has a variety of formal realizations, as the composition in permutation groups, in transformation groups, in abstract groups, in monoids, in categories, and in semigroups. In each of these cases the formal properties include the associative law, valid for any composition of functions.

In this chapter we have followed this development from functions and transformations through transformation groups (geometric and otherwise) to abstract groups and some of their remarkably rich properties. One is led to ask why group theory, starting with such a simple system of axioms, turns out to be so deep. The explanation may depend on several features of the axiomatics for groups, such as the following:

(a) Multiple origins (geometric transformations, algebraic operations)
(b) Decisive uses (Galois theory; geometry; number theory)
(c) Additional applications (Quantum Theory, crystal structure)
(d) Exact representation in terms of the origins: Every abstract group is isomorphic to (at least one) transformation group
(e) Multiple models: Many, but not too many, finite groups.

Listing these features does not provide a real explanation of the depth of group theory, unless we could show that other axiom systems with these five features also had deep consequences. This we can do only tentatively in a few cases. Thus monoids do have properties (a) and (d), but do not have a wealth of additional applications–and have "too many" finite models with "too little" structure. In set theory, properties of intersection, union, and complement can be abstracted in the axioms for a Boolean algebra. Here (d) holds, in view of the Stone representation for Boolean algebras, (and (a) and (c) hold as well). However, the finite models of Boolean algebra are dull: For each natural number n, there is just one model with 2^n elements–namely, the Boolean algebra of all subsets of an n element set. There are in Mathematics many other axiomatic notions–but few of them enjoy all five of the properties (a)–(e) above.

The material of this chapter may be summarized in the following network, which indicates, only in part, how these ideas are all interrelated.

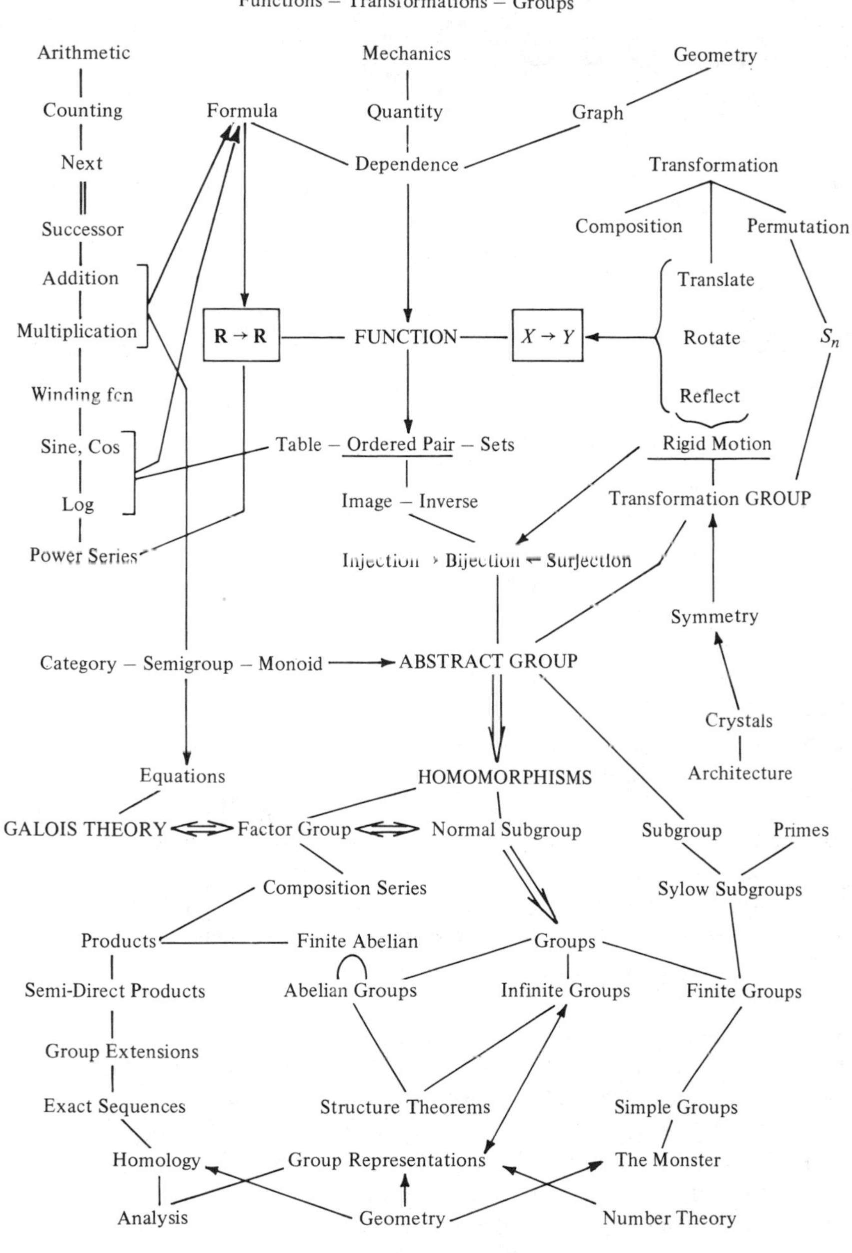
Functions – Transformations – Groups
Arithmetic
Counting
Next
Successor
Addition
Multiplication
Winding fcn
Sine, Cos
Log
Power Series
Formula
R → R
Mechanics
Quantity
Dependence
FUNCTION
Table – Ordered Pair – Sets
Image – Inverse
Injection → Bijection ← Surjection
Geometry
Graph
Transformation
Composition
Permutation
Translate
Rotate
Reflect
X → Y
S_n
Rigid Motion
Transformation GROUP
Symmetry
Crystals
Architecture
Category – Semigroup – Monoid
ABSTRACT GROUP
HOMOMORPHISMS
Equations
GALOIS THEORY
Factor Group
Normal Subgroup
Subgroup
Primes
Sylow Subgroups
Composition Series
Products
Finite Abelian
Groups
Semi-Direct Products
Abelian Groups
Infinite Groups
Finite Groups
Group Extensions
Exact Sequences
Structure Theorems
Simple Groups
Homology
Group Representations
The Monster
Analysis
Geometry
Number Theory

CHAPTER VI

Concepts of Calculus

1. Origins

Many sorts of calculations press themselves upon us. Thus, given a piece of surface, how does one calculate its area? Or, given a section of a curve, how does one calculate its length or the direction of its tangent line at some point? More generally, how does one calculate the rate at which this or that variable quantity is changing with time? The striking discovery (by Newton and Leibniz) that there were systematic methods to calculate *all* these things, and many more like them, had a major influence on the directions and structure of Mathematics. For a considerable period, more practical calculations of such things tended to dominate conceptual understanding, in a way that emphasizes the observation that Mathematics takes its origin in human activities.

The calculation of areas began in Euclidean geometry, with formulas for the areas of triangles and squares. Next in line was the area of a circular disc. This could be determined by inscribing and circumscribing regular polygons in the disc; as the number of sides of these polygons increased, the area inside the circle was "pinched" between the (calculable) areas inside the larger and the smaller polygon–and this pinching process produced the area inside the disc as a sort of limit. This method, neatly adapted to the case of the circle, suggested to Archimedes and others a search for extensions of the method to calculate other areas–that of an elliptical disk, and that for more irregular figures. Similarly the measurements of length began with the Pythagorean theorem used to determine the length of a slanted line and hence the perimeter of any polygon. The circumference of a circle could then be found (§III.2) through successive approximation by inscribed polygons. These and other problems of measurement (of volumes, weights, centers of gravity, and the like) emphasized the ubiquitous role of approximation and may have indicated the need for a systematic understanding of the way in which such successive approximations can converge to a desired limit–an understanding then provided in the calculus by the general definition of the integral as the limit of a sum.

Also, how does one construct tangents? Drawing the tangent line to a circle at some point P is hardly problematical, since the line through P and perpendicular to the radius there is evidently the correct tangent–just touching the circle evenly on both sides of P. For an ellipse, such as a simple construction will succeed only for tangents perpendicular to the major or minor axis at one end of that axis. At other points, here and on the hyperbola, the parabola or on other curves in the plane, more sophisticated methods, some known to the Greeks, are needed to draw an exact tangent; with time, it became plausible that the tangent line to a curve at a point P might be well approximated by the line (the secant) determined by two points P' and P'' on the curve taken close (and then closer) to P. The need to locate such tangent lines is suggested not just by the geometrical aspects of plane or spatial curves, but also by mechanical situations where moving things fly off "on a tangent"–from a bobsled or a roller coaster, for example. Thus the determination of tangent lines again can suggest their calculation by the limit of some systematic sequence of approximations.

Observations of motion raise analogous questions of approximation. Terrestrial and heavenly bodies move about, but with velocities which often do not stay constant. There still should be some way of calculating a velocity for such motion–an instantaneous velocity which might, on reflection, be measured by using as approximations suitable average velocities over shorter and shorter time intervals. Similar thoughts might then arise for other measurable quantities which vary with time, but not at a uniform rate. In this way attention might be directed to the calculation in general of such instantaneous rates of change–again by some sort of scheme of successively better approximations.

Such problems can also occur for "rates" which are not time rates but rates relative to some other changing quantity. An elementary example might be the rate at which the area of a circle changes as the radius alters. Decisive examples arise in Mathematical Economics. There, in analysing the costs of producing more, it is wiser not to consider just the average cost for the production of one widget out of a total of n such, but to use the marginal cost–meaning roughly the additional cost of producing the last one of those n widgets or, more conceptually, the rate at which the cost of production changes relative to the number n of widgets produced. This example of the use of derivatives may have developed historically much later than the original discovery of the calculus; however, it is intrinsically one of the appropriate origins for the idea of a systematic calculation of relative rates of change for variable quantities.

This summary description provides a conceptual (though not a specifically historical) account of the multiple origins of the calculus.

2. Integration

It is remarkable that so many different processes of approximating total measured quantities by adding together little bits of these quantities can all be subsumed under *one* process, that of integration. Area, volume, length, pressure, moment of inertia, weight, and the like all can be managed by such sums. To be sure, the usual formal definition of the Riemann Integral is usually presented as the calculation of an area—specifically the area below a curve $y = f(x)$, above the x axis, and bounded left and right by the ordinates $x = a$ and $x = b$. This area is broken up into the usual thin vertical and rectangular strips of width dx running from $x = a$ to $x = b$, as in Figure 1. The area of such a strip is altitude $f(x)$ times base d, hence is written $f(x)dx$ and the sum of them all—and hence the total area desired—is the definite integral

$$\int_a^b f(x)dx. \tag{1}$$

On one reading, the width dx of each strip is an infinitesimal increment in x (hence the notation dx), there is an infinite number of strips, and so the integral sign (an elongated *S*) represents an infinite sum of infinitesimal quantities, giving the desired area. Another reading, devoid of any such uncertain appeal to the infinitely small, would approximate the desired area by a finite sum of the areas of rectangles of finite width. To this end, subdivide the interval from $x = a$ to $x = b$ in n intervals with successive endpoints $a = x_0 < x_1 < x_2 < \cdots < x_n = b$; call such a subdivision σ. In the ith interval from x_{i-1} to x_i the function f (if continuous) will take on a maximum $\text{Max}_i f$ and a minimum $\text{Min}_i f$ (Figure 2), and the presumptive true area A_i under this portion of the curve will lie between the areas of the two rectangles which have width $x_i - x_{i-1}$ and height $\text{Min}_i f$ and $\text{Max}_i f$, respectively, as expressed by the inequalities

$$(\text{Min}_i f)(x_i - x_{i-1}) < A_i < (\text{Max}_i f)(x_i - x_{i-1}).$$

Adding up these rectangles gives for our subdivision σ a "lower sum" $L(f,\sigma)$ and an "upper sum" $U(f,\sigma)$

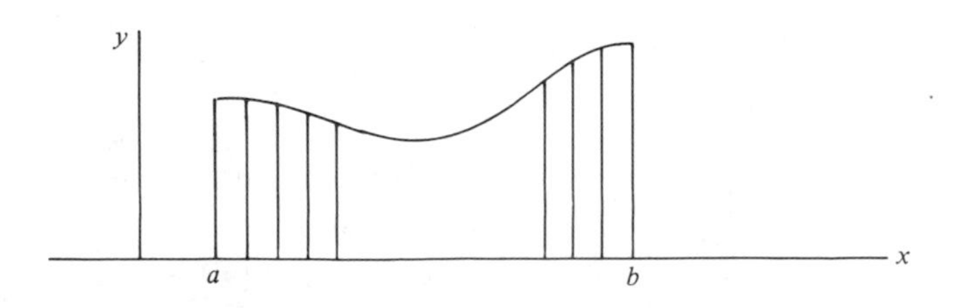

Figure 1. The Riemann integral.

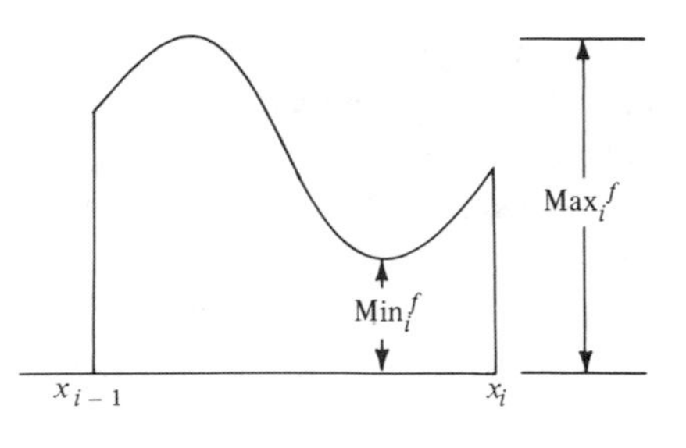

Figure 2

$$L(f,\sigma) = \sum_{i=1}^{n} (\mathrm{Min}_i f)(x_i - x_{i-1}) < U(f,\sigma) = \sum_{i=1}^{n} (\mathrm{Max}_i f)(x_i - x_{i-1}); \quad (2)$$

the desired area under the whole curve must be squeezed between these two sums. To actually express this area, we must then take a limit of these sums for successively finer subdivisions σ, so let the size of σ be measured by $|\sigma|$, the maximum of all the interval lengths $|x_i - x_{i-1}|$. When f is continuous (on the whole interval $a \leqslant x \leqslant b$) it is then the case (as we will soon see) that for each measure $\epsilon > 0$ of approximation there is a $\delta > 0$ such that, whenever $|\sigma| < \delta$, then $|U(f,\sigma) - L(f,\sigma)| < \epsilon$. This implies that L and U have a common limit as $|\sigma|$ approaches 0,

$$\lim_{|\sigma| \to 0} L(f,\sigma) = \lim_{|\sigma| \to 0} U(f,\sigma) = \int_a^b f(x)dx\,; \quad (3)$$

this limit is the definite integral (3) of the function f over the interval $a \leqslant x \leqslant b$. By its construction, it represents what the area under the curve over this interval should be; if you wish, it is the definition of this area; for those applications such as pressures and volumes it can also serve as a definition. Such definite integrals, so defined as limits of sums, are still written in the classical notation (3) suggesting the infinite sum of infinitesimals; in fact that intuitive view of the matter allows us to easily set up the integral representing all sorts of other quantities: The weight of a thin slab of known but variable density, the water pressure on a slab-like portion of a dam, or the volume enclosed by a surface of revolution. Much of the instruction in elementary integral calculus consists of repetitive exercises practising the formulations of such integrals. When they cannot be done by slices or slabs, they can be managed (with more technique but in a wholly similar spirit) by multiple integrals.

This formal definition of the *Riemann integral* of a continuous function replaces the intuitive idea of successive approximations by the use of limits and of the standard logical *quantifiers* (for all $\epsilon > 0$ there exist a $\delta > 0$ such that . . .) needed for the exact description of these limits. Various general properties of the integral also follow directly from this definition. Linearity is an example: the definite integral from a to b of the sum of two continuous functions is the sum of the integrals of these functions separately. Again, if $a < b < c$, the integral of a given continuous function from a to c is the sum of the integrals from a to b and from b to c. This property is really using the idea of "composing" a path (of integration) from a to b with a path from b to c—an idea already present in group theory, and one which will recur in more elaborate cases of line integrals.

So much for the initial formulation of measurement by integration. However since we do not really carry out infinite sums or limits of an infinite number of successive finite sums, the real "calculus" of such integrals rests on the connection with differentiation, to which we now turn.

3. Derivatives

The derivative of a variable quantity y with respect to another such quantity x on which it depends is to be the instantaneous rate of change of y relative to the change in x. This description has intuitive appeal, especially in the special case when x is taken to be time. Instantaneous rates are modeled on average rates: For a value of y_1 at time x_1, changed to a value y_2 at a different time x_2, the average rate of change is the ratio $(y_2 - y_1)/(x_2 - x_1)$ of change in y to change in x. The "instantaneous" aspect might again be formulated by an infinitesimal change dx from x to $x + dx$. Then if y depends on x by a function $y = f(x)$, the instantaneous rate of change is the ratio $[f(x + dx) - f(x)]/dx$ or, in an evident extension of the dx notation, dy/dx, the quotient of two infinitesimals. Such infinitesimals serve for quick calculation. For example, if $f(x) = x^2$, then the instantaneous rate is

$$\frac{dy}{dx} = \frac{(x + dx)^2 - x^2}{dx} = \frac{2xdx + dx^2}{dx} = \frac{2xdx}{dx} = 2x, \tag{1}$$

since the square $(dx)^2$ of an infinitesimal surely vanishes, at least in comparison with its first power. Other rates easily follow. Moreover, if y depends on x and x in turn on time t, cancellation of the infinitesimal dx gives

$$\frac{dy}{dx} \cdot \frac{dx}{dt} = \frac{dy}{dt}, \tag{2}$$

and one has (but hasn't quite proved) the important (because necessarily most useful) "chain rule" for the derivative of a composite function. Thus calculus with infinitesimals is intuitive and an efficient means of calculation.

But what *are* these infinitesimals? The archimedean law for the real numbers (§IV.4) says that a positive number, no matter how small, has arbitrarily large multiples–in effect, that a positive number cannot be infinitesimal. With Bishop Berkeley, one may conclude that "infinitesimals are the ghosts of departed quantities."

What remains are limits. For each value $x = a$ and each actual finite increment $x = a + h$ with $h \neq 0$ one may form from a given function $y = f(x)$ the very same ratio $[f(a + h) - f(a)]/h$ as before. The derivative, when it exists, is then defined to be the limit of this ratio as $h \neq 0$ approaches 0; with either standard notation dy/dx or $f'(x)$, this means that the derivative is

$$\left[\frac{dy}{dx}\right]_{x=a} = f'(a) = \lim_{h \to 0} \frac{f(x + h) - f(a)}{h}. \tag{3}$$

The necessary meticulous $\epsilon - \delta$ definition of the limit in this description must again use the quantifiers (for all ϵ there exists a δ) to achieve precision. With this precision (as careful students of the calculus know) one can then prove that the derivatives of x^2 and of x^n are what they should be and that the desired rule (2) for the derivative of a composite function does hold for suitable differentiable functions–not by simple cancellation of infinitesimals, but by limits taken after cancellation.

This discussion of alternative views of the calculus may serve to support our thesis that Mathematics is not just formalism and is not just empirically convenient ideas, but consists in formalizable intuitive or empirical ideas. Calculus fits this thesis as to the nature of Mathematics because it starts from problems (such as the calculation of areas or of rates of change) and from these problems develops suggestive ideas which ultimately can be fully formalized. For calculus, the initial formalization dealt chiefly with practical rules for finding and manipulating derivatives and integrals. This is not yet a complete formalization, which was first provided in the 19th century by the rigorous $\epsilon - \delta$ treatment by limits.

Another thesis asserts that the same intuitive idea can be variously formalized. For the calculus, we now know that this is the case. Indeed, the earlier and originally wholly speculative use of infinitesimals can be rigorously formalized in at least two different ways: By using Abraham Robinson's non-standard model of the reals, as presented for example in Keisler's text [1976] on calculus, or by using Lawvere's proposals to use "elementary topoi" in which the real line **R** is presented not as a field but as a ring in which there is a suitable infinitesimal neighborhood of 0 (see Kock [1981]).

4. The Fundamental Theorem of the Integral Calculus

The essential connection between differentiation and integration is provided by the theorem of the title of this section. This theorem goes back to a simple intuitive idea: That the total change in a quantity ought to be exactly the sum of the successive small instantaneous changes—an idea which in other forms arises early in geometry, with the rule that the whole is the sum of its parts. Specifically, if the quantity in question is a function $F(t)$ of time t, the total change from time $t = a$ to a later time $t = b$ is just the different $F(b) - F(a)$. Now suppose this function $F(t)$ has at each time t a derivative $F'(t) = f(t)$. The instantaneous change in an infinitesimal interval dt of time is then the product of this interval dt by the instantaneous rate of change $F'(t)$, and so the sum of all the successive instantaneous changes is the definite integral $\int_a^s f(t)dt$. The suggested theorem then reads

Theorem. *If the function $F(t)$ has a continuous derivative $f(t) = F'(t)$ on the interval $a \leqslant t \leqslant b$, then*

$$\int_a^b f(t)dt = F(b) - F(a). \tag{1}$$

The expression on the right is often written as $[F(t)]_a^b$, while the function F is called the *indefinite integral* $\int f(t)dt$.

Before we discuss a rigorous version of the "infinitesimal" motivation of this theorem, let us consider its utility. It provides a formula for calculating the (Riemann) integral $\int_a^b$, provided the function $f(t)$ inside the integral is known to be the derivative $f(t) = F'(t)$ of some other function F. If we have differentiated enough such functions F and have prepared a "Table of Integrals" giving the results, we may then hope to find to the given "inside" function $f(t)$ a suitable "primitive" $F(t)$. When this is the case, the definite integral (representing an area, a volume, or some other quantity to be determined) may be calculated as the difference $F(b) - F(a)$ of two values of this function F. Thus the fundamental theorem conceptually ties the process of differentiation to that of integration, and provides means of (sometimes) calculating definite integrals.

But alas, our table of integrals may not contain any function F with the desired derivative f. For instance, if we have so far differentiated only polynomials, powers, square roots and the like, we will have no function of t with derivative the square root $(1 - t^2)^{1/2}$. Progress on the corresponding definite integral $\int_a^b (1 - t^2)^{1/2}dt$, with well chosen limits a and b (i.e., with $0 \leqslant a \leqslant b \leqslant 1$) is then possible only when we use trigonometric functions; set $t = \sin\theta$ for some angle θ, determine that the derivative of $\sin\theta$ is $\cos\theta$, and then use the standard rules for "change of variables"— $dt/d\theta = \cos\theta$ or $dt = \cos\theta d\theta$ —to get

$$\int_a^b (1 - t^2)^{1/2}dt = \int_\alpha^\beta (1 - \sin^2\theta)^{1/2}\cos\theta d\theta = \int_\alpha^\beta \cos^2\theta d\theta,$$

where the new limits α and β for the integral over θ are chosen so that $\sin\alpha = a$, $\sin\beta = b$. Now, if we have noticed that the derivative of $\sin\theta\cos\theta$ is $2\cos^2\theta - 1$, we have a primitive, so we may again use the fundamental theorem to calculate

$$\int_\alpha^\beta \cos^2\theta d\theta = (1/2)[\theta + \sin\theta\cos\theta]_\alpha^\beta.$$

In particular,

$$\int_0^1 (1 - t^2)^{1/2} dt = \int_0^{\pi/2} \cos^2\theta d\theta = \pi/4. \tag{2}$$

This is a small indication of the way in which integration forces the consideration of derivatives of new classes of functions–here the trigonometric functions; later elliptic functions. A still simpler example is the integral $\int dx/x$, which leads to the function $\log_e x$. In the present case, the entry of trigonometric functions is no accident–after all, the integral on the left of (2) is just a representation of the area $\pi/4$ of the quadrant of a unit circle, and the trigonometric functions serve to provide rectangular coordinates for points on the circle.

Now return to consider a possible proof of the fundamental theorem. We can derive it from two more primitive facts; as follows.

Lemma A. *If a continuous function $F(t)$, defined for all t with $a \leqslant t \leqslant b$, has derivative 0 for every such t, then $F(t)$ is constant for all t with $a \leqslant t \leqslant b$.*

This is intuitively plausible: If the rate of change of F is zero everywhere then it doesn't change at all, hence is constant. Later we will return to examine a rigorous proof of this lemma from the Law of the Mean (§7).

Lemma B. *If the function $f(t)$ is bounded between two constants m and M, so that $m \leqslant f(t) \leqslant M$ for all t with $a \leqslant t \leqslant b$, then the definite integral of $f(t)$ satisfies the inequalities*

$$(b - a)m \leqslant \int_a^b f(t)dt \leqslant (b - a)M. \tag{3}$$

If one considers the definite integral as a measure of the area under the curve $y = f(x)$ and between the ordinates $x = a$ and $x = b$, these two inequalities are evident, since $(b - a)M$ is the area of a rectangle including all that area under the curve, while $(b - a)m$ is the area of a rectangle wholly under the curve. An exact proof of (3) from the limit definition of the integral in §2 is straightforward: In the sum (2.2), each term is by hypothesis bounded as in

$$m(x_i - x_{i-1}) \leqslant (\mathrm{Min}_i f)(x_i - x_{i-1}) \leqslant (\mathrm{Max}_i f)(x_1 - x_{i-1}) \leqslant M(x_i - x_{i-1}).$$

The whole sum is then squeezed between $m(b - a)$ and $M(b - a)$, so the same must be true for the limit of this sum; that is, for the definite integral.

As for the fundamental theorem (1), consider the definite integral

$$G(t) = \int_a^t f(t)dt$$

with a variable upper limit t. To get at its derivative, change t to $t + h$. By Lemma B

$$hm \leqslant G(t + h) - G(t) = \int_t^{t+h} f(t)dt \leqslant hM,$$

where m and M are now lower and upper bounds for $f(t)$ on the short interval from t to $t + h$. Since f is continuous at t, both bounds can be made close to $f(t)$ by taking h small. Therefore $G'(t) = f(t)$. (Geometrically, the rate of change of the area under the curve is the length of the ordinate $f(t)$.) This means that $F(t) - G(t)$ has derivative zero. By Lemma A, it is constant; since $G(a) = 0$, this constant must be $F(a)$. Thus

$$G(b) = \int_a^b f(t)dt = F(b) - F(a),$$

which is the fundamental theorem.

5. Kepler's Laws and Newton's Laws

The problems presented by mechanics have had a decisive influence on the development of calculus. In mechanics, one needed to analyse the ways in which bodies move–both heavenly bodies and terrestrial ones, falling freely under gravity, sliding down inclined planes, or propelled by other forces. Thus Isaac Newton developed the calculus in order to account for Kepler's laws describing the motions of the planets. Kepler, on the basis of extensive earlier empirical observations, asserted that each planet moves around the sun in a plane containing the sun, that the orbit of the planet is an ellipse in that plane, with the sun at one focus, and that the variable speed of the planet in its orbit is such that the radius vector from the planet to the sun would sweep out equal areas in equal times.

In the calculus, one had to hand the conceptual description of velocity and acceleration. For a body moving in a straight line, with position given at any time t by its distance s from some origin on the line, the velocity v, as we have already seen, is just the rate of change ds/dt of s with respect to time; Newton wrote this rate (this derivative) as $\dot{s}$, and called it the fluxion of the fluent s. One also needs to consider acceleration, described as the rate of change of velocity with respect to time; this acceleration is a second derivative $a = d^2s/dt^2$, also written as $\ddot{s}$. For bodies moving in three-dimensional space, referred to coordinates x, y, and z, the velocity is a vector with components $\dot{x}$, $\dot{y}$, and $\dot{z}$, while the acceleration is the vector $\ddot{x}$, $\ddot{y}$, $\ddot{z}$. In this way, the concept of the derivative enters in the description of the kinematics of motion.

Newton's laws of motion describe the effect of external forces on the motion of a body (or better, of a particle). When there are no such external forces, the body continues to move in a straight line at a constant velocity; in other words, its acceleration is zero. When there are external forces, the acceleration is proportioned to the force. More exactly, the acceleration vector is in the same direction as the force; specifically mass times acceleration is equal to force. Thus for one-dimensional motion along a straight line under a constant force there will be a constant acceleration, call it g, and the law reads $a = \ddot{s} = g$. Here we are given the (constant) derivative g; one function with this derivative is gt and by Lemma A of §4 any other function with this derivative can differ from gt by most a constant, call it v_0, so that

$$\dot{s} = gt + v_0 . \tag{1}$$

This constant v_0 represents the initial velocity (that at time $t = 0$). By a similar argument, one finds

$$s = (1/2)gt^2 + v_0 t + s_0 \tag{2}$$

where s_0 is a constant, the coordinate (initial position) of the body at time $t = 0$. Thus Newton's laws and a simple integration easily suffice to derive the familiar formula (2) for a body falling under (constant) acceleration of gravity.

The corresponding derivation of Kepler's laws for the motion of planets rests also on the assumption that the only force acting on the planet is the force of gravitational attraction to the sun. This force is directed from the planet to the sun, and its magnitude is given by the inverse square law and so is proportional to the product of the masses of the sun and the planet and the square reciprocal of the distance r from sun to planet. At any one time the sun, the planet and this velocity vector of the planet relative to the sun determine a plane (Figure 1). The force of gravity is directed toward the sun and in this plane. Hence the acceleration of the planet perpendicular to that plane is zero—and remains zero. Thus Newton's laws easily prove Kepler's first law, that the planet moves in a plane. For coordinates x and y in that plane, with origin at the sun one may now express the acceleration $\ddot{x}$ and $\ddot{y}$ in terms of the corresponding polar coordinates r and θ and the components of force as

$$\ddot{x} = A \cos \theta/r^2, \qquad \ddot{y} = A \sin \theta/r^2, \tag{3}$$

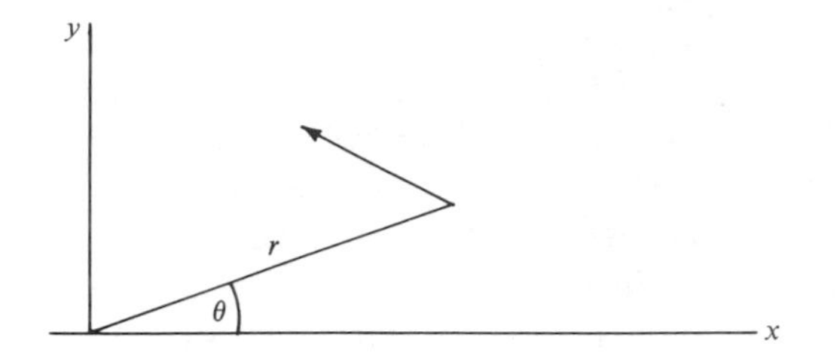

Figure 1. Planetary orbit.

where A is a proportionality constant. A suitable integration of these equations (Chapter IX) proves Kepler's other laws. This represents the initial triumph of the calculus–a clear derivation of the empirically based laws for the motion of the planets from more conceptual laws of motion.

This triumph is all the more remarkable because it shows that the *same* laws of motion apply both to the planets and to motions here on earth. We have already noted the derivation of the equation of motion for a freely falling body. For a projectile much the same argument works. If a body is projected from the origin in the (x,y)-plane, with the vertical axis y positive upwards, and with initial velocities v_x and v_y along the two axes, the accelerations of the body are given by Newton's laws as

$$\ddot{y} = -g, \qquad \ddot{x} = 0, \tag{4}$$

where g is the (constant) acceleration of gravity. Integrating twice gives

$$y = gt^2/2 + v_y t, \qquad x = v_x t.$$

Thus, substituting x for t in the first equation, the path of the body is the locus of the equation

$$2v_x^2 y = gx^2 + 2v_x v_y x;$$

This is a parabola in the (x,y)-plane.

The case of a projectile illustrates in an especially simple way one of the characteristics of formal mathematical calculations. One need not watch the whole path of the projectile; rather the Mathematics predicts this path from physical laws (here Newton's laws) and initial observations–here the initial position and velocity of the body. The calculation then proceeds without further attention to the actual observation of the body, but comes out with a (more or less accurate) description of the successive positions of the body. The same formal character applies to more careful or accurate calculations, such as those taking air resistance into account.

6. Differential Equations

This discussion of the motion of planets and of projectiles is a typical instance of the general method of describing physical and other phenomena by differential equations. A typical differential equation, now of the first order and not "second order", will determine some unknown function x of the variable t by giving the first derivative of x,

$$dx/dt = h(x,t), \tag{1}$$

in terms of a known function h of the two quantities x and t. The underlying way of determining a solution is then the same basic notion: That the total change in a variable quantity x ought to be the sum of the successive instantaneous changes $h(x,t)dt$. The expression of this idea in the fundamental theorem of the integral calculus can be rewritten (see (4.1)) as

$$F(b) = F(a) + \int_a^t f(t)dt.$$

In words, the final values of the variable F (or x) is determined by the initial value $F(a)$ and *all* the intermediate values $f(t) = F'(t)$ of the derivative on the interval $a \leqslant t \leqslant b$. This underlies the use of "initial values" in getting solutions of differential equations. In explicit terms, this amounts to a search for some function $x = g(t)$ defined for a suitable range of t and such that this function with its derivatives satisfies the given differential equation (1):

$$g'(t) = h(g(t),t) \tag{2}$$

for all the intended t. To do this, we must again canvass functions g whose known derivatives might satisfy (2); elementary treatments provide various rules and tricks to do this, for example by changing the variables x or t to other more manageable quantities. As with elementary integration, the known functions may not suffice to get the solution. At this point one might use numerical methods, or might try to invent a new function g to do the job. Such invention might seem a chancy matter, but this need not be so. Given the general definition of a function, it is possible to prove "existence theorems" which specify conditions under which a solution g to equation (2) must exist and (perhaps) conditions which make the solution unique. One such existence theorem is the Picard theorem.

To formulate this theorem, we consider the *initial condition* that $x = x_0$ when $t = t_0$ and assume for the equation (1) that the function h on the right is continuous, say in the square D consisting of all (x,t) such that $|x - x_0| < a$ and $|t - t_0| < a$. A *solution* of (1) with the given initial conditions is then a function $g(t)$, defined for t in some interval

$| t - x_0 | < \delta$ about t_0, and with $| g(t) - x_0 | < a$ there, such that $g(t_0) = x_0$ and that (2) holds for all $| t - t_0 | < \delta$. One version of the Picard theorem then asserts that there is a $\delta > 0$ for which such a solution exists and is unique, provided that the given function h satisfies the *Lipschitz condition*: There is a constant $M > 0$ such that, for all (x,t_1) and (x,t_2) in the square D

$$| h(x,t_2) - h(x,t_1) | < M | t_2 - t_1 | . \tag{3}$$

Incidentally, this Lipschitz condition will certainly hold if the function h has a continuous partial derivative with respect to t everywhere in the square D.

This result will illustrate the care necessary to formulate exactly the intuitive idea that a solution ought to exist and ought to be unique—at least ought to if the differential equation formulates correctly a physical situation known to have a physical solution.

Our illustrations from mechanics had to do with a second order differential equation such as $\ddot{x} = f(x,t)$. In this case the appropriate initial conditions amount to giving *both* x_0 and its first derivative $\dot{x}_0$ at the initial time $t = t_0$. This one may see by reducing the single second order equation to a pair of first order equations; use the velocity $v = \dot{x}$ as another variable; the equations are then

$$\dot{v} = f(x,t), \qquad \dot{x} = v$$

and there is a corresponding existence theorem. For a particle moving in three dimensions the corresponding second order differential equations involve three second derivatives $\ddot{x}$, $\ddot{y}$, and $\ddot{z}$, and reduce to first order equations in the six variables x, y, z and $\dot{x}$, $\dot{y}$, $\dot{z}$. The space of these six variables—called a *phase space*—is a simple example of the need for higher dimensional spaces in the understanding of mechanics (Chapter IX). For k moving particles, one needs a phase space of $6k$ dimensions! Mechanics motivates higher geometry.

7. Foundations of Calculus

The differential and integral calculus, as we have seen, starts with problems of calculating various quantities and the invention of uniform methods of making these calculations. These methods rest initially on vague but persuasive ideas of infinitesimals, infinite sums, and of rates of change; their very success and rapid development forcibly raises the question of formulating a rigorous foundation for these methods. This foundation, developed in the 19th century, must start from a clear notion of a function (the thing which might have a derivative). The foundation then rests on the notion of a limit, the definition of derivative and of integral in terms of limits, and the careful proof of a number of necessary facts about these processes. In particular, this involves the proofs from the axioms for **R** of the following sequence of intuitively plausible properties of any real valued function f, continuous on the unit interval [0,1], which consists of all x, $0 \leqslant x \leqslant 1$:

$$f\colon [0,1] \to \mathbf{R}, \text{ continuous}. \tag{1}$$

(i) The function f is *bounded* above and below; that is, there are real numbers m and M such that, for all x in [0,1],

$$m \leqslant f(x) \leqslant M.$$

(ii) The function f takes on its maximum; that is, there is a point x_0 in [0,1] such that $f(x) \leqslant f(x_0)$ for all $x \in [0,1]$.

The maximum can be determined from property (i) and continuity as the least upper bound of the set of all the real numbers $f(x)$ for $x \in [0,1]$. This is typical of the use of the least upper bound property of the reals.

(iii) The function f takes on its minimum in the interval [0,1].

(iv) The function f takes on all intermediate values. It is enough to prove that if $f(0) < 0$ and $f(1) > 0$, then there is an $x \in [0,1]$ with $f(x) = 0$. That is, f takes on the value of 0.

(v) The function f is uniformly continuous in [0,1]; that is, for each $\epsilon > 0$ there is a $\delta > 0$ such that, for all x_1, x_2 in [0,1],

$$|x_1 - x_2| < \delta \Rightarrow |f(x_1) - f(x_2)| < \epsilon. \tag{2}$$

Compare this with ordinary continuity in the interval. That continuity requires that for all x_1 and all $\epsilon > 0$ there is a $\delta > 0$ (perhaps depending on x_1 and ϵ) such that, for all x_2 in [0,1], the implication (2) holds. Thus for ordinary continuity one must produce for each x_1 and ϵ a δ which works in (2) for that one value of x_1; for uniform continuity one must produce for each ϵ a single δ which works for all points x_1 in the interval. Thus uniform continuity is (2) with prefix the sequence of quantifiers

$$(\forall \epsilon)(\exists \delta)(\forall x_1)(\forall x_2)$$

($\forall \epsilon$ here is short for "for all $\epsilon > 0$"), while ordinary continuity throughout the interval is the same statement (2) preceded by the sequence of quantifiers

$$(\forall \epsilon)(\forall x_1)(\exists \delta)(\forall x_2).$$

In other words, interchanging a universal quantifier $(\forall x_1)$ and an existential quantifier $(\exists \delta)$ makes a difference (examples below)—while interchanging two universals such as $(\forall \epsilon)$ and $(\forall x_1)$ makes no difference in meaning.

We continue with the properties of the continuous function f of (1).

(vi) (Rolle's Theorem). If $f(0) = f(1)$ and f has a first derivative at every point x with $0 < x < 1$, then there is a point ξ, $0 < \xi < 1$, with $f'(\xi) = 0$. (In other words, the graph of f has a horizontal tangent somewhere.)

For ξ one can take a point where f attains its maximum as given by property (ii), and then show from the definition of the derivative that f' must be zero there. This argument fails if this ξ is 0 or 1; one may then choose a point ξ where f attains its minimum.

(vii) (Law of the Mean). If f has a first derivative at every point x with $0 < x < 1$, then there is a point ξ, $0 < \xi < 1$, where $f'(\xi) = f(1) - f(0)$. Put differently, the graph of f has somewhere a tangent parallel to the longest secant.

This law follows from Rolle's theorem applied to the function

$$g(x) = f(x) - x[f(1) - f(0)]$$

which does satisfy the hypothesis $g(0) = g(1)$ of Rolle's theorem. The law of the mean also proves Lemma A of §4.

All of these theorems apply at once to real-valued functions f continuous on any finite closed interval, say the interval $[a,b]$ of all x with $a \leqslant x \leqslant b$. In that case, for example, the law of the mean states that

$$f(b) - f(a) = f'(\xi)\,(b - a)$$

for some point ξ with $a < \xi < b$: The total change in the function is the length of the interval times the rate of change at some intermediate point ξ. This law is another way of expressing one of the basic ideas of calculus: The first derivative gives a linear approximation for the function.

Uniform continuity, as formulated in (v), is powerful. For example, it can be used in connection with our description in §2 of the definite integral $\int_a^b f(x)dx$. Given $\epsilon > 0$, uniform continuity yields a $\delta > 0$ so that $|x_1 - x_2| < \delta$ implies $|f(x_1) - f(x_2)| < \epsilon/(b - a)$. Thus if we subdivide the interval of integration into pieces $x_i - x_{i-1}$ of length less than δ, the maximum and the minimum of f in any subinterval differ by less than $\epsilon/(b - a)$, and this in turn implies that the lower sum $L(f,\sigma)$ and the upper sum $U(f,\sigma)$, defined as in §2, differ by less than $(\epsilon/(b - a))\Sigma x_i - x_{i-1}) = \epsilon$. Hence they do approach the same limit, as we asserted in (2.2) in defining this limit to be the definite integral.

The assertion in (v) of uniform continuity depends essentially upon the hypothesis that f is continuous in the *closed* interval [0,1]—the interval including its endpoints. For example, the function $g(x) = 1/x$ is continuous on the *open* interval (0,1), consisting of all real numbers x with

$0 < x < 1$, but g is not *uniformly* continuous on this open interval–given $\epsilon > 0$, as x gets closer to 0 even smaller δ is required in (2). In fact the uniform continuity property (v) is a consequence of the following property of the topological space $X = [0,1]$.

Lemma (Heine–Borel alias Borel–Lebesgue). *If the interval* $[0,1]$ *is contained in the union of open intervals* U_i *(for i in any index set I), then* $[0,1]$ *is contained in the union of a finite number of these* U_i.

One says that $[0,1]$ is covered by the U_i when it is contained in their union.

PROOF. Use least upper bounds! Let [0,1] be covered by some collection $\{U_i \mid i \in I\}$ of open intervals. Consider the set S consisting of all those real numbers $x \in [0,1]$ such that the closed interval $[0,x]$ from 0 to x can be covered by a finite number of the U_i. This set S is bounded by 0 and 1, hence has a least upper bound, call it ξ. If ξ is less than 1, this ξ belongs to one of the U, say to U_j–and U_j contains numbers $\xi' > \xi$. Since ξ is a least upper bound of S, U_j also must contain some $x \in S$. Then the interval $[0,x]$ is covered by a finite number of the U_i, and this number, together with U_j, cover the interval $[0,\xi']$ stretching beyond ξ, a contradiction to the choice of ξ as the least upper bound of S. Therefore $\xi = 1$, so the interval [0,1] is covered by a finite number of U's.

One should note that this is not a "constructive" argument. It does not produce the finite cover, but argues by contradiction.

Heine–Borel suffices to prove uniform continuity (property (v) above). For let $\epsilon > 0$ be given. Since f is continuous at each point x_0 there is for each x_0 an open interval U_{x_0} (with center x_0) such that $|f(x_1) - f(x_2)| < \epsilon$ whenever x_1 and x_2 are in U_{x_0}. Now (a small trick) consider the intervals $V_{x_0} \subset U_{x_0}$ with the same center x_0 and half the radius. Since each $x_0 \in V_{x_0}$, these smaller intervals cover [0,1]. By Heine–Borel, a finite number of them will cover. Take δ to be the minimum radius of this finite number of intervals. Clearly $\delta > 0$. Now consider any two points x_1 and x_2 in [0,1] at distance apart less than δ. Since the intervals cover, x_1 is in some interval V_{x_0}. By the trick, x_1 *and* x_2 lie together in the larger interval U_{x_0} so $|f(x_1) - f(x_2)| < \epsilon$, as desired.

With this result, we have completed the announced proof of the existence of the definite integral. However, for double integrals we will need Heine–Borel for a square. The proof of uniform continuity, however, involves nothing really new. To understand it, think of the interval (or the square) as a metric space and hence a topological space and call such a space *compact* when it satisfies Heine–Borel. In other words X is compact if every covering of X by open sets U_i contains a finite subcovering. Then the proof we have just given for (v) really proves

Theorem 1. *Any continuous real valued function $f\colon M \to \mathbf{R}$ on a compact metric space M is uniformly continuous.*

There are many compact spaces. For example, any product of compact spaces is compact; in particular the unit square or cube is compact. Any closed subset C of a compact space X is compact; here a subset $C \subset X$ is said to be *closed* when its compliment $X - C$ is open in X, while the open subsets of C (which define its topology) are just the intersections $U \cap C$ with open subsets U of X. A subset of the line or the plane is compact if and only if it is both closed and bounded. On the other hand, the open interval of all real x with $0 < x < 1$ is not compact; when it is covered by expanding proper open subintervals, no finite number of them will suffice to cover. Similarly the whole real line $\mathbf{R}$ is not compact, and it carries functions such as x^2 which are continuous but not uniformly continuous.

Compactness is related to convergence. One readily proves that in a compact space X every infinite sequence of points x_n has an infinite subsequence converging to some point of X; here the convergence of a sequence of points is defined just as was the convergence of a sequence of numbers (§IV.4). This indicates again why $\mathbf{R}$ is not compact, because the sequence of natural numbers has no convergent subsequence. A metric space X is compact if and only if every infinite sequence of points in X has an infinite subsequence converging to a point in X–but this result fails for general topological spaces with convergence defined using neighborhoods. The recognition of the importance of compactness and of its description by coverings is a major step in the understanding of topological spaces. It developed only slowly–and was not really codified until Bourbaki, in his influential 1940 volume on topology, insisted.

This concept of compactness is only one of the many issues arising from the development and foundation of the calculus. Geometry and mechanics had led to the intuitive ideas of rates of change, area, and summation. The somewhat vague formulation of these ideas by means of infinitesimals proved to be powerful tools in the 18th century, but various difficulties developed. For example (Titchmarsh [1932], 1.75), the sequence of functions $f_n(x) = n^2x(1-x)^n$ converges to zero for all x in the interval $0 \leqslant x \leqslant 1$, but the definite integral of the $f_n(x)$ from 0 to 1 is $n^2/(n+1)(n+2)$, so the sequence of integrals does not converge to 0! (The original convergence is not "uniform".) This is typical of problems arising with the interchange of two infinite processes (here convergence and integration). Such difficulties eventually required a more sophisticated formulation of the calculus, based on careful development of the concept of a limit. This concept, in its turn, involves a careful use of logical quantifiers and suggests the consideration of limits in $\mathbf{R}^2$, in $\mathbf{R}^3$, and in much more general topological spaces. The study of the definite integral inevitably involves new notions of uniform continuity and then compactness, and the proofs must rest upon a careful axiomatization of the real numbers–an arithmetization of one-dimensional geometry. The axioms alone do not suffice–there is a whole list of subtle consequences. Also, compactness in its turn must be disentangled from the more pictorial notions "closed and bounded". There are many more developments, especially of more general integrals, such as the Lebesgue integral. Thus, all told, the initial intuitive ideas and problems, and their extensions, and applications, lead to notions which in their turn demand more subtle concepts internal to Mathematics.

8. Approximations and Taylor's Series

Ideas of successive approximation underly the calculus. Areas are approximated from inside and outside the simple areas, and integrals are approximated by finite sums. Tangent lines to curves are approximated by secants, instantaneous velocities are approximated by average velocities, and similarly for higher derivatives. For example, the first derivative of a function $y = f(x)$ allows an approximation of that function by a linear function, as in

$$f(x) \sim f(a) + (x - a)f'(a), \tag{1}$$

where the linear function on the right agrees with f and with f' at $x = a$; in other words, it is the linear function whose graph is the tangent line to $y = f(x)$ at $x = a$. Moreover, one can estimate the error in such a linear approximation. If the first derivative $f'(x)$ is continuous on the closed interval from a to x, while the second derivative $f''(x)$ exists there, a simple iteration of the Law of the Mean (§7) shows that there is a real number ξ between a and x such that

$$f(x) = f(a) + (x - a)f'(a) + f''(\xi)\frac{(x - a)^2}{2}. \tag{2}$$

In other words, if the second derivative f'' is small in the interval, the linear approximation (1) is good.

This formula (2) is a special case of Taylor's theorem. If a function $f(x)$ has a continuous $(n - 1)$st derivative $f^{(n-1)}(x)$ on the closed interval from a to x and if the nth derivative exists on the open interval from a to x, then there is a real number ξ between a and x such that

$$f(x) = f(a) + \sum_{k=1}^{n-1} f^{(k)}(a)\,\frac{(x-a)^k}{k!} + f^{(n)}(\xi)\,\frac{(x-a)^n}{n!}, \tag{3}$$

where $k! = 1 \cdot 2 \ldots k$. In other words, the first $n - 1$ derivatives of f provide coefficients for an approximation to f by a polynomial of degree

$n-1$. The polynomial is uniquely determined as the one of that degree which at $x = a$ agrees with f and with the first $n - 1$ derivatives of f. The error in this approximation is then measured by a value of the nth derivative, as in the last term of (3); there are also other (integral) formulas for this "remainder term", and there are other types of polynomial approximations.

For a function $f(x)$ which has derivatives of all orders, this process suggests the formation of an infinite power series

$$\sum_{k=0}^{\infty} f^{(k)}(a)\frac{(x-a)^k}{k!} \tag{4}$$

called the *Taylor's series* for f. The explicit meaning of such infinite sums is then given by the convergence of the partial sums, as discussed in §IV.4. Often the remainder formula given by Taylor's theorem (3) can be used to show that the Taylor series converges for some (or even all) values of x, or that it converges "uniformly" for some closed interval of values of x. This leads to the familiar power series for e^x, $\sin x$, and $\cos x$. They—and their variants—are then at hand for the computation of tables of trigonometric functions. Indeed one can use the power series to define analytically the function $\sin x$, thereby avoiding the winding function of §IV.2—and obscuring the reasons for the periodicity of $\sin x$. The numerous other applications of such power series indicate that the initial elementary idea of approximation, in particular linear approximations, does indeed have extensive consequences. This does not justify the overuse of merely linear approximations in the multiple regression methods (least squares) so popular in econometrics.

9. Partial Derivatives

For elementary problems in algebra, it is always vital to know whether to add or to multiply. For example, the total change due to two separate causes clearly ought usually to be the sum of the separate changes. On the other hand, one must multiply to get a composite rate of change: For example, given the exchange rate from pounds to dollars and the rate from dollars to francs, the product will be the exchange rate from pounds to francs.

These two simple observations of the practical meaning of addition and multiplication appear formally in the *chain rule* for differentiation in the calculus. This we have already used: If $z = g(y)$ and $y = h(x)$ are two functions with continuous derivatives, then in the relevant range $z = g(h(x))$ is a function of x and has derivative

$$z'(x) = g'(y)h'(x), \qquad \text{or} \quad \frac{dz}{dx} = \frac{dz}{dy}\frac{dy}{dx}. \tag{1}$$

The proof requires a little care with the limits entering in the definition of the derivatives involved; the limits involve the formulas for the corresponding finite increments $\Delta x = x - x_1$, for then

$$\frac{\Delta z}{\Delta x} = \frac{\Delta z}{\Delta y}\frac{\Delta y}{\Delta x}.$$

Thus (1) expresses the underlying reason why one should multiply rates.

The corresponding chain rule is more striking for functions of several variables, such as a quantity z given as a function $z = f(x,y)$ for all points (x,y) in some open set U of the cartesian (x,y)-plane. There is no problem in finding what derivatives might mean here; if one holds y fixed, the quantity z remains just a function of x; its derivative, when it exists, is called the *partial derivative* with respect to x. Thus at a point (x,y) in U this derivative, for $h \neq 0$, is

$$\frac{\partial z}{\partial x} = f'_x(x,y) = \lim_{h\to 0}\frac{f(x + h,y) - f(x,y)}{h}. \tag{2}$$

Holding x fixed, there is a similar partial derivative $f'_y(x,y)$. This is the more explicit notation, specifying both variables and indicating which is "held fixed" and which is variable. The notation $\partial z/\partial y$ for the same partial derivative is incomplete unless an indication is added to show *which* other variable is at hand and is held fixed. (The observation is important, say in thermodynamics, where several alternative pairs of independent variables may be to hand.)

For $z = f(x,y)$ these two partial derivatives give the rate of change of z when the point (x,y) in the plane moves horizontally (in the x direction) or vertically (in the y direction). As such, they do not give complete information about all the possible rates of change of z. One would at least also want the rate at which z changes when the point (x,y) moves off in another direction–say in a direction making an angle θ with the positive x axis. A linear change in this direction is then given, in terms of a parameter t, by

$$x(t) = x + (\cos\theta)t, \qquad y(t) = y + (\sin\theta)t.$$

The intuitive principle that the total change is the *sum* of the separate changes in x and in y then suggests that the derivative of z with respect to the parameter t in the direction θ should be the linear combination

$$\frac{dz}{dt} = \left[\frac{\partial z}{\partial x}\right]\cos\theta + \left[\frac{\partial z}{\partial y}\right]\sin\theta, \tag{3}$$

usually called the *directional* derivative. When both the partial derivatives of z are continuous, this formula holds; it is a special case of the following chain rule. If the functions $x = g(t)$ and $y = h(t)$ have continuous derivatives, giving values x,y in the set $U \subset \mathbf{R}^2$ where the function $z = f(x,y)$ also has two continuous first partial derivatives, then $z = f(g(t),h(t))$ is a function of (suitable values of) t with the continuous derivative

$$\frac{dz}{dt} = \frac{\partial z}{\partial x}\frac{dx}{dt} + \frac{\partial z}{\partial y}\frac{dy}{dt}. \tag{4}$$

This clearly includes the motivating case (3). A similar formula applies when z is given as a function of more than two variables or when these variables x and y depend not on one but on several parameters.

This chain rule (4) has several different aspects.

First, think of $dx = (dx/dt)dt$ as an infinitesimal change in x, caused by the (equally) infinitesimal change dt in t. Then, multiplying (4) by dt and cancelling gives

$$dz = \frac{\partial z}{\partial x}dx + \frac{\partial z}{\partial y}dy. \tag{5}$$

This expression is called the *total differential* of z; for given values of x and y it gives the total change in z due to infinitesimal changes dx and dy; we will soon give a less "infinitesimal" interpretation.

Starting from a point x_0, y_0 with finite changes $x - x_0$ and $y - y_0$, the formula (5) suggests a linear approximation $z - z_0$ to the change in z:

$$(z - z_0) = \left[\frac{\partial z}{\partial x}\right]_0 (x - x_0) + \left[\frac{\partial z}{\partial y}\right]_0 (y - y_0). \tag{6}$$

This suggests (and correctly) that there is a Taylor's formula and also a Taylor series, each valid for functions z of two variables x and y which are sufficiently "smooth". Here and later we will use the term *smooth* for functions with enough continuous derivatives, in cases where we do not wish to specify in detail how many such derivatives are in fact needed.

In the chain rule (4), dz/dt can be regarded as an "inner product" of two "vectors", as follows

$$\frac{dz}{dt} = \frac{\partial z}{\partial x}\frac{dx}{dt} + \frac{\partial z}{\partial y}\frac{dy}{dt} = \left[\frac{\partial z}{\partial x}, \frac{\partial z}{\partial y}\right] \cdot \left[\frac{dx}{dt}, \frac{dy}{dt}\right]. \tag{7}$$

The first factor on the right is called the *gradient* of $z = f(x,y)$; it is defined at each point (x_0,y_0) of the plane, and is written

$$(\nabla f)_0 = \left[\frac{\partial f}{\partial x}, \frac{\partial f}{\partial y}\right]_{x=x_0,\ y=y_0}. \tag{8}$$

This vector "points" in the direction of the maximum rate of increase of the function f, and has that rate as its length. The function f determines one such vector at each point of the plane. At each point, all such vectors for all f form a two-dimensional vector space, called the *cotangent space* attached to the plane at that point.

The second vector in the product (7) depends on the functions $x = g(t)$, $y = h(t)$. They describe a continuous *path* passing through the point x_0, y_0; such a path is also called a *parametrized curve*, consisting of the points $(g(t),h(t))$ each labelled with the corresponding value of the *parameter* t. Such a curve is the *trajectory* of a moving point. At time t_0, where $x = x_0$, and $y = y_0$, the velocity of this moving point is the second factor of (7)

$$\left[\frac{dx}{dt}, \frac{dy}{dt}\right]_{t=t_0} = (g'(t_0), h'(t_0)). \tag{9}$$

It is called the *tangent vector* to the path at the point. All the tangent vectors to trajectories through the point (x_0,y_0) form a two dimensional vector space, called the *tangent space* T_0 to the plane at this point.

Now the chain rule, in the form (7), gives dz/dt as the product of the gradient vector of z by the tangent vector to the path. If one considers both vectors to lie in the same two-dimensional space, this product is just the inner product as described in §IV.9. However, it is preferable to consider the tangent and cotangent spaces as conceptually distinct. The "product" in (7) is then a real-valued function of two vectors, one from each space. This function is linear in each vector when the other is held constant, so is said to be *bilinear*; in Chapter VII we will indicate why it makes the cotangent space "dual" to the tangent space.

The cotangent space may be constructed formally as follows: Take all smooth functions $f(x,y)$, each defined in some neighborhood of (x_0,y_0); they form an (infinite dimensional) vector space under addition of values and multiplication by real constants. Call two such smooth functions $z = f(x,y)$ and $w = k(x,y)$ *cotangent* (or equivalent) at the point (x_0,y_0) when they have the same first partial derivatives there:

$$\left[\frac{\partial z}{\partial x}\right]_0 = \left[\frac{\partial w}{\partial x}\right]_0, \qquad \left[\frac{\partial z}{\partial y}\right]_0 = \left[\frac{\partial w}{\partial y}\right]_0. \tag{10}$$

The equivalence classes of these functions under this relation then form the desired two-dimensional space, called the *cotangent* space at the point (x_0,y_0). This construction works not just for the plane, but for other curved surfaces such as the sphere, when the coordinates (x,y) in the plane are replaced by suitable coordinates, such as latitude and longitude for the sphere.

On the other hand, every smooth function f has a gradient ∇f as in (8). In particular, the coordinates x and y are smooth functions, with gradients $\nabla x = (1,0)$ and $\nabla y = (0,1)$. Hence every gradient can be expressed at each point as a linear combination of these two gradients, in the form

$$\nabla f = \left[\frac{\partial f}{\partial x}\right]\nabla x + \left[\frac{\partial f}{\partial y}\right]\nabla y.$$

Except for notation, this is just the definition

$$df = \left[\frac{\partial f}{\partial x}\right]dx + \left[\frac{\partial f}{\partial y}\right]dy$$

of the *total differential*. Thus the differential, born as an infinitesimal, may be defined to be the gradient ∇f—a vector in the cotangent space. This is why the cotangent space differs from the tangent space.

The tangent vector at t_0 to the path $g(t)$, $h(t)$ also determines the usual tangent line to that path, with the parametric equations

$$x - x_0 = g'(t_0)(t - t_0), \qquad y - y_0 = h'(t_0)(t - t_0), \tag{11}$$

where $x_0 = g(t_0)$ and $y_0 = h(t_0)$.

The chain rule (4) also has a three-dimensional interpretation. The function $z = f(x,y)$ represents a height z above (or below) the point (x,y) in the plane, and so may be pictured by a smooth surface S at these heights above some portion of the plane. The *tangent plane* π to this surface at a point $p = (x_0, y_0, z_0 = f(x_0, y_0))$ is by definition the plane (if there is one) containing all the tangent lines at p to all the smooth curves on S passing through p. Such a smooth trajectory is given by $x = g(t)$, $y = h(t)$ and $z = f(g(t),h(t))$; its tangent line at p is given by the parametric equation (11) plus the corresponding equation for z:

$$z - z_0 = \left[\frac{dz}{dt}\right]_0 (t - t_0). \tag{11$'$}$$

But these three equation (11) and (11$'$) together satisfy the linear equation (6) for the approximate change $z - z_0$ in z. This linear equation (6) represents a plane in 3-space; since it is satisfied by (11), it must be the tangent plane to the surface S according to our definition.

In this way, the chain rule combines ideas from geometry (tangent planes), from mechanics (velocity vectors), from calculus (linear approximation), and from algebra (dual spaces), with the results appropriately added or multiplied. It gives meaning to the "total differential".

Some of these ideas are more vivid in pictures. Thus the gradients of a function $f(x,y)$ defined in the whole (x,y)-plane give a vector at each point in the plane—hence a *vector field* in the plane (Figure 1). Alternatively the loci where $f(x,y) =$ constant give a family of curves in the plane—the *contour lines* for f. Their use is suggested by topographic maps, picturing in the plane the varying heights above sea-level (Figure 2). When f is smooth the gradient vectors, if non-zero, are orthogonal to the contour lines; for topography, they represent the direction of fastest ascent. For Mathematics, their exploitation has proved decisive in topology and Morse theory.

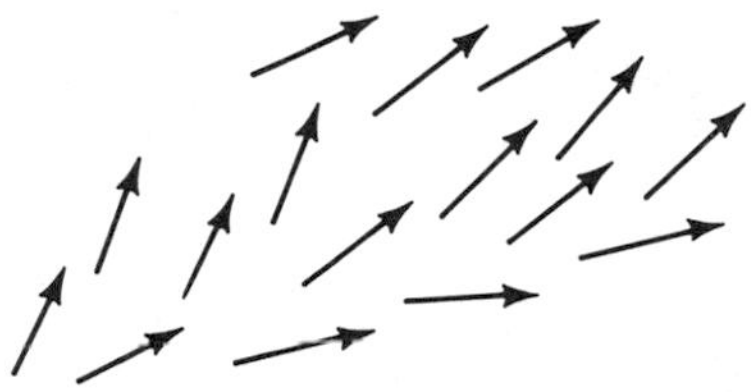

Figure 1. A vector field.

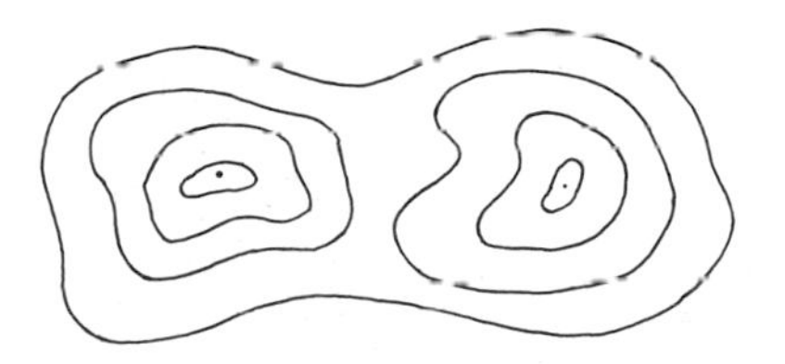

Figure 2. Contour lines.

10. Differential Forms

The one dimensional definite integral

$$\int_a^b f(x)dx$$

taken along the x-interval from a to b adds up successive increments dx in x, each weighted by the value $f(x)$ of a function f. In two dimensions the analogous "line" integral

$$\int_{(a,b)}^{(c,d)} (P(x,y)dx + Q(x,y)dy)$$

is taken along some path in the plane and adds up infinitesimal increments both in x and in y, each weighted by a function $P(x,y)$ or $Q(x,y)$. These ideas will be clarified if we begin by thinking first just about the gadget which sits under the integral sign.

A first order exterior *differential form* ω in an open set U of the x,y plane is an expression

$$\omega = P(x,y)dx + Q(x,y)dy \tag{1}$$

where P and Q are smooth functions of a point (x,y) in U. Such a differential form can also be described as a function which assigns to each point (x,y) in U, in a smooth way, a cotangent vector $Pdx + Qdy$ at that point. For each smooth function $z = f(x,y)$ the total differential $dz = f'_x dx + f'_y dy$ is such a differential form ω. However, there are many differential forms which are *not* total differentials. This is because of the important theorem which states that the order of partial differentiation can be interchanged, at least for a function $f(x,y)$ with continuous second partial derivatives:

$$\frac{\partial^2 f}{\partial y\, \partial x} = \frac{\partial^2 f}{\partial x\, \partial y}. \tag{2}$$

There is a standard proof, using the mean value theorem, of this intuitively plausible result. It means that a differential form ω in (1) which is a total differential must satisfy the condition

$$\frac{\partial P}{\partial y} = \frac{\partial Q}{\partial x}; \tag{3}$$

such a differential form is said to be closed.

A variety of examples, some to appear below, suggest the following formal definition of a *line integral* $\int_L \omega$. Here the "line" L means a smooth curve in the region U of the (x,y)-plane where ω is defined. The integral is again a limit of a sum: Subdivide the curve L at points $p_0, p_1, \ldots, p_n$, with $p_i = (x_i, y_i)$, into n "short" pieces and form the sum

$$\sum_{i=1}^{n} P(x_i,y_i)(x_i - x_{i-1}) + Q(x_i,y_i)(y_i - y_{i-1}), \tag{4}$$

each summand is then the value of the differential form ω at one of the points p_i (or, alternatively, at some other point on the part of the curve from p_{i-1} to p_i), with the differentials dx and dy replaced by the actual increments $x_i - x_{i-1}$ or $y_i - y_{i-1}$ in the coordinates. Under suitable hypotheses, this sum approaches a limit as $n \to \infty$ and the length of each piece of the curve approaches zero; this limit is the line integral $\int_L \omega$. The idea here is just that of the ordinary integral; the identity in ideas can be expressed formally. Represent the line L by smooth parametric equations $x = g(t)$, $y = h(t)$; the line integral is then equal to an "ordinary" definite integral between suitable limits in the parameter t,

$$\int_L \omega = \int_{t_1}^{t_0} [P(g(t),h(t))g'(t) + Q(g(t),h(t))h'(t)]dt;$$

moreover the standard formulas for change of variable in an integral show that the integral on the right (between suitable limits in t) is independent of the choice of a smooth parametrization.

Examples are numerous. In physics, the *work* done by a (vector) force F in a (vector) displacement D is defined to be the product of the force by the displacement in the direction of the force; that is, to be the inner product $F \cdot D$ of these two vectors. If the force in question varies with position in the plane as $F = (P(x,y), Q(x,y))$, then the work done under a small displacement $\Delta x, \Delta y$ is approximately the inner product $P(x,y)\Delta x + Q(x,y)\Delta y$, so that the sum (4) will approximately measure the total work done along the line L–and the line integral, in the limit, measures the exact amount of work.

The flow of fluid in a plane provides another example of a line integral. If the fluid is flowing everywhere with the same constant vector velocity V, with x and y components M and N, then the amount of fluid crossing a position vector $(\Delta x, \Delta y)$ in unit time will be the length of that position vector times the component of velocity V perpendicular to it; in other words, it will be the inner product of V and the vector $(-\Delta y, \Delta x)$ perpendicular to the position vector: The same formula $-M(x,y)\Delta y + N(x,y)\Delta x$ applies where the fluid velocity $V = (M(x,y), N(x,y))$ varies with the point (x,y). Adding up all the contributions and taking the limit, as in the line integral, shows that the integral $\int_L (N dx - M dy)$ measures the total flow of fluid across the line L in unit time. This example (and others) shows how the line integral is another embodiment of the general idea of measuring a total quantity by adding up all the little bits (intuitively, the infinitesimal bits).

A *double integral* $\int\int_A f(x,y) dx dy$ of a smooth function f over a bounded area A of the (x,y) plane is again a suitable limit of a double sum $\sum_{ij} f(x_i, y_j)(x_{i+1} - x_i)(y_{j+1} - y_j)$. There are multiple interpretations. Thus when $f(x,y) = 1$ the double integral is just the area A, while for a slab A with density $f(x,y)$ varying from point to point, the double integral is the weight of the slab. Here the fundamental theorem of the integral calculus reappears in the form of the Gauss lemma ("the divergence theorem"). Suppose that the area A is bounded by a smooth closed curve C, called the *boundary* $C = \partial A$ of A, and let ω be a smooth differential form defined throughout A and on C. Then

$$\int_C [P(x,y)dx + Q(x,y)dy] = \int\int_A \left[\frac{\partial Q}{\partial x} - \frac{\partial P}{\partial y} \right] dx dy; \qquad (5)$$

the line integral of ω is reduced to a double integral (which in the case of fluid flow can be interpreted as a total "divergence"). The outline of the proof of (5) is straightforward: Suppose that $P = 0$, so that only Q is present, and suppose also that each horizontal line meeting the curve C

meets C in just two points–one, (x_l, y_l), at the left and another, (x_r, y_r), at the right, as in Figure 1. Then the double integral of $\partial Q/\partial x$ can plausibly be replaced by an iterated integral–first on x, then on y, and the integral on x along a horizontal line (i.e., the sum along a horizontal strip) is by the fundamental theorem just $[Q(x_r, y_r) - Q(x_l, y_l)]dy$; the second integration on y then produces the line integral of Qdy in two pieces–left side and right side of A. As the arrows in the figure may indicate, it is here important to integrate along the curve C in the appropriate direction, so that the area A always lies to the left of C as traversed. (This requires the geometric ideas of orientation, as discussed in Chapter III.) Our concern here is not with the necessarily careful details of rigorous proof of (5) (say, for more convoluted boundaries C) but with the more sweeping observation that what appears in the Gauss lemma (6) is another (necessarily formal and careful) realization of the idea that the total change in a quantity (the total along the boundary curve) is exactly the sum of the infinitesimal changes–just the idea which was already expressed (§4) in the fundamental theorem.

The same idea reappears in higher dimensions. An integral over a two-dimensional surface S in Euclidean 3-space has the general form of a double integral

$$\int\int_S [L(x,y,z)dydz + M(x,y,z)dxdz + N(x,y,z)dxdy],$$

where L, M, and N are smooth functions of the coordinates x, y, and z and the integrand is a second order differential form. Such an integral may be defined as an appropriate limit of a sum or by reduction to a double integral taken in the plane of two parameters s and t when the surface S is given by parametric equations $x = g(s,t)$, $y = h(s,t)$, $z = k(s,t)$. If the surface S is the total boundary $S = \partial V$ of some volume V, Green's theorem (also called Gauss' lemma) asserts that

$$\int\int_{\partial V} [Ldydz + Mdxdz + Ndxdy] =$$

$$\int\int\int_V \left[\frac{\partial L}{\partial x} - \frac{\partial M}{\partial y} + \frac{\partial N}{\partial z}\right] dxdydz. \tag{6}$$

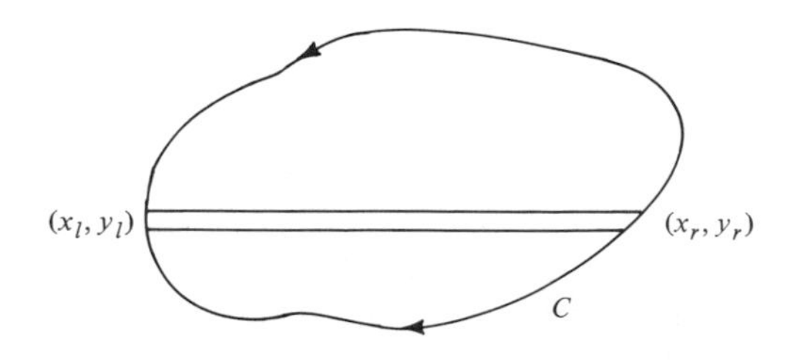

Figure 1

Similarly Stokes' theorem deals with the integral of a first order form taken over a curve ∂S bounding a piece S of surface:

$$\int_{\partial S} [Pdx + Qdy + Rdz] = \iint_S \left[\left[\frac{\partial Q}{\partial x} - \frac{\partial P}{\partial y}\right] dxdy + \right.$$
$$\left. + \left[\frac{\partial R}{\partial x} - \frac{\partial P}{\partial z}\right] dxdz + \left[\frac{\partial R}{\partial y} - \frac{\partial Q}{\partial z}\right] dydz \right]. \qquad (7)$$

In each of these versions (5), (6), and (7) of the fundamental theorem, the differential form ω which on the left is integrated over the "boundary" determines another differential form which appears as the integrand on the right; the latter is called the *exterior derivative* of ω, and is written as $d\omega$:

$$d[P(x,y)dx + Q(x,y)dy] = \left[\frac{\partial Q}{\partial x} - \frac{\partial P}{\partial y}\right] dxdy,$$

$$d[Pdx + Qdy + Rdz] = \left[\frac{\partial Q}{\partial x} - \frac{\partial P}{\partial y}\right] dxdy +$$
$$\left[\frac{\partial R}{\partial x} - \frac{\partial P}{\partial z}\right] dxdz + \left[\frac{\partial R}{\partial y} - \frac{\partial Q}{\partial z}\right] dydz,$$

$$d[Ldydz + Mdxdz + Ndxdy] = \left[\frac{\partial L}{\partial x} - \frac{\partial M}{\partial y} + \frac{\partial N}{\partial z}\right] dxdydz.$$

With this description of the exterior derivative of a form, all of these versions of the "fundamental theorem" of the calculus can be written in a uniform way. Consider a bounded smooth portion V (volume, surface, area) of some space with a smooth boundary ∂V. Then

$$\int_{\partial V} \omega = \int_V d\omega,$$

where the smooth differential form ω (and thus it exterior derivative) must be defined throughout the region V *and* on its boundary. Note that the formulas above for these exterior derivatives can all be obtained from a simple memnonic: For a function P or L, dP (or dL) is the usual total differential; for a product, $d(PQ) = (dP)Q + PdQ$, for a variable x, $d(dx) = 0$, and the differentials dx, dy of the variables are multiplied with the understanding that every square is zero. Then $(dx)(dx) = 0$, while $(dx + dy)^2 = 0$ and hence $dxdy = -dydx$. From the rule (2) for interchanging the order of two successive smooth partial derivatives, it then follows that a second exterior differential $dd\omega$ is always zero.

These various observations form the starting point of many important developments in topology and geometry (see Chapter VIII). These techniques are also heavily used in classical Mathematical Physics in the language of vectors rather than that of forms.

11. Calculus Becomes Analysis

At about this point in the conceptual development the subject of calculus–rules for the effective calculation of magnitudes by instantaneous rates and infinite sums–is gradually transformed into analysis. Here "real" analysis deals with functions on the reals **R** to **R** and examines their deeper properties as to integration, limits, and derivatives. It has many aspects not covered in this book save in the following brief notes.

The (Riemann) integral which we have discussed involves a variety of unpleasant problems such as those of the termwise integration of a convergent series of functions or the interchange of order of two successive integrations. Many of these become smooth when the integral is suitably generalized to the Lebesgue integral $\int_L f(x)dx$, available for a larger class of functions f. One approach to the development of this integral is by step functions. So consider bounded functions f on an interval of the real line and sets S of points on that interval. Cover S by disjoint open intervals and take the sum of the lengths of these intervals. The greatest lower bound of all these sums is the "outer measure" of S. A set E is called *measurable* if for each $\epsilon > 0$ there is an open set U containing E and such that the outer measure of the difference $U - E$ is less than ϵ. This also gives the measure (the "width") of E. A step function is one which is constant on each of a number of disjoint measurable sets E_i and zero elsewhere. The integral of a step function f is an evident sum of the terms "width of each step times the value of f on that step"; when $f(x)$ is approximated from below by a step function $s(x)$ and from above by another such $t(x)$, and when the least upper bound of the lower approximations equals the greater lower bound of the upper ones, this common value is the Lebesgue integral $\int_L(f(x))dx$, and the function f is said to be *Lebesgue integrable*. There are evident algebraic properties for the measures of intersection and union of sets; indeed such measures are essential to the conceptual axiomization of probability theory. There is an especially simple definition of a set of measure zero and it is useful to assert that two functions are equal *almost everywhere* when they are equal everywhere except on a set of measure zero. Now identify two functions f which are equal almost everywhere and consider the set L^2 of all (complex-valued) functions f on the interval whose absolute value squared is Lebesgue integrable. To each function f in L^2 we can assign a *norm* $\|f\|$, defined by the integral over the interval

$$\|f\|^2 = \int |f(x)|^2 dx .$$

The "distance" between two functions f and g in L^2 is then the norm $\|f - g\|$. This distance makes L^2 into a metric space, and it proves to be very useful to apply geometric ideas about spaces to the analytically described set L^2. It is an especial virtue of the Lebesgue integral that this metric space is complete–every Cauchy sequence in L^2 has a limit there. Any two functions f and g in L^2 also have an "inner product" $\int f(x)g(x)dx$, so L^2 is an infinite dimensional inner product space (§VII.5).

The Lebesgue integral, as distinguished from the Riemann integral, shows that the same primitive idea of "adding up the little pieces" can have more than one formal expression (in fact, more than just these two!).

For differential equations, a central idea is that a smooth function $f(t)$ can be determined completely by its initial value at $t = 0$ and knowledge of all of its first derivative. The same idea appears for smooth functions $f(x,y)$ of two or more variables: They can be determined by "boundary conditions" and knowledge of the partial derivatives. Such P.D.E. (*partial differential equations*) arise not just from the conceptual analogy of these ideas but from many different cases in theoretical physics where exactly this kind of data is at hand for quantities depending on several independent variables.

For example, let a one-dimensional wave at time $t = 0$ be represented by a height y (above each point x) given by a smooth function $y = f(x)$ as at the left of Figure 1. If this wave keeps the same form as it moves to the right at a constant velocity c (in units distance per time), then the height at position x at time t is given by $u = f(x - ct)$ (try it out!). This function u satisfies the first order P.D.E. $\partial u/\partial t = -c\, \partial u/\partial x$ and (from Rolle's theorem) *any* smooth function u satisfying this P.D.E. must have the form $u(x,t) = k(x - ct)$ for some smooth function k. Thus the solution $k = k(x,t)$ of the P.D.E. is determined by its initial values $k(x,0)$. Similarly a wave moving steadily at velocity c to the left is given by $v = g(x + ct)$. Both u and v satisfy the second order P.D.E.,

$$\frac{\partial^2 u}{\partial t^2} = c^2 \frac{\partial^2 u}{\partial x^2}, \tag{1}$$

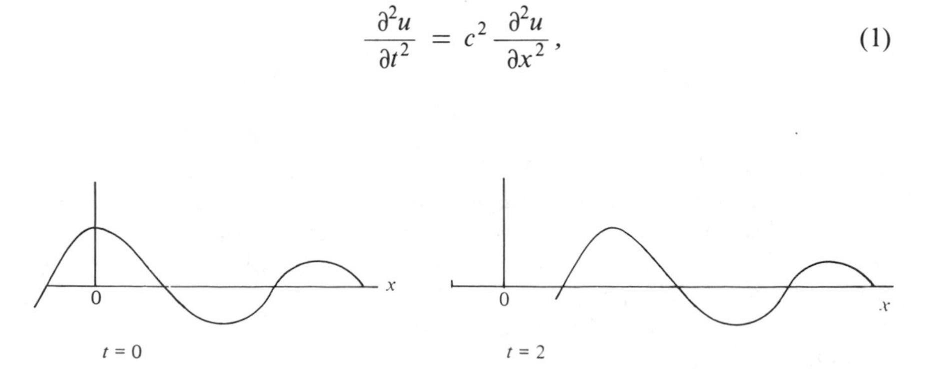

Figure 1. Motion of a one-dimensional wave.

called the *wave equation*; the most general smooth solution of this equation have the form $f(x - ct) + g(x + ct)$ for arbitrary smooth f and g.

In three space dimensions x, y, and z the wave equation reads

$$\frac{\partial^2 u}{\partial t^2} = c^2 \left[\frac{\partial^2 u}{\partial x^2} + \frac{\partial^2 u}{\partial y^2} + \frac{\partial^2 u}{\partial z^2} \right]; \tag{2}$$

it applies to many sorts of quantities u varying in time and space.

Potential. Under Newton's law inverse square for gravitational attraction, the potential energy of a unit mass at distance r from an attracting body at the origin is the work done by gravity in bringing the unit mass up "from infinity". Since the force of gravity is proportional to $1/r^2$, this potential (the integral over r) is proportional to $1/r = (x^2 + y^2 + z^2)^{-1/2}$. This function $u = 1/r$ of x, y, and z satisfies (except at the origin) the P.D.E.

$$\frac{\partial^2 u}{\partial x^2} + \frac{\partial^2 u}{\partial y^2} + \frac{\partial^2 u}{\partial z^2} = 0. \tag{3}$$

By addition, the potential due to several attracting bodies or due to a smooth distribution of such bodies will also satisfy this P.D.E., called *Laplaces's equation* for the potential u.

A smooth function u which satisfies the equation (3) in an open set V of $\mathbf{R}^3$ is said to be *harmonic* in V. For example, an electric charge distributed over a closed surface S in space will define a potential u constant on S and harmonic inside S. This and related physical situations suggests the following Dirichlet problem: Let V be an open set in $\mathbf{R}^3$ bounded by a suitable surface S, while f is a smooth real-valued function defined on S. Extend f to a function u defined on S plus V and harmonic in V. Dirichlet intended to solve this problem by considering, among all functions u extending f, one which minimized the triple integral

$$\int\int\int_V \left[\frac{\partial^2 u}{\partial x^2} + \frac{\partial^2 u}{\partial y^2} + \frac{\partial^2 u}{\partial z^2} \right] dxdydz\,; \tag{4}$$

he could then show that this minimizing function is harmonic. The integral (4) is non-negative, so has a greatest lower bound over all u, but it is not clear that there is an actual function taking on these greatest lower bound as the minimum. The clarification of this situation led to major advances in analysis (Monna [1975]).

Harmonic functions (of two variables) also arise in complex analysis (Chapter X).

Harmonic motion is governed by the ordinary differential equation

$$\frac{d^2x}{dt^2} = -k^2x \tag{5}$$

for (one-dimensional) position x as a function of time t. The most general solution of this equation has the form

$$x = A \cos kt + B \sin kt \tag{6}$$

for constants A and B. This function x is periodic in t, of period $2\pi/k$. Moreover, (6) asserts that the vector space (over the reals) of all the solutions is a two dimensional vector space, with basis the two solutions $\cos kt$ and $\sin kt$. Thus harmonic motion reduces to trigonometry; these two basic solutions can be regarded as the two projections on the axes of a point moving at uniform velocity around a circle.

In determining the solutions of P.D.E. it is often convenient to first solve the P.D.E. for specified simple boundary values, then suitably adding up the results. To do this one must express the finally desired boundary value as a sum (or a series) of simpler functions. For a function $f(x)$ with period 2π in x a typical such expression is the Fourier series

$$f(x) \sim \frac{1}{2}a_0 + \sum_{k=1}^{\infty} a_k \cos + kx + b_k \sin kx. \tag{7}$$

The study of such series expansions had a major historical role in the consideration of the question of Chapter V: "What is a function"?

Now suppose that the series (7) does converge to $f(x)$ for all x and try to determine the coefficients a_k and b_k by multiplying f and the series by $\sin mx$ or $\cos mx$ and then integrating from 0 to 2π. One would like to integrate the resulting series term-by-term. This is a typical example of the problem of double limits (interchange convergence and integration). This is a general question, with many ramifications in analysis. When it can be justified in the present case, it leads to simple formulas for the coefficients in (7):

$$a_m = \frac{1}{\pi}\int_0^{2\pi} f(x)\cos mx\,dx, \qquad b_m = \frac{1}{\pi}\int_0^{2\pi} f(x)\sin mx\,dx. \tag{8}$$

The study of the actual convergence of such series makes effective use of the Lebesgue integral, and leads to many deep theorems. There is also a uniqueness question: If two such series (7) converge to the same function $f(x)$, are their coefficients all the same? This question reduces at once to the question: If such a series (7) converges to 0 for all x, $0 \leqslant x \leqslant 2\pi$, are the coefficients all zero? The answer is yes, and also yes if one assumes convergence to 0 for all x except for a finite number of points. In fact, the question of what additional exceptions are possible is what first led George Cantor to consider abstract sets: The sets S such that convergence to 0 for all $x \in S$ implies zero coefficients. It is striking that such specific questions about convergence were the principle origin of the general concepts of set theory! The relevance of this origin has been emphasized by quite recent research by Solovay in set theory.

These Fourier series deal with functions $f(x)$ periodic of period 2π; such a function can alternatively be described as a function $f(\theta)$ defined for all points θ on the circle (of radius 1); here the *additivity* of angles enters, so the *circle* is a topological group both compact and abelian. General aspects of this subject of "harmonic analysis" (this term is derived from harmonic motion) concern functions defined on other topological groups–there are remarkable ramifications.

At a maximum or a minimum, a smooth function $y = f(x)$ necessarily has derivative zero; this fact is used in the elementary (and useful) determination of maximum and minimum points. There are many more sophisticated minimum problems: Along which curve will a moving particle descend from A to B in the least possible time? Which simple closed curve of given length will enclose the greatest area? In these cases one tries to find not the values of x which make a given function $y = f(x)$ a minimum but the values of a given function $f(x)$ (the form of that curve of descent) which make a given quantity a minimum. Typically, that quantity is usually measured by an integral whose integrand is some expression F involving both x, values of the function $y = f(x)$ at interest and the values of its derivative–say an integral

$$\int_a^b F(y,y',x)dx, \qquad y = f(x).$$

Such questions of choosing a minimizing function $f(x)$ arise in the "least action" principles of Mechanics (Chapter IX) as well as in the Dirichlet problem (4) above. These minimum (or maximum) problems constitute the *Calculus of Variations*–a subject with many conceptual questions, sharp need for a careful $\epsilon - \delta$ foundation, a number of (recently developed) uses in practical problems of optimization and finally many striking conceptual ideas. For example, what happens to the elementary observation that between any two minimum of a smooth function $y = f(x)$ there lies a maximum? (The answer leads to the so called "Morse relations" on the numbers of minima, of saddle points, and of maxima, say for functions of two variables).

In the Calculus of Variations one maximizes or minimizes an integral over "all" smooth functions $f(x)$; it turns out to be suggestive to think of the collection of all of these functions as a "space". The study of such spaces of functions is one of the reasons historically for the introduction of the general idea of a metric space, as in the case of the space L^2.

12. Interconnections of the Concepts

Algebra (e.g., group theory) and geometry (e.g., vector spaces) could each be described as the study of models of a simple formal axiomatic system. The calculus is not easily so described; it is rather the development, from intuitive methods of calculation, of a variety of formal concepts (limit, derivatives, integrals, differential forms) all found within one overarching axiomatics (that for the real numbers and sets of reals and sets of functions on the reals). Nevertheless it has the same basic character as do algebra and geometry: It begins with practical problems and intuitive ideas useful for calculations which turn out to be strictly formalizable–and which have been so formalized, both with $\epsilon - \delta$ and with actual infinitesimals. Moreover, the various ideas involved are tightly connected with each other, as suggested in the following diagram.

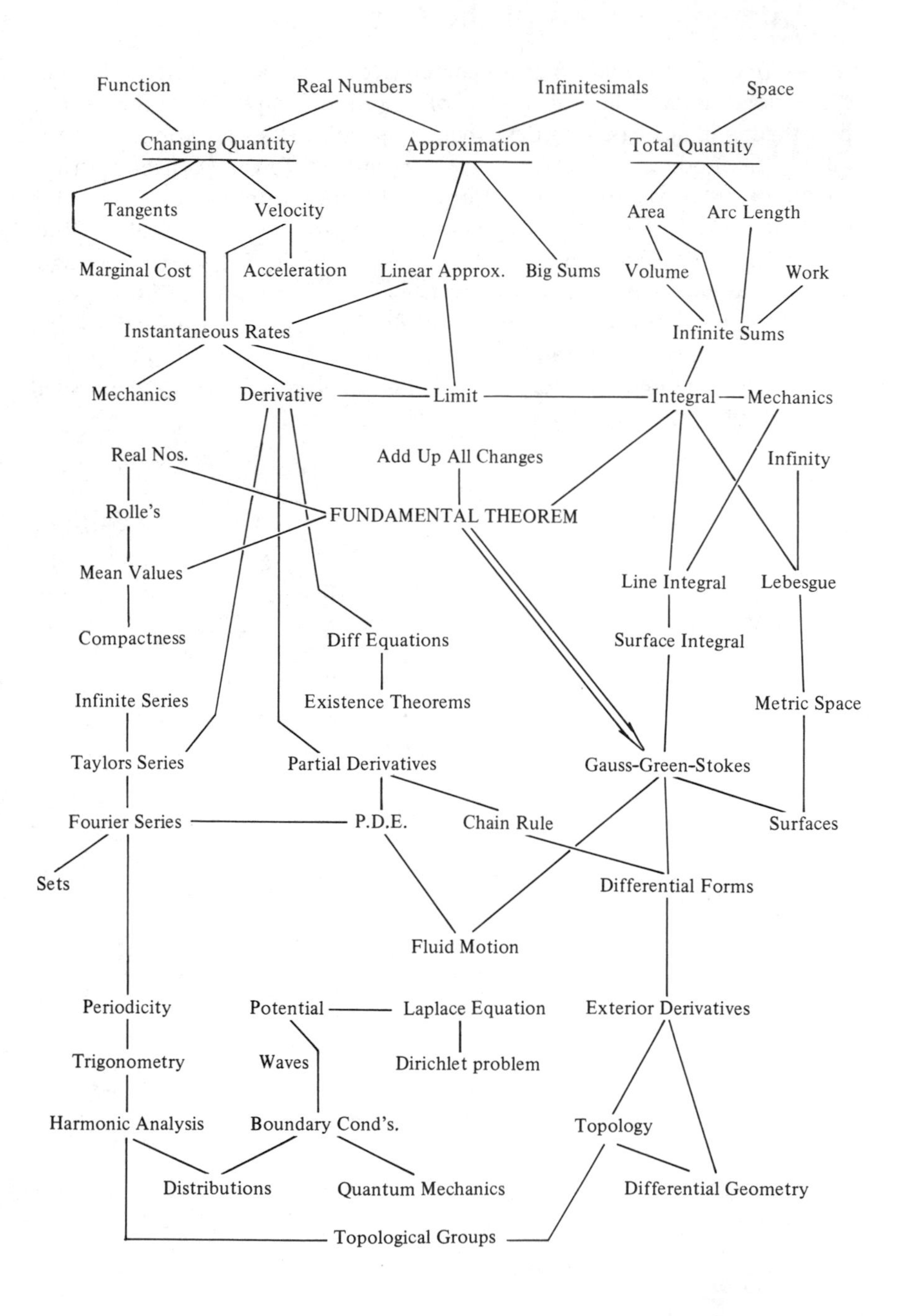
Function
Real Numbers
Infinitesimals
Space
Changing Quantity
Approximation
Total Quantity
Tangents
Velocity
Area
Arc Length
Marginal Cost
Acceleration
Linear Approx.
Big Sums
Volume
Work
Instantaneous Rates
Infinite Sums
Mechanics
Derivative
Limit
Integral
Mechanics
Real Nos.
Add Up All Changes
Infinity
Rolle's
FUNDAMENTAL THEOREM
Mean Values
Line Integral
Lebesgue
Compactness
Diff Equations
Surface Integral
Infinite Series
Existence Theorems
Metric Space
Taylors Series
Partial Derivatives
Gauss-Green-Stokes
Fourier Series
P.D.E.
Chain Rule
Surfaces
Sets
Differential Forms
Fluid Motion
Periodicity
Potential
Laplace Equation
Exterior Derivatives
Trigonometry
Waves
Dirichlet problem
Harmonic Analysis
Boundary Cond's.
Topology
Distributions
Quantum Mechanics
Differential Geometry
Topological Groups

CHAPTER VII

Linear Algebra

An algebraic approach to problems of plane geometry led us in §IV.7 to introduce two-dimensional vector spaces over the field **R** of real numbers, while in the calculus gradients and tangent lines lead to cotagent and tangent vector spaces. Three dimensional vector spaces are standard in physics, while the algebra of vectors is an effective way of handling geometrical ideas in dimensions higher than 3. Analysis soon produces infinite-dimensional spaces such as L^2 (§VI.11). This chapter will summarize the properties of such linear vector spaces over an arbitrary field, not necessarily **R** or **C**.

1. Sources of Linearity

To say that an effect is "linear" means that the effect respects proportions and that the effect of a sum is the sum of the separate effects. The formal description of such linearity can be stated in the context of a "linear" vector space defined much as in §IV.7. Such a space, defined over a given field F of "scalars", is a set V of "vectors" equipped with two operations: addition $(v,w) \mapsto v + w$ of two vectors v and w, and multiplication $(a,v) \mapsto av$ of a vector v by a scalar $a \in F$. The axioms require that the vectors form an abelian group under addition, and that multiplication by scalars satisfies the identities

$$a(v + w) = av + aw, \qquad 1v = v, \tag{1}$$

$$(a + b)v = av + bv, \qquad (ab)v = a(bv) \tag{2}$$

for all vectors v and w in the space V and for all scalars a, b in F. These two operations of addition and scalar multiple yield the more general *linear combinations*, such as the combination

$$a_1 v_1 + a_2 v_2 + \cdots + a_n v_n = \sum a_i v_i$$

of n vectors $v_i \in V$ with corresponding scalar coefficients $a_i \in F$. All the linear combinations of n given vectors $v_1, \ldots, v_n$ form a subspace of V, the subspace *spanned* by V. *Linear independence* of the $v_1, \ldots, v_n$ means that $\Sigma a_i v_i = 0$ only when all the coefficients a_i are 0. Moreover, the $v_i, \ldots, v_n$ form a *basis* of the finite dimensional space V when every vector in V can be expressed in just one way as a linear combination $\Sigma a_i v_i$. The vectors of a basis are necessarily linearly independent. Any two bases of the same space have the same number of elements, and this number n is the *dimension* of the space, dim V. All these concepts depend on the use of linear combinations.

Just as for the plane, the geometry of a vector space is expressed in terms of its transformations—more exactly, its linear transformations—into itself or into some second vector space W (over the same field F). Such a transformation is by definition a function T from V to W which preserves all linear combinations, in the sense that always

$$T\left[\sum_{i=1}^{n} a_i v_i\right] = \sum_{i=1}^{n} a_i T(v_i). \tag{3}$$

It is enough to require just the two simpler identities

$$T(v_1 + v_2) = Tv_1 + Tv_2, \qquad T(av) = aT(v) \tag{4}$$

stating that T preserves sums (is *additive*) and that T preserves scalar multiples (is *homogeneous*). If both $S: W \to U$ and $T: V \to W$ are linear, so is their composite $S \cdot T: V \to U$. A linear transformation $T: V \to V$ of V into itself is a linear *endomorphism*. It is said to be *non-singular* when it has a two-sided inverse $T^{-1}: V \to V$. This inverse (unique as always) is necessarily linear.

This "transformation" language serves to codify the many cases of Mathematical operations which are additive and homogeneous.

In calculus, the operations of differentiation and integration are linear. Thus on an interval of the real axis, say the unit interval $I = \{x \mid 0 \leqslant x \leqslant 1\}$, consider the set $C^\infty = C_I^\infty$ of all those functions $f: I \to \mathbf{R}$ which have continuous derivatives of all orders. This set C^∞ is a vector space over $\mathbf{R}$ with addition and scalar multiple defined in "termwise" fashion by the equations

$$(f + g)(x) = f(x) + g(x), \qquad (af)x = a(f(x)) \tag{5}$$

for all x in I. The operation of differentiation, say $DF = df/dx$, is then a transformation $D: C^\infty \to C^\infty$ which is linear according to elementary rules of the calculus ("The derivative of a sum is the sum of the derivatives"). The operation of integration, sending each f to $\int_0^x f(x)dx$, is also linear ("The integral of a sum"). This space C^∞ is clearly very large; it has no finite basis and so is *infinite dimensional*. It is just one of many examples of a *function space* (a space whose elements are functions of some specified type). It has turned out that it is very illuminating to use the geometric language of vector spaces for this and many other such function spaces.

For harmonic motion, we observed in §VI.11 that all solutions for the differential equation $d^2x/dt^2 = -k^2x$ were uniquely linear combinations of two solutions, sin kt and cos kt. This differential equation for harmonic motion is a homogeneous *linear differential equation* with constant coefficients (i.e., a linear combination, over **R**, of derivatives (d^nx/dt^n , . . . , dx/dt, x). For such an equation of order n, it is similarly the case that all C^∞ solutions are linear combinations of n linearly independent solutions.

In the calculus for functions of two variables x and y, we have seen in §VI.9 that the chain rule suggests two different vector spaces attached to each point in the (x,y)-plane: The tangent space, consisting of all the vectors tangent to the various possible paths through the point, and the cotangent space, consisting of all the differentials of smooth functions at this point. The relation between these two spaces is one source of a notion of "dual" vector spaces (§4), and is a background on which much higher-dimensional geometry is developed. If we think of a path as the trajectory of a moving particle, it becomes clear that the tangent space is suggested by the development of mechanics, as we will see in more detail in Chapter IX below.

For algebra, vector space dimensions provide a measure of size. Thus when a polynomial $f(x)$ with coefficients in F is *irreducible* over F (has no proper polynomial factors over F), it is always possible (§10) to construct a field $K = F(\alpha)$ generated by F and by one root α of $f(x)$; this field is unique up to an isomorphism of fields leaving all elements F fixed. This field K can be viewed as a vector space over F (vectors are elements of K, added as there, and multiplied by elements of F). As such, its vector space dimension is precisely the degree of the given irreducible polynomial $f(x)$ (Special case: The complex numbers **C** arise from **R** by "adjoining" a root of $x^2 + 1$, and form a two dimensional vector space over **R**). This use of dimensions has already been noted in §V.7 for the Galois group, whose order is the dimension of the splitting field. Similar dimensions occur throughout the study of fields (for example, in the determination of all finite fields) and in the study of quaternions and other rings containing a field (often called *linear algebras* over that field).

The notion "vector space" thus organizes diverse examples. Given all these examples it is remarkable that the axioms for a vector space (though stated by Peano in 1888 for spaces over **R**) were not generally recognized by Mathematicians until Weyl's 1913 publication of his book on relativity (where he needed to use vector and affine geometry).

Our formulas (1) and (2) describe V as a *left* vector space over the field F, in the sense that the scalar multiples, $a,b,\ldots$ appear to the left of the vectors. The same vector spaces can be described with scalar multiples on the right. It will be convenient to use such *right* vector spaces in the next sections.

2. Transformations versus Matrices

There are two styles of doing linear algebra: Geometrically (or "invariantly") with linear combinations of vectors, or explicitly, with coordinates. These coordinates depend on a choice of basis. In detail, if the (right) vector space V has a finite basis of vectors $u_1, \ldots, u_n$, then every vector v in V has a unique expression as a linear combination

$$v = u_1x_1 + u_2x_2 + \cdots + u_nx_n$$

of these basis vectors u_i with scalar coefficients x_i (here written on the right). These scalars are the *coordinates* $X = (x_1, \ldots, x_n)$ of the vector v relative to the basis u_i. One can say that the vector "is" the n-tuple X of numbers x_i, so that V is the space F^n of these n-tuples. One can then express all properties of vectors in term of these numbers. Thus a linear endomorphism T: $V \to V$ is completely determined by giving the images

$$Tu_j = \sum_{i=1}^{n} u_i a_{ij} \tag{1}$$

of the n basis vectors u_j; hence T is determined by the array

$$A = \begin{bmatrix} a_{11} & a_{12} & \ldots & a_{1n} \\ \cdot & & & \\ \cdot & & & \\ \cdot & & & \\ a_{n1} & a_{n2} & \ldots & a_{nn} \end{bmatrix} \tag{2}$$

of coefficients—an $n \times n$ *matrix*. By linearity $T(\Sigma u_j x_j) = \Sigma_{i,j} u_i a_{ij} x_j$ so the coordinates x'_i of the transformed vector $v' = Tv$ are given by the equations

$$x'_i = \sum_{j=1}^{n} a_{ij} x_j, \qquad i = 1, \ldots, n. \tag{3}$$

This is the familiar way in which the square matrix X also represents the transformation T in terms of coordinates (or "variables") x_i; in particular, the jth basis vector u_j is sent into the jth column of the matrix A.

The coordinate representation (3) and the basis representation (1) of T are equivalent. however, the matrix A for T depends on T and on the choice of basis. This leads to a central question: When do two different square matrices represent the same linear transformation (§12 below)?

The practical issue of solving simultaneous linear equations is another source of linearity: Given $x'_1, \ldots, x'_m$ and the m equations (3)—this time with $i = 1, \ldots, m$—to find all the x_j, if any, to the given x'_i. These m equations now represent a linear transformation $T: V \to W$ from the n-dimensional space V of the $(x_i, \ldots, x_n)$ to the m-dimensional space W of the x'_i. Familiar facts about the solutions now translate to observations about this transformation and its rectangular matrix. Thus the *image* of T is the subspace Im T of W consisting of all the vectors Tv for v in V; that is, the set of all resulting values $(x'_1, \ldots, x'_m)$; its dimension is the *rank* of T (and of its matrix A). The *null-space* of T is the subspace Null T of V consisting of all those vectors v of V with $Tv = 0$; its dimension is the *nullity* of T (and of its matrix A). An elementary construction of bases shows that these dimensions ("dim") satisfy

$$n = \dim(\text{domain } T) = \dim(\text{Im } T) + \dim(\text{Null } T). \tag{4}$$

This equation in dimensions subsumes the facts about solutions of n simultaneous linear equations. For given x'_j, the equations (3) have a solution if and only if the vector $(x'_1, \ldots, x'_m)$ lies in Im T. In particular, if $n = m$ and rank $A = n$ (i.e., the rows of A are linearly independent), then the solution always exists and is unique—and the solution is given, via the usual determinants, by Cramer's rule. For n homogeneous equations (i.e., when all $x'_i = 0$) the number of linearly independent solutions (the nullity) is n minus the rank. These results provide another example of the translation of algebraic calculations into geometric concepts.

Products of matrices arise from composites of the corresponding transformations. If the square $n \times n$ matrix B represents a second linear $S: V \to V$ by $Su_k = \Sigma u_j b_{jk}$, then calculation with (2) shows that the composite transformation $T \cdot S$ (first apply S, then T) has

$$TSu_k = \sum_{i=1}^{n} u_i \sum_{j=1}^{n} a_{ij} b_{jk}. \tag{5}$$

So the matrix C of the composite transformation has entries $c_{ik} = \Sigma_j a_{ij} b_{jk}$—the ith row of A times the kth column of B. This explains the familiar row by column definition of the product $C = A \cdot B$ of two square matrices. This matrix product is associative *because* the composition of transformation is associative, but this can also be verified directly from the row by column definition. In this way the "idea" of composition is realized for computations by the matrix formalism—especially handy for numerical calculations, for example for "sparse" matrices (those with only a few non-zero entries).

Rectangular matrices have corresponding products. A linear transformation S from an l-dimensional space to an m-dimensional one plus a choice of basis in each space yields an $m \times l$ matrix B (m rows, l columns). When T is a transformation from this m-dimensional space to one of dimension n, with an $n \times m$ matrix A, the matrix product AB representing the composite $T \cdot S$ is defined row by column, as in (5). Again a triple product of matrices, *when* defined, is associative; and there is a suitable distributive law for the addition of matrices. For these calculations it is convenient to think of a vector X as a column—an $n \times 1$ matrix, then the action (3) of a matrix A on a vector X can itself be written as a matrix product $X' = AX$. (Here the matrix A appears on the left, just as the corresponding transformation and as for functions generally; this is why we used scalars on the right, to balance functions on the left.) The columns of a matrix A can be changed to rows by defining the *transpose* matrix A^T to be the matrix with entries $(A^T)_{ij} = A_{ji}$. Then $(AB)^T = B^T A^T$.

The whole subject can be developed exclusively in terms of n-tuples X (though this popular approach misses the geometric invariance, and so loses some understanding). Then the typical finite dimensional vector space F^n consists of all such n-tuples (all $n \times 1$ columns) with the evident addition and multiplication by scalars, and a subspace is a collection of n-tuples closed under these operations. An $n \times n$ matrix A is *non-singular* if and only if its rows are linearly independent. It is equivalent to require that the columns be linearly independent, or that there exists a matrix B with product $BA = I$, the identity matrix. Then B is unique, hence is the inverse A^{-1} of A, and $A^{-1}A = I = AA^{-1}$. When A is non-singular, the linear equations (3) have a unique solution X for given X'; to wit, $X = A^{-1}X'$. The (usual) *determinant* $|A|$ of a matrix A satisfies $|AB| = |A| \; |B|$, so is a homomorphism (of multiplication) to F.

Under composition, all the non-singular matrices with entries in a given field F clearly form a group, the *general linear group* $GL_n(F)$. Equivalently, it can be described as the group of all non-singular linear endomorphisms $T: V \to V$ of an n-dimensional vector space over F. Geometrically, it consists of all the non-singular affine transformations of space which leave a point (the origin) fixed—it contains contractions, expansions, shears, reflections, and rotations about the origin. Linearity is a common formal property of these quite various transformations.

Matrix multiplication is not commutative, because the composition of linear transformations is not necessarily commutative. In the 20th century, matrix multiplication had triumphant (and unexpected) applications in quantum mechanics. However, the ideas underlying matrix theory first developed from the study of elasticity or at least from the use of many-variable calculus to get linear approximations. Thus consider a small smooth deformation $\mathbf{R}^3 \to \mathbf{R}^3$ of 3-space, keeping the origin fixed. Each point with coordinates x, y, z is deformed to a point x', y', z'; these latter quantities, as functions of x, y, and z, have Taylor expansions beginning with linear terms

$$x' = a_{11}x + a_{12}y + a_{13}z + \cdots,$$

$$y' = a_{21}x + a_{22}y + a_{23}z + \cdots,$$

$$z' = a_{31}x + a_{32}y + a_{33}z + \cdots,$$

Properties of the resulting deformation then depend in the first approximation on the behavior of the array of coefficients a_{ij}, though it was a long time before Cayley in England and other algebraists on the continent took the step of detaching the coefficients $A = (a_{ij})$ from the environment. Again in this case, an abstract idea of matrix manipulation arises from more concrete problems.

3. Eigenvalues

A matrix description of any mathematical object depends on a choice of a basis, and so requires attention to the effect of a change in that basis—a change which replaces old coordinates x_i by new coordinates y_i, where $y_i = \Sigma P_{ij} x_j$ and the square matrix P with entries P_{ij} is non-singular (because one can change back). Two matrices A and B which represent the same endomorphism of V relative to (possibly) different bases are said to be *similar*. Thus in old (X) and new (Y) coordinates the same transformation reads

$$X' = AX, \qquad Y' = BY.$$

The change of coordinates is expressed by the matrix equations $Y = PX$ and $Y' = PX'$, so an easy calculation with matrix products gives $B = PAP^{-1}$. In words, square matrices A and B are similar if and only if there is a non-singular matrix P with $B = PAP^{-1}$. This is the formal setting; we will return later (in §12 below) to the explicit question of determining when two given matrices are similar.

What is the simplest matrix similar to a given matrix A? In particular, can A be similar to a diagonal matrix D—one with entries (say) $\lambda_1, \ldots, \lambda_n$ along the main diagonal,

$$D = \begin{bmatrix} \lambda_1 & & & & 0 \\ & \lambda_2 & & & \\ & & \cdot & & \\ & & & \cdot & \\ & & & & \cdot \\ 0 & & & & \lambda_n \end{bmatrix} \tag{1}$$

and zeros elsewhere? The corresponding linear transformation then simply multiplies the ith basis vector by λ_i. Such a scalar λ_i is called an eigenvalue of A. Specifically, λ is an *eigenvalue* of an endomorphism T if and only if $Tv = v\lambda$ for some non-zero vector v; equivalently, in matrix language, λ is an eigenvalue of a matrix A if and only if $AX = X\lambda$ for some $X \neq 0$. An *eigenvector* X for A is then any vector (zero or not) with $AX = X\lambda$ for a scalar λ. The notion of an eigenvalue is manifestly invariant under change of coordinates, and the above discussion proves the

Theorem 1. *An $n \times n$ square matrix A is similar to a diagonal matrix if and only if A has n linearly independent eigenvectors.*

In terms of the identity matrix I (diagonal with 1's down the diagonal) the equation $AX = X\lambda$ can be written as $(A - \lambda I)X = 0$. Hence a scalar λ is an eigenvalue if and only if this system of n homogeneous linear equations in X has a solution $X \neq 0$; that is, if and only if $A - \lambda I$ is a singular matrix. But a square matrix is singular if and only if its determinant is 0, and the determinant $|A - \lambda I|$ of $A - \lambda I$ for A $n \times n$ is a polynomial of degree n in λ,

$$(-1)^n\lambda^n + (-1)^{n-1}(a_{11} + \cdots + a_{nn})\lambda^{n-1} + \cdots + |A| = 0. \tag{2}$$

This is called the *characteristic polynomial* of A; we have proved the

Theorem 2. *The eigenvalues of A are the roots of its characteristic polynomial.*

Therefore by the fundamental theorem of algebra, a real (or a complex) matrix A does have eigenvalues.

The notions of eigenvector and eigenvalue dominate much of the use of linear algebra, especially in analysis and physics. The initially simple ideas of linearity have remarkable ramifications. As an elementary sample, consider a system of simultaneous first order linear (and homogeneous) differential equations

$$\frac{dx_i}{dt} = \sum_{j=1}^{n} a_{ij}x_j, \qquad i = 1, \ldots, n \tag{3}$$

in n (real) variables x_i. The real matrix A with entries a_{ij} has n eigenvalues $\lambda_1, \ldots, \lambda_n$, possibly not all different and real or possibly complex. If A is similar to a diagonal matrix, the corresponding change of coordinates to $y_1, \ldots, y_n$ then transforms the equations (3) to the form

$$\frac{dy_i}{dt} = \lambda_i y_i, \qquad i = 1, \ldots, n.$$

In this form, some solutions are evident: $y_i = C_i e^{\lambda_i t}$, for $i = 1, \ldots, n$ and for arbitrary constants (constants of integration) C_i. If some eigenvalue λ_i is real and positive, the solution y_i grows exponentially. However, λ_i may be complex. This is a typical case in which a "real" problem in real numbers leads to formulas involving complex numbers—here requiring the exponential function e^z for complex values of the exponent z.

The characteristic polynomial (2) of a matrix A has an unexpected property. In (2), substitute the matrix A for the (scalar) variable λ. the resulting (matrix) expression is always zero. This is the Cayley–Hamilton theorem.

4. Dual Spaces

Vector spaces come naturally in pairs, acting on each other. Thus the matrix product of a row by a column is a scalar; by this product, the "row vectors" may be regarded as linear functions of the "column vectors", and vice-versa. In §VI.10 the directional derivative of a function $z = f(x,y)$ along a path turned out to be a "product" of the gradient of f by the tangent vector to the path, so the gradients (cotangent vectors) are thereby linear functions of the tangent vectors, and vice-versa. To develop this idea, notice first that scalar-valued functions on *any* set X constitute a vector space. Specifically, if F is a field, the collection F^X of all functions $f,g\colon X \to F$ is a vector space when the vector operations are defined by the equations

$$(f + g)(x) = f(x) + g(x), \qquad (af)(x) = a(f(x)) \tag{1}$$

for all x on X and all scalars $a \in F$. One says of (1) that the operations are defined *pointwise*. For an infinite set X, this vector space F^X is infinite dimensional, but when X is finite, say with n elements, F^X is isomorphic to the familiar vector space F^n of n-tuples of elements of F.

If $X = V$ is itself a vector space over F, it is natural to consider just those functions $f\colon V \to F$ which are linear. Under the pointwise operations (1) they constitute a vector space (a subspace of F^V)

$$V^* = \hom(V,F) = \{f \mid f\colon V \to F \quad \text{linear}\} \tag{2}$$

called the *dual space* of V. This space is conceptually different from V, but when V is finite dimensional, it has the same "size" as V. Specifically, if V has a basis of n elements $u^1, \ldots, u^n$, then the value of the function f on any vector with coordinates x_i may be calculated as

$$f\left(\sum_{i=1}^{n} u^i x_i\right) = \sum_{i=1}^{n} \left(f u^i\right) x_i . \tag{3}$$

Thus the linear function f is completely determined by its n values $y^i = fu^i$ on the n given basis vectors of V. This suggests that these y^i ought to be coordinates of f. Indeed, one can define n special vectors $u_1, \ldots, u_n$ of the dual space V* by setting

$$\begin{aligned} u_i(u^j) &= 1, \qquad i = j, \\ &= 0, \qquad i \neq j, \end{aligned} \tag{4}$$

where the values $u_i(u^j) = \delta_{ij}$, called the *Kronecker* δ, are the entries of the $n \times n$ identity matrix. Then $f = \Sigma y^i u_i$, so the u_i do form an n-element basis of the dual space V^*. This basis $u_i, \ldots, u_n$ of V^* is called the *dual basis* to $u^1, \ldots, u^n$. (The latter, we note, is written with indices upstairs, all so arranged that sums $\Sigma y^i u_i$, $\Sigma u^i x_i$ over an index i have one index up, the other down.) In any event, we conclude for the dimensions

$$\dim V \text{ finite} \quad \text{implies} \quad \dim V^* = \dim V. \tag{5}$$

Despite this comparison, it is not in order to identify the space V with its dual V^*, because they are conceptually different and because the difference appears explicitly for infinite dimensional spaces. For example, take V to be the space of all finite sequences $(x_0, x_1, \ldots, x_n, 0, 0, \ldots)$ of elements of F—equivalently, all functions $\mathbf{N} \to F$ with only a finite number of non-zero values. Every vector in this space is a *finite* linear combination $\Sigma x_i u^i$ of the sequences u^i, where u^i is zero except for the scalar 1 in position i. These vectors u^i thus form a (denumerable) basis for V, so any linear function f: $V \to F$ is determined by its values $fu^i = y_i$ on these basis vectors. This means that the dual vector space V^* can be represented as the space $F^{\mathbf{N}}$ of all infinite sequences $(y_0, y_1, \ldots)$ of elements $y_i \in F$. This space V^* is much "bigger" than V; for example, if F is the field of rational numbers, V is denumerable but V^* is not. This illustrates why V^* is not V.

We have "constructed" V^* from V, but the two spaces can better be handled symmetrically. The process of *evaluating* a function $f \in V^*$ at a vector $v \in V$ is a function

$$e: V^* \times V \to F, \qquad e(f,v) = f(v) \tag{6}$$

of two variables, f in V^* and v in V. For fixed f, this function is linear in V—because f was defined to be a linear map f: $V \to F$. For fixed v, this function is linear in f—because this linearity is required in the vector space operations defined by (1) on the dual space V^*. All told, the above function e of two variables is *bilinear*—i.e., linear in each variable.

In general, for any two vector spaces W and V over the same field F, a function b: $W \times V \to F$ of two variables is said to be *bilinear* if and only if $b(w,v)$ is linear in v for each fixed $w \in W$ and linear in w for each fixed

v. This in turn means that, for each fixed w_0, $b(w_0, -)\colon V \to F$ is an element of the dual space V^*, so that there is a map

$$W \longrightarrow V^*, \qquad w_0 \mapsto b(w_0, -), \tag{7}$$

linear from W to V^*. Similarly, b determines a linear map

$$V \longrightarrow W^*, \qquad v_0 \mapsto b(-, v_0). \tag{8}$$

Theorem. *Let W and V be finite dimensional vector spaces over F. If $b\colon W \times V \to F$ is bilinear, and if for all w_0 and v_0*

$$b(w_0, -) = 0 \textit{ implies } w_0 = 0, \qquad b(-, v_0) = 0 \textit{ implies } v_0 = 0, \tag{9}$$

then b determines isomorphisms $W \cong V^*$ and $V \cong W^*$ by (7) and (8).

(A bilinear b with property (9) is called a *dual* pairing; the theorem then says that dually paired spaces are duals.)

Proof. The hypothesis (9) insures that each map (7) and (8) has null space zero; by using the basic fact that rank plus nullity equals dimension of domain it follows that both linear maps are isomorphisms.

This abstract formulation recaptures the geometric relation between the cotangent and the tangent spaces (§VI.9). At that point, the chain rule for functions of two variables involved the product of the cotangent vector (the gradient of a function) by a tangent vector (to a path). This product is bilinear and moreover satisfies (9), so is a dual pairing. Therefore, by this theorem, it makes the cotangent space the dual of the tangent space (and vice versa). Thus calculus ties in to algebraic duality.

The matrix product of a row $Y = (y_1, \ldots, y_n)$ by a column $X = (x_1, \ldots, x_n)$—an $n \times 1$ matrix—is a scalar $\Sigma y_i x_i$; this is a dual pairing; thus the one-row matrices are the *duals* of the one-column matrices.

Neither picture of duality provides room for a double dual V^{**} or for higher duals. For finite dimensions there is no such need. The evaluation $v \mapsto e(-, v)$ of (6) turns each vector v into a linear function on the dual space V^*; that is, into an element of the double dual space V^{**}. This map

$$v \mapsto e(-, v)\colon V \longrightarrow V^{**} \tag{10}$$

is an isomorphism when V is finite dimensional. Put differently, the evaluation e of (6) *is* a dual pairing of V and V^*, so by the theorem produces an isomorphism $V \cong (V^*)^*$. This is called a *natural* isomorphism because it is defined directly in (10), using only the given structure and not any special choices, such as a choice of basis. For a finite dimensional

V there is also an isomorphism $V \cong V^*$, say that sending each basis vector u^i of V to the corresponding dual basis vector u_i of V^*—but this isomorphism depends on a choice of basis, so is "unnatural".

Matrices work backwards *or* forwards. Thus a rectangular $m \times n$ matrix A gives on column vectors X a linear transformation $X \mapsto AX$ of n space to m-space. Equally well, on row vectors Y it yields a different linear transformation $Y \to YA$ of m space to n space. Alternatively, A for basis vectors comes on the right in (2.1), for coordinates, on the left in (2.2). This may be explained by duality. If T: $V \to W$ is a linear transformation (say, the transformation on column vectors given by the matrix A), then each vector f in the dual space W^* is really a linear map f: $W \to F$, hence the composite fT: $V \to F$ is an element of the dual space V^*, so that $f \mapsto fT$ is itself a linear mapping

$$T^*\colon W^* \longrightarrow V^* \tag{11}$$

—with direction opposite that of T. Moreover $(TS)^* = S^*T^*$; one says that the formation of the dual applies to spaces $V \mapsto V^*$ and to their maps, as $T \mapsto T^*$, is a *contravariant* functor (of spaces and maps).

The construction of V^* can be formulated more generally. The addition (term by term) of two rows of the same length suggests the addition, entry by entry, of two matrices of the same size. More conceptually, for given vector spaces V and W, two linear transformations T, T': $V \to W$ have a sum $T + T'$: $V \to W$ by the (pointwise) formula

$$(T + T')v = Tv + T'v.$$

With pointwise multiplication of each T by scalars, this makes the collection of all the "homomorphisms" T: $V \to W$ into a vector space, called

$$\hom(V,W) = \{T \mid T\colon V \longrightarrow W \text{ linear}\}. \tag{12}$$

In case W is just the base field F, regarded as a vector space over itself, this space $\hom(V,F)$ is just the dual space V^*. If V and W have dimensions m and n, a choice of basis in each replaces each T by a matrix and so proves the vector space $\hom(V,W)$ isomorphic to the space of all $n \times m$ matrices with entries in F. Its dimension is then nm, with a basis (say) those matrices which have one entry 1 and all other entries zero.

5. Inner Product Spaces

For many geometric purposes, linear vector spaces are not enough. For example, a linear transformation does carry straight lines to straight lines, but it can distort angles, change circles into ellipses, and generally alter distances. However, in discussing trigonometry and the rotations of the plane, we already saw in §IV.9 the utility of the inner product of two plane vectors in describing angles and distances. It turns out here (as in so many other cases) that the two-dimensional case is typical: One can get the appropriate "Euclidean" geometry in an n-dimensional space simply by adjoining to the vector space structure the additional structure of an inner product.

An *inner product space* E is thus defined to be a vector space E over the field $\mathbf{R}$ of reals with an inner product–a real valued function $u \cdot v$ of two vectors u and v–which (as in equations (10), (11), and (12) of §IV.9) is bilinear, symmetric, and positive definite. Such spaces are at hand. Thus for each dimension n the space $\mathbf{R}^n$ of n-tuples $(x_1, \ldots, x_n)$ of real numbers has the "standard" inner product

$$(x_1, \ldots, x_n) \cdot (y_1, \ldots, y_n) = x_1 y_1 + \cdots + x_n y_n . \tag{1}$$

There is a Hilbert space l^2 consisting of all those infinite sequences of $(x_1, \ldots)$ of reals for which the sum of the squares is convergent. Here Σx_i^2 and Σy_i^2 convergent imply the convergence of the inner product

$$(x_1, x_2, \ldots) \cdot (y_1, y_2, \ldots) = x_1 y_1 + x_2 y_2 + \cdots . \tag{2}$$

Similarly, analysis uses the space L^2 of equivalence classes of all those real-valued functions f (say on the unit interval) which are Lebesgue square integrable (§VI.11). The inner product in this space is the Lebesgue integral

$$f \cdot g = \int_0^1 f(x) g(x) dx . \tag{3}$$

With any inner product, the *length* or the *norm* $|u|$ of each vector can be defined by the equation

$$|u|^2 = u \cdot u, \qquad |u| \geqslant 0, \tag{4}$$

while the angle θ between two vectors u and v is determined by a variant of the law of cosines as

$$u \cdot v = |u| \; |v| \cos \theta ; \tag{5}$$

in particular, u is perpendicular (orthogonal) to v when $u \cdot v = 0$). If this angle θ is to make sense, one must have the *Schwarz inequality*: For any two vectors the norm of the inner product is

$$|u \cdot v| \leqslant |u| \; |v| . \tag{6}$$

This can be proved from the axioms of an inner product space. Moreover one can define a distance between two vectors u and v–that is, the distance between their end points–as $|u - v|$. From the Schwarz inequality it follows that this distance satisfies the standard triangle inequality. Hence, with this distance, the inner product space is a metric space–as it should be. In fact, in $\mathbf{R}^n$ with the standard inner product (1), the distance between two n-tuples X and Y is just

$$((x_1 - y_1)^2 + \cdots + (x_n - y_n)^2)^{1/2};$$

we have returned to the starting point of geometry with the Pythorgorean theorem–now in n dimensions.

For an arbitrary basis $u^1, \ldots, u^n$ of a finite dimensional inner product space, the inner product of any two basis vectors $u^i \cdot u^j$ is a scalar g^{ij}, and these scalars determine the inner product of any two vectors in terms of their coordinates as

$$\left[\sum u^i x_i\right] \cdot \left[\sum u^j y_j\right] = \sum_{i,j} g^{ij} x_i y_j ; \qquad (7)$$

the right hand side is the coordinate expression of the inner product as a "bilinear form" in the x's and the y's. However, for the standard inner product the matrix $G = (g^{ij})$ of this form is just the $n \times n$ identity matrix. This can be arranged in any finite dimensional inner product space, by choosing c *normal orthogonal* basis–one in which the inner product $u^i \cdot u^j$ is the Kronecker δ_{ij}. This requirement means geometrically that each basis vector has length 1 and any two are perpendicular (orthogonal). Any given basis $v^1, \ldots, v^n$ can be made orthogonal as follows: First tip v^2 to be perpendicular to v^1 (by adding to v^2 an appropriate multiple of v^1). Then tip v^3 to be orthogonal to v^1 and v^2, and so on. Finally shrink each of the resulting vectors to a length 1. This is called the Gram–Schmidt process. Similar uses of normal and orthogonal bases crop up for infinite dimensional spaces in Fourier series, Hilbert spaces, and elsewhere. The idea is transportable from geometry to analysis.

The inner product is automatically a dual pairing of the space with itself. Hence each finite dimensional space with an inner product is isomorphic to its dual space–because each vector v is also a linear function "inner product with v". Hence for *such* spaces one can drop the distinction between a space and its dual.

6. Orthogonal Matrices

In a space with an inner product one has all the concepts of Euclidean geometry: One can define spheres and rigid motions. For such a space E, the appropriate endomorphisms are the orthogonal transformations; those functions $T: E \to E$ which are linear *and* which preserve the inner product, in the sense that $Tu \cdot Tv = u \cdot v$ for all pairs of vectors u, v. This means also that T preserves distances and angles, and carries orthonormal vectors to such. Indeed, if T carries any one normal orthogonal basis of E into a normal orthogonal basis, then T is an orthogonal transformation—because the inner product is computed by the standard formula (5.1) from the coordinates relative to this basis.

A square matrix A is *orthogonal* if and only if the corresponding transformation is orthogonal (relative to the standard inner product (5.1)). Since this transformation on column vectors X is $X \mapsto AX$, it carries the successive basis vectors into successive columns of A. Hence the matrix A is orthogonal if and only if its columns are normal and orthogonal—each of length 1 and any two orthogonal. If A^T is the transposed matrix, this is exactly the statement that the matrix product A^TA is the identity matrix I, so that the transpose A^T of A is its left inverse. But such a left inverse is also a right inverse. Hence a square matrix A is orthogonal if and only if

$$A^TA = I = AA^T; \tag{1}$$

its transpose is its inverse. Since the determinant of a product is the product of the determinants, this equation implies that an orthogonal matrix has determinant ± 1 And if the orthogonal matrix A has a real eigenvector $X \neq 0$, the equation $AX = X\lambda$ and preservation of length implies that $\lambda = \pm 1$—the only possible real eigenvalues of an orthogonal matrix.

In two dimensions, an orthogonal transformation is a rigid motion leaving the origin fixed, so must be either a rotation or a rotation followed by a reflection. Hence the only 2×2 orthogonal matrices are $\pm A_\theta$ where

$$A_\theta = \begin{bmatrix} \cos\theta & -\sin\theta \\ \sin\theta & \cos\theta \end{bmatrix} \tag{2}$$

is the matrix of a rotation through the angle θ about the origin, just as in equation (IV.9.6). The characteristic polynomial of this matrix is $\lambda^2 - 2\lambda\cos\theta + 1$; hence the eigenvalues are $\lambda = \cos\theta \pm i\sin\theta$—non-real complex numbers unless $\theta = \pi k$ for some integer k. Thus, except for the identity rotation and the half-turn, there are no (real) eigenvectors—no vectors rotated into a multiple of themselves. The eigenvalues are complex numbers of absolute value 1.

This fact is general. The eigenvalues of an orthogonal matrix A (and hence of an orthogonal transformation) are always complex numbers of absolute value 1. To see this, we treat the real-number entries in A as complex numbers, and let A act on the complex vector space $\mathbf{C}^n$ of columns X of n complex numbers. In this space $\mathbf{C}^n$, the inner product should be given by the formula

$$(x_1, x_2, \ldots, x_n)\cdot(y_1, \ldots, y_n) = x_1y_1{}^* + \cdots + x_ny_n{}^* \tag{3}$$

where y^* denotes the complex conjugate of y. This inner product is bilinear and positive definite, because $yy^* \geqslant 0$. However, it is not symmetric. Instead it is antisymmetric, in the sense that complex conjugation gives

$$(v \cdot u) = (u \cdot v)^*$$

for any two vectors u and v. A vector space over $\mathbf{C}$ with such an inner product (bilinear, positive definite, and antisymmetric) is called a *unitary* space. The theory of these "unitary" spaces over the complex numbers is wholly analogous to the theory of inner product spaces over the reals. In particular, a linear transformation is *unitary* if and only if it preserves the inner product, and an $n \times n$ matrix A of complex numbers is unitary if and only if its conjugate transpose is its inverse. It follows that the eigenvalues of a unitary matrix (a transformation) all have absolute value 1. In particular a real orthogonal matrix is necessarily unitary–and hence the result cited above about its eigenvalues. This is again a case where problems about real numbers require complex numbers for a solution.

From the definition it follows that the $n \times n$ orthogonal matrices form under multiplication a group, called the *orthogonal group* 0_n. Equivalently, it can be described as the group of all orthogonal endomorphisms of an n-dimensional inner product space. This group includes reflections which invert the orientation of space, where orientation can be described as in §III.8. The matrices of such transformations have determinant ± 1. The *proper* orthogonal matrices (those with determinant $+1$) form a subgroup $S0_n$ of 0_n, called the *special orthogonal group*.

An orthogonal transformation may also be described as a rigid motion which leaves the origin (the vector 0) fixed. Just as in the plane, the most general rigid motion of an n-dimensional inner product space is an orthogonal transformation followed by a translation.

To summarize, we see that the geometric ideas of space, transformation, and rigid motion extend naturally beyond two and three dimensions (and, for analysis, into infinite dimensions) and have an effective algebraic formalization by vector spaces, linear orthogonal transformations, and their representations by matrices.

7. Adjoints

As we have seen, any linear transformation $T: E \to E$ has a dual $T^*: E^* \to E^*$. But if E is an inner product space, E can be identified with E^* by regarding each vector v of E as the linear function $v \cdot -$, "inner product with v". Then T^*v, by definition of the dual T^*, is just the composite function $v \cdot T$. In other words,

$$(T^*v) \cdot u = v \cdot Tu \qquad (1)$$

for all vectors u in E. Since a vector is determined by giving its inner product with all other vectors u, this equation defines T^*v and hence defines the linear transformation T^*. It is called the *adjoint* of T. The definition (1) is often written with a bracket notation for inner product, as

$$<T^*v,u> \; = \; <v,Tu>. \tag{2}$$

Now suppose that T is given by a matrix A with entries a_{ij}, relative to some normal orthogonal basis u_i. Then, as in (2.1), $Tu_j = \Sigma u_i a_{ij}$. According to the definition (1)

$$(T^*u_k)\cdot u_j = u_k \cdot Tu_j = u_k \cdot \sum_i u_i a_{ij} = a_{kj},$$

because $u_k \cdot u_i = \delta_{ki}$. For the same reason,

$$\left[\sum_i u_i a_{ki}\right] \cdot u_j = a_{kj}, \qquad j,k = 1, \ldots, n.$$

Therefore $T^*u_k = \Sigma u_i a_{ki}$. In other words, T^* is given by the transposed matrix A^T whose ij entry is a_{ji}. In short, adjoint transformations have transposed matrices (relative to orthonormal bases). This is the intrinsic meaning of a transposed matrix.

Adjoints probably appeared first in the study of homogeneous second order linear differential equations. Such an equation may be written as $L(y) = 0$ where y is a (smooth) function of x, $'$ denotes d/dx and L is the differential operator

$$L(y) = (py')' + ry' + qy,$$

with p, q, and r suitable functions of x. Classical texts then simply decree that the "adjoint" operator on a function z is (Frank–von Mises [1930], p. 363)

$$M(z) = (pz')' - (rz)' + qy.$$

More conceptually, regard L and M as linear operators on a suitable space of functions in which the inner product of functions y and z is the integral, over some x interval from a to b, of the product function yz. Then if y and z vanish at the ends of the integral, integration by parts shows that

$$\int_a^b L(y)z = \int_a^b yM(z).$$

Thus L is indeed adjoint to M in exactly the sense (1). Corresponding notions for partial differential operators involve Gauss–Green–Stokes. These ideas turned out to be essential in the applications of Hilbert space to quantum mechanics.

We return to the case of a finite-dimensional inner product space E.

A linear endomorphism $T\colon E \to E$ is *self-adjoint* when $T^* = T$; that is, when its matrix A relative to any normal orthogonal basis is symmetric (equal to its transpose). Thus, for an inner product space E:

Theorem. *All the eigenvalues of a self-adjoint $T\colon E \to E$ are real.*

The proof must use complex numbers (to prove a theorem about real vector spaces!); it will be convenient to formulate it in terms of a (real) symmetric matrix A representing T. By the fundamental theorem of algebra, the characteristic polynomial $|A - \lambda I|$ of A has (possibly) complex roots; they are the eigenvalues of A. Take one such eigenvalue λ. There must then be a non-zero and (possibly)-complex column vector X which is an eigenvector, so that $AX = X\lambda$. Let * denote the operation which transposes each matrix *and* takes the complex conjugate of each entry; then $X^*A^* = \lambda^*X^*$, where X^* is a row and $A^* = A$ because A^* is real and symmetric. By the associative law for matrix multiplication (row by matrix by column)

$$(X^*A)X = X^*(AX),$$

or, evaluating each by $AX = X\lambda$ and $X^*A = \lambda^*X^*$,

$$\lambda^*(X^*X) = \lambda(X^*X).$$

But $X \neq 0$, and so $X^*X \neq 0$. Hence $\lambda^* = \lambda$, so that the eigenvalue λ is real, as asserted.

In this proof the matrix A originally gave an endomorphism $X \mapsto AX$ of the real vector space E, but we changed it to give an endomorphism, written also $X \mapsto AX$, of a complex vector space E' of the same dimension. This change of field was done *in terms of* a basis of E. It is desirable to describe the change invariantly *without* using a basis. This we will soon do, in §9.

8. The Principal Axis Theorem

In the analytic geometry of the plane, ellipses and hyperbolas can be described as the loci of equations

$$\frac{x^2}{a^2} \pm \frac{y^2}{b^2} = 1 \tag{1}$$

(+ for ellipses, − for hyperbolas). What about the locus of a more general quadratic equation

$$ax^2 + 2bxy + cy^2 = 1 ? \tag{2}$$

This equation can also be written in matrix form as

$$(x,y)\begin{bmatrix} a & b \\ b & c \end{bmatrix}\begin{bmatrix} x \\ y \end{bmatrix} = 1 \tag{3}$$

with a 2×2 symmetric matrix, call it A, in the middle. Can the equation be so simplified that the matrix A becomes diagonal, as in (1), just by rotating the coordinate axes? This amounts to multiplying the column vector of coordinates by an orthogonal matrix P; this will evidently replace the matrix A in (2) by the matrix P^TAP. But P is orthogonal, so its transpose P^T is its inverse P^{-1}, and thus A has been replaced by the similar matrix (§2) $P^{-1}AP$. This means that we now think of A not as the coefficients in the quadratic form (2) but as the matrix of a (self-adjoint) transformation. In a moment we will prove

Theorem 1. *In an inner product space, each self-adjoint $T: E \to E$ has a diagonal matrix relative to a suitable orthonormal basis of E.*

For the equation (2) this theorem means that with new (rectangular) coordinates x' and y' the equation reads

$$\lambda x'^2 + \mu y'^2 = 1,$$

where λ and μ are the (real) eigenvalues of the 2×2 symmetric matrix A. Then if both are positive, the equation represents an ellipse (taken relative to its "principal" axes); if $\lambda > 0$ and $\mu < 0$, it represents a hyperbola, again relative to "principal" axes. If both λ and μ are negative or zero, there is no (real) locus, while if $\mu = 0$ and $\lambda > 0$ the locus is a line $x' = \lambda^{-1/2}$. In each case, the theorem has reduced the curve to principal axes by a rotation (possibly followed by a reflection) of the coordinate axes.

To prove the theorem for a space E of any finite dimension n, take a real eigenvalue λ of the self-adjoint transformation T. The corresponding eigenvector can be multiplied by a suitable scalar to have length 1. We can then make this vector the first vector of an orthonormal basis of E. The matrix A for the transformation T relative to this new basis must then carry the eigenvector with coordinates $(1,0,\ldots,0)$ into $(\lambda,0,\ldots,0)$, so the latter vector is the first column of the matrix, and the matrix has the form

$$A = \begin{bmatrix} \lambda & \text{---} \\ \cdot & \\ \cdot & B \\ \cdot & \\ 0 & \end{bmatrix},$$

where B is some $(n - 1) \times (n - 1)$ matrix. But A is symmetric, so the first row here is really $(\lambda, 0, \ldots, 0)$ and the lower corner B is also symmetric. Induction applied to this B then gives the result desired.

This theorem is sometimes called the *principal axis theorem* because of its application to the geometry of ellipses and hyperbolas, and also to the corresponding result in three dimensions for ellipsoids and hyperboloids of one or two sheets. The theorem can be stated simply as a theorem about matrices, in two different ways:

For every real symmetric matrix A there is an orthogonal matrix P with $P^{-1}AP$ diagonal, or

For every real symmetric matrix A there is an orthogonal matrix P with P^TAP diagonal.

In both cases, the diagonal entries are the eigenvalues.

For the first statement, A is considered as the matrix of a self adjoint transformation T; for the second A is considered as the matrix of a *quadratic form* $X^TAX = \Sigma x_i a_{ij} x_j$; it then amounts to the assertion that every such quadratic form can be made diagonal by an orthogonal change in the coordinates X. This is a striking example of the fact that a formal result (here, one about matrices) has two different geometric interpretations—one for transformations, and the other for quadratic forms.

There is still another formulation of Theorem 1. For each eigenvalue λ of $T: E \to E$ consider the subspace E_λ consisting of *all* the corresponding eigenvectors. This subspace has at least the dimension 1, and in fact its dimension is the "multiplicity" of the eigenvalue λ; that is, its multiplicity as a root of the characteristic polynomial. The theorem can then be read to say that every vector v in E can be written uniquely as a sum $\Sigma_\lambda v_\lambda$ of vectors v_λ in E_λ. This is said to make the space E the *direct sum* of its subspaces E_λ. In the infinite dimensional case a more complicated statement of this character will apply; it is called the *spectral theorem*.

Suitable versions of all these results apply also to spaces over the complex numbers. They deal not with matrices which are symmetric but with *hermitian* matrices—those equal to their conjugate transpose. This case is specifically relevant in analysis.

9. Bilinearity and Tensor Products

For three vector spaces U, V, and W over the same field F, a function

$$B: U \times V \to W \tag{1}$$

is said to be *bilinear* when the values $B(u,v)$ are linear in $u \in U$ for each fixed vector v and also linear in $v \in V$ for each fixed u. For example, the inner product (5.7) is bilinear. It turns out that *all* of the bilinear functions B described under (1) can be represented by linear functions and a single bilinear function $\otimes$ from $U \times V$ to a new space, called the *tensor product* of U and V. The properties of this space are very useful for a wide range of geometric and algebraic problems. Its existence is asserted in the following theorem.

Theorem 1. *Given two vector spaces* U and V *over a field* F *there is a third vector space* $U \otimes V$ *and a bilinear function*

$$\otimes : U \times V \to U \otimes V \tag{2}$$

with the following property: Whenever B *is a bilinear function (1) on* U *and* V *to any third space* W *over* F, *there is a unique linear transformation* $T\colon U \otimes V \to W$ *such that*

$$B(u,v) = T(u \otimes v) \tag{3}$$

holds for all vectors $u \in U$ *and* $v \in V$.

This property (3) may also be formulated in a diagram

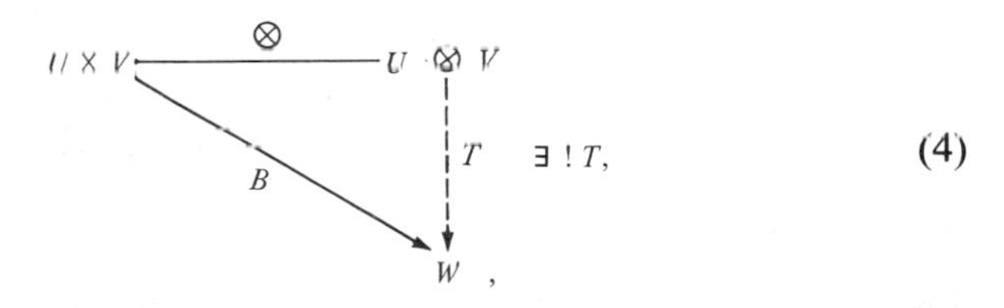

to express the idea that the new bilinear function $\otimes$ *has the property that to any old bilinear* B *there is exactly one linear* T *(dotted arrow) which makes* $T\otimes = B$*; that is, which makes this diagram "commute".*

The force of this theorem is to reduce as much as possible about bilinear functions B to linear functions T—on the grounds that we know more how to handle the linear ones. Since all the bilinear functions B are obtained—and uniquely—from the one bilinear $\otimes$ it is appropriate to call this one bilinear function $\otimes$ a *universal* one; specifically, universal for the given spaces U and V.

Here is one employment for the tensor product space. Just as we have constructed the dual space $V^* = \hom(V,F)$ and the space $\hom(V,W)$ of linear transformations (or matrices), so we need a space

$$\text{Bilin}(U,V;W)$$

of all bilinear functions $B\colon U \times V \to W$. But by the theorem the bijection $B \mapsto T$ of the diagram (4) reduces this space to one previously constructed, by an isomorphism of vector spaces

$$\mathrm{Bilin}(U, V; W) \cong \hom(U \otimes V, W). \tag{5}$$

This result is sometimes formulated by saying that bilinear functions are *represented* by $U \otimes V$; in some sense they are "objectified" by $U \otimes V$. There are similar representations for trilinear functions.

Before we indicate how to construct the desired universal bilinear function $\otimes$ of (4), let us observe that it will not matter *how* it is constructed, provided only that it has the property stated under (4)–in other words, the formal property matters, and not the explicit elements of the desired space $U \otimes V$. More exactly, suppose that we had two such universal bilinear functions $\otimes$ and $\square$ into (say) two apparently different spaces such as $U \otimes V$ and $U \square V$. Then we have a diagram

$$\begin{array}{ccc} U \times V & \xrightarrow{\quad\otimes\quad} & U \otimes V \;\circlearrowleft\; S\cdot T \\ \Vert & & S \uparrow\,\downarrow T \\ U \times V & \xrightarrow{\quad\square\quad} & U \square V \end{array} \tag{6}$$

with both $\otimes$ and $\square$ bilinear and universal. Now since $\otimes$ is universal, it is universal for $\square$, and so there is a linear T as shown, with $T \otimes = \square$. Vice versa, since $\square$ is universal, there is a linear S as shown, with $S\square = \otimes$. This means that the composite $S \cdot T$: $U \otimes V \rightarrow U \otimes V$ has $(S \cdot T)\otimes = \otimes$. But the identity map 1 of $U \otimes V$ also has this property $1 \otimes = \otimes$, while the definition of the universality of $\otimes$ said that there is only one such linear map. Therefore $S \cdot T$ is the identity of $U \otimes V$. By exactly the same argument $T \cdot S$ is the identity (of the second space $U \square V$). In other words, S is a two-sided inverse for T, so T is an isomorphism. The two spaces $U \otimes V$ and $U \square V$ are isomorphic–and by an isomorphism which carries the universal bilinear $\otimes$ in the first space into the $\square$ for the second space.

Note first that this is really not an argument just about the universal bilinear function, but about a universal what-not. Another example has already appeared. In §V.8, the free product $A * G$ of two groups was defined to have as elements certain words $a_1 g_1 \ldots a_n g_n$ in elements of A and of G; then there were two group homomorphisms k_1: $A \rightarrow A * G \leftarrow G$ and k_2 which were "universal" among all pairs of homomorphisms from A and G to a third group H, as in (V.8.4)

$$A \longrightarrow H \longleftarrow G.$$

The argument we have just given for the tensor product shows that the free product $A * G$ with the homomorphism k_1 and k_2 is determined uniquely, up to an isomorphism respcting k_1 and k_2, by this universal property. In other words, what matters about the free product $A * G$ is not its specific construction by means of words, but its universal property.

Recall also that the (direct) product $A \times G$ of two groups with its projections $A \leftarrow A \times G \rightarrow G$ was also shown to have a universal property—see (V.8.2). In other words, the product of groups (or of spaces) need not be constructed from ordered pairs of elements. Any other construction will do, provided only that it yields the universal property!

Now we can return to formulate our construction for the tensor product $\otimes$ of spaces, as in (2). Suppose for instance that U and V are finite dimensional, with bases say $u_1, \ldots, u_m$ and $v_1, \ldots, v_n$. First take mn symbols $u_i \otimes v_j$ for $i = 1, \ldots, m$ and $j = 1, \ldots, n$ and manufacture a new vector space $U \otimes V \cong F^{mn}$ over F with these symbols as basis. Its vectors are then formal linear combinations $\Sigma a_{ij}(u_i \otimes v_j)$. Now consider any bilinear B to any space W. By its bilinearity, the values of B are all determined once one knows its values $B(u_i, v_j)$ on the basis vectors. Now if we define T by

$$T(\sum a_{ij}(u_i *X\ u_j)) = \sum_{i,j} a_{ij} B(u_i, v_j)$$

we get a linear map $T\colon U \otimes V \rightarrow W$ with the desired property that $T\otimes = B$; clearly this is the only such T with this property.

This construction used arbitrarily chosen bases, but by earlier remarks, the same construction with any other bases would give an isomorphic result. The construction also gives the dimension for the tensor product (of finite dimensional spaces) as

$$\dim(U \otimes V) = (\dim U)(\dim V). \tag{7}$$

The construction also shows that the universal bilinear function $\otimes$ yields "product" vectors $u \otimes v$ in the tensor product. These product vectors certainly *span* $U \otimes V$, but there are in general many vectors in $U \otimes V$ which are not just products $u \otimes v$, but linear combinations of such products.

For an infinite dimensional space U (or V) much the same proof will give the existence of the tensor product $U \otimes V$—once one has proved (using Zorn's lemma) that every vector space has a (possibly infinite) basis—one such that each vector in the space is uniquely a linear combination of a finite number of basis vectors. A more invariant construction of $U \otimes V$, done without any use of bases, will appear in the next section.

The tensor product allows us to "change the field" (of scalars). For example $\mathbf{C}$ is a two-dimensional vector space over the reals $\mathbf{R}$. For any real vector space V the tensor product $V \otimes \mathbf{C}$ has scalar multiples (on the right) by $\mathbf{C}$. Thus $V \mapsto V \otimes \mathbf{C}$ changes the real vector space V to a complex one (of the same dimension)—as needed in §7 above.

Starting from one vector space V and using also its dual space V^* one may now construct all sorts of tensor spaces, such as the space $V^* \otimes V \otimes V$—of mixed tensors with one contravariant and two covariant indices. Once given a basis $u^1, \ldots, u^n$ of V one then has the dual basis $u_1, \ldots, u_n$ of V^* and therefore, in the above case, a basis for $V^* \otimes V \otimes V$ consisting of n^3 vectors $u_i \otimes u^j \otimes u^k$. One may then exhibit a tensor by its coordinates relative to such a basis. One may readily derive the (many indexed) formulas for changing coordinates from one basis to another. They were once used to *define* what a tensor is, but this we have avoided by the invariant definition of the tensor product itself. For an inner product space $V = E$, there are also formulas for shifting upstairs indices downstairs–on the basis of the canonical isomorphism $E \cong E^*$.

There are many other useful properties of tensor products. For example, if $T: U \to U'$ and $S: V \to V'$ are two linear transformations, then $\langle u,v \rangle \mapsto Tu \otimes Sv$ is a bilinear function $U \otimes V \to U' \otimes V'$; hence there is by universality a corresponding linear function

$$T \otimes S: U \otimes V \to U' \otimes V' \tag{8}$$

with $(T \otimes S)(u \otimes v) = Tu \otimes Sv$. In other words, the tensor product operation applies not just to spaces, but also to the linear transformations between them. It is said to be *functorial* (Chap. XII).

The triple tensor product $U \otimes (V \otimes W)$ of three vector spaces produces a *trilinear* function $(u,v,w) \mapsto u \otimes (v \otimes w)$ which is clearly universal among trilinear functions on $U \times V \times W$. Now, reassociating parentheses, $(u \otimes v) \otimes w$ is also a univeral trilinear function to the space $(U \otimes V) \otimes W$. Since the universal property determines the space up to an isomorphism, there is an isomorphism

$$U \otimes (V \otimes W) \cong (U \otimes V) \otimes W, \qquad u \otimes (v \otimes w) \mapsto (u \otimes v) \otimes w \tag{9}$$

It is convenient to identify these two spaces by this isomorphism–then the tensor product of vectors has the handy property of associativity.

10. Collapse by Quotients

In this chapter, our subject is linear maps. One needs means to construct them; one such is the process which collapses a vector space to a quotient space.

The study of congruence of integers modulo an integer m leads (as described in §II.8) to the construction of the ring $\mathbf{Z}_m$ of integers modulo m, with the corresponding homomorphism (or projection) $\mathbf{Z} \to \mathbf{Z}_m$ from the ring $\mathbf{Z}$ of integers. The effect of this homomorphism is to map all the multiples of m to zero–in effect, to collapse the set of all multiples of m. Similarly for groups: The factor group G/N collapses the elements of any given normal subgroup N.

A corresponding process applies to a vector space V. If one wishes to ignore a particular vector u_0 and its multiples, one may map V into a newly constructed space in which exactly the subspace consisting of these multiples is mapped to zero. For example, in cartesian 3-space the vertical projection onto the $x-y$ plane collapses exactly all the multiples of the unit z-vector. The same process works more generally to collapse any subspace S of V. There is a new space V/S, called the *quotient space*, and a linear transformation P: $V \to V/S$ with $P(S) = 0$ and the following universal property: Any linear transformation T: $V \to W$ into a third vector space W for which $T(S) = 0$ can be written as a composite $T = T' P$ for a unique linear transformation T': $V/S \to W$, as in the following commutative diagram:

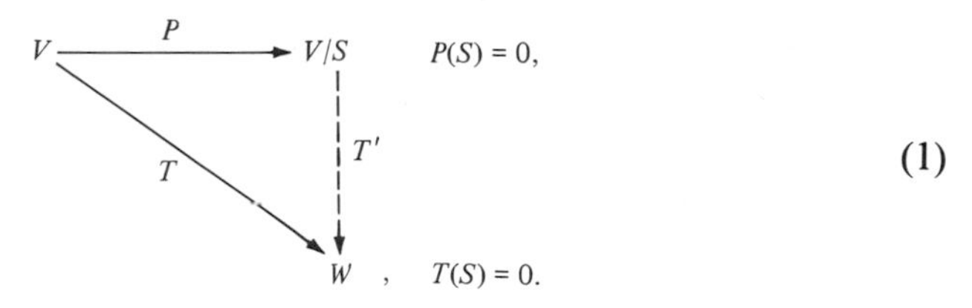

This may be proved by taking P to map each vector v of V into the "coset" (or hyperplane) $S + v$ consisting of all vectors $s + v$ for s in S; these cosets themselves form a vector space under operations such as the addition given by $(S + v) + (S + w) = S + (v + w)$. Essentially, these cosets are just like the congruence classes $C_m(a)$ modulo m used to construct the integers modulo m, because each coset $S + v$ consists of all vectors v' in V with $v' - v$ in S (in effect, with $v' \equiv v \pmod{S}$). However, once the universal property (1) is established, it no longer matters how the elements of this *quotient space* V/S are described, because it is this universal property which formalizes everything about the quotient (and which proves that the quotient is unique up to an isomorphims of vector spaces).

This result allows one to collapse any given collection of vectors $s_1, \ldots, s_k$ of a space V—simply take the subspace S to be the set of all linear combinations $a_1 s_1 + \cdots + a_k s_k$ of these vectors; that is, the subspace *spanned* by the s_i; clearly any transformation T which maps all s_i to zero must map to zero all of the subspace.

As a typical example of the use of this process, let us describe another construction of the tensor product $U \otimes V$ of two vector spaces U and V. First take all pairs $\langle u,v \rangle$ of vectors $u \in U$ and $v \in V$, and write each pair as if it were a product $u \cdot v$. Form the (very large) vector space L which has all these pairs $u \cdot v$ as its basis; thus L consists of all finite linear combinations of symbols $u_i \cdot v_i$ with coefficients in the field F. Now $\langle u,v \rangle \mapsto u \cdot v$ is a function F: $U \times V \to L$, but it is by no means bilinear. One can force it to be bilinear simply by collapsing in L all the things which would be zero if F were bilinear—thus collapsing all elements

$$(a_1 u_1 + a_2 u_2)\cdot v - a_1(u_1 \cdot v) - a_2(u_2 \cdot v) \tag{2}$$

and similarly for linearity in the second factor v. Now F becomes a bilinear functions $U \times V \to L/S$ into the collapsed space. One may verify that it is universal, so that this construction L/S is indeed a tensor product.

This example illustrates a use of infinite dimensional spaces to construct finite dimensional ones–and also shows that there are many different formal ways to construct what is effectively the same tensor product.

Mathematics uses many other types of "collapse" for other structures. In each case one must note first what can be collapsed; that is, what can be mapped to zero by a homomorphism: For vector space, any subspace; for groups any normal subgroup. For a ring R, define an *ideal* of R to be a subset $K \subset R$ such that k_1 and k_2 in K imply $k_1 + k_2$ in K, while k in K and r in R imply that rk and kr are in K. In any homomorphism $f\colon R \to S$ of rings, the subset of R consisting of all $k \in R$ with $fk = 0$ is necessarily an ideal. This ideal is called the *kernel* of the homomorphism. Conversely, given any ideal K in a ring R, there is a ring R/K and an homomorphism $R \to R/K$ with kernel K which is universal among homomorphisms from R with kernel containing K. This *quotient ring* may be constructed by taking its elements to be cosets $K + r$ of K. This construction includes the case of integers modulo m, because the set of all multiples of m is an ideal in the ring $\mathbf{Z}$ of integers. It also includes one of the constructions of the complex numbers $\mathbf{C}$ from the field R of real numbers. For this, first form the ring R of all polynomials in a symbol x with real coefficients. In this ring, the set of all multiples of the polynomial $(x^2 + 1)$ is an ideal K, and the quotient ring R/K is (isomorphic to) the complex numbers–because the collapse of $x^2 + 1$ to 0 forces x (when collapsed) to satisfy the equation $x^2 = -1$ defining the basic complex number i.

This process also makes Galois theory possible without any recourse to complex numbers. Over any (abstract) field F the polynomials in x form a ring $R = F[x]$ in which the multiples of any one polynomial $f(x)$ form an ideal (f). If f is irreducible, the quotient ring $R/(f)$ is a field containing F and an element ξ (the coset of x) with $f(\xi) = 0$. In short, $R/(f)$ is a field $F(\xi)$ generated by F and one root of f; the remaining roots can be adjoined in the same way to give an (abstract) splitting field and its Galois group.

This notion of ideal, useful for collapses, will turn out to have other uses in number theory and arithmetic (§XII.3).

11. Exterior Algebra and Differential Forms

For differential forms and the relations (Gauss–Green–Stokes) between surface integrals and volume integrals we needed to manipulate differentials dx and dy according to the formal rules

$$(dx)(dy) = -(dy)(dx), \qquad (dx)^2 = 0. \tag{1}$$

These are rules intended to apply to the vectors dx, dy, dz in the cotangent space at a point, say, in Euclidean 3-space. They need not be mysteious, because they are algebraic rules which can be achieved by suitable collapse of formal products taken over any vector space V. This collapse leads to *exterior algebra*.

Starting with V, the successive tensor products yield a string of vector spaces

$$F,\ V,\ V \otimes V,\ V \otimes V \otimes V, \ldots \tag{2}$$

Their elements are called (covariant) *tensors*; those in the n-fold tensor product T_n being tensors of *rank n* while (for completeness) the scalars in F count as tensors of rank 0. Taken together, all these tensors form an algebraic system with the following rules of operation: Any tensor can be multiplied by a scalar; any two tensors of the same rank can be added; any two tensors can be multiplied by using the product $\otimes$; for example, the product of the tensors u and $v \otimes w$ is

$$u \cdot (v \otimes w) = u \otimes v \otimes w. \tag{3}$$

All these tensors, with these algebraic operations, form the so-called *tensor algebra* $T(V)$. This algebraic system isn't quite a vector space, because we do not (and do not need to) add two tensors of different ranks; for the same reason it is not quite a ring under sum and product. Technically, it is called a *graded* algebra; i.e., both a *graded* vector space and a *graded* ring (with grading by ranks); see *Algebra* XVI.4). The reader will find it straightforward to write down the axioms for such a graded system, noting that the convention (8.9) about triple tensor products ensures that the multiplication is associative. However, the multiplication is *not* commutative.

Every tensor is a sum of products of vectors, so that $T(V)$ is generated by the vector space V; it is the "most general" graded algebra so generated, and so is called the *free* graded algebra over V. It thus has a "universal" property, as follows: For any graded algebra A, each linear transformation $V \to A$ into the vector space of elements of grade 1 in A can be extended uniquely to a morphism $T(V) \to A$ of graded algebras. In consequence, any such A generated by V can be obtained by suitably collapsing the tensor algebra $T(V)$.

For the graded algebra of differentials (1) we need a product with $u \otimes v = -v \otimes u$ and $u \otimes u = 0$. These identities do not hold in the tensor algebra $T(V)$. Hence we simply take the appropriate quotient of $T(V)$. This can be done by collapsing in $T(V)$ all the elements $t \otimes t$ (for any tensor t of positive rank) and all the multiples (left and right) of any such $t \otimes t$ by another tensor. The resulting quotient algebra $E(V)$ is called the *exterior algebra* of V. We will write its elements as e, f, g and the product of two such as $e \wedge f$. By its construction, this algebra satisfies

$$e \wedge e = 0 \tag{4}$$

for all e of positive rank. This in its turn implies

$$0 = (e + f) \wedge (e + f) = e \wedge e + e \wedge f + f \wedge e + f \wedge f.$$

Since $e \wedge e$ and $f \wedge f$ are 0, this gives the desired equation

$$e \wedge f = -f \wedge e \tag{5}$$

again for e and f of positive rank.

To pin this algebra down, write $E_k = E_k(V)$ for its elements of rank k, so that E_k is a vector space quotient of the k-fold tensor product $T_k(V) = V \otimes \cdots \otimes V$. Since we have divided out only by "squares" $t \otimes t$ with t of positive rank, nothing in rank 0 or 1 has been collapsed, so that $E_1(V)$ is just $T_1(V) = V$. Now suppose that V is finite dimensional, with some basis $u_1, \ldots, u_n$. Then $T_k(V)$ has as a basis all k-fold tensor products of these basis vectors, possibly with repetitions. In $E_k(V)$ all products with a repeated u_i become 0 by (4), while any two different u_i can be interchanged, by (5). Hence in $E_k(V)$ every element is a linear combination of the k-fold "exterior" products

$$u_{i_1} \wedge u_{i_2} \wedge \cdots \wedge u_{i_k}, \qquad i_1 < i_2 < \cdots < i_k \tag{6}$$

of basis vectors with indices in ascending order. For $k > n$ there can be no such product, so $E_k(V)$ is zero for $k > 0$. For smaller k, it turns out that the $(k,n-k)$ possible products of (6) do form a basis. This one may see more easily for $n = 3$, say. Here we have the list of basis elements suggested:

$$\begin{aligned}
E_0&: \quad 1, \\
E_1&: \quad u_1, u_2, u_3, \\
E_2&: \quad u_1 \wedge u_2, u_1 \wedge u_3, u_2 \wedge u_3, \\
E_3&: \quad u_1 \wedge u_2 \wedge u_3 .
\end{aligned} \tag{7}$$

All the products of these elements can be computed by the rules (4) and (5); with these product rules one verifies that any $t = a_1 u_1 + a_2 u_2 + a_3 u_3$ has $t \wedge t = 0$. Hence the required collapse already happened when all the elements in the list (7) are taken to be linearly independent. Therefore they are indeed linearly independent in $E(V)$.

For $u_1 = dx$, $u_2 = dy$, and $u_3 = dz$ and so $V = \mathbf{R}^3$ this exterior algebra then does provide a formal setting for the calculations with differential forms in three variables (and similarly for higher dimensions!). From this geometric origin of exterior algebra there also flows an unexpected byproduct. For any $n \times n$ matrix A with entries a_{ij} in a field think of the i-th row R_i as an element $R_i = \Sigma a_i u_j$ of rank 1 in the exterior algebra. Then the exterior product of all n rows computes out to

$$R_1 \wedge R_2 \wedge \cdots \wedge R_n = |A| \, u_1 \wedge u_2 \wedge \cdots \wedge u_n \tag{8}$$

where $|A|$ is the usual determinant of the matrix A (try it out for $n = 3$). Indeed, the rules (4) and (5) simply encode the formal rules about calculating determinants–when two columns are equal, the determinant is 0 (rule (4)); when two columns are interchanged, the sign of the determinant changes (rule (5)). The consequene is that exterior algebra can be used to provide a definition of determinants and a complete development of their properties (*Algebra*, §XVI.7). This can be done in invariant style, defining the determinant not of a matrix A but of the (more geometric) linear endomorphism represented by A. Here again the tension between geometry and matrices reappears and is resolved.

Observe also that linear transformations $T: V \to W$ will induce corresponding homomorphism $E(T)$: $E(V) \to E(W)$ on the associated exterior algebras; in brief, this contruction from V is *functorial*.

12. Similarity and Sums

Now we return to the central question of linear algebra: When do two square matrices represent the same linear endomorphism $T: V \to V$, relative to possibly different bases? Or, given such a T, can one find a simplest matrix representation? In case $V = E$ is Euclidean and G is self-adjoint, the principal axis theorem of §8 gives the answer: The matrix may be chosen to be diagonal, with diagonal entries the (real) eigenvalues of T. This answer will not extend to all other cases. For example, the 2×2 matrix

$$\begin{bmatrix} 0 & 0 \\ 1 & 0 \end{bmatrix} \tag{1}$$

has zero as its only eigenvalue, but it is not similar to the matrix of all zeros because it does not represent the zero endomorphism. Hence finding a "simplest" matrix for a general T will require something much more elaborate than just diagonal matrices.

One approach is to consider V not just as a vector space but as a "richer" algebraic system (a "module") given by the additional presence of T and its iterates. What this means is that any vector v of V can be multiplied not just by scalars in the ground field F, but also by polynomials in T with coefficients in F, as in

$$(a_0 + a_1 T + \cdots + a_k T^k)v = a_1 v + a_1 Tv + \cdots + a_k T^k v. \quad (2)$$

We call this the "product" of v by the polynomial $a_0 + a_1 x + \cdots + a_k x^k$. The set of all these polynomials in x forms a commutative ring, the *polynomial ring* $F[x]$. the multiplication of the vector v by such polynomial scalars satisfies all the usual rules for such multiples. Hence the definition (1) makes V a " module" over $F[x]$ in the sense of the following definition.

A *module* M over a ring R is an abelian group under addition with an operation $(a,v) \mapsto av$ of multiplication for each element $v \in M$ and each scalar a in R such that

$$a(v + w) = av + aw, \qquad 1v = v. \quad (3)$$

$$(a + b)v = av + bv, \qquad (ab)v = a(bv). \quad (4)$$

Higher geometry and analysis present many other examples of such "modules"–abelian groups with actions by various rings of "operators". For instance, differential forms constitute a graded module with an operator d where $d^2 = 0$.

The module axioms (3) and (4) are identical with those for a vector space over a field–except that now the scalars a in the ring R need not have multiplicative inverses, and the multiplication of two scalars need not be commutative. Much of the elementary theory of vector spaces carries over unchanged to modules; in particular, the definition of a linear map $T: M \to N$ of one module into another. Given a submodule S of M (a subset of M closed under all the operations) one can again construct the quotient module M/S and a linear map $P: M \to M/S$ which sends S to zero and is universal with this property. Given two modules M and N over the same ring R one can form their *direct sum* $M \oplus N$, consisting of all ordered pairs of elements (v,w) with $v \in M$, $w \in N$, with the operations

$$a_1(v_1,w_1) + a_2(v_2,w_2) = (a_1 v_1 + a_2 v_2, a_1 w_1 + a_2 w_2). \quad (5)$$

This is a module over R; it has linear maps

$$M \leftarrow M \oplus N \rightarrow N$$

to the factors, given by $v \leftarrow\!\shortmid \langle v,w \rangle \mapsto w$. They are "universal", so make $M \oplus N$ a product of M and N in the sense used in §V.9 for groups. It also has linear maps

$$M \rightarrow M \oplus N \leftarrow N$$

from the factors, given by $v \mapsto (v,0)$ and $w \mapsto (0,w)$. They are also universal, so make it a "coproduct"–the construction which for groups was called a free product in §V.9.

For vector spaces over F, the direct sum of $F \oplus \cdots \oplus F$ of a number (say n) of copies of the field F is simply the standard vector space of n-tuples of scalars in F. Every finite dimensional vector space has this form, once a basis is chosen. This last fails for modules: There can be many modules over a ring R which are in no wise direct sums of copies of R. For example, take the quotient of the polynomial ring $F[x]$ by the ideal (x^2) consisting of all multiples of x^2. This quotient $F[x]/(x^2)$ is still a module over $R = F[x]$ but it is much "smaller" than R. Every element comes from one of the polynomials $a + bx$, for a and b in F. Multiplication of this element by x is a linear transformation T sending a to ax and bx to 0 ($x^2 = 0$!). In other words, the module $F[x]/(x^2)$ is just a two-dimensional vector space over F with a linear endomorphism T acting on the basis vectors by $u_1 \mapsto u_2$ and $u_2 \mapsto 0$—and so with exactly the matrix (1) noted above. In this way every module over the polynomial ring $F[x]$ is just a vector space V over F together with a linear endomorphism $V \to V$, given by multiplication by x.

Now in the polynomial ring $F[x]$, just as in the ring $\mathbf{Z}$ of integers, there is a division algorithm, from which one may prove that every ideal K in $F[x]$ is 0 or is an ideal $K = (f(x))$ which consists of the multiples of a single polynomial $f(x)$, which might as well be taken to be *monic* (i.e., with leading coefficient 1). The corresponding quotient module $F[x]/(f(x))$ is a vector space V over F with dimension the degree m of the polynomial $f(x)$. Take as basis of this vector space the successive powers $1, x, x^2, \ldots, x^{m-1}$. Then the corresponding endomorphism $T: V \to V$, "multiply by x" sends each basis vector into the next one, but the last x^{m-1} into

$$x^m = -a_0 - a_1 x - \cdots - a_{m-1}x^{m-1},$$

where the a_i are the coefficients of $f(x) = a_0 + \cdots + a_{m-1}x^{m-1} + x^m$. Therefore the matrix of T, relative to this basis, has the appearance (for $m = 4$)

$$\begin{bmatrix} 0 & 0 & 0 & -a_0 \\ 1 & 0 & 0 & -a_1 \\ 0 & 1 & 0 & -a_2 \\ 0 & 0 & 1 & -a_3 \end{bmatrix} \qquad (6)$$

It is called the *companion matrix* (of the polynomial $f(x)$). These matrices will be the building blocks for the analysis of similarity.

The analysis uses modules over special kinds of rings. An *integral domain* is a commutative ring in which $bc = 0$ implies $b = 0$ or $c = 0$. A *principal ideal domain* is a domain D in which *every* ideal K is an ideal $K = (d)$ consisting of all the multiples (in D) of some one element d.

Both $F[x]$ and $\mathbf{Z}$ are principal ideal domains. For modules over such a D one can prove (with considerable trouble; *Algebra*, Chap. X) the

Theorem. *If a module M over a principal ideal domain D is spanned by a finite number of its elements, then there is a number r and non-zero elements $d_1, \ldots, d_k$ of D, each a multiple of the next, such that M is the direct sum*

$$M \cong D \oplus \cdots \oplus D \oplus D/(d_1) \oplus \cdots \oplus D/(d_k). \tag{7}$$

The integer r and the ideals (d_i) are uniquely determined by M.

This result does include vector spaces (as modules over a field F), since F is a principal ideal domain whose only ideals are (0) and F itself. However, the result (7) exhibits the complexity of modules relative to vector spaces; for a finite dimensional vector space V over F one has only $V \cong F \oplus \cdots \oplus F$–just the first r summands of (7), with r the dimension. For general D the additional modules $D/(d_i)$ here are called *cyclic* modules because, like cyclic groups (§V.6) they are generated by one element–the 1 of D.

What does this theorem mean when D is the polynomial ring $F[x]$? A module M is then just a vector space V together with a linear endomorphism, "multiply by x". In case V is finite dimensional, there can be no factor $D = F[x]$ in the decomposition (7), because each such factor D would correspond to an infinite dimensional vector space $F[x]$. hence there is only the string of cyclic modules $D/(d_i)$. Each of these corresponds to a transformation with matrix a companion matrix (6). Hence the

Corollary. *For any linear endomorphism T: $V \to V$ of a finite dimensional vector space V over a field F there is a basis of V in which the matrix of T, displayed as*

$$\begin{bmatrix} C_1 & 0 & \cdots & 0 \\ 0 & C_2 & & 0 \\ \cdot & & & \\ \cdot & & & \\ \cdot & & & \\ 0 & 0 & & C_k \end{bmatrix}, \tag{8}$$

consists of companion matrices C_i of polynomials $d_i \in F[x]$ down the diagonal. Moreover, each d_i is a multiple of d_{i+1}, $i = 1, \ldots, k-1$.

The number k and the polynomials d_i, chosen monic, are uniquely determined by the transformation T.

These monic polynomials are called the *invariant factors* of T. They form a *complete* system of invariants under similarity: Two square matrices are similar if and only if they have the same invariant factors. Correspondingly, the representation (8) is called a *canonical* form for a (square) matrix, under similarity—every square matrix is similar to exactly one such canonical form. Much more can be said (but not here) about this "rational canonical form" and other canonical forms for matrices, such as the Jordan canonical form.

What does the theorem mean when the principal ideal domain D is the ring $\mathbf{Z}$ of integers? Consider any (additive) abelian group A. In A a repeated sum (say three times, as $a + a + a$) amounts to the multiplication of $a \in A$ by an integer, here by 3, as $3a = a + a + a$. This "scalar" multiple enjoys the properties (3) and (4) used to define a module. In other words, an abelian group A is just the same thing as a $\mathbf{Z}$-module.

Corollary. *Any finitely generated abelian group A can be represented as a direct sum*

$$A = \mathbf{Z} \oplus \cdots \oplus \mathbf{Z} \oplus \mathbf{Z}/(m_1) \oplus \cdots \oplus \mathbf{Z}/(m_k) \tag{9}$$

of a number r of infinite cyclic groups $\mathbf{Z}$, *plus cyclic groups of orders* $m_1, m_2, \ldots, m_k$ *where each* m_i *is a multiple of* m_{i+1}, $i = 1, \ldots, k-1$. *The numbers r and* $m_1, \ldots, m_k$ *are invariants of A.*

This result, which describes all finitely generated abelian groups by 'invariants" r and m_i, is especially useful in algebraic topology. There each "homology group" A of a polyhedron is abelian and finitely generated, so is determined by its "Betti numbers" r and its "torsion coefficients" m_i.

In the canonical form (8), each invariant factor d_i can be written as a product of powers of irreducible polynomials, called the *elementary divisors* of T; there is a corresponding decomposition of the companion matrix C_i. For abelian groups, for example, each m_i can be written as a product $p_i^{e_1} \ldots p_s^{e_s}$. of powers of primes p_i. Then just as $\mathbf{Z}_6 \cong \mathbf{Z}_2 \otimes \mathbf{Z}_3$, there is an isomorphism

$$\mathbf{Z}/(m_i) \cong \mathbf{A}/(p_1^{e_1}) \oplus \cdots \oplus \mathbf{Z}/(p_s^{e_s}). \tag{10}$$

Combined with (9) this gives the decomposition of finite abelian groups already announced in (V.8.4)—the direct product there *is* the same as the direct sum here.

Our main observation now is that this result (9) for abelian groups and that for companion matrices are essentially the same. Separate proofs (often presented by mysterious manipulations of matrices) would come down to the use of essentially the same devices, and these devices are codified—and hence better understood—by using the concept of a module, generalized from that of vector space.

13. Summary

Linear algebra starts with geometrical pictures of vectors and with elementary ideas about "linear" operations–those which preserve sum and proportion. The exact formulation of these ideas is possible only with the notion of a linear transformation between vector spaces. The manipulation of such transformations involves both conceptual formulations and calculations with matrices and presents a central problem, that of similarity: When do two matrices represent the same linear endomorphism? Geometric and analytic considerations introduce various constructions on spaces–dual spaces and inner product spaces. Linear functions lead to bilinear functions and more general products, using the tensor product of spaces. In the presence of an inner product, the similarity problem for such symmetric matrices can be solved by the use of eigenvectors and eigenvalues. In the general case, one of the canonical forms for matrices under similarity requires that vector spaces over a field be generalized to consider modules over a suitable ring. Further developments of linear algebra, not summarized here, involve the same pattern–an interaction between geometry and analysis, leading to successive formal generalizations helpful in formulating and understanding aspects of linearity and its manifold uses, for linear approximations and analytic operations.

References. Survey of Modern Algebra was the first text in [1941] in English to effectively combine the Anglo-Saxon matrix methods with the conceptual view of vector spaces (Hermann Weyl). This connection with Hilbert space (von Neumann) is emphasized in the classic text of Halmos [1942]. The use of modules is emphasized in *Algebra* and in Hungerford [1974]. Modern computers have made big matrices manageable, but unfortunately many current texts of elementary linear algebra bury the ideas under a morass of muddled matrix manipulations with no understanding of the concepts.

CHAPTER VIII

Forms of Space

Perceptions of space and of motions in space have led mathematicians to describe a wide variety of formal geometrical structures. In this chapter we will introduce a few of these structures, beginning with the description of arc length and of various curvatures, and going on to topological spaces, sheaves, manifolds, and the like. It will appear that the role of intuitive ideas is very important in the analysis of such geometric structures—and that it often is a long time before evident geometric intuitions are brought to a clear formal expression. These expressions provide many different forms for the elusive idea of "space".

1. Curvature

In Euclidean and Non-Euclidean geometry, the phenomena of space were analysed in terms of lines, triangles, angles, and congruence; in brief, such a geometry is primarily linear. Many other geometrical phenomena, however, involve curved lines in the plane and twisted curves and curved surfaces in three-dimensional space. By approximating these curves by straight lines, the methods of calculus come into play, leading to the subject of differential geometry, to which we now turn. The resulting elementary methods of analysing curvature lead inevitably to a study of the intrinsic geometry of surfaces and to problems in the classification of surfaces and higher-dimensional manifolds.

A smooth *path* (that is, a *parametrized curve*) in the x,y plane is given, as in §VI.9, by a pair of smooth functions

$$x = g(t), \qquad y = h(t) \tag{1}$$

defined for some interval $t_1 \leqslant t \leqslant t_2$ of the real parameter t. The tangent vector at any point $p = (g(t_0),h(t_0))$ of the path is the vector

$$(g'(t_0),\ h'(t_0)) \tag{2}$$

in the tangent space T_p of the plane at that point. If the tangent vector is never zero, the path is said to be *regular*. The parameter t can be changed, say to $u = k(t)$, provided the smooth function k has $k'(t) \neq 0$ throughout the interval in t. Such a change alters the length of the tangent vector (2) but not its direction—and gives the same set of points (x,y). This collection of points is the *curve* traced out by the path (1). Thus a change of parameter gives the same curve, traversed at a (possibly) different speed and with points labelled with (possibly) different parameter values.

One first wants the length of such a curve. To measure the length of this curve from t_1 to t_2 one inscribes a polygon in the curve, calculates the length of the polygon by the Pythagorean theorem for each piece, and takes suitable successive such polygons with shorter and shorter sides. In the limit, this gives the length expressed as the (Riemann) integral

$$\int_{t_1}^{t_2} [g'(t)^2 + h'(t)^2]^{1/2} dt = \int (dx^2 + dy^2)^{1/2}. \tag{3}$$

(More generally, a possibly not-so-smooth curve is said to be *rectifiable* when this limiting process leads to a definite result.) The length s is a function of t; that is, the length from t_1 up to t is then the integral

$$s = \int_{t_1}^{t} (dx^2 + dy^2)^{1/2}; \tag{4}$$

then s determines t, so one may use this *arc length* s as a parameter in place of t. The differential of this arc length s is then given by the quadratic expression

$$ds^2 = dx^2 + dy^2; \tag{5}$$

it is the Pythagorean theorem in infinitesimal form. It can be regarded as an inner product $dx\, dx + dy\, dy$ in the tangent plane T_p. This inner product structure on the tangent plane T_p applies to all the curves through p.

For curvature, the circle provides a typical example. The smaller the radius r, the greater the curvature. Here one defines the *curvature* of the circle to be $\kappa = 1/r$. For a general curve the natural approach is then to approximate the curve at each point p by a circle through that point. Specifically, take p and two nearby points p' and p'' on the curve. Through them there passes a (unique) circle C. Because the curve is smooth, this circle C will approach a unique limiting position C_p, except when p is a point of inflection. The curvature of this limiting circle, called the *osculating* circle, is by definition the curvature of the given curve at the point p in question. From this description as a limit one can obtain an analytic formula giving the curvature κ at each point as a function of the parameter value t at that point:

$$\kappa = \frac{g'(t)h''(t) - h'(t)g''(t)}{(g'(t)^2 + h'(t)^2)^{3/2}}. \tag{6}$$

Here the term in the denominator is non-zero because the parametrization has been assumed to be regular, as defined above. For a curve given in the familiar form of an equation $y = f(x)$ in the coordinates x and y, one may regard the x coordinate as the parameter; in that case the formula (6) for the curvature κ simplifies to

$$\kappa = \frac{y''(x)}{[1 + y'(x)^2]^{3/2}}. \tag{7}$$

However, the choice of a coordinate as the parameter has no intrinsic geometric meaning, since the same geometric path may be referred to different coordinates. The arc length s, measured from some starting point, provides an intrinsic parameter, and the curvature κ at each point is more naturally expressed as a function $\kappa = \kappa(s)$ of the arc length s up to that point.

Now in view of the "intrinsic" geometric definition of arc length and of curvature, a rotation or a translation of the curve will not alter its arc length between two points or the curvature as a function of this length. Hence these quantities are said to be Euclidean *invariants* of the curve, in the sense that they are unaltered by the stated (Euclidean) transformations. Much more is true: The quantities arc length and curvature form a *complete* system of invariants in the following sense. If two smooth curves have the same curvature function $\kappa = \kappa(s)$ when the arc lengths are measured from points p_1 and p'_1 on the curves, then there is a rigid motion of the plane which carries the first curve exactly onto the second in such a way that p_1 coincides with p'_1.

For curves in space there is an additional invariant, the *torsion*, which measures the rate at which the curve deviates from one lying wholly in one plane. The definition again involves a suitable use of limits to formalize the intuitive notion, while the curvature and the arc length of a space curve are defined much as in the plane. It is then true that arc length, curvature, and torsion together provide a complete set of Euclidean invariants for a smooth space curve.

In space, a smooth path is given, in terms of a parameter u, by three smooth functions of u

$$x = g(u), \qquad y = h(u), \qquad z = k(u).$$

It is often convenient to write the coordinates (x,y,z) as a single vector $\mathbf{p}$, in bold face type. Again, the differential of arc length at p is $ds^2 = dx^2 + dy^2 + dz^2$. If the arc length s is taken as the parameter, the tangent vector $\mathbf{t}$ at each point has length 1. This means that the inner product $\mathbf{t}\cdot\mathbf{t} = 1$ is constant. Differentiating this equation with respect to the parameter s gives $2\mathbf{t}\cdot\mathbf{t}' = 0$, so that the vector $\mathbf{t}'$ representing the rate of change of the tangent vector (with respect to arc length) is perpendicular to the tangent vector $\mathbf{t}$. Now the unit *normal* vector of the space curve is defined to be the vector **n** in the direction of $\mathbf{t}'$, but with length 1. This means that the derivative $\mathbf{t}'$ is proportional to **n**; indeed, $\mathbf{t}' = \kappa\mathbf{n}$ because the curvature κ (of a plane curve or a space curve) represents the rate at which the tangent is turning. For a space curve one may take a third unit vector perpendicular to both the normal **n** and the tangent **t** and suitably directed; it is called the *binormal* **b**. Thus at each point of the space curve one has an orthonormal basis **t**, **n**, **b** (for the tangent space at that point). The curvature and torsion together specify how these basis vectors (the "frame") changes as one moves along the curve–as in the equation $\mathbf{t}' = \kappa\mathbf{n}$ and other such equations constituting the "Frenet formulas". This method of "moving frames" has an immediate pictorial content (think of the three first fingers of a hand stretched out to be orthogonal and moving along the curve). This method applies not just to curves, but to surfaces and beyond, and is a powerful formal tool in the use of calculus to understand geometry.

In brief, the lines, circles and planes of elementary Euclidean geometry can be used to approximate the curves (and later the surfaces) in space, thereby making the ideas of the calculus apply in a geometric context.

2. Gaussian Curvature for Surfaces

Just as two dimensions (the plane) are decisive for the understanding of Euclidean geometry, so two dimensions (a surface) are decisive for the understanding of curvature. Various surfaces, variously curved, are at hand in three-space: Spheres, ellipsoids, cylinders, paraboloids, hyperboloids, toruses and the like. Some of these surfaces may be described as the locus of a single smooth equation, just as the sphere of radius r with center at the origin is (by Pythagoras again!) the locus of the equation $x^2 + y^2 + z^2 = r^2$. The coordinates x, y, z of these points P on the sphere may also be represented as functions of suitable parameters. The most familiar parameters are latitude θ and longitude ϕ (the names recall the problems of navigation along the long direction in the Mediterranean). They are described by the familiar Figure 1 or by the corresponding parametric equations read off from this figure:

$$\begin{aligned} x &= r\cos\theta\cos\phi\,, \\ y &= r\cos\theta\sin\phi\,, \\ z &= r\sin\theta\,. \end{aligned} \tag{1}$$

In the case of a curve, the choice of a parameter amounts to the choice of a continuous map of an interval into the (set of points of) the curve. As for the sphere with these parameters, we note that the longitude ϕ is

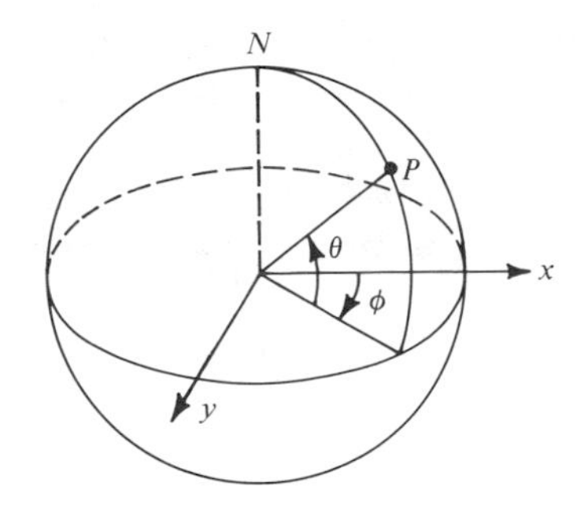

Figure 1. Spherical coordinates.

ambiguous at the north and south poles of the sphere; moreover it is periodic, with period 2π. Hence we had better say that the spherical coordinates ϕ and θ in the equations (1) describe a continuous bijection of the open square I^2

$$-\pi/2 < \theta < \pi/2, \qquad -\pi < \phi < \pi \tag{2}$$

in the $\theta-\phi$ plane onto all of the sphere except the north pole, the south pole and the "international date line" (the half great circle where $\phi = \pi$). This map $I^2 \to S^2$ is called a *chart* on the sphere. Since the sphere is manifestly *not* the same figure as the rectangle, it is impossible to cover the whole sphere with one such chart. Two charts will do—this one and a similar one where the "poles" are taken on the present equator. One may then say that the sphere is obtained by pasting together two such charts (where they overlap). Other surfaces can be similarly described by pasting together several charts, where each chart is a map defined on an open square like (2) (or more generally on an open set in the plane). For example, the surface of a torus can be covered by three such charts, but no two will cover the whole torus.

Often, a piece of a surface is given by a smooth equation $z = f(x,y)$ in rectangular coordinates x, y, and z; and this is just a chart with the $x-y$ coordinates as parameters.

The curvature of such a piece of surface should somehow measure how rapidly the tangent planes to the surface change direction as one moves from point to point. This change can be observed equally well by the "tipping" of the normal vector (the vector orthogonal to the tangent plane). Specifically, using the calculus and an equation $z = f(x,y)$ as in §VII.9 one can determine at each point p of the surface S a plane tangent to S and hence a vector $\mathbf{n}$ of length 1 orthogonal to the tangent plane; this vector is called a *unit normal vector* to S. If we choose these vectors smoothly at nearby points of the chart we get a *field* of such vectors, one attached to each point of this piece of the surface. (A reversal of sign, from $\mathbf{n}$ to $-\mathbf{n}$, gives a second choice of such a local field.) Now the curvature of the surface ought to be measured by the way in which these unit normal vectors vary (tip about) from point to point on the surface. The idea is much the same as that for a curve, where the curvature measures the rate at which the tangent vector (or equally well the normal vector) turns. This intuitive idea for measuring curvature for a surface can be formalized in several ways.

First, one may try to reduce the question to one dealing directly with the curvature of plane curves. To do this, consider all the planes containing the point p and the unit normal vector $\mathbf{n}$ at this point. They will all be orthogonal to the tangent plane to the surface at p, and each will cut out from the surface a plane curve which has some definite curvature κ at p. If the surface is a sphere, these "sectional" plane curves will all be circles of the same curvature, but in general the sectional curvature κ will vary as the plane through $\mathbf{n}$ rotates about the axis $\mathbf{n}$. This is the case, for example, with the different curves cut out in this way at one vertex of an ellipsoid. This example suggests that it is reasonable to concentrate attention on the maximum curvature κ_1 and the minimum curvature κ_2 of these "sectional" plane curves. These two are called the *principal curvatures* of the surface at the point p and their directions (in the tangent plane at that point) are called the *principal directions* at p. (In fact the principal curvatures can be calculated as eigenvalues of a suitable matrix, and the two principal directions turn out to be eigenvectors orthogonal to each other.) At any point on a convex surface (such as an ellipsoid) both the principal curvatures will have the same sign (say, positive). However, at a saddle point p of a surface (for example at the origin for the surface $z = x^2 - y^2$) the principal curvatures have opposite signs, because one of the sectional curves bends up, while the other bends down at p. If one wants a single measure of the curvature of the surface at the point, one may take a suitable combination of the two principal curvatures–for example, the *mean curvature* $(\kappa_1 + \kappa_2)/2$. But notice that this mean curvature might come out to be zero at a saddle point, although the surface is surely "curved" there.

A different and deeper concept was developed by Gauss. On a little piece A of the surface about the point p, take *all* the unit normal vectors $\mathbf{n}$ at points of A and translate them so that they all start from the origin 0. The ends of these unit vectors then trace out a region B on the unit sphere about the origin. Moreover, the more sharply the surface curves at p, the larger the region B. Hence, assuming that areas can be measured, one defines the *Gaussian curvature* of the surface at p to be a limit

$$K = \operatorname{Lim} \frac{\text{area } B \text{ (on the unit sphere)}}{\text{area } A \text{ (on the surface)}}, \tag{3}$$

where the limit is taken as all the dimensions of A approach zero. (Later, we will assign a sign to this limit.)

This intuitive description can be formalized in various ways. One such method uses a process of "covariant differentiation", which amounts in this case to differentiating along a vector $\mathbf{v}$ in the tangent plane at the point p on the surface. Specifically, consider each of the three components of the normal vector $\mathbf{n}$, and its derivative (a real number) in the direction $\mathbf{v}$. These three derivatives are the components of a vector $\nabla_{\mathbf{v}}\mathbf{n}$, called the derivative of the vector field $\mathbf{n}$ in the direction $\mathbf{v}$. Since all the vectors of the field $\mathbf{n}$ have unit length, the inner product $\mathbf{n}\cdot\mathbf{n}$ is 1; therefore the derivative $\nabla_{\mathbf{v}}\mathbf{n}$ will be perpendicular to $\mathbf{n}$—and so will be a vector in the tangent plane at the point p. In other words, the correspondence

$$\mathbf{v} \mapsto \nabla_{\mathbf{v}}\mathbf{n}$$

is a function mapping the tangent plane T_p into itself; one then sees that it is a linear transformation

$$L\colon T_p \longrightarrow T_p.$$

With careful formulation (see for example Barrett O'Neill [1966]) it turns out that this transformation is self adjoint (i.e., has a symmetric matrix relative to any basis of T_p). This transformation is sometimes called the "shape mapping" for the surface in question.

This shape mapping clearly contains information as to how rapidly the unit normal is turning, for motions in any direction from the point p. Hence it is not surprising that it can be shown (see e.g. O'Neill *loc. cit.*) that L contains all the information about the curvature of the surface at the point p. Specifically, the determinant of the shape mapping (calculated as the determinant of any one of the 2×2 matrices representing that mapping) is the Gaussian curvature K of the surface, complete with the desired sign. Like every self-adjoint transformation, this particular transformation L can be brought to principal axes. It turns out that these axes are exactly the principal directions of curvature, as defined above, while the eigenvalues (the diagonal entries) are the two principal curvatures κ_1 and κ_2. Since the principal axes of a symmetric matrix are orthogonal, this shows that the principal directions of curvature are orthogonal—as one may readily see in particular surfaces such as the ellipsoid. Moreover the Gaussian curvature (the determinant of the matrix) is the product $K = \kappa_1\kappa_2$ of the two principal curvatures (this defines the sign of K). This indicates that the Gaussian curvature is positive when the surface is convex or concave at the point in question (i.e., when the surface near the point lies all on one side of the tangent plane there). On the other hand, it also indicates that the Gaussian curvature is negative at a saddle point, where the surface lies on both sides of the tangent plane.

The detailed demonstration of these results requires time and care and can be done by various techniques, not necessarily formulated in terms of the "shape transformation" L. However, the use of this transformation does illustrate in striking form the close relation which obtains between the algebraic properties of symmetric matrices and the geometric study of curvature.

3. Arc Length and Intrinsic Geometry

On a sphere, the length of an arc of any smooth curve can be computed by an integral like (1.4), where the differentials dx, dy, and dz are computed in a suitable chart directly from the expression for the curve in the coordinates of latitude θ and longitude ϕ. The formal equations (2.1) for the cartesian coordinates x, y, and z give (with r held constant)

$$dx = -r\sin\theta\cos\phi\; d\theta - r\cos\theta\sin\phi\; d\phi\,,$$

$$dy = -r\sin\theta\sin\phi\; d\theta + r\cos\theta\cos\phi\; d\phi,$$

$$dz = r\cos\theta\; d\theta\,.$$

Hence the three-dimensional element $ds^2 = dx^2 + dy^2 + dz^2$ of distance, when restricted to the sphere, can be expressed formally as

$$ds^2 = r^2 d\theta^2 + r^2\cos^2\theta d\phi^2\,. \tag{1}$$

One may also motivate this formula by drawing on the sphere a small "spherical triangle" made up of arcs of circles with hypotenuse ds, horizontal side $r\cos\theta\; d\phi$ and vertical side $rd\theta$; then (1) is the Pythagorean theorem for this infinitesimal "right triangle". What is more (and more than we will prove here), this formal calculation gives the correct integrand ds for the arc length of a spherical path written in latitude and longitude coordinates for a parameter t as $\theta = g(t)$, $\phi = h(t)$. On the practical side, such length calculations are needed in ocean navigation, where the great circle joining two points provides the shortest possible (smooth!) voyage between those two points. On the theoretical side, such calculations of length form the starting point of much of Riemannian geometry (§10 below).

There are similar formulas for the length of arc applicable to paths on other surfaces in 3-space. One assumes that a piece of the surface is given in terms of two parameters u and v by smooth functions

$$x = f(u,v), \qquad y = g(u,v), \qquad z = h(u,v)\,. \tag{2}$$

This is to be a chart for a piece of the surface, defined for u,v in some open set of the u,v plane, so that (2) is to be a bijection of this open set onto an open subset of the surface. The representation is to be regular, in the sense that the matrix

$$\begin{bmatrix} \dfrac{\partial f}{\partial u} & \dfrac{\partial g}{\partial u} & \dfrac{\partial h}{\partial u} \\[2ex] \dfrac{\partial f}{\partial v} & \dfrac{\partial g}{\partial v} & \dfrac{\partial h}{\partial v} \end{bmatrix} \tag{3}$$

has rank 2 everywhere (i.e., has its rows always linearly independent). This means that at each point in the u,v plane the functions (2) maps the tangent plane of the (u,v)-plane linearly to the tangent plane at the corresponding point on the surface–and that (3) is the matrix of this linear mapping. Then the same sort of formal calculations as those for the sphere yields a formula for arc length (that is, for a ds^2) of the form

$$ds^2 = E\,du^2 + 2\,F\,dudv + G\,dv^2, \tag{4}$$

where E, F, and G are smooth functions of the parameters u and v in the open set at issue; these functions are expressible in terms of the given functions f, g, and h of (2) and their derivatives. Moreover, this formal expression can again be used to determine actual lengths of arc for smooth paths on the surface. This "quadratic differential form" (4) is called the *first fundamental form* for the surface. There is also a "second" such form, which we will not discuss here.

Now there arise questions about the "intrinsic" geometry of the surface–geometric properties of figures in the surface which can be determined entirely in terms of measurements in the surface which make no reference to the ambient space–the rest of the space surrounding the surface. These properties would include quantities which could be computed just from the first fundamental form ds^2, because this form depends only on measurements of arc length for paths in the surface. Not all measurable quantities for a surface are "intrinsic" in this sense. For example, the two principal curvatures are not intrinsic. This can be seen in the example of a circular cylinder–such a cylinder can be "rolled out" onto a plane without changing the lengths of any arc on the cylinder–but clearly changing one of the two principal curvatures of the cylinder. An amazing theorem found by Gauss asserts, however, that the Gaussian curvature (unlike the principal curvatures) *is* intrinsic–it depends only on the coefficients in the first fundamental form (4).

The proof of this result is beautiful but subtle. Initially, one can express the Gaussian curvature K in term of the coefficients of the *two* fundamental forms; a calculation then shows how to express it just in terms of E, F, and G and their derivatives.

Another version of this argument uses small "triangles" ABC on the surface. Each edge of the triangle is to be a geodesic–so that the length of AB, for example, represents the shortest distance from A to B. From the metric one can also measure each angle of the triangle, in radians, and hence determine the *excess* $E(ABC)$–the sum of the three angles, less π. One may then prove that the Gaussian curvature K at each point p is given by the limit

$$K_p = \lim_{A,B,C \to p} \frac{E(ABC)}{\operatorname{area}(\Delta ABC)}$$

with limit taken as the three vertices approach the point p. Thus, just as on the sphere, positive curvature means angle sum in a triangle larger than π, and larger positive curvature corresponds to larger angle sum. This result makes the intrinsic character of curvature visible. It also recalls the results on angle-sums in non-Euclidean geometry (Chapter III).

The discovery of this result raised in explicit form the problem of studying the geometry of the surface *itself*, intrinsically and without reference to the ambient space. This leads to the problem of describing surfaces and higher-dimensional varieties without using *any* ambient space. This is the origin of the idea of an (intrinsically) "curved space", as used in relativity theory. Thus it is that the Euclidean viewpoint for geometry is transcended.

4. Many-Valued Functions and Riemann Surfaces

There is a second important origin for the idea of considering surfaces "intrinsically". In solving quadratic equations and elsewhere one comes up against "functions" such as $u = \pm\sqrt{x}$ which are two-valued (or many-valued). At first, mathematicians were inclined to treat such a formulas as an ordinary "function" $u = f(x)$, with an ambiguity indicated by the "$\pm$", and indeed this is the usual practice in writing the standard formula for the solution of a quadratic equation. But the two-valued function $u = \pm\sqrt{x}$ can also be displayed by a graph in the (x,u) plane which gives both the values of u for given x as the coordinates of points on the parabola $u^2 = x$ of Figure 1. In this way, one can avoid the hopelessly ambiguous two-valued function $u = \pm\sqrt{x}$ by observing that both u and x are *single* valued functions, not of the *number* x, but of the *point* p on the parabola. A similar analysis applies to more involved cases of two-valued functions. Thus for the function $u = \pm\sqrt{(x^2-1)(x^2-4)}$ one may say that both u and x are single-valued functions of a point p on the quartic curve $u^2 = (x^2-1)(x^2-4)$, a curve which has the general form shown in Figure 2.

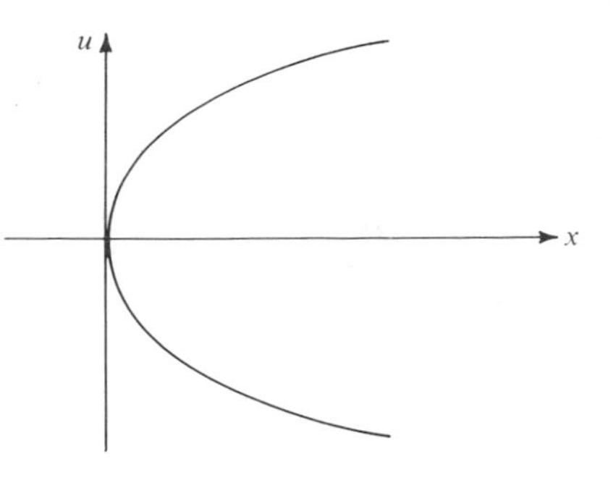

Figure 1

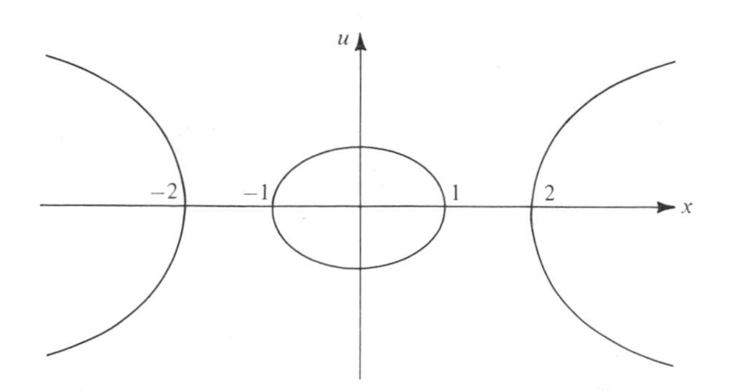

Figure 2. Quartic curve.

In brief, these many-valued functions of a real variable can be made into single-valued functions, not of a number but of a point on a curve. In effect, such a curve displays (in a continuous fashion) all the values of the function as the projections on the axes u and x, with nearby values corresponding to nearby points.

This geometric process becomes more decisive and more illuminating when it is used not for functions of a real variable but for functions of a complex variable. Such a variable $z = x + iy$ has two real dimensions x and y, so instead of a curve one must use a surface. As an example, consider the two values $w = \pm\sqrt{z}$ for the square root function. This is to be represented by the surface S whose points are all pairs (w,z) of complex numbers with $w^2 = z$, where two points (w_1,z_1) and (w_2,z_2) count as neighboring when both $|w_1 - w_2|$ and $|z_1 - z_2|$ are small. Thus the surface is the geometrical "manifold" of all solutions of $w^2 = z$. The correspondence $(w,z) \mapsto z$ is a continuous projection $p: S \to Z$ of this manifold into the complex plane Z. Now the point (w,z) with $w^2 = z$ on S is really completely determined just by the number w, so the surface S is "really" just the complex w-plane W, while the projection $W = S \to Z$ is continuous. Since each $z \neq 0$ has two complex square roots w_1 and w_2, the projection $W \to Z$ sends two points (w_1,z) and (w_2,z) to each point $z \neq 0$ of Z. In this sense, W consists of two copies of Z.

To see this in more detail, we use polar coordinates r and θ in the z-plane, so that $z = r(\cos\theta + i\sin\theta) = re^{i\theta}$. Since the multiplication of complex numbers multiples their absolute values and adds their arguments, the two square roots of $z \neq 0$ are

$$w_1 = \sqrt{r}\,e^{i\theta/2}, \qquad \sqrt{r}\,e^{i(\theta/2+\pi)}, \qquad 0 \leqslant \theta < 2\pi$$

when $\sqrt{r}$ is taken real and positive. The values w_1 of the first square root cover all the *upper* half of the w-plane, except for the negative real axis $0B$ (Figure 3), while the values w_2 cover the *lower* half-plane, omitting the positive axis $0A'$. Each of these half-planes is the image under $\sqrt{z}$ of the whole of the z-plane, so we need two copies of the z-plane, as in Figure 4.

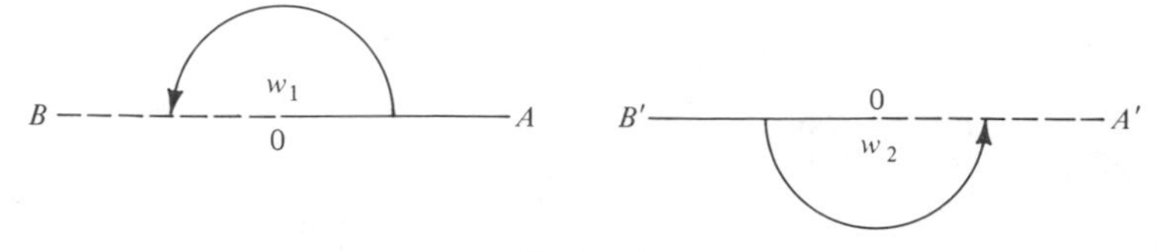

Figure 3

To make this image continuous, each copy of the z-plane should be cut apart by a slit along the positive z-axis. Now the whole surface S' is obtained by pasting together the two halves of the w-plane from Figure 3–ray $0B$ in the top half pasted to ray $0B'$ in the bottom half and then ray $0A'$ in the bottom half pasted (point for point) to the ray $0A$ in the top half. Now exactly the same surface S can be described by pasting together the two slit copies of the z-plane (Figure 4): The lower slit $0B$ in the top plane is pasted to the upper slit $0B'$ in the bottom plane while the upper slit $0A$ in the top plane is pasted (point for point) to the lower slit $0A'$ in the bottom plane. The result, seen end-on from the point at ∞ on the positive real axis, is illustrated in Figure 5 which is slightly inexact, because the ray 0∞ represents two separate rays, meeting only at the origin 0. All told, the surface S on which the function $\sqrt{z}$ becomes single-values is thus represented by pasting together two copies of the slit z-plane, so that each copy carries one of the two possible values (one of the "branches") of $\sqrt{z}$. Note that the location of the slit could be changed by replacing the positive real axis by any other smooth curve running from 0 to ∞. Different such choices of a slit give the same topological surface, which is described intrinsically as the manifold of all pairs (w,z) with $w^2 = z$.

This first example illustrates a general method for turning a many-valued "function" w of a complex variable z into a single-valued function not of z but of a point on a surface S, a *Riemann Surface*. First decompose the z-plane by slits so that w becomes single-valued when it is followed continuously along paths constrained not to cross the slits. This yields several single valued functions w_j which may be called the *branches* of w. For each such branch w_j, take one copy of the slit z-plane to carry these values w_j. Then paste these copies together along the slits appropriately matched so as to get the desired surface S. This is a process which visualizes the surface S, which is described intrinsically as the manifold of all the pairs (w,z) involved.

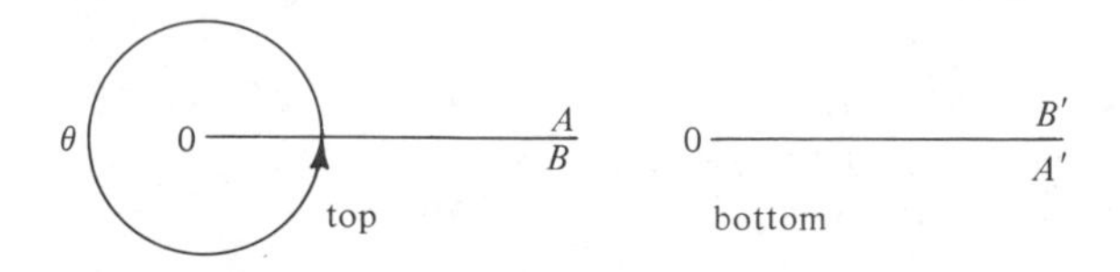

Figure 4

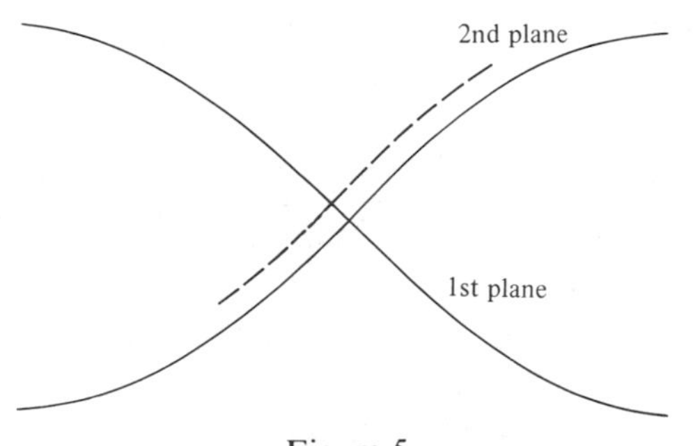

Figure 5

For this function $w = \sqrt{z}$ it is also natural to say that $z = \infty$ yields $w = \infty$, and so to use instead of the z-plane the *Riemann sphere* both for z and for w (see §IV.11). Then the surface is constructed by taking two copies of the Riemann z-sphere, slitting each along a great circle (the real axis) from 0 to the north pole, and then joining one edge of each slit to the other edge of the slit on the other sphere, much as in Figure 5 above. This process yields a two-dimensional closed surface on which both w and z are single-valued, including values ∞.

There is another way to visualize this surface. First distort each sphere continuously so that the opened slit becomes a circular hole; the needed connection of the edges of the slit can then be visualized by connecting these two holes by a cylinder, as in Figure 6 below. A further distortion of this figure shows that our Riemann surface is "really" just the surface of a single sphere (that is, is homeomorphic to the sphere).

A more striking case is that of the two-valued function $w = \pm\sqrt{(z^2-1)(z^2-4)}$ which occurs in the integrand of a certain "elliptic" integral which cannot be integrated in elementary terms. Again we wish to construct a "Riemann surface" S whose points are all pairs (w,z) with $w^2 = (z^2-1)(z^2-4)$, so that both w and z will be single-valued functions of a point (w,z) on this surface. Now for each z except $z = \pm 1$ or $z = \pm 2$ there are two corresponding values of w, so the projection $(w,z) \mapsto z$ makes this surface cover the z-plane twice. One may write w as a product

$$w = \pm\sqrt{z-1} \quad \sqrt{z+1} \quad \sqrt{z-2} \quad \sqrt{z+2}.$$

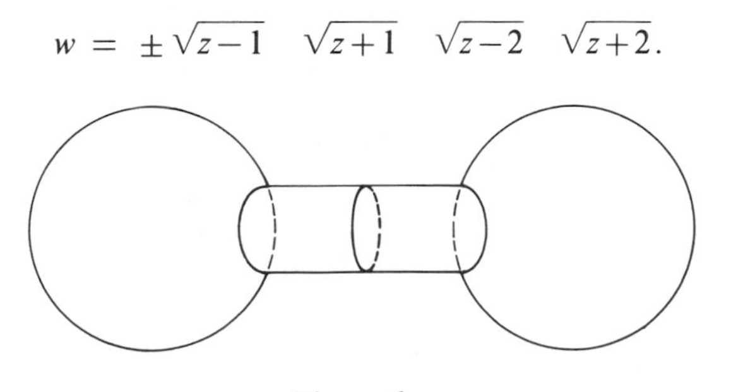

Figure 6

We do not need explicit formulas for the two roots w_1 and w_2, but we note that for values of z very near $z = 1$ the values w_1 and w_2 will interconnect much as did the values of the square root $\sqrt{z-1}$ (just like $\sqrt{z}$). In other words, as the point z moves once around $z = \pm 1$ in a small circle, the first root w_1 will change into the second root w_2, and w_2 into w_1. There will be a similar interchange near the three points $z = -1$, $z = +2$, and $z = -2$, and our two copies of the z-plane should be connected correspondingly. This can be done if we take two copies of the z-plane and slit each copy along the real x-axis from $+1$ to $+2$ and from -1 to -2. These slits will insured that a circle or other closed path *not* crossing the slits cannot go around just one of the points ± 1, ± 2, but must (Figure 7) go around two (or four) of them–and so following such a closed path will not change one value w_i into the other. Now to follow what happens to w_i across the slits we paste the two z-planes together, so that the upper edge of a slit on the top z-plane attaches to the lower edge on the bottom, and vice versa, much as in the previous Figure 5. Also the two copies of each point ± 1 and ± 2 are identified. Now the two copies of the z-plane can carry the two values w_1 and w_2 so that they fit together across the slits. The result is a Riemann surface such that both z and w are single-valued and continuous functions of a point on the surface.

This description of the surface is not really intrinsic, since we could have used differently placed slits, say slits along semicircles from -2 to $+2$ and -1 to $+1$. However, any such description provides a picture of the intrinsic manifold ("algebraic curve") of all solutions of the equation $w^2 = (z^2-1)(z^2-4)$. The previous Figure 2 displays just the "real" points on this curve.

As before, one may make a more "geometric" picture of this surface by starting with two slit Riemann spheres instead of two slit z-planes, then distorting each of the slits into an open circular hole. The result is then two spheres, each with two circular holes which are to be joined rim to rim. This join can be done by using two cylindrical tubes, as in the Figure 8. A distortion of the resulting surface shows that it is "like" the surface of a doughnut–a torus. For these and more elaborate Riemann surfaces one wishes to formulate an intrinsic description.

Figure 7. Slits.

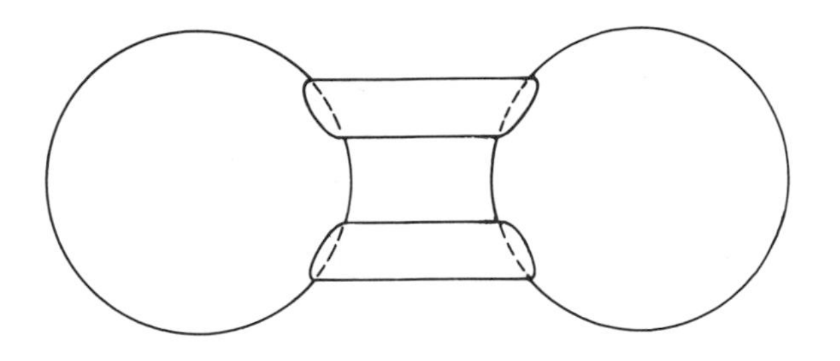

Figure 8. Distorted torus.

5. Examples of Manifolds

The sphere can be described as the "manifold" of all solutions (x,y,z) of the equation $x^2 + y^2 + z^2 = 1$. In many other cases the set of all solutions of some problem or the collection of all "things" with some observable property can be regarded as a geometric entity–curve, surface, or solid–in which "nearby" solutions are pictured as "nearby" points.

One simple example is the manifold of all possible quadratic polynomials $ax^2 + bx + c$ with real coefficients. Since such a polynomial is fully determined by the three real numbers a, b, and c, it can be represented by the points in Euclidean 3-space with these coordinates a, b, c. Hence this manifold of quadratic polynomials *is* just the space $\mathbf{R}^3$.

In mechanics one has occasion to consider a double pendulum in the plane, with a second bob B hanging by a string from the end of a first bob A, the latter hung from a fixed point P (Figure 1). To study the possible motions, one needs first of all the "manifold" of all possible positions of this double pendulum. Now a position is described completely by giving two angles θ and ϕ; to wit, the angle made by each bobstring with the vertical. Here the possible values of the angles θ and ϕ range from 0 to 2π radians, with 2π counting the same as zero (Figure 2). Hence the possi-

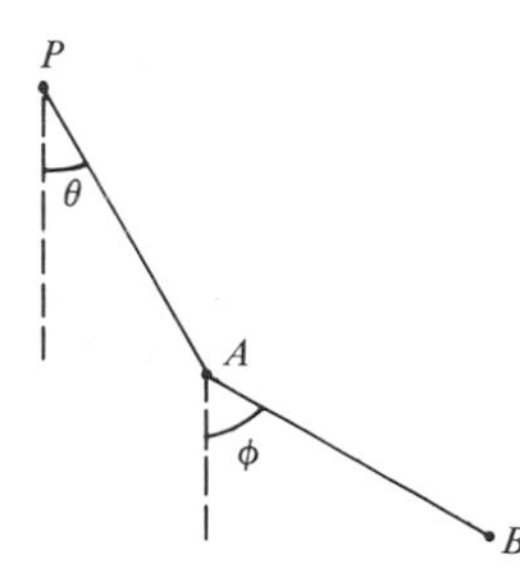

Figure 1. Double pendulum.

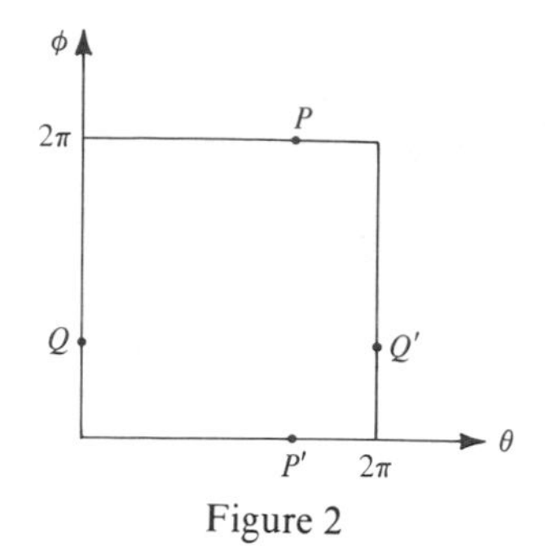

Figure 2

ble positions can be considered to be the points of a $2\pi \times 2\pi$ square in which each point P of the top edge is to be identified with the corresponding point P' of the bottom edge, while the left-hand edge is similarly identified point by point (Q with Q') with the right-hand edge. If one replaces the square by a flexible piece of paper, making the identification by pasting, one may see that the top to bottom identification by itself produces a cylinder which on left-to-right identification becomes a torus. In other words, the torus *is* the geometric manifold of possible positions of the double pendulum; on this torus one may picture various periodic motions of the pendulum. It is striking that the *same* torus arises both from the pendulum problem and as a Riemann surface. This illustrate the ambiguity of geometric form.

The Möbius band is another example. Take a long rectangle and then identify points on the left-hand edge—but in upside down order—with points on the right-hand edge, as in Figure 3. The resulting surface or band has a "twist" and is one-sided. Moreover, it has an edge (a "boundary") which is a single twisted circle.

A related example is the "projective plane"—described as the manifold of all possible lines passing through the origin 0 in 3-space $\mathbf{R}^3$, with the obvious meaning of "nearby" lines. In brief, it is the manifold whose "points" are the lines through 0. Since each line through the origin meets the unit sphere about the origin in *two* diametrically opposite points, this manifold may also be described as the manifold whose points are *pairs* of diametrically opposite points on a sphere. More simply we might use just

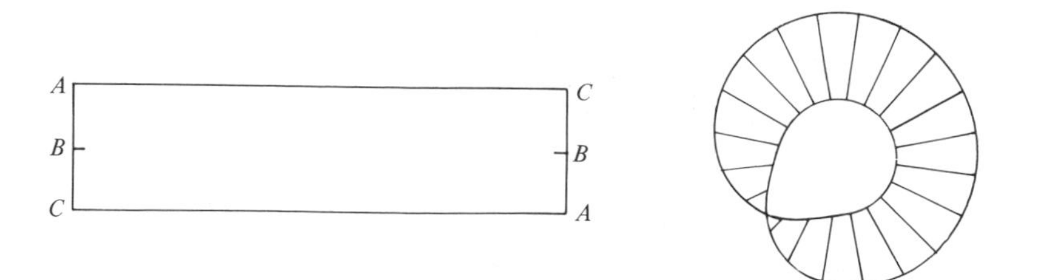

Figure 3. Möbius band.

single points–say the points in the lower-half hemisphere, except that then each pair of diametrically opposite points on the equator of the hemisphere must be identified. If we think of this as a circular disc with these diametrical points identified, the picture is almost like that of the torus above. However, if we try to carry out the identifications by pasting together corresponding points from the equator of the hemisphere we arrive at a figure like 4, with A to be pasted to A', B to B' and so on. One cannot actually carry this out in space without making one edge PAQ' pass through $P'BQ'$ where it really should not–but one can imagine the resulting manifold existing, though not in familiar space $\mathbf{R}^3$. This manifold is clearly a surface–hence two-dimensional.

This manifold may also be pictured by projecting the lower hemisphere from the center (of the hemisphere) onto the plane π tangent to the hemisphere at the south pole. The points below the equator then project onto all the points of the plane π, while the points on the equator are projected off to "infinity". The manifold is thus pictured as an ordinary plane *plus* a stock of points at infinity, and one may show that any two parallel lines in the ordinary plane meet at exactly one of these points at infinity. This is why this manifold is called the *projective plane*. There are many more such manifolds, such as a projective 3-space or the manifold of planes through the origin in $\mathbf{R}^4$. They arise intrinsically and not just as subsets of Euclidean space.

The projective plane of Figure 4 is a "one-sided" surface–one may move from the inside (say, at A') to the outside (say, at A) without going through the surface. In this respect, it is like the Möbius band. Indeed, if we cut off the bottom (the disc below the circle $TSVR$ of Figure 4) what is left *is* just the Mobius band! To see this, project the rest of Figure 4 onto the plane through $TSVR$. After some distortion this gives the ring on the left of Figure 5, where $Q'PQ$ is still to be pasted (reversed) on $QP'Q'$. To do this, cut the ring apart vertically along RQ and $Q'S$, and turn the right-hand half over. Now $Q'PQ$ can be pasted to $QP'Q$–while RQS is still to be pasted to $RQ'S$–again with a reversal. But this (as on the right of Figure 5) is just a distorted Möbius band, with the arc RQS pasted in

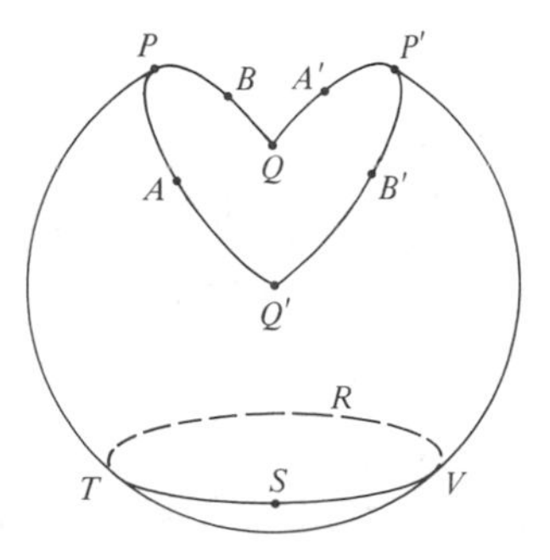

Figure 4. Projective plane.

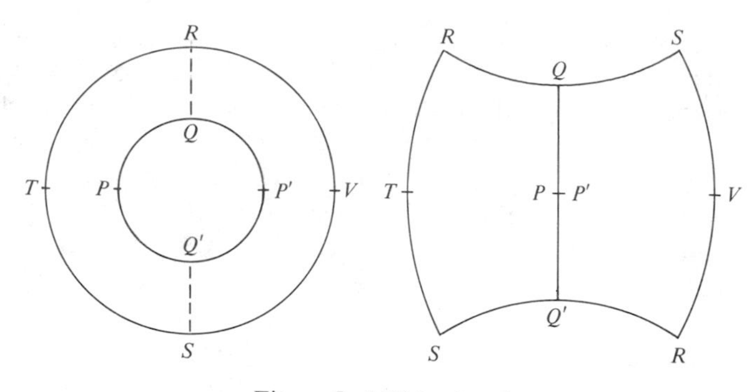

Figure 5. Möbius band.

reverse on the arc $RQ'S$–and a single free edge $STRVS$ (a circle) as boundary. In other words, the projective plane arises from the Möbius band by pasting in a disc (the bottom disc of Figure 4) along the circular edge of the Möbius band of Figure 3!

6. Intrinsic Surfaces and Topological Spaces

Examples have indicated the need to describe a two-dimensional surface intrinsically, without thinking of the surface S as something embedded in ordinary 3-space. Thus the idea of Gaussian curvature is intrinsic, while the various Riemann surfaces and other manifolds all illustrate this need–most dramatically for the projective plane, which simply cannot be embedded in 3-space without some singularity. It turns out that the intrinsic description of a surface proceeds in several stages, each stage a formalism which describes more of the structure of the surface.

First, a surface S is at least a topological space (Chapter I). This requirement provides points and a sense in which two points on the surface may be regarded as "nearby". A "distortion" of a surface is just a homeomorphism of spaces.

Second, a surface S is two-dimensional. This is a "local" property of the surface, which will require that each point has a neighborhood which "looks like" a neighborhood of a point in the Euclidean plane. "Local" properties are to be contrasted with "global" properties of the whole surface.

Third, a surface S should normally be smooth; this means that one should know which curves on S are smooth and which functions on the surface are differentiable. In turn, this provides means to construct tangent and cotangent planes to the surface.

Fourth, the surface may be equipped with a metric, which will make it possible to measure the length of smooth curves on the surface as well as the angles between such curves–thus providing for the basic operations of mensuration on the surface. This metric structure on the surface is usually expressed by giving a suitable differential form ds^2, where ds is the "element of arc", used in integration to get the length of arcs on the surface.

To fill in the details, we turn first to the topology. A surface S is at least a topological space. This means that certain subsets of S are specified as "open", and that these open sets satisfy the axioms given in §I.10. Then we can define when a function f on the surface is continuous–for any function f on S with values in S, or in the reals $\mathbf{R}$, or in any other topological space (as always, f continuous means that the inverse image of each open set is open). For example, if S and S' are two surfaces, a *homeomorphism* (see §V.5) $S \to S'$ is a function f which is a bijection of the set S onto the set S' such that both f and its inverse f^{-1} are continuous. Thus the Riemann surface of $z = \sqrt{w}$ pictured as two spheres joined by a tube is homeomorphic to the sphere; similarly the two-spheres picture (Figure 4.8) of the Riemann surface of $z = \sqrt{(w^2-1)(w^2-4)}$ is homeomorphic to the torus (but a torus is *not* homeomorphic to a sphere). For a homeomorphism $f\colon S \to S'$ it is not enough to require that f be continuous and a bijection. For example the map $\theta \mapsto e^{2\pi i\theta}$ is a bijection of the "half open" interval $[\theta \mid 0 \leqslant \theta < 2\pi]$ of the real axis onto the unit circle in the complex plane–but this interval is clearly *not* homeomorphic to this circle, because the inverse function f^{-1} is not continuous at 1. On the other hand, the circle *is* homeomorphic to the ellipse or to the periphery of the square.

The definition of a topological space is very general. Any surface with fins and filaments attached or with holes punched out is a topological space. Any subset Y of the plane $\mathbf{R}^2$ is a topological space, provide one takes as the open sets U of Y the intersections $V \cap Y$ with Y of arbitrary open subsets V of $\mathbf{R}^2$. In fact, this procedure makes a subset Y of *any* topological space X into a topological space (with the so-called *induced* or "subspace" topology). This choice of a topology on Y makes the inclusion $Y \subset X$ of Y in X into a continuous function $Y \to X$; moreover, it provides the fewest open sets in Y necessary to have such continuity. The indicated generality in the definition of a topological space is useful not just for the study of "pathological" spaces (of which there are many) but also to allow a flexible way of handling many mathematical objects as spaces. One example is the consideration of "spaces of functions". Thus the set of all continuous real-valued functions f on the unit interval $[0 \leqslant x \leqslant 1]$ of the reals is a metric space with the distance $\rho(f,g)$ between two functions f and g taken to be the maximum value attained by the quantity $| f(x) - g(x) |$ on the interval (recall from §I.10 that every metric space is also a topological space). Other ways of specifying when two functions are close to each other can be similarly used to construct topological spaces, as for example with the space L^2 of Lebesgue square integrable functions.

Any set X whatever can be regarded as a topological space when all the subsets of X are taken to be open; this gives the so-called *discrete* topology on X. For an infinite set X there is a topology in which the empty subset and all subsets U with a finite complement are taken to be open. This clearly satisfies the axioms for open sets, but is far from the usual geometric pictures of a space.

In view of this generality, it has been useful to consider various restricted classes of topological spaces. As in §I.10 a space X is said to be *Hausdorff* when it satisfies the following "separation" axiom:

Hausdorff Axiom. To each pair $p \neq q$ of distinct points of X there exist disjoint open sets U and V in X with $p \in U$ and $q \in V$.
(In other words, different points p and q have disjoint open neighborhoods U and V.) All the subspaces of $\mathbf{R}^2$ or of $\mathbf{R}^n$ are clearly Hausdorff spaces, as is any metric space. There are, however, many non-Hausdorff spaces; the simplest is the Sierpinski space which has exactly two points p and q, with open sets the empty set $\varnothing$ and the sets $\{p\}$ and $\{p,q\}$ (the point p is open, but not the point q!).

For many purposes one may wish to put a space together out of smaller pieces, just as a sphere may be described by two overlapping charts or a circle by two or more overlapping intervals. This pictorial idea (scissors and paste!) can be formalized in the notion of a "covering". An *open covering* of a space X indexed by a set I is a family U_i, for indices $i \in I$, of open sets U_i of X, such that X is the union of the U_i (i.e., such that every point of X lies in U_i for some index $i \in I$). The space X is thus completely determined by knowing all the subspaces U_i and how they overlap. For example, a subset V of X is open if and only if each intersection $V \cap U_i$ is open in the (induced) topology of the space U_i; for the same reason a function f on X to some other space is continuous if and only if the restriction of f to each of the open sets U_i is continuous. This result is the reason why it is useful to consider coverings by *open* sets–and why one describes surfaces by overlapping open charts.

Coverings are also used in the definition of compactness (§VI.7).

For many purposes, one studies not general topological spaces, but the more special compact Hausdorff spaces–in particular surfaces which are both compact and Hausdorff.

It is also appropriate to ask when a space is connected. The intuitive idea is that a space is connected when it does not "fall apart" into two or more pieces. This can be turned into a formal definition in at least two ways. One definition reads: A space is *connected* when it is not the union of two disjoint, non-empty open sets U and V. For example, the topological space consisting of the whole real line minus the origin is not connected, because it is the union of two (disjoint) half lines, each open–because the origin has been omitted.

A second definition of connectivity uses *paths*, where a path in any space is defined just as for paths in the plane: A *path* in X is a continuous map $f: I \to X$ of the unit interval $I = \{t \mid 0 \leqslant t \leqslant 1\}$ into the space X; it is of course called a *path* from $f(0)$ to $f(1)$. Then a space X is said to be *path connected* (or, sometimes, "arcwise connected") when there is at least one path joining any two points of the space. One may readily show that any path-connected space is connected in the previous sense. Indeed, if it is not so connected, and "falls apart" into disjoint open sets U and V, join a point of U by a path f to a point of V; then the open sets $f^{-1}U$ and $f^{-1}V$ are disjoint open sets which disconnect the interval I—but the interval *is* a connected space. However, there *are* connected spaces which are not path-connected! Here again we have an example of two different formalizations of the same intuitive idea!

7. Manifolds

The second aspect of the "intrinsic" description of surfaces is the sense in which any surface is "two-dimensional". This should mean that each point of the surface has a neighborhood which looks like (i.e., is homeomorphic to) a neighborhood of some point in the Euclidean plane. We can express this condition in terms of the "charts" which have already been used (in navigation and) in our description of the intrinsic geometry of the sphere. Specifically, if S is any topological space a (2-dimensional) chart on S is a homeomorphism

$$\phi: U \longrightarrow V, \qquad U \subset S, \qquad V \subset \mathbf{R}^2 \tag{1}$$

of an open subset U of S to an open set V in the Euclidean plane. The open set U is called the *domain* of the chart. An *atlas* for S is a family of charts $\phi_i: U_i \to V_i$, indexed by the elements i of some set I and such that S is the union of the domains of these charts (i.e., such that every point of S belongs to at least one chart). A two-dimensional topological *manifold* (a *surface*) is a topological space which has such an atlas; usually one also requires that the surface be connected in the sense described above; this will make it likely that there are charts which overlap. Here two charts ϕ and ϕ' have an *overlap* when the intersection $U \cap U'$ of their domains is nonempty. In this event, since ϕ^{-1} and ϕ'^{-1} exist and are continuous, the images $\phi(U \cap U') = W$ and $\phi'(U \cap U') = W'$ of the overlap are open sets W and W' in the plane, and the relevant restrictions ϕ_1 and ϕ'_1 of ϕ and ϕ' yield a diagram (see Figure 1)

$$W \xleftarrow{\phi_1} U \cap U' \xrightarrow{\phi'_1} W', \tag{2}$$

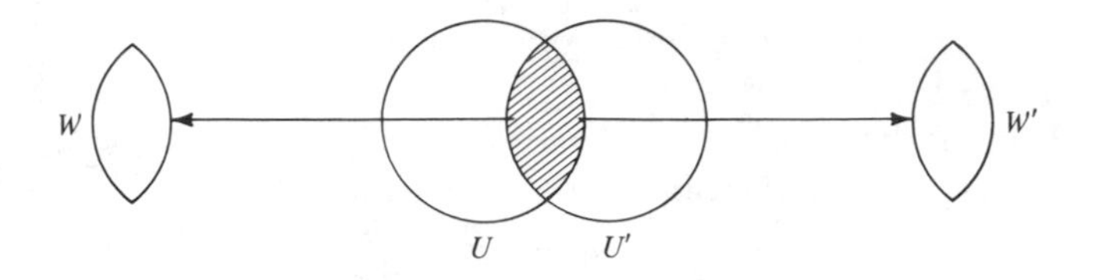

Figure 1. Overlap of charts.

which is a homeomorphism

$$\phi'_1 \cdot \phi_1^{-1}: W \to W'; \tag{3}$$

this is informally called the *overlap* (or *patching*) map for the two charts, because it codes the way in which the two charts are to be pasted together on the manifold. In other words, a surface is a topological space obtained by pasting together open sets (more explicitly, open discs) cut from the Euclidean plane.

This description of a surface S is not yet an invariant one, because one and the same surface may have many different atlases. It is for this reason that the definition reads "A surface is a topological space for which there exists an atlas" (e.g., there can be many different atlases for the sphere!). Alternatively one might take *all* the possible charts for S; they form an atlas called the *maximal* atlas because it contains every other atlas. Then a surface would be defined as a topological space which is covered by the domains of all of its two-dimensional charts.

As an example, consider the projective plane, defined to be the manifold of all lines through the origin of $\mathbf{R}^3$. As we have seen in §5, it can also be described as a circular disc in which diametrically opposite pairs of points on the circumference have been identified. It can be covered by three charts, for example, by three sectors of the disc, each extended a bit over its boundary so as to be an open set (see Figure 2). In this atlas, the charts U and U' overlap right side up along their extensions on $A0$ and also along BC—but upside down there because CB is pasted upside down

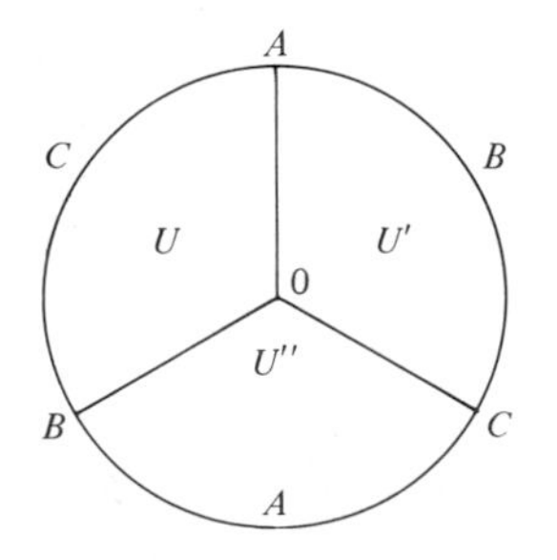

Figure 2. Projective plane with three charts.

on BC. Thus $U \cap U'$ falls into two connected pieces W_i and W_2. If we choose the overlap map on the first piece so as to preserve the orientation, then on the second piece it must reverse the orientation. We could have made the intersection of the charts connected by taking more and smaller charts as in the string of beads in Figure 3—but it still turns out that at least one of the overlap maps $W \to W'$ must reverse the orientation of the open sets W and W' of the Euclidean plane—because CB goes upside down onto BC. In other words, there is no way to pick "matching" orientations on all the charts of an atlas for the projective plane. This surface is therefore said to be non-orientable (and this property also means that the projective plane turns out to be a "one sided" surface whenever it is embedded in an (orientable) Euclidean space.

These observations about orientability can be formalized so as to apply to any surface. First, the basic axioms of plane geometry (§III.8) provide a choice of orientation for the Euclidean plane $\mathbf{R}^2$; this choice may be pictured as a choice of one of the possible directions of rotation as the "clockwise" one. Then any overlap map (3) between two charts sends an open set W of $\mathbf{R}^2$ to another open set W' of $\mathbf{R}^2$. We can assume that W and W' are both nonempty and connected. At any one point of W, the overlap map will either preserve the chosen orientation (i.e., preserve the chosen sense of rotation) or reverse it. Since W is connected, orientation preserved at one point of W will mean orientation preserved at all points of W. Now if a surface has an atlas in which every nonempty overlap map preserves the orientation, we say that the surface is *orientable*, and that it is *oriented* by a choice of orientation in one chart (and, by continuation, in every chart). This formal definition matches the intuitive notion. Indeed, the chosen orientation in $\mathbf{R}^2$ may be pictured by a small circle with a sense of rotation (a "clock"). Each chart $\phi : U \to W$ transfers (by ϕ^{-1}) this picture to the piece U of the manifold; when the overlap maps can all be made invertible, these directed small circles can be moved continuously along the overlap from the domain of one chart to the next—and this cannot be done on the projective plane. However, the plane itself, the sphere and the torus are all orientable in the sense of this definition. This development shows how the elementary Euclidean notion

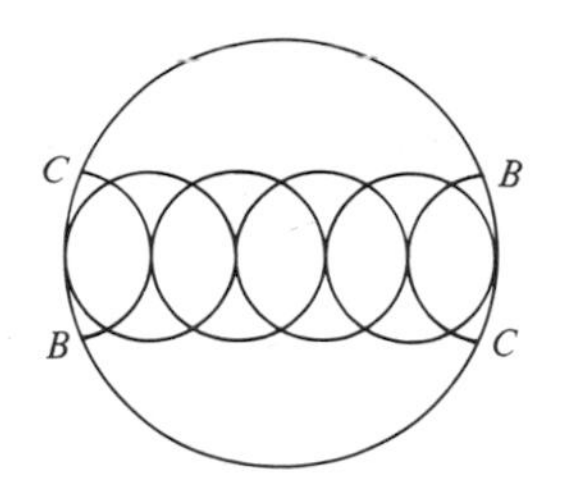

Figure 3

of orientation (Chapter III) extends to a much more general geometric context.

One may also readily prove that a surface is connected if and only if it is path-connected.

With these definitions in hand, we may ask what surfaces there may be, and in particular: What are *all* the compact connected orientable surfaces? Here two surfaces are to count as the same if they are homeomorphic; thus the surfaces of the sphere and of an ellipsoid are the same, but they are evidently different from the surface of the torus. The torus, in its turn, is different from the pretzel with two holes (Figure 4) and so on. This sequence "and so on" gives all the possibilities. Any compact connected orientable surface is homeomorphic to the sphere or to a pretzel with g holes, where the natural number $g \geqslant 1$ is called the *genus* of the surface. Two such surfaces of different genera cannot be homeomorphic. The proof of this theorem (to be found in many elementary texts on algebraic topology) uses a description of surfaces in which the edges of a closed polygon are suitably pasted together. For example, the torus has been described above as a square in which the four edges in cyclic order are pasted according to $bab^{-1}a^{-1}$ (Figure 5). Similarly an octagon with the edges paired up in cyclic order in the pattern $aba^{-1}b^{-1}cdc^{-1}d^{-1}$ will produce a pretzel with two holes (just think of two tori, connected along the circle e of Figure 6). The proof of the general theorem about surfaces amounts to showing that any such pasting of

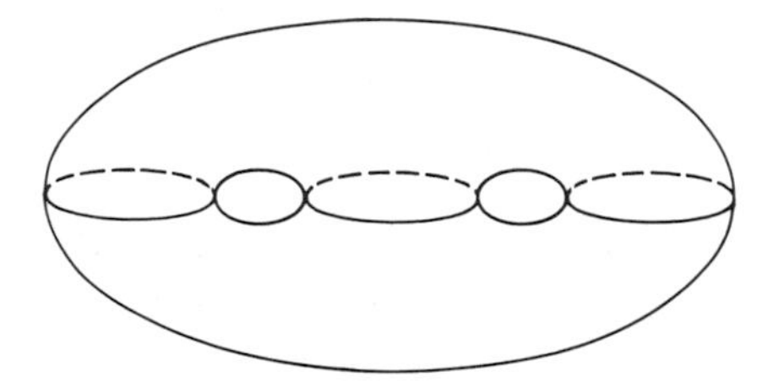

Figure 4. Pretzel with two holes.

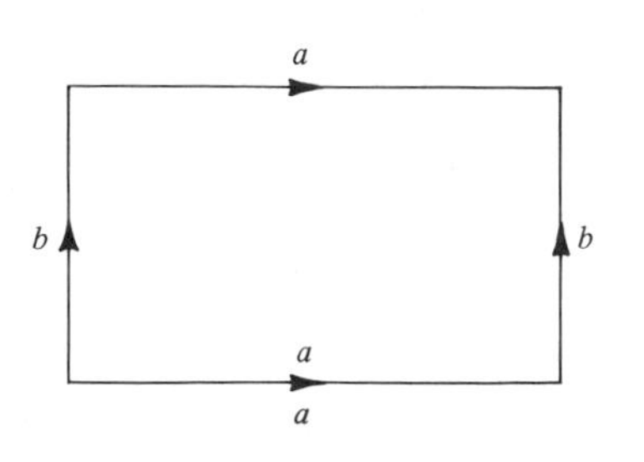

Figure 5. Torus as pasted square.

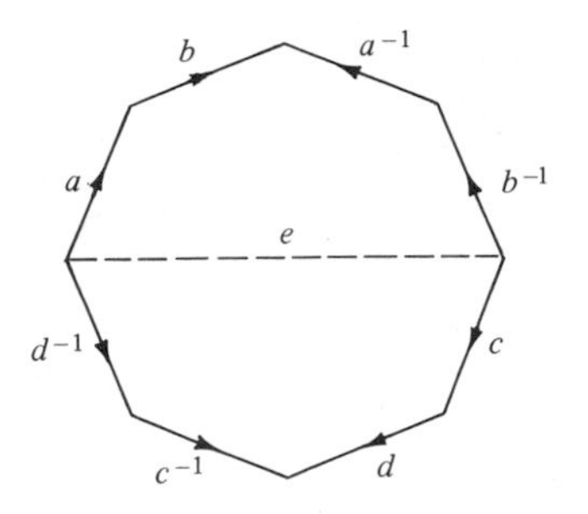

Figure 6. Pastings for a pretzel.

polygons to form a connected orientable surface can be reduced to one of the standard forms for a sphere or a pretzel. The result has evident meaning for the Riemann surfaces we have already described; indeed, every compact Riemann surface for a suitable function $w = w(z)$ is such a surface, and the genus is a significant invariant.

For non-orientable surfaces there is an exactly similar theorem. It can be formulated in terms of the Möbius band, which, as we have seen, is a non-orientable surface with a single circular arc as boundary. So take any orientable surface, cut out a circular disk with a circular arc as boundary, and along the boundary paste in a Möbius band—boundary to boundary. The resulting surface is evidently non-orientable, and the general theorem states that any compact connected non-orientable surface is homeomorphic to one of those so constructed. For example, this construction on the sphere will produce the projective plane (Möbius band plus disc!).

For noncompact surfaces the result cannot be so easy—for example one can produce such a surface by removing a finite or infinite number of points or slits from a compact surface.

For higher dimensions, our formal definition of a surface (a two-dimensional topological manifold) carries over at once to a definition of an n-dimensional topological manifold; the only change is that charts are now homeomorphisms ϕ: $U \to V \subset \mathbf{R}^n$ into open subsets V of $\mathbf{R}^n$. For each chart ϕ the cartesian coordinates x_i of $\mathbf{R}^n$, regarded as functions $\mathbf{R}^n \to \mathbf{R}$, yield by composition n functions $x_i\phi$: $U \to \mathbf{R}$, $i = 1, \ldots, n$, which are called the *local coordinates* for that chart. A point in the domain U of the chart is completely determined by its local coordinates, while the overlap functions describe the possible changes in local coordinates. Some more old-fashion treatments of manifolds operated with local coordinates only—but the full notion of a manifold requires both its global description as a topological space and its local description by charts.

For $n \geqslant 3$ the classification of all possible manifolds is a difficult problem. There are also infinite-dimensional analogs of manifolds (Lang [1967]).

8. Smooth Manifolds

The surface of a rectangular box (a parallelopiped) is a two-dimensional manifold according to our definitions–it is a compact and connected topological space which is, moreover, homeomorphic (say by radial projection) to the surface of a sphere. However, a curvature can hardly be defined at the corner points of such a box, and at these points there is also no tangent plane. In order to describe tangent planes, intrinsic curvature, and the like we must restrict attention to *smooth* surfaces.

The idea, as in §VI.9, is that a function is smooth when it has enough derivatives. To expand this, consider first a function f of two real variables x_1 and x_2 which is defined in some open set W of the plane $\mathbf{R}^2$, thus f is a function

$$\mathbf{R}^2 \supset W \xrightarrow{f} \mathbf{R}. \tag{1}$$

Since W is open, each point (x_1,x_2) in W has an open neighborhood also contained in W; this allows the usual definition of the partial derivatives of f at that point. Then f is said to be *smooth* of class C^∞ when f has partial derivatives of all orders at each point of its domain W, while f is said to be of class C^k when it has continuous partial derivatives of all orders up through order k.

Consider next a function ψ

$$\mathbf{R}^2 \supset W' \xrightarrow{\psi} W \subset \mathbf{R}^2 \tag{2}$$

between two open sets W' and W in the plane $\mathbf{R}^2$. Here each point p in W' is determined by its coordinates, say x_1 and x_2, while each point in W is determined by its coordinates, say u_i and u_2; they can be regarded as functions u_1,u_2: $W \to \mathbf{R}$. In this way the function ψ of (2) is described by two composite functions $u_1 \cdot \psi$ and $u_2 \cdot \psi$: $W' \to \mathbf{R}$

$$u_1 = f_1(x_1,x_2), \qquad u_2 = f_2(x_1,x_2). \tag{3}$$

This is the expression of ψ in terms of (local) coordinates. The function ψ is defined to be *smooth* if both these coordinate functions f_1 and f_2 are smooth (have continuous partial derivatives of all orders).

This definition has three basic properties:

1^0 *Composition.* The composite of two smooth functions is smooth. In particular, if ψ: $W' \to W$ in (2) and f: $W \to \mathbf{R}$ in (1) are both smooth, so is their composite $f \cdot \psi$: $W' \to \mathbf{R}$–because its partial derivatives are given by the chain rule.

2^0 *Restriction.* A smooth function remains smooth under any restriction of its domain. Specifically, if ψ: $W' \to W$ in (2) is smooth, while V' is any open subset of W', then the restriction of ψ to V' is a smooth function $\psi \mid V': V' \to W$.

3^0 *Patching.* A function put together from smooth pieces is smooth. For example, if $f: W \to \mathbf{R}$ in (1) is smooth and if the open set W is covered by a family of open sets V_i, for i is some index set I, then if all the restrictions $f \mid V_i: V_i \to \mathbf{R}$ are smooth, so is f—because each derivative of f at a point of W can be calculated from the values of f in one of the sets V_i which contains that point. In this case, one says that f is "patched together" from the smooth pieces $f \mid V_i$.

A smooth surface is now described by requiring that there be an atlas of charts with smooth overlaps. Indeed, recall from §7 that any two charts of a surface S,

$$S \supset U \overset{\phi}{\to} V \subset \mathbf{R}^2, \qquad S \supset U' \overset{\phi'}{\to} V' \subset \mathbf{R}^2$$

have on the intersection $U \cap U'$ of their domains an "overlap" map

$$\mathbf{R}^2 \supset W' = \phi(U \cap U') \overset{\phi'_1\phi_1^{-1}}{\to} W = \phi'(U \cap U') \subset \mathbf{R}^2$$

given as a composite of (suitable restrictions of) ϕ' and ϕ^{-1}. This overlap map $W' \to W$ goes between subsets of $\mathbf{R}^2$, so, as in (2), we can determine when it is smooth. Hence one can say that the charts ϕ and ϕ' have a smooth overlap if the composite $\phi'_1\phi_1^{-1}$ is smooth. If we write u_1 and u_2 for the local coordinates in V' and x_1 and x_2 for the local coordinates in V, this amounts exactly to saying that the local coordinates u_1, u_2 in the overlap are smooth functions (on $\phi(U \cap U')$) of the local coordinates x_1, x_2.

Now define a smooth surface S to be a (topological) surface S together with an atlas A of charts $\phi_i: U_i \to V_i$ for $i \in I$ such that any two charts ϕ_i and ϕ_j in the atlas A have a smooth overlap. As before, this atlas may not be maximal, but we can construct the maximal atlas by the following details. So consider any other chart $\phi: U \to V$ for S which is "smooth for A" in the sense that ϕ has a smooth overlap with every chart ϕ_i in the atlas A. We want to consider the collection A^* of *all* such charts ϕ. We claim that if $\phi': U' \to V'$ is another such chart, smooth for A, then ϕ and ϕ' have a smooth overlap. Indeed, for each chart ϕ_i of the given atlas A, there is a (possibly empty) intersection $V_i = U \cap U' \cap U_i$ of the domains and a diagram (Figure 1). Then, for suitable restrictions of the charts ϕ we have smooth overlaps

$$\phi_i\phi^{-1}: \phi(V_i) \to \phi_i(V_i); \qquad \phi'\phi_i^{-1}: \phi_i(V_i) \to \phi'(V_i),$$

the first because ϕ has smooth overlaps with all ϕ_i, and the second similarly for ϕ'. By composition (principle 1^0 above) the map

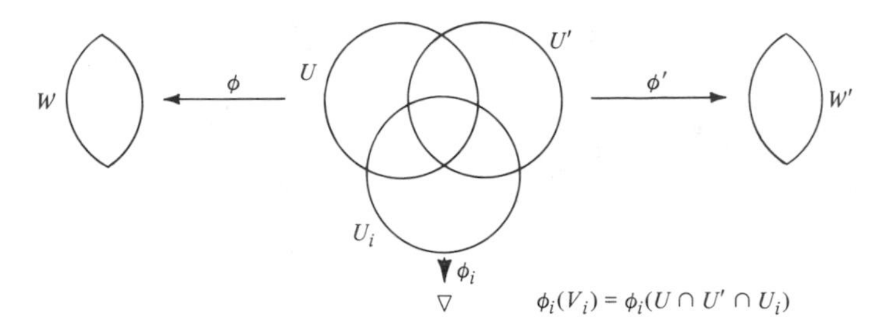

Figure 1. Overlapping charts.

$\phi'\phi^{-1}$: $\phi(V_i) \to \phi'(V_i)$ is smooth. On the other hand the domains U_i of the given charts cover the whole surface S, so the intersections $\phi(U_i \cap U \cap U')$ cover the whole of $W = \phi(U \cap U')$. Therefore, by the patching principle, $\phi'\phi^{-1}$: $W \to W'$ is smooth on the whole of W. In other words, the atlas A^* of all charts smooth for A has the property that any two of its charts ϕ and ϕ' have a smooth overlap. Among all smooth atlases for the smooth surface S, it is the maximal one—and every smooth surface can be described in an invariant way by such a *maximal atlas*. The description evidently depends on the three basic properties (composition, restriction, and patching) for smooth maps.

The same ideas will describe smooth manifolds M of any dimension. One also has the notion of a smooth map

$$\theta: M \to M' \tag{4}$$

between two such manifolds M and M' (of the same or of different dimensions). Indeed, for each point p of M choose charts for M and M' with domains U and U' such that $p \in U$ and $\theta(p) \in U'$. Then if U has dimension k and local coordinates $x_1, \ldots, x_k$ from the given chart, while U' has local coordinates $u_1, \ldots, u_n$, the map θ is expressed, much as in (3), via the local coordinates by n real-valued functions

$$u_1 = f_1(x_1, \ldots, x_k), \ldots, u_n = f_n(x_1, \ldots, x_k). \tag{5}$$

The map θ is then defined to be *smooth* if these n functions are smooth; this does not depend on the choice of the charts because the overlap functions between any two charts are themselves smooth, and because the composite of smooth functions is smooth.

In the simple case of a real-valued continuous function

$$f: M \to \mathbf{R} \tag{6}$$

one needs only one chart for the reals $\mathbf{R}$, considered as a one-dimensional manifold. Thus the definition above states that such a function f is smooth if on the domain U of each chart of M with local coordinates

$x_1, \ldots, x_k$ the corresponding real-valued function $f(x_1, \ldots, x_k)$ is of class C^∞. For any open set V in M, this also describes what is meant by a smooth function $V \to \mathbf{R}$, since V itself is a smooth manifold (with charts the intersections with V of all the charts of M). It is then straightforward to prove that a map θ: $M \to M'$ between smooth manifold M and M' is smooth if and only if, for every smooth chart of M' with local coordinates u_i, all the real-valued composites $u_i\theta$ are smooth in the sense described under (6). By the patching property, it is enough to require this just for the charts of some atlas covering M'–but the notion of smoothness is intrinsic, since it is independent of any particular choice of charts or of coordinates.

The plane, the sphere, the torus and the projective plane, with the respective atlases which we have described, are all smooth two-dimensional manifolds, with the expected smooth maps. There are similar smooth manifolds in higher dimensions. For that matter, the circle, the open interval and the whole real line are smooth one-dimensional manifolds.

9. Paths and Quantities

Next we study maps between manifolds. The calculus for a smooth surface S involves pairs of smooth maps

$$I \xrightarrow{h} S \xrightarrow{f} \mathbf{R}. \tag{1}$$

Here I is an interval, say the interval $-1 < t < 1$ of $\mathbf{R}$, with t as local coordinate. For suitable coordinates x_1, x_2 on S these two functions h and f are expressed by formulas such as

$$t \mapsto h(t) = (h_1(t), h_2(t)); \qquad (x_1, x_2) \mapsto f(x_1, x_2);$$

we may suppose that $h(0) = (0,0)$ is the origin (of local coordinates) on S.

The map h is a *path* through $h(0)$, parametrized by t, while the map f is a (real-valued) *quantity* (e.g., a physical quantity) defined everywhere on the surface S. Such a quantity on S may be pictured by its *level lines* (the loci $f(x_1, x_2) =$ constant) on S, as in Figure 1. At the origin $t = 0$ the path has a tangent vector with coordinates

$$\left(\frac{dh_1}{dt}, \frac{dh_2}{dt}\right)_{t=0} \tag{2}$$

while the quantity f has a gradient (or differential) given (as in §VI.9) by coordinates

$$df = (\nabla f)_0 = \left(\frac{\partial f}{\partial x_1}, \frac{\partial f}{\partial x_2}\right)_{x_1 = x_2 = 0}; \tag{3}$$

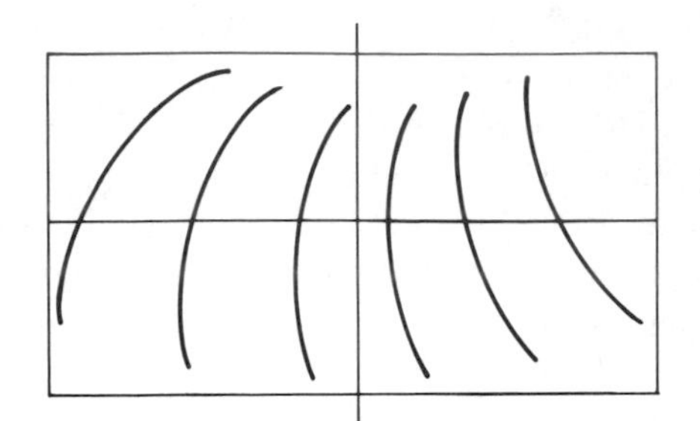

Figure 1. Level lines = countour lines.

this is its "cotangent" vector at the origin; it "points" in the direction of the maximum rate of increase of f at the origin. Indeed the *directional derivative* of f at the origin in the direction of the path h is defined to be the derivative there of the composite function fh. It can be expressed in coordinates by the usual chain rule as

$$\frac{d(fh)}{dt}\Big|_{t=0} = \left[\frac{\partial f}{\partial x_1}\frac{dh_1}{dt} + \frac{\partial f}{\partial x_2}\frac{\partial h_2}{\partial t}\right]\Big|_{t=0}. \tag{4}$$

This formula exhibits the cotangent vector $(\text{grad } f)_0$ of (3) as a linear function of the tangent vector (h'_1, h'_2) of (2), and it indicates that two quantities f and g have the *same* cotangent vector at this origin when they have the same directional derivatives along all paths–and dually that two paths through the origin have the same tangent vector there if and only if they give equal directional derivatives there for all quantities.

This last remark suggests how one can define tangent vectors (or cotangent vectors) without any use of coordinates. So consider all smooth paths through the origin and all smooth quantities f there. Under addition and multiplication by reals a (real scalars $a \in \mathbf{R}$) these quantities do form a (high-dimensional) vector space. Without using coordinates, each smooth composite function fh has a derivative

$$D_0(fh) = \frac{d(fh)}{dt}\Big|_{t=0} \tag{5}$$

at the origin; moreover, this derivative is linear as a function of the quantity f. Hence the assignment $h \mapsto D_0(fh)$ makes each path h a real-valued linear function on the vector space of all quantities. Now call two paths h and k through 0 *tangent* at 0 (and write $h \sim_0 k$) when they determine in this way the same linear function; that is when $D_0(fh) = D_0(fk)$ for every quantity f. This relation $h \sim_0 k$ is reflexive, symmetric, and transitive. Hence we can introduce for each path h the equivalence class

$$\tau_0 h = \{\text{all smooth paths } k \mid h \sim_0 k\} \tag{6}$$

and call it *the* "tangent vector" defined by the path h at 0. In much the same way, one may call two quantities $f,g\colon S \rightarrow \mathbf{R}$ "cotangent" at 0 (in symbols, $f \sim_0 g$) when $D_0(fh) = D_0(gh)$ for all smooth paths through 0. This relation is again reflexive, symmetric, and transitive, so we may introduce for each f the class

$$d_0 f = \{\text{all } g \mid g \text{ smooth and } f \sim_0 g\} \tag{7}$$

and call it the gradient (or the differential or the cotangent vector) for f at 0.

Now $f_1 \sim_0 g_1$ and $f_2 \sim_0 g_2$ imply that

$$(f_1 + f_2) \sim_0 (g_1 + g_2), \qquad af_1 \sim_0 af_2$$

for each $a \in \mathbf{R}$. Hence the vector operations of addition and multiplication by real scalars carry over from the vector space of functions to the classes $d_0 f$, so these cotangent vectors, as constructed, do form a vector space over $\mathbf{R}$. It is a two-dimensional space, since the chain rule shows that each $d_0 f$ is determined by two coordinates $(\frac{\partial f}{\partial x_1})_0$ and $(\frac{\partial f}{\partial x_2})_0$ relative to the basis $d_0 x_1$, $d_0 x_2$. This completes the invariant description of the cotangent space $T^\circ(S)$ to the surface S at the point 0. Moreover, each path h through 0 gives by the assignment

$$d_0 f \mapsto D_0(fh)$$

a linear function $T_\circ(S) \rightarrow \mathbf{R}$ on the cotangent space. This function, determined by h, actually depends only on the equivalence class $\tau_0 h$ of the path h. Now the linear functions $T^\circ(S) \rightarrow \mathbf{R}$ are the elements of the dual vector space $T^\circ(S)^*$. Hence each tangent vector $\tau_0 h$ can be taken to be an element of the dual space $T^\circ(S)^*$. The sum of two tangent vectors can then be defined to be their sum as elements of this dual space–and the sum agrees with that which would be calculated from the sum of their components in a local coordinate system. Hence the dual space $T^\circ(S)^*$ provides an invariant (coordinate-free) description of the tangent space $T_0(S)$ to the surface S at the point 0. In this way, both for surfaces and for manifolds of higher dimension, the tangent and cotangent spaces arise naturally and together from the smooth structure and from the smooth paths and quantities defined by that structure. Many texts do not explain this natural origin.

Each point of the surface S has its cotangent space, a vector space of dimension 2. Now consider the set of all cotangent vectors at all points of S; they form a geometric object called the *cotangent bundle* $T^{\bullet}S$. It is also a smooth manifold; for a given chart of S, each cotangent vector df is determined by four real numbers: The local coordinates (x_1,x_2) of the point of cotangency, plus the values $\frac{\partial f}{\partial x_1}$ and $\frac{\partial f}{\partial x_2}$ of the partial derivatives at that point of some quantity f. These four coordinates transform smoothly on the overlap between two charts on S. Hence the cotangent bundle $T^{\bullet}S$ for S is a four dimensional manifold. Moreover, the function $p\colon T^{\bullet}S \to S$ which sends each cotangent vector $d_0 f$ to the point of cotangency ($d_0 f \mapsto 0$) is a smooth map, called the *projection* of the cotangent bundle on its *base* S. A *cross section* ω of this projection is a smooth map $\omega\colon S \to T^{\bullet}S$ such that $p\omega\colon S \to S$ is the identity; in other words, a cross section is a smooth map which sends each point (x_1,x_2) of the surface S to a vector $d^{(x_1,x_2)}f$ cotangent at that point. In local coordinates x_1 and x_2 the differentials dx_1 and dx_2 of the coordinates form a basis of each cotangent space, hence each cross section ω has (in local coordinates) an expression of the form

$$\omega = g_1(x_1,x_2)dx_1 + g_2(x_1,x_2)dx_2\,, \tag{8}$$

where g_1 and g_2 are smooth functions. Expressions such as (8), usually called *differential forms*, can be described intrinsically as cross sections of the cotangent bundle. In particular, each smooth function $f\colon S \to \mathbf{R}$ on the surface determines such a differential form; in local coordinates

$$df = \left(\frac{\partial f}{\partial x_1}\right)dx_1 + \left(\frac{\partial f}{\partial x_2}\right)dx_2\,. \tag{9}$$

This idea of a cotangent bundle also arises from mechanics (Chapter IX).

The *tangent bundle* $T_{\bullet}S$ is similarly described; its points are all the tangent vectors at all the points of the surface S; they form a four dimensional smooth manifold with the evident coordinates, and with a smooth projection $T_{\bullet}S \to S$. A smooth cross section of this projection is called a *vector field* on S. Such a field assigns to each point of S a tangent vector at that point—and this in a smooth way. For example, if S is the plane with coordinates x, y, a pair of simultaneous differential equations

$$\frac{dx}{dt} = f(x,y), \qquad \frac{dy}{dt} = g(x,y)$$

with smooth f and g is in effect a vector field in the plane.

These few definitions are just the starting point for the study of the calculus on manifolds. For example, integration of differential forms can be defined, and the fundamental theorem of calculus has higher dimensional analogs in the theorems of Gauss, Green, and Stokes. Higher derivatives of smooth quantities also enter. For example, the *k-jet* of a quantity $f\colon S \to \mathbf{R}$ is the object determined by the collection of all the partial derivatives of f up through order k, with respect to some set of local coordinates.

The topological structure of a surface or a manifold is given in a very simple way, by specifying the open subsets. A smooth structure requires a more elaborate description, in terms of an atlas. Familiar topological manifolds, like $\mathbf{R}^n$, also have a familiar (standard) smooth structure–but just recently it has been discovered that $\mathbf{R}^4$ has a second smooth structure–not equivalent to the standard one. (This cannot happen for $\mathbf{R}^n$, $n \neq 4$.) However, consider the 7-sphere S^7; that is, the locus in $\mathbf{R}^8$ of

$$x_1^2 + x_2^2 \quad \cdots \quad + x_8^2 = 1.$$

It has a standard smooth structure–with charts given (say) by projection onto coordinate hyperplanes of $\mathbf{R}^8$. However, it has also another, essentially different, smooth structure.

10. Riemann Metrics

The final step in the intrinsic description of a surface S is the metric, defined by specification of the lengths of curves on S. This will be done by integrating a formula for the differential ds of arc length. For the sphere (and other surfaces) we have seen that ds^2 can be given in terms of suitable local coordinates u and v by a formula of the form

$$ds^2 = E(u,v)du^2 + 2F(u,v)dudv + G(u,v)dv^2,$$

where E, F, and G are smooth functions. More generally, one may express such a Riemann metric in terms of local coordinates x_i as a "quadratic differential form"

$$ds^2 = \sum_{i,j=1}^{2} g^{i,j}(x_i,x_j)dx_i dx_j, \tag{1}$$

where the g^{ij} are smooth functions of the local coordinates and are symmetric ($g^{ij} = g^{ji}$) and positive definite. The latter means, as usual, that $(a_1,a_2) \neq (0,0)$ implies $\Sigma\, g^{ij}(x_i,x_j)a_i a_j > 0$. A *Riemann metric* on a surface means a metric (1) on each chart, such that the metrics agree (under change of coordinates) on each overlap of charts.

To interpret this, consider a path $h\colon I \to S$ on the surface for which the image $h(I)$ lies in the domain of the chart involved. We have already seen that each differential form

$$\omega = g^1(x_1,x_2)dx_1 + g^2(x_1,x_2)dx_2$$

represents a linear function on each tangent vector $h'(t)$, by the formula

$$\omega(\tau_0 h) = g^1(x_1,x_2)\frac{dh_1}{dt} + g^2(x_1,x_2)\frac{dh_2}{dt}.$$

In just the same way, the quadratic ds^2 of (1) is a quadratic function of tangent vectors, as in

$$Q(\tau_0 h) = \sum_{i,j=1}^{2} g^{ij}(h_1(t),h_2(t)) \frac{dh_i}{dt} \frac{dh_j}{dt}. \quad (2)$$

By the assumption about overlaps of charts, the value of $Q(\tau_0 h)$ at any value of t does not depend on the choice of a chart. Finally, the length of the path h from $h(t_1)$ to $h(t_2)$ is defined by the integral

$$\int_{t_1}^{t_2} Q(\tau_0 h) dt. \quad (3)$$

The metric may also be used to describe corresponding formulas for areas.

The meaning of ds^2 can be formulated in terms of vector space concepts. At each point of the surface S the form ds^2 involves a positive definite 2×2 matrix $g^{ij} = g^{ij}(x_1,x_2)$. If v and w are two tangent vectors to S, expressed in the same local coordinates as $v = (v_1,v_2)$ and $w = (w_1,w_2)$, then the formula (2) gives a bilinear, symmetric, and positive definite inner product

$$(v,w) = \sum_{i,j=1}^{2} g^{ij} v_i w_j. \quad (4)$$

Thus specifying a Riemann metric on the surface S amounts exactly to specifying an inner product in the tangent plane at each point of S, in such a way that this specification is smooth (i.e., smooth as a function of the local coordinates). This description (and the formula above) apply equally well for smooth manifolds of higher dimensions. It is the starting point of an extensive theory of Riemannian manifolds.

We have thus completed the intrinsic definition of surfaces. A surface is a topological space covered by charts with smooth overlaps and with a Riemann metric. On this basis there is a remarkable further development of geometric properties of surfaces—and similarly of higher dimensional manifolds. In particular, the Gaussian curvature of a surface may be defined directly in terms of the metric. With this, the intrinsic definition of a surface is complete.

11. Sheaves

Continuous functions on a topological space X can be described just in terms of the open subsets of X, since a function $f: X \to Y$ is continuous when the inverse image under f of each open subset of Y is open in X. Smooth functions on a manifold, on the other hand, require more for their definition: A whole array of charts. There is a different approach which describes directly just the collections of continuous or of smooth functions (call them the "good" functions). For this approach one must consider not just the good functions defined on all of X, but those defined on each open subset U of X, together with the way in which these functions restrict to smaller open subsets. Such collections of functions (or of other objects) are called "sheaves". They play a major role in higher geometry.

In the case of two topological spaces X and Y, the topology determines for each open set $U \subset X$ the set

$$C(U, Y) = \{f: \mid f: U \to Y \text{ continuous}\} \tag{1}$$

of all continuous functions on U to Y. Moreover, if $V \subset U$ is a smaller open set in X, each continuous f on U has a restriction $f \mid V$ to V which is continuous there; this gives an operation "restriction"

$$f \mapsto f \mid V, \qquad \Upsilon_V^U: C(U, Y) \to C(V, Y).$$

For fixed Y, we now examine the way in which the set $C(U, Y)$ depends on U.

A *presheaf* P on a topological space X is a function which assigns a set $P(U)$ to each open set U of X and a restriction map

$$\Upsilon = \Upsilon_V^U: P(U) \to P(V) \tag{2}$$

to every inclusion $V \subset U$ of open sets, but in such a way that Υ_U^U is always the identity and also so that restrictions compose. This means that for each nested array

$$W \subset V \subset U \tag{3}$$

of open sets in X one has the composition rule for restrictions,

$$\Upsilon_W^U = \Upsilon_W^V \Upsilon_V^U: P(U) \to P(W). \tag{4}$$

In general, think of $P(U)$ as the set of all "good" functions on the open set U, together with the operation Υ_V^U of restricting any good function to an open subset V of U. For example, for each fixed Y, the sets $C(U, Y)$ of (1) constitute a presheaf on X because, given a nested array (3), the restriction of a continuous function f from U to W is the same as the restriction of f to V, further restricted to W. If X is a smooth manifold, the sets $S(U, \mathbf{R})$ of smooth functions $f: U \to \mathbf{R}$ is also an example.

Cross-sections provide another example. If $p: E \to X$ is any continuous map of topological spaces, a *cross section* of p over an open set U is a function $\omega: U \to E$ such that the composite $p\omega: U \to X$ is the identity on U. Each cross section on U has an evident restriction to each open subset

$V \subset U$, and with this restriction the cross sections of p form a presheaf $\chi(U,p)$.

Still another example, for any topological space X, is the set

$$\Omega(U) = \{W \mid W \subset U \text{ is open in } X\} \tag{5}$$

of all open subsets W of the given open set U. Here the restriction map is to be the intersection

$$\Omega(U) \to \Omega(V) \quad \text{by} \quad W \mapsto W \cap V. \tag{6}$$

The next observation about continuous functions is that a continuous function on a large open set can be "pieced together" from values on smaller open sets when these given values match. Specifically, given open sets $U_1, U_2 \subset X$ consider continuous function $f_1: U_1 \to Y$ and $f_2: U_2 \to Y$ which match on the overlap $U_1 \cap U_2$ (i.e., the restrictions of f_1 and f_2 to $U_1 \cap U_2$ are the same). Then there is a unique continuous function $f: U_1 \cup U_2 \to Y$ which restricts to f_i on each U_i—as for example when U_1 and U_2 are overlapping intervals on the line, as in Figure 1.

The idea here is that good functions which match on all overlaps can be pieced together. When this applies to a presheaf P, that presheaf is called a sheaf.

Formally, a *sheaf* S on a topological space X is a presheaf with the following property: For every covering

$$U = \bigcup_{i \in I} U_i \tag{7}$$

of an open set U of X by open sets U_i and for every list of elements $f_i \in S(U_i)$ which "match", in the sense that any two f_i and f_j have the same restriction to $S(U_i \cap U_j)$, there is a unique $f \in S(U)$ with restriction f_i on each U_i. In other words, given (7) and $f_i \in S(U_i)$ with

$$\Upsilon_V^{U_i} f_i = \Upsilon_V^{U_j} f_j, \qquad V = U_i \cap U_j \tag{8}$$

for all i and j, there exists a unique $f \in S(U)$ with

$$\Upsilon_{U_i}^{U} f = f_i, \qquad i \in I. \tag{9}$$

This property: Things that match as in (8) can be pieced together as in (9), turns out to be a central characteristic of collections (sheaves) of "good" functions, and of other geometric objects.

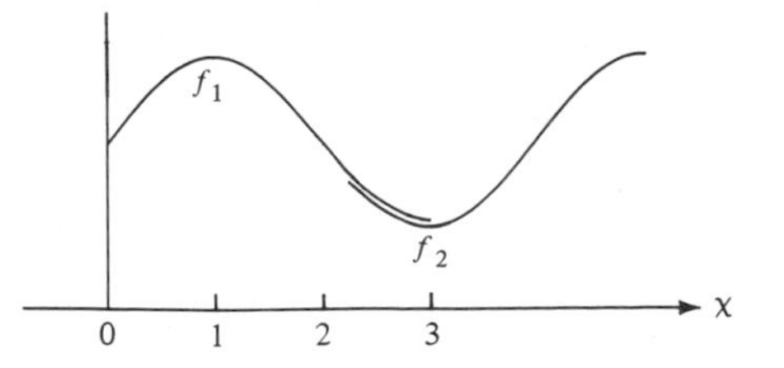

Figure 1. Pasting two curves.

The examples of presheaves given above are all also examples of sheaves. For instance, the matching condition clearly holds for the presheaf $C(U, Y)$ of continuous functions and for the presheaf of cross sections of a map $p: E \to X$. The matching conditions all hold for the presheaf Ω of open sets described under (5). For, given an open covering (7) of an open set U and given open subsets $W_i \subset U_i$ for each $i \in I$ which satisfy the matching conditions

$$W_i \cap U_j = W_j \cap U_i$$

for all i and j, the union $W = \cup W_i$ is the unique open subset of U with $W \cap U_i = W_i$ for each i. Hence Ω is a sheaf.

The matching–piecing condition applies also to smooth functions. Thus for an open set U on a smooth manifold and a covering $U = \cup V_i$, a function $f: U \to \mathbf{R}$ such that its restrictions $f_i: U_i \to \mathbf{R}$ are all smooth is necessarily smooth itself–because the derivative of f at a point of U depends only on the values of f in any specific neighborhood of the point. Hence for any smooth manifold M the sets

$$C_M^\infty(U) = \{f \mid f: U \longrightarrow \mathbf{R} \text{ smooth}\} \tag{10}$$

define a sheaf on M, sometimes called the "structure sheaf" on M.

The notion of a structure sheaf can also be used to give an alternative definition of what is meant by a smooth structure on an n-dimensional manifold. For this approach, one starts with certain model spaces–the open sets V of $\mathbf{R}^n$, each V equipped with its structure sheaf S_V of smooth real-valued functions on open subsets of V. An n-dimensional smooth manifold is then defined to be a topological space X equipped with a sheaf S and which has a covering by open sets U_i, for i in some I, such that S restricted to any one U_i is "isomorphic" to the structure sheaf S_{V_i} of some open set V_i of $\mathbf{R}^n$. We omit the details, which need a description of isomorphisms and of more general morphisms from one "space X with a sheaf S" into another such object. However, the general intent should be clear: A manifold can be described as a topological space on which the "good" functions are specified, by giving the sheaf of such functions. There are corresponding sheaf definitions for other types of manifolds–for example, for C^k manifolds and for complex analytic manifolds. The same approach has recently been especially successful in describing "algebraic" manifolds. In other words, the notion of a sheaf has proved to be an effective conceptual tool in describing a considerable variety of sorts of spaces; the elementary notion of space thus acquires (and requires) a sophisticated conceptual development.

One may also need algebraic structures on sheaves. For example, the structure sheaf C^∞ of a manifold is actually a sheaf of rings: Each set $C_M^\infty(U)$ of smooth functions is a commutative ring under the evident addition and multiplication of smooth functions, and the corresponding restriction maps Υ^U_V of (2) are homomorphism of rings. Similarly, sheaves of abelian groups on a topological space may be used to describe the connectivity properties of that space. Sheaves also enter in certain independence proofs in set theory (Chapter XII).

12. What Is Geometry?

Within Mathematics, geometry plays a significant role, as we have seen; however we have not yet seen just what it is that makes a given piece of Mathematics "geometry". The starting point for geometry is the idea that there should be a "science of space", but the subsequent development of this science goes in many directions, some of them unexpected and some not strictly "geometrical". Space, to start with, is flat, three-dimensional, and extended. Indeed it is this unlimited "extension" of space which introduces infinity into geometry, formally by way of the Archimedean axiom (though this axiom also has a strictly arithmetic formulation with real numbers). Similarly geometry involves the consideration of rigid motions, in part as a means of comparing differently situated but similar figures and in part because the very idea of geometric extent arises from the experience of motion. This use of rigid motions implicitly involves the compositions of motions and hence groups of motions—even though this idea of group was formalized only late in the historical development of geometry.

On the other hand, geometry employs simple physical instruments, the straight-edge and the compass, and then idealizes these instruments as perfectly straight lines and perfectly round circles. Using these idealized forms the raw empirical facts of geometry are organized as systematic consequences of a system of postulates. Amazingly, most of these postulates are formulated just for the geometry of the plane, even though such an ideal plane is not initially present in the rough original experience of space. Moreover, this Euclidean plane has not just one structure, but two: One structure not oriented, the other oriented and hence equipped with a smaller group of symmetries.

The Euclidean plane is an effective formal object for expressing many geometric phenomena, and moreover provides a handy tool for managing three-dimensional space—but geometry does not stop short at the three dimensions of our ambient physical space. All manner of appearances require geometrical formulations with more than three dimensions. Linear phenomena in these higher dimensions can be managed readily, partly with algebraic means via coordinates and partly by extending suitable insights from 2 and 3 dimensions. The resulting study of n-dimensional geometry then appears in a double light, both as linear algebra and as vector geometry—and the tension resulting from this double classification reappears repeatedly, as for example in the alternative between matrices and linear transformations. Such double billing means that any answer to our question "What is geometry?" must inevitably mix geometry with other sources of Mathematics. Similarly, elementary geometrical notions of angle, mensuration, and trigonometry reappear in higher dimensions in the guise of an inner product structure for Euclidean vector spaces. Here again the principal axis theorem, alias the spectral theorem, formulates both the geometric properties of the axes of conic sections and the eigenvalues appropriate to linear differential operators. Therefore geometry is not just a subdivision or a subject within Mathematics, but a means of turning visual images into formal tools for the understanding of other Mathematical phenomena.

But geometry is not just one form of space. The possibility of different geometries was first indicated by the development of non-Euclidean geometry. It turned out that the Euclidean parallel axiom is just not a consequence of the other axioms of Euclidean geometry; instead it represents one possible choice among several: Through a point not on a given line in a plane there may be no parallel lines, one parallel line, or many parallels, and the axioms expressing these varying possibilities can be supported by geometric models–in which the "lines" mentioned in the axioms are not the expected Euclidean lines. What then matters is the coherent development of angle sums and other metric (and group-theoretic) properties from the axioms. At this point in the historical development it becomes clear that geometry is not the science of a unique form of "space", but the extended study of many space-like forms.

Within any one of these spaces one then comes to consider all sorts of figures–not just the lines and triangles of linear algebra. Curves, surfaces, and curved solids appear both analytically as the loci of equations and geometrically as configurations in space. Curvature and global form is of their essence. Locally such figures can be specified when point-coordinates are given by functions of suitable parameters, such as time, latitude, and longitude–but the full nature of these figures requires a subtle global description. At first, they may appear to be just figures embedded in some standard Euclidean space, but this is not adequate. Their properties, in particular their curvature properties, have an intrinsic existence, independent of any Euclidean embedding. In fact there are many "manifolds", such a manifold of lines, which are originally given directly as manifolds and not in some Euclidean space. In this chapter we have been concerned primarily with the intrinsic description of such manifolds. This description has required quite a number of separate steps, involving separate kinds of "space" and "manifolds". It turns out to be convenient to begin with the very general notion of a topological space, where everything is described by the notion of a neighborhood (via open sets). Such spaces provide a satisfactory generality and a way of using geometrical language for analytic purposes, as with function spaces. For the much more specific notion of a manifold the essential "local" idea is that of a "chart"–an idea derived perhaps from the use of maps in navigating the globe. At any rate, the global manifold is pieced together from overlapping local charts, each a bijective image of a chunk (open set) of Euclidean space (or of some other model space). Specifying the overlap maps between adjacent charts is then an essential feature; according as the overlap is continuous, or continuous and smooth, or complex analytic one has a topological, or a differentiable, or a complex-analytic manifold. In this case again a simple geometric object–for example a torus–can be viewed as with different structures–simply as a topological manifold (or space), or as a smooth manifold, or as a complex one (e.g., a Riemann surface). These alternatives, in cases more general than that of the torus, later involve a remarkable variety of problems. (Can one topological manifold have different smooth structures?)

For a manifold, the charts are primarily a means to an end–in this case to the determination of those functions on the manifold which are continuous, or differentiable, or complex-analytic, as the case may be. If one instead considers directly such sets of "good" functions and their behavior under restriction of the domain one is led to the notion of a "sheaf". This concept, with its subsequent extensive use elsewhere in Mathematics, provides a striking example of the way in which pictorial ideas from geometry turn into conceptual idealizations.

This was the last item of this chapter, but it is by no means the end of geometry. A few (among many) other examples will appear later. In the next chapter we will see how certain problems in mechanics naturally lead to the study of tangent bundles of manifolds. Then complex analysis (Chapter X) will make use both of topology (connectivity) and manifolds (Riemann surfaces). Algebraic varieties (loci described in coordinates by polynomial equations) play an important role, for example in number theory (Chapter XII). Here again geometric intuition aids mightily in the understanding of parts of Mathematics which initially are not at all geometric.

Beyond these few examples lie many other parts of geometry. They are enormously varied; they do not really provide a simple answer to our original question "What is geometry?". It would seem that the original and pictorial study of space and motion, pursued systematically and imaginatively, leads to the study of many different kinds of Mathematical forms. The resulting development is partly geometrical and partly conceptual, and uses geometric insights to provide both organization and understanding. Geometry is not so much a subdivision of Mathematics as one source of Mathematical form.

CHAPTER IX

Mechanics

There is a remarkable interaction between theoretical constructions in Science and conceptual notions in Mathematics: the same idea–or the same idea, disguised, may arise both in Science and in Mathematics. This interaction is especially visible in the relations between theoretical Physics and pure Mathematics, where it first became apparent in mechanics at the hands of Galileo and in the invention of the calculus by Newton. One may say that the calculus (which dominated the development of Mathematics for at least two centuries) was developed by Newton in order to tackle problems of mechanics, especially those of celestial mechanics. This is only one of the striking interactions we have in mind. Today there is a amazing confluence of the gauge theories in Physics (for the Yang–Mills equations) and the geometrical theory of connections on fiber bundles. It is the aim of this chapter to sketch some of these developments. Though we cover only a small number of items, they are intended as samples of many decisive interactions between Science and Mathematics.

1. Kepler's Laws

Fascination with the motions of the planets and the stars is endemic. For the Greeks, a precise description of planetary motion was provided by ptolomaic astronomy. Given that the circle was for Greek geometry the dominant representation of repetitive or periodic motion, there was an initial inclination to think that the planets move in circular orbits with the earth as center. However, such an orbit did not suffice to explain the appearances–in particular the observation that at times the planets seemed to move backwards, in a retrograde motion. To account for this, Ptolemy provided epicycles (small circles superimposed on the original circular orbits–and with enough epicycles most of the motions (perhaps almost all) could be explained. Such an explanation required a considerable number of epicycles; it served until Copernicus introduced a heliocentric theory. It was then Kepler who used many careful measurements of the positions of the planets (i.e., their orbits) relative to the sun. After elaborate calculations he was able to propose his three laws about these orbits of the planets. His laws read

(1) The planets describe orbits in a plane containing the sun, in such a way that the areas swept out in equal times are equal. (This refers to the area swept out by the radius vector from the sun to the planet.)
(2) Each planetary orbit is an ellipse, with the sun at one focus.
(3) The square of the period of each planet in its orbit is proportional to the length of the major axis of the ellipse.

In this form, Kepler's laws are a summary (in geometrical terms) of facts of observation. Newton, using the calculus for this purpose, was able to deduce these laws from more basic principles of motion. In particular, Newton's second law of motion states that the force on a particle is its mass m times its acceleration. Here the force $\mathbf{F}$ and the acceleration $\mathbf{a}$ are three dimensional vectors, written with bold face letters as is the custom in theoretical physics. Thus Newton's second law takes the form of a vector equation

$$m\mathbf{a} = \mathbf{F}, \tag{1}$$

where the acceleration $\mathbf{a}$ is measured relative to some "inertial" base. Such a vector equation is independent of the choice of coordinates; given a choice of coordinates, it is equivalent to three scalar equations in the three (rectangular) components of $\mathbf{s}$ and $\mathbf{F}$.

The calculus enter essentially because the acceleration of the particle is defined to be the second time derivative of its position vector $\mathbf{s}$ (the vector from the origin of the inertial base to the planet). Using Newton's notation of one dot for one time derivative, this definition can be written as $\mathbf{a} = \ddot{\mathbf{s}}$. In particular, if the force $\mathbf{F}$ on the particle is zero, so is the acceleration $\mathbf{a}$. From this it follows that the vector velocity $\mathbf{v} = \dot{\mathbf{s}}$ is constant. This is Newton's first law: A particle under the action of no forces remains at rest or moves in a straight line with constant velocity. Newton also formulated a universal law of gravitation: The force of gravitational attraction between two point masses m_1 and m_2 is directly proportional to the product of these two masses, inversely proportional to the square of the distance r between them, and directed along the line joining the bodies. In other words the magnitude F of the force is given by the *inverse square law*

$$F = \gamma\frac{m_1 m_2}{r^2}, \tag{2}$$

where γ is a constant (the gravitational constant) in suitable units. This is called a "universal" law because it applies effectively both to the motions of planets and stars, and to the more local motions of bodies (such as projectiles) in the earth's gravitational field. Newton presented both kinds of applications in his famous "Mathematical Principles of Natural Philosophy" (1687), but the presentation there did not explicitly use the calculus. The full development of the subject of "Newtonian Mechanics", with its use for extended bodies, fluid mechanics, and the like, required more than a century and was advanced by many other noted mathematicians, especially Leonard Euler (1707–1783).

Newton did deduce Kepler's laws from his; let us summarize the process. Consider the motion of a planet of mass m relative to the sun. We neglect the gravitational forces from the other planets, so that we have a *two-body* problem, with sun and planet as the two bodies. At a chosen initial time, the vector force on the planet is directed toward the sun, and the planet has some initial vector velocity. If this velocity happens to be directed toward (or away from) the sun, all the subsequent motion (by Newton's second law) must be along the line joining the planet to the sun. Leaving aside this exceptional case, the initial velocity and the line from planet to sun together determine a plane containing both the velocity and acceleration vectors. Hence, by Newton's second law, all the subsequent motion takes place in that plane. This is the derivation of that part of Kepler's first law which asserts that the orbit is planar. The proof did not use the inverse square law.

To get the shape of the orbit, we describe the plane of the orbit by rectangular coordinates x and y relative to the sun 0 as origin. The gravitational force F on the planet P is then proportional to $1/r^2$ and directed from P to the origin 0, where r is the distance $0P$, This suggests the introduction of polar coordinates (r,θ), as in Figure 1. These polar coordinates convert to rectangular coordinates (x,y) by the familiar equations

$$x = r\cos\theta, \qquad y = r\sin\theta \tag{3}$$

which embody the definitions of sine and cosine. Therefore the force $\mathbf{F}$ is a vector of magnitude F with components $-F\cos\theta$ and $-F\sin\theta$ along the x and y axes. Thus Newton's second law, written in components, becomes

$$m\ddot{x} = -F\cos\theta, \qquad m\ddot{y} = -F\sin\theta; \tag{4}$$

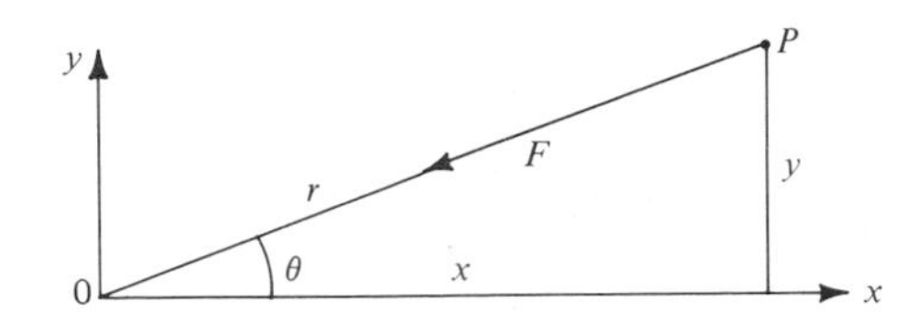

Figure 1. Coorinates for an orbit.

here the double dots represent the second derivatives of the (variable) coordinates with respect to time. These two differential equations can be combined to give the equation

$$m(\ddot{x}y - \ddot{y}x) = 0$$

which states that the time derivative of $\dot{y}x - \dot{x}y$ is zero, so that

$$\dot{y}x - \dot{x}y = k \tag{5}$$

is a constant k.

This formal deduction has substance. The quantity $\dot{y}x - \dot{x}y$ on the left of (5), when multiplied by the mass m, is called the *angular momentum* of the planet P about the point 0. So the proof has shown that the angular momentum about 0 is constant when the force is directed toward (or away from) 0. In polar coordinates, the equation (5) for angular momentum becomes, by (3),

$$mr^2\dot{\theta} = mk\,; \tag{6}$$

this displays the relation of angular momentum about 0 to the angular velocity $\dot{\theta}$ about 0. In rectangular coordinates, consider the position and velocity vectors as three dimensional vectors with components $(x,y,0)$ and $(\dot{x},\dot{y},0)$. Their vector product (§IV.14) is then $(0,0,\dot{y}x - \dot{x}y)$; multiplied by m, it is the momentum, considered as a vector through 0 perpendicular to the plane of the motion.

This is also related to the area $A = A(t)$ swept out by the radius vector $0P$ in time t; Figure 2 indicates that the increment in area $A(t)$ due to an increment $\Delta\theta$ in the polar angle is approximately the area $\frac{1}{2}r^2\Delta\theta$ of a circular sector of radius r and angle $\Delta\theta$. Hence, by the usual limiting procedures, the rate of change of area is $\dot{A} = (1/2)r^2\dot{\theta}$. Thus the equation (6) contains that part of Kepler's first law which states that area is swept out at a constant rate; again the conclusion does not depend on the inverse square law!

To get the equation of the orbit, we must use the inverse square law in the form $F = m\mu/r^2$, where μ is some constant. By (6), this can be written $F = (m\mu/k)\dot{\theta}$; then the second order differential equations (4) become

$$\ddot{x} = -(\mu/k)\cos\theta\,\dot{\theta}, \qquad \ddot{y} = -(\mu/k)\sin\theta\,\dot{\theta}.$$

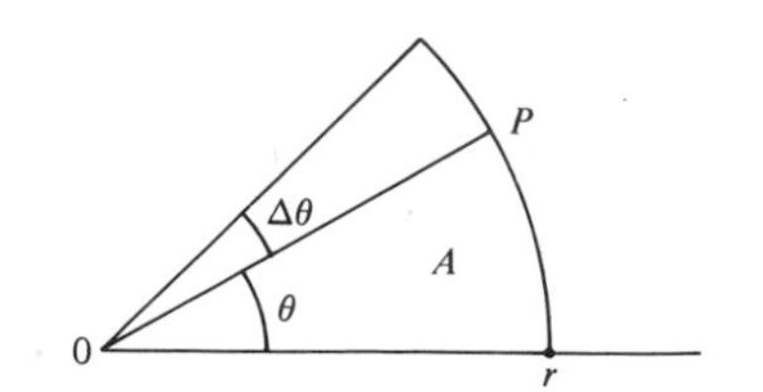

Figure 2. Area swept out.

One integration then yields

$$\dot{x} = -(\mu/k)(\sin\theta + A), \qquad \dot{y} = (\mu/k)(\cos\theta + B),$$

where the constants of integration A and B are to be determined by the initial velocity at the initial state. When these two equations are entered into the expression (5) for the constant k of angular momentum, we obtain

$$k^2 = \mu(x\cos\theta + y\sin\theta) + \mu(xB + yA).$$

Since $r = x\cos\theta + y\sin\theta$, this equation can be written as

$$r = k^2/\mu - xB - yA. \tag{7}$$

By putting in the polar coordinate values (3) of x and y, this is the equation of the orbit written in polar coordinates. More directly, the right hand side in (7) is proportional to the distance from the line whose equation is that right hand side set equal to 0. It is easier to see this by choosing the initial conditions at a point where $\theta = 0$ and $\dot{x}$ is zero. Then $A = 0$ and the equation (7) reads

$$r = B(k^2/\mu B - x). \tag{8}$$

Now $k^2/\mu B - x$ on the right is clearly the horizontal distance from the planet P to the (vertical) line $x = k^2/\mu B$. Then equation (8) states that the planet moves so that its distance r from the sun at 0 is always a fixed proportion B of its distance from a vertical line. But a conic section (Figure 3) can be defined, in terms of a point 0 as *focus* and a line D as *directrix*, as the locus of all those points P with distances from 0 and D so that $P0 = ePD$, where the constant e is the *eccentricity* of the conic. The orbit must then be an ellipse ($e < 1$), a parabola ($e = 1$), or a hyperbola ($e > 1$); in the case of the planets we know by observation of their recurrence that the orbit must be an ellipse (with the sun at one focus). As one knows, an ellipse can also be defined in terms of two foci, as the locus of points P such that the sum of the distances from P to the two foci is a constant $2a$; then $2a$ is the length of the "major axis" of the ellipse. A straightforward analysis will then give Kepler's third law about the square of the periods of different planets. But we have already established the essential point: When velocity and acceleration are represented by deriva-

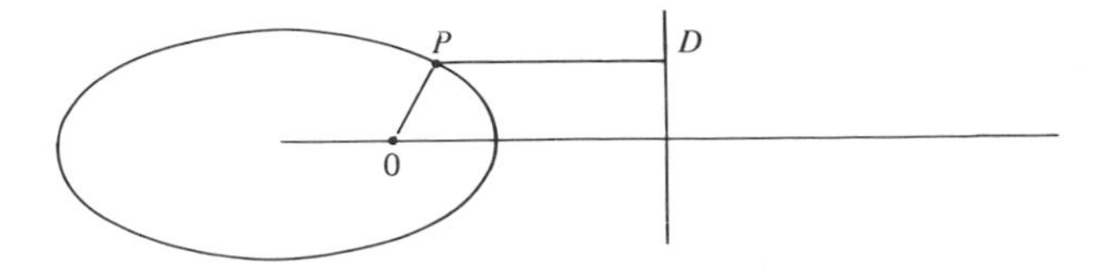

Figure 3. An ellipse as a locus.

tives, as defined in the Calculus, then Kepler's laws, summarizing the observations of the positions of the planets, can be deduced from Newton's second law of motion and the inverse square law of gravitation. Moreover, the deduction leads naturally to physical quantities such as momentum and to Mathematical ideas—the choice of suitable coordinate systems and the process of integrating differential equations, using initial conditions to determine the inevitable constants of integration. The further scope of these ideas is tremendous!

2. Momentum, Work, and Energy

Newton's third law of motion states that action equals reaction. This means, say, that in a mechanical system consisting of just two particles the force exerted by the first particle on the second will be equal and opposite (equal in magnitude and opposite in direction) to the force exerted by the second particle on the first. It is this law which gives rise to the principle of the *conservation of momentum* of an isolated mechanical system.

The (linear) momentum of a mass particle is a vector defined as the mass times the vector velocity. Thus, for a particle of mass m with coordinates x_1, x_2, and x_3 the momentum is the vector $m\mathbf{v}$ with coordinates $m\dot{x}_1$, $m\dot{x}_2$, and $m\dot{x}_3$. Since the acceleration is the time derivative of the vector velocity $\mathbf{v}$, Newton's second law (1.1) becomes the statement that the vector force on a particle is the time rate of change of momentum; in particular, when the force is zero, the momentum is conserved. Also, for an isolated system of two particles, exerting equal and opposite forces on each other, the total momentum of the system is constant. The same conclusion holds for systems of n particles.

Next comes the notion of *work*—specifically, the work done by a vector force $\mathbf{F}$ in moving a particle along a path. In the simplest case, when a constant force F moves a particle by a distance d along the direction of that force, it is said to do work of magnitude the product Fd. More generally, the work done by a vector force $\mathbf{F}$ in a vector displacement $\mathbf{w}$ depends only on the component of $\mathbf{F}$ along the direction $\mathbf{w}$, so is the product of that component by the length of $\mathbf{w}$; in more invariant terms, it is the inner product $\mathbf{F}\cdot\mathbf{w}$ of these vectors. (The inner product is often called the scalar product, to distinguish it from the vector product which is also used in mechanics.) This formula, work equals $\mathbf{F}\cdot w$, suggests that force acts as a linear function of displacement, so that a variable force will appear as the dual of a vector field; that is, as a differential form.

This leads to the following formulation. In an open set U in $\mathbf{R}^3$, with rectangular coordinates x^1, x^2, and x^3, a field of force $\mathbf{F}$ has three components F_1, F_2, and F_3 depending on position, as $F_i = F_i(x^1,x^2,x^3)$, and so determines a (smooth) differential form

$$\omega = F_1 dx^1 + F_2 dx^2 + F_3 dx^3 \tag{1}$$

on U. We want the work done in moving along a path, given as usual by a smooth map $u: I \to U$ of a time interval, say $I = \{t \mid 0 \leqslant t \leqslant t_1\}$. This work will be approximately the sum of small bits of work along short pieces of the path; hence, by the main idea of integration, the work done by the force F along the path u is (defined to be) the line integral

$$\begin{aligned} \int_u \omega &= \int_u (F_1 dx^1 + F_2 dx^2 + F_3 dx^3) \\ &= \int_0^{t_1} (F_1 \dot{x}^1 + F_2 \dot{x}^2 + F_3 \dot{x}^3) dt. \end{aligned} \tag{2}$$

This definition of work as an integral does include the special case (of a constant force), as introduced above. What we call a "differential form" in (1) is in physics texts usually called the differential δW of work, but the idea is the same–it is the thing which is to be integrated (along paths) to get the work done along these paths; the symbol δW does not necessarily mean that there is a function W of position with this differential.

Now use Newton's second law for a particle of mass m moving under this force field $\mathbf{F}$ along this path u, given by smooth functions $x^i(t)$. Then

$$F_i \dot{x}^i = m \frac{d\dot{x}^i}{dt} \dot{x}^i, \qquad i = 1,2,3.$$

The indefinite integral of this product is $(1/2)(m(\dot{x}^1)^2)$, so that the work done along the path is

$$\int_u \omega = (1/2)m[(\dot{x}^1)^2 + (\dot{x}^2)^2 + (\dot{x}^3)^2]_0^{t_1}, \tag{3}$$

where the notation at the right means the difference of the values of the bracketed expression at t_1 and 0. This leads to the definition of the *kinetic energy* T of the particle as

$$T = (1/2)m[(\dot{x}^1)^2 + (\dot{x}^2)^2 + (\dot{x}^3)^2] \tag{4}$$

and to the theorem that the work done along the path is equal to the change (3) in the kinetic energy along that path. In brief, if v is the magnitude of the velocity, then the kinetic energy is

$$T = (1/2)mv^2.$$

Sometimes the differential form $\omega = \delta W$ of work may be the differential $\omega = -dV$ of an actual function $-V$ of position. This means that the force F_i in each direction x^i is the partial derivative

$$F_i = -\frac{\partial V}{\partial x^i}, \qquad i = 1,2,3. \tag{5}$$

This implies that the work done along any smooth path from a point b to a point c is just the difference $V(c) - V(b)$–and so is independent of the choice of path. This function V—unique up to a constant—is then called the *potential energy* and the forces are said to be *conservative*. For example, under the inverse square law for a particle moving under the gravitational attraction of another particle at distance r there is a potential energy of the form

$$V = -\gamma m/r; \tag{6}$$

indeed, the partial derivative with respect to the coordinate r is the desired inverse square force $-\gamma m/r^2$. (The minus sign indicates that the force is directed toward the origin.) The potential energy per unit mass—in this case $V = -\gamma/r$—is often just called the *potential*. For several attracting sources there can be similar potentials; the study of these potentials is extensive, with profound connections to partial differential equations and to probability theory.

When the forces are conservative, the work done along a path, as expressed in the integral (3), is exactly the change in its potential energy along that path. It follows that in such a system the sum of potential and kinetic energy remains constant. This is a first form of the famous physical principle of the *conservation of energy*.

The harmonic oscillator is a simple example of a conservative system. A particle moves along a line under a force of attraction directed toward the origin and proportional to the distance x from the origin. Thus by Newton's law, the motion is governed by the second order differential equation

$$\frac{d^2x}{dt^2} = -kx \tag{7}$$

for some constant k. This force $-kx$ may be derived from a potential $kx^2/2 + C$, for any constant C (potential energy is determined only "up to" an additive constant). The equation (7) has the well known integral $A \cos kt + B \sin kt$ or

$$x = A \cos(kt + \phi), \tag{8}$$

where A and B or A and ϕ are two constants of integration. In the form (8) the constants have an immediate interpretation, because A is the largest value of x (at $kt = -\phi$), so is the *amplitude* of the motion, while the angle ϕ measures the "phase" by which the motion is translated from the simple cosine function $A \cos kt$. The formula (8) also yields the velocity as a function of time,

$$\dot{x} = -kA \sin(kt + \phi).$$

The motion may be visualized by plotting x and $\dot{x}$ together in a plane, called the *phase plane*. When the constant k is 1, the plot (Figure 1) is a

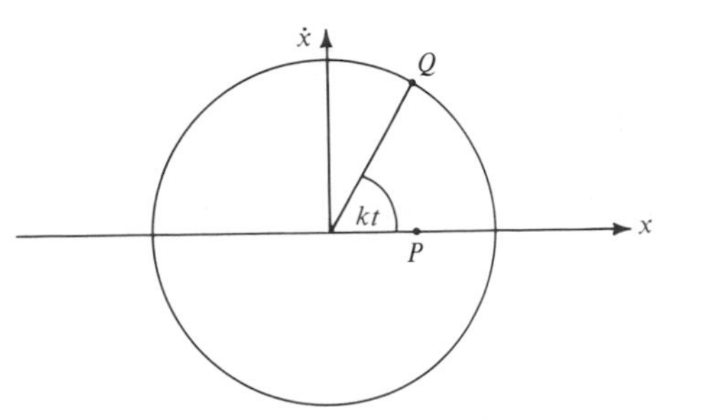

Figure 1. Phase plane.

circle. *A* point Q moves around the circle at uniform speed; its projection P on the x axis is the point executing the original harmonic motion. Such "harmonic oscillators" are important because general periodic motions (oscillations) can be built up (via Fourier series) out of linear combinations of these "simple" harmonic oscillations. The reappearance of the geometrically defined trigonometric functions $\cos kt$ and $\sin kt$ in this connection with mechanics is a remarkable (and classical) instance of the interrelation of mathematical constructs.

The idea of using the phase plane with coordinates *both* position and velocity is relevant, because the initial conditions for the motion (position and velocity) fix a point in this phase plane. This point (and thus these conditions) determine the constants of integration A and ϕ; geometrically, this point determines the circle above. Put differently, this point is a position on the real axis *and* a velocity there, hence is a point on the tangent bundle for the x axis. In this form, as we will soon see, the idea generalizes.

3. Lagrange's Equations

The calculation of planetary orbits in §1 started with the usual rectangular coordinates x and y in the orbital plane of the planet, but soon switched to polar coordinates in that plane. It would have been convenient to have the equations of motion written directly in terms of polar coordinates. This can be done, and done in a way to suggest the form of such equations in any other coordinate system.

The transformation from rectangular to polar coordinates r and θ reads

$$x = r\cos\theta, \qquad y = r\sin\theta. \tag{1}$$

From these we can express the rectangular components of velocity as

$$\begin{aligned} \dot{x} &= -r\sin\theta\,\dot{\theta} + \cos\theta\,\dot{r}, \\ \dot{y} &= r\cos\theta\,\dot{\theta} + \sin\theta\,\dot{r}; \end{aligned} \tag{2}$$

hence the kinetic energy becomes

$$T = (1/2)m(\dot{x}^2 + \dot{y}^2) = (1/2)m(r^2\dot{\theta}^2 + \dot{r}^2). \tag{3}$$

Differentiating the equations (2) again with respect to time gives the rectangular components $\ddot{x}$ and $\ddot{y}$ of the acceleration vector **a**. The radial and angular components are then (Figure 1)

$$\begin{aligned} a_r &= \ddot{x}\cos\theta + \ddot{y}\sin\theta = \ddot{r} - r\dot{\theta}^2, \\ a_\theta &= -\ddot{x}\sin\theta + \ddot{y}\cos\theta = r\ddot{\theta} + 2\dot{\theta}\dot{r}. \end{aligned} \tag{4}$$

The force has two corresponding components, F_r along the radial direction and F_θ perpendicular to this direction, but for this "angular" direction it is more natural to use the *torque* T_θ about the origin, defined as F_θ times the lever arm r from 0. For these two components, Newton's second law then reads

$$F_r = m(\ddot{r} - r\dot{\theta}^2), \qquad T_\theta = m(r^2\ddot{\theta} + 2r\dot{\theta}\dot{r}). \tag{5}$$

The right hand side at first looks mysterious; for example, the term $-r\dot{\theta}^2$ on the right in the first equation is often shifted to the left side, where it is called the *centrifugal force* due to the angular velocity $\dot{\theta}$. (Swing a horse chestnut on a rope and you can feel the centrifugal force!) However, this right hand side can be explained more systematically in terms of the partial derivatives of the polar coordinate kinetic energy function T of (3). This T is a function of r, $\dot{r}$, and $\dot{\theta}$; calculating its partial derivatives, one finds

$$F_r = \frac{d}{dt}\left(\frac{\partial T}{\partial \dot{r}}\right) - \frac{\partial T}{\partial r}, \tag{6}$$

$$T_\theta = \frac{d}{dt}\left(\frac{\partial T}{\partial \dot{\theta}}\right). \tag{7}$$

We can even make these two equations look alike by subtracting $\partial T/\partial\theta$ on the right of (7)—a zero term because the kinetic energy T is in fact independent of θ.

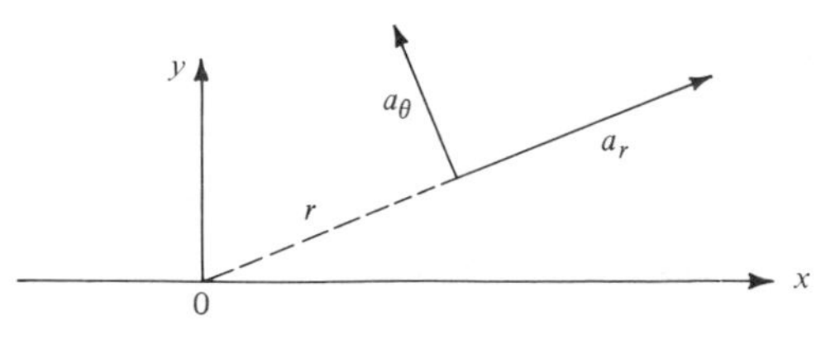

Figure 1. Angular acceleration.

As a result of these apparently mechanical computations, a miracle now happens: Equations like these hold in *any* coordinate system. Specifically, if q is any one coordinate and T the kinetic energy, as it depends on q and the q-component $\dot{q}$ of velocity, Newton's equation for this coordinate becomes

$$\frac{d}{dt}\left[\frac{\partial T}{\partial \dot{q}}\right] - \frac{\partial T}{\partial q} = Q_q, \tag{8}$$

where Q_q (like the torque in (7)) is a "generalized" force in the q-direction, and the partial derivatives with respect to q hold constant all the other coordinates of the system. This is the Lagrange equation for the coordinate q.

This Lagrange equation (8) does include the polar coordinate equations (6) and (7) for the coordinates r and θ. The miracle is that this one form of equation applies to any choice of coordinate system. This exemplifies the observation that one Mathematical form has many realizations.

We will derive the equations (8) and the definition of generalized force for the general case of N particles, where the position in space of each one is given by the usual three cartesian coordinates; the position of the whole system is then determined by $k = 3N$ coordinates x^i, for $i = 1, \ldots, k = 3N$; in effect we are considering the motion of a single point in the space $\mathbf{R}^k$. Then Newton's law makes

$$m_i \frac{d^2x^i}{dt^2} = F_i, \qquad i = 1, \ldots, k, \tag{9}$$

where each F_i is the appropriate force while m_i is the appropriate mass (thus $m_1 = m_2 = m_3$ is the mass of the first particle). The kinetic energy T depends on the velocities according to the standard formula

$$T = (1/2)(m_1\dot{x}_1^2 + \cdots + m_n\dot{x}_n^2). \tag{10}$$

Then $m_j\dot{x}_j = \dfrac{\partial T}{\partial \dot{x}_j}$, so that Newton's law (9) can be written as

$$\frac{d}{dt}\left[\frac{\partial T}{\partial \dot{x}^j}\right] = F_j, \qquad j = 1, \ldots, k. \tag{11}$$

Since T does not depend on the x^i, this is a special case of the Lagrange equation (8)

Now replace the x^j by new coordinates. These new coordinates are traditionally denoted by q's, so that the x^j are given by smooth functions h^j,

$$x^j = h^j(q^1, \ldots, q^n), \qquad j = 1, \ldots, k, \tag{12}$$

in terms of the new coordinates. The partial derivatives $\partial h^j/\partial q^i$ are traditionally written as $\partial x^j/\partial q^i$, thereby avoiding the use of a letter h for the function, and using the letter x^j to denote both the quantity x^j (the coordinate) and the function h^j of the q's. The components of velocity $dx^i/dt = \dot{x}^i$ along a path are then, by the chain rule,

$$\dot{x}^j = \frac{\partial x^j}{\partial q^1}\dot{q}^1 + \cdots + \frac{\partial x^j}{\partial q^n}\dot{q}^n, \qquad j = 1, \ldots, k. \tag{13}$$

Consequently the kinetic energy T becomes a function of both the q^i's and the $\dot{q}^i$'s. In these coordinates we will show that Newton's second law takes the form of the Lagrange equations

$$\frac{d}{dt}\left[\frac{\partial T}{\partial \dot{q}^i}\right] - \frac{\partial T}{\partial q^i} = Q_i, \qquad i = 1, \ldots, n, \tag{14}$$

where the Q_i are suitable "generalized" forces; for examples, torques.

Since the Lagrange equations hold for the case of rectangular coordinates we can prove that they hold generally by showing that they remain true under any change (12) from old coordinates x^j (not necessarily rectangular) to new coordinates $q^1, \ldots, q^n$. It is convenient to think of the x^j as coordinates of a point in one *configuration space* D, say the space $D = \mathbf{R}^{3N}$, while the q's are coordinates in some possibly different configuration space C of dimension n; the change of coordinates (12) is then interpreted as a smooth map $\phi : C \to D$ of spaces; such a map is often called a *change of base* (e.g., in algebraic geometry).

In order to transfer Lagrange's equations from C to D we need to change both paths (trajectories) and functions (kinetic energy). The diagram

$$I \xrightarrow{u} C \xrightarrow{\phi} D \xrightarrow{f} \mathbf{R}, \tag{15}$$

with I an interval of time, suggests what happens. Each path u in the space C becomes by composition with ϕ a path $\phi \cdot u$ in D, while each function f or quantity defined on D becomes by composition with ϕ a function $f\phi$ on C; one says that f is *pulled back* to C. Also the differential form ω which determines the work (and thus the forces) on D can also be pulled back along the smooth map ϕ. First, each of the k coordinates x^j on D becomes by a composition a quantity $x^j\phi$ on C, with differential there given by the usual formula

$$dx^j = \sum_{i=1}^{n} \frac{\partial x^j}{\partial q^i} dq^i, \qquad j = 1, \ldots, k.$$

(Here $\partial x^j / \partial q^i$ are just the partial derivatives $\partial h^j / \partial q^i$ of (13).) By these equations, the differential form $\omega = \Sigma F_j dx^j$ for work ω on the original manifold D becomes a differential form

$$\phi^*\omega = \sum_j F_j dx^j = \sum_j \sum_i F_j \frac{\partial x^j}{\partial q^i} dq^i$$

on the second configuration space C; collecting the coefficients of each differential dq^i, it has the expression

$$\phi^*\omega = \sum_{i=1}^{n} Q_i dq^i, \qquad Q_i = \sum_{j=1}^{k} F_j \frac{\partial x^j}{\partial q^i}. \tag{16}$$

The quantities Q_i so obtained on C are by definition the *generalized* forces in the direction q^i. The essential point of this definition is that the work done by given forces, obtained as the integral of the differential form ω over a path, is preserved; by using the definition of the line integral (straightforward, as specified by (2.2)) one shows, for each path $u: I \to C$, that

$$\int_{\phi \cdot u} \omega = \int_u \phi^*\omega. \tag{17}$$

The components $\dot{x}^j$ of velocity are expressed as functions of the q's and the $\dot{q}$'s by the equations (13), and by these equations the partial derivatives of these functions $\dot{x}^j(q^1, \ldots, q^n, \dot{q}^1, \ldots, \dot{q}^n)$ are

$$\frac{\partial \dot{x}^j}{\partial \dot{q}^i} = \frac{\partial x^j}{\partial q^i}, \qquad \frac{\partial \dot{x}^j}{\partial q^i} = \sum_l \frac{\partial^2 x^j}{\partial q^i \partial q^l} \dot{q}^l. \tag{18}$$

If I is a time interval, with parameter t, then along each path $u: I \to C$ all the coordinates q^i, $\dot{q}^i$ and x^j become, by composition with u or with $\phi \cdot u$, functions on I (i.e., functions of t). By the chain rule for differentiating a composite,

$$\frac{d}{dt}\left(\frac{\partial x^j}{\partial q^i}\right) = \sum_l \frac{\partial^2 x^j}{\partial q^i \partial q^l} \frac{dq^l}{dt}.$$

Since $\dot{q}^l = dq^l/dt$, the right hand side is the same as in the second equation of (18). hence

$$\frac{d}{dt}\left(\frac{\partial x^j}{\partial q^i}\right) = \frac{\partial \dot{x}^j}{\partial q^i}, \tag{19}$$

valid along any path u. This means that the partial derivatives here are those of functions of $q^1, \ldots, q^n, \dot{q}^1, \ldots, \dot{q}^n$; they then become functions of t alone via u.

With these preparations, we can now deduce the Lagrange equations for the coordinates q by proving that: A path u: $I \to C$ satisfies these equations whenever the composite path $\phi \cdot u$: $I \to D$ satisfies the Lagrange equations

$$\frac{d}{dt}\left[\frac{\partial T}{\partial \dot{x}^j}\right] - \frac{\partial T}{\partial x^j} = F_j, \qquad j = 1, \ldots, k$$

in the x-coordinates on D. Indeed from the definition (16) of the generalized forces Q_i one has

$$\begin{aligned} Q_i &= \sum_j F_j \frac{\partial x^j}{\partial q^i} = \sum_j \left[\frac{d}{dt}\left[\frac{\partial T}{\partial \dot{x}^j}\right]\frac{\partial x^j}{\partial q^i} - \frac{\partial T}{\partial x^j}\frac{\partial x^j}{\partial q^i}\right] \\ &= \sum_j \left[\frac{d}{dt}\left[\frac{\partial T}{\partial \dot{x}^j}\frac{\partial x^j}{\partial q^i}\right] - \frac{\partial T}{\partial \dot{x}^j}\frac{d}{dt}\left[\frac{\partial x^j}{\partial q^i}\right] - \frac{\partial T}{\partial x^j}\frac{\partial x^j}{\partial q^i}\right], \end{aligned}$$

by the rule for the differentiation of a product. After substituting (19) in the second term here, the second and third terms combine to give $-\ \partial T/\partial q^i$. As for the first term, the chain rule gives

$$\frac{\partial T}{\partial \dot{q}^i} = \sum_j \left[\frac{\partial T}{\partial \dot{x}^j}\frac{\partial \dot{x}^j}{\partial \dot{q}^i} + \frac{\partial T}{\partial x^j}\frac{\partial x^j}{\partial \dot{q}^i}\right];$$

here $\partial x^j / \partial \dot{q}^i = 0$, while $\partial \dot{x}^j / \partial \dot{q}^i$ is $\partial x^j / \partial q^i$ by (18) so that the whole becomes

$$Q_i = \frac{d}{dt}\left[\frac{\partial T}{\partial \dot{q}^i}\right] - \frac{\partial T}{\partial q^i}, \qquad i = 1, \ldots, n.$$

These are Lagrange's equations for the path u on C, as desired.

The Lagrange equations apply also to the case of motion under constraints–for example a so called "holonomic" constraint, where a particle or particles is required (i.e. is constrained) to move only along some curve or surface in Euclidean space. An example is a sphere rolling along a smooth table or a pendulum where the pendulum's bob is suspended by a weightless string from some fixed point 0. Then the motion of the bob is just the motion of a point on a sphere, and so can be described in terms of coordinates on the sphere–say the latitude and the longitude. In this case the configuration space C is a manifold (the sphere) of dimension $n = 2$, while the original configuration space D is $\mathbf{R}^3$, Euclidean space of dimension $k = 3$. The argument above applies directly to this case and gives the Lagrange equations which describe the motion in terms of latitude and longitude. Notice here that the dimension 2 of C is smaller than that of D (hence $k \neq n$ in the formulas above) and that the configuration space C is a manifold, included in the Euclidean space D by the inclusion map $\phi: C \to D$. Moreover the coordinates q^i must be regarded as local coordinates, valid in some chart–the longitude is not defined at the north pole of a sphere!

In such a case there are forces F^c_j of "constraint" which hold the particle to the submanifold, as well as "external" forces F^e_j. Thus in the Euclidean space the jth component of the total force is $F_j = F^e_j + F^c_j$. But the forces of constraint act orthogonally to the submanifold C; as a result the pullback to C of the differential form for work done by the constraints is $\phi^*(\Sigma F^c_j dx^j) = 0$. More simply, "the forces of constraint do no work". What this means is that the required generalized forces Q_i on C may be calculated by pulling back only the external forces F^e.

Lagrange's equations apply also to more general ("non-holonomic") constraints, such as the case of a ball rolling on a rough table. At any one time, the ball is constrained to move on a three dimensional manifold, with local coordinates the two rectangular coordinates of the point of contact in the plane (the table) plus one angle (the angle of rotation of the sphere about the diameter through this point of contact). However, by rolling suitably forward and backwards, one may see that the ball can reach any position in a four dimensional manifold (two coordinates in the plane and two for rotation of the sphere). The Lagrange equation in such a case may be derived much as in the argument above, replacing the holonomic constraint $\phi: C \to D$ by a map $\phi: C \times I \to D$ which at each time $t \in I$ gives the position $\phi_t = \phi(-,t)$ of the constraint to C at the time t. The generalized forces Q_i for the Lagrange equations of motion on C are then obtained by pulling back the work ω along ϕ^*_t; this is classically called a "virtual" displacement.

Lagrange's equations have a simpler form when the forces are conservative. There is then a potential V, which is a quantity defined on D such that the forces F_j are given by the partial derivatives $-\partial V/\partial x^j$. This potential is then also a quantity defined on C (as the composite $V \cdot \phi$). The generalized forces Q_i on C then are

$$Q_i = \sum_j F_j \frac{\partial x^j}{\partial q^i} = -\sum_j \frac{\partial V}{\partial x^j} \frac{\partial x^j}{\partial q^i} = -\frac{\partial V}{\partial q^i}:$$

In other words, they are also derivatives of the potential V in the new coordinates. One may then introduce $L = T - V$, a quantity defined on D and on C, and called the *Lagrangian*; the Lagrange equations for conservative forces then take the form, for coordinates q_i on the configuration space C,

$$\frac{d}{dt}\left[\frac{\partial L}{\partial \dot{q}^i}\right] - \frac{\partial L}{\partial q^i} = 0, \qquad i = 1, \ldots, n. \tag{20}$$

There are many practical applications of the Lagrange equations. In the case of a simple pendulum, a particle of mass m is constrained to move on a vertical circle of radius r under the force of gravity, with the standard acceleration g. There is one generalized coordinate, the angle θ (of the pendulum arm) to the vertical. The force is conservative, with potential energy $V = -mgr \cos \theta$. The kinetic energy, transformed from rectangular coordinates, is $mr^2\dot{\theta}^2/2$. Lagrange's equation in the coordinate θ then becomes

$$\frac{d}{dt}(\dot{\theta}) = -(g/r)\sin \theta .$$

An exact integration requires elliptic functions. If the angle θ is small, sin θ may be approximated by θ, and the equation becomes that of a simple harmonic oscillation.

The spherical pendulum consists of a particle of mass m constrained to move on a sphere of radius r. The appropriate coordinates are longitude ϕ and latitude θ (measured from the south pole of the sphere). By the standard equations transforming rectangular to spherical coordinates one obtains

$$T = mr^2(\dot{\theta}^2 + \sin^2 \theta\, \dot{\phi}^2)/2, \qquad V = -mgr \cos \theta .$$

The two Lagrange equations in the coordinates θ and ϕ then are

$$\frac{d}{dt}(r^2\dot{\theta}) = r^2 \sin \theta \cos \theta\, \dot{\phi}^2 - rg \sin \theta .$$

$$\frac{d}{dt}(r^2 \sin^2\theta\, \dot{\phi}) = 0 .$$

The second one integrates once to give $\dot{\phi} = h/sin^2\theta$, for h a constant of integration. The first equation then takes the form

$$\ddot{\theta} = \frac{h^2 \cos \theta}{\sin^3\theta} - \frac{g}{r} \sin \theta .$$

If $h = 0$, this reduces to the case of the simple pendulum. The integration in other cases is more complex.

4. Velocities and Tangent Bundles

The Lagrange equations,

$$\frac{d}{dt}\left[\frac{\partial T}{\partial \dot{q}^i}\right] - \frac{\partial T}{\partial q^i} = Q_i , \qquad i = 1, \ldots, n .$$

with Q_i a function of position, are n second order differential equations in the n coordinates q_i, as one may exhibit by the expansion

$$\sum_{j=1}^{n}\left[\frac{\partial^2 T}{\partial \dot{q}^i\,\partial \dot{q}^j}\frac{d^2 q^j}{dt^2}+\frac{\partial^2 T}{\partial \dot{q}^i\,\partial q^j}\frac{dq^j}{dt}\right]-\frac{\partial T}{\partial q^i}=Q_i. \tag{1}$$

In general, a solution will be determined by giving the $2n$ initial values of the position and velocity coordinates q^i and $\dot{q}^i$. Alternatively, we can think of these $2n$ quantities as $2n$ independent variables. Then we replace d^2q^j/dt^2 by $d\dot{q}^j/dt$ and rewrite (1) as

$$\sum_{j=1}^{n}\left[\frac{\partial^2 T}{\partial \dot{q}^i\,\partial \dot{q}^j}\frac{d\dot{q}^j}{dt}+\frac{\partial^2 T}{\partial \dot{q}^i\,\partial q^j}\frac{dq^j}{dt}\right]-\frac{\partial T}{\partial q^i}=Q_i. \tag{2}$$

Since $\dot{q}^i$ is now a coordinate and not a derivative, we must add the equations

$$\frac{dq^i}{dt}=\dot{q}^i,\qquad i=1,\ldots,n. \tag{3}$$

We now have $2n$ *first* order differential equations in the $2n$ variables q^i and $\dot{q}^i$ for position and velocity (phase space), with coefficients functions of these variables.

For rectangular coordinates, the equations (2) have the simpler form

$$m_i\frac{d\dot{x}^i}{dt}=F_i\,;$$

In general, the matrix with coefficients $\partial^2 T/\partial\dot{q}^i\,\partial\dot{q}^j$ is usually nonsingular; when this is the case, the equations (2) can be solved explicitly for the derivatives $d\dot{q}^j/dt$. If we label the $m=2n$ variables $y^1,\ldots,y^m$, the equations (2) and (3) then have the general form

$$\frac{dy^k}{dt}=G_k(y^1,\ldots,y^m),\qquad k=1,\ldots,m \tag{4}$$

with each G_k a smooth function of the y's. A trajectory of the given mechanical system is then a solution $y^k=y^k(t)$, $k=1,\ldots,m$, of these equations. Here the m quantities G_k can be considered as the m components of a tangent vector at the point with coordinates $y^1,\ldots,y^m$, so that the G_k describe a *vector field* (see §VIII.9) on (part of) the m-dimensional space $\mathbf{R}^m$. A solution of the equations is then a smooth path in $\mathbf{R}^m$ whose tangent vector at each point is the vector given by the field at that point. More informally, a solution is a path which threads through the given vector field, as suggested by Figure 1 for $m=2$. In these terms, one can readily visualize the corresponding "existence theorem", which

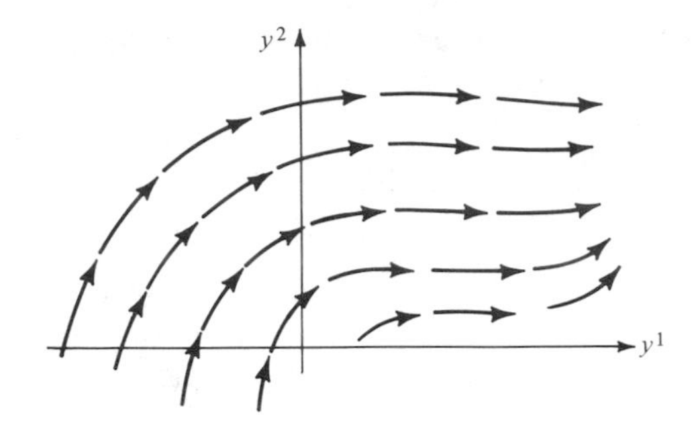

Figure 1. Paths through a vector field.

asserts that such a smooth system always has, for each given initial point, a smooth solution defined for some sufficiently small interval of time—so solutions exist, even though they may not be expressed in terms of elementary functions.

The phase space for simple harmonic motion (§2) exhibited both position and velocity. Similarly the quantities q^i (position) and $\dot{q}^i$ (velocity) are the coordinates for the *phase space* of the motion. A point in this space is a point in the configuration space C plus a tangent vector at that point. In other words, it is a point on the tangent bundle $B.C$ to C—which we write with the letter B, to avoid confusion with the use of the letter T for kinetic energy. Thus the device of thinking of a point moving in such a phase space is really equivalent to the geometric idea of using a tangent bundle to a manifold.

Our proof that the Lagrange form of the equations is preserved under change of base can be illuminated in these terms. Each smooth map $\phi : C \to D$ of manifolds carries tangent vectors (to paths) in C into tangent vectors to D and hence induces a map $B.\phi : B.C \to B.D$ on the tangent bundles. On each tangent space, it is the linear map given in coordinates by the equations (3.13). The determinant of this linear map, with entries $\partial x^j / \partial q^i$, depending on the q's, is called the *Jacobian* of the transformation ϕ.

In these terms, the diagram (3.15) can be expanded to represent all the coordinates involved in the change of base (q_i coordinates on C, x_j on D):

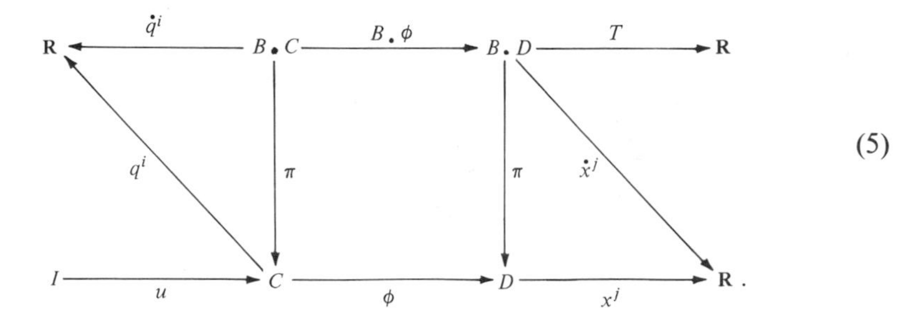

Here the map π is the projection of each tangent bundle (phase space) onto its base, and the coordinates q^i (or x^j) on the base become, by composition with π, coordinates $q^i \cdot \pi$ on the bundle. This diagram also displays the other functions involved. Thus the kinetic energy T, originally given as a quantity on the initial phase space (i.e., a function T on $B.D$ to the reals) becomes by composition with $B.\phi$ a quantity on the new phase space. By convention, the letter T usually stands for either (or both) of these functions. The whole proof of §3 can then be reformulated in these terms.

For iterated constraints one has successive smooth maps

$$E \xrightarrow{\Psi} C \xrightarrow{\phi} D$$

of configuration spaces. In this case, successive images of the tangent vectors of E are just the images under the composite map $\phi \cdot \Psi$; in other words, $B.(\phi\Psi) = (B.\phi)(B.\Psi)$. The fact that application of $B.$ respects composition in this way is sometimes formulated by saying that the "tangent bundle is a functor $B.$ on the category of manifolds"; see §XI.9.

5. Mechanics in Mathematics

We pause to emphasize the many and remarkable exchanges of ideas from mechanics to Mathematics and back. To begin with, the very notion of the calculus was found by Newton in order to formulate the mechanics of planetary motion. The idea of rate of change (derivative) was needed in order to get at the ideas of velocity and acceleration necessary for mechanics. This in turn led to the consideration of (ordinary) differential equations and their solutions (exact or approximate) subject to initial conditions. Since these differential equations are typically of second order, their initial conditions involved initial position and velocity and so led to the use of first order differential equations on tangent bundles (phase spaces) where the location of a point is determined by position and velocity. For several particles, the configuration spaces often required spaces of more than three dimensions. Both the configuration spaces and the phase spaces were originally described just in terms of coordinates, but the effective formulation of the differential equations of motion called attention to the need for some invariance (say in the form used for the Lagrange equations) under changes of coordinates. Once this is considered, one is really dealing with a conceptually described idea of a smooth differentiable manifold. Although the explicit general definition of such a manifold was not formulated till the 1930's, it was implicitly present for at least a century before that in the minds of both geometers and physicists.

This is not all. The physicist's notion of the "differential of work" is the Mathematician's "differential form". The total work in a process is obtained by summing up differentials, and this amounts to the definition of line integrals. Other such connections of ideas will appear later in this chapter. They are discussed here as examples of the very many cases of ideas connected from theoretical Physics to Mathematics. They cannot all be listed in any one book, but we can mention the recent exciting parallels which have appeared between gauge theories (the Yang–Mills equations in quantum field theory) and the Mathematical study of connections in fiber bundles. The interaction between tensor analysis and relativity theory is another such case (Weyl [1923]), as is the use of matrices and of Hilbert spaces in quantum mechanics.

In such cases, a particular idea may arise first in Physics or first in Mathematics or often with apparent independence in both. At issue is not the question of which comes first, but the remarkable fact that both come—that ideas from physical problems and from apparently pure Mathematical speculation come together.

6. Hamilton's Principle

The form of the Lagrange equations in a configuration space C is independent of the choice of coordinates in that space. This fact needs explanation, and has one: Hamilton's principle asserts that the solutions of Lagrange's equations are exactly those paths in C which "minimize" the integral of the Lagrangian function L along the path. We consider a conservative system, and a Lagrangian L which depends not only on position and velocity, but also on time t. Such a "time dependent" Lagrangian is thus a smooth function $L\colon B.C \times I \to \mathbf{R}$, where I is the time interval, say all t with $0 \leqslant t \leqslant 1$. Thus for each choice of coordinates $q^1, \ldots, q^n$ on C, L appears as a function $L(q^1, \ldots, q^n, \dot{q}^1, \ldots, \dot{q}^n, t)$. Given two fixed points a and b in the space C, we consider smooth paths from a to b; that is, smooth functions $u\colon I \to C$ with $u(0) = a$ and $u(1) = b$. Each point of the path has a tangent vector at that point, so $u(t)$ together with these tangent vectors defines to a path $u_B\colon I \to B.C$ in the tangent bundle; we say that u_B *lifts* u over π because the projection $\pi\colon B.C \to C$ will again give u as the composite $u = \pi \cdot u_B$. Also, sending each time t to itself yields by $t \mapsto (u_B.t, t)$ a map $u_\# : I \to B.C \times I$.

The integral J to be minimized is the integral of L along this path,

$$J(u) = \int_0^1 L(q^1, \ldots, q^n, \dot{q}^1, \ldots, \dot{q}^n, t)dt = \int_0^1 (L \cdot u_\#)dt\,; \tag{1}$$

the second version of this formula indicates that one is simply integrating the composite function $L \cdot u_\#$ on I. We want to compare this integral $J(u)$ with the same integral along other smooth paths in C with the same endpoints (at the same starting and ending times $t = 0$ and $t = 1$), we aim to find those paths u for which $J(u)$ is a minimum or at least is "stationary" (see below).

Now to minimize (or to maximize) a smooth function f of a real variable x one first finds where the derivative of f vanishes. Let us say that f is *stationary* at some value x_0 if $df/dx = 0$ for $x = x_0$. Thus a function is stationary both at a maximum and at a minimum—and also at a horizontal inflection of the curve $y = f(x)$; to distinguish these cases one needs more than just the vanishing of the first derivative. Now we wish to find similar stationary positions not for a function like $f(x)$ of a number x, but for a function $J(u)$ of a curve u; now u can vary over the set of all curves from a to b. Such a problem is said to belong to the *calculus of variations* (§*VI.11*). Here "variation" refers to the variation of the path; this we interpret as an embedding of the given path $u(t)$ in a one-parameter family $U(t,\epsilon)$ of paths, where ϵ is a real parameter, say $\epsilon \in I$. Thus U is a smooth function U: $I \times I \to C$, such that $U(t,0) = u(t)$ is the given path, while all the "varied" paths have the same endpoints a and b, so that $U(0,\epsilon) = a$ and $U(1,\epsilon) = b$ for all ϵ in I. The situation may be pictured as in Figure 1 (for a two dimensional space C). For each such family U, one can again form the integral $J(U)$; it will be a smooth function of ϵ. We say that the original integral $J(u)$ is *stationary* if $dJ(U)/d\epsilon$ is zero for $\epsilon = 0$ whenever u is embedded in a smooth family U of paths.

Hamilton's Principle. For each time-dependent smooth Lagrangian function L on a manifold C, a path u: $I \to C$ from a point a to a point b satisfies Lagrange's equations for L if and only if the corresponding integral $J(u)$ of the Lagrangian is stationary at u.

Since the integral $J(u)$ depends only on the function L and the path u and not on the coordinates, this theorem explains the fact that the Lagrange equations have the same form, independent of the choice of coordinates.

Note that this principle compares the actual path u (which does satisfy the Lagrange equations) with other nearby paths, which do not in general satisfy the Lagrange equations—but which do have the same starting and ending points and times.

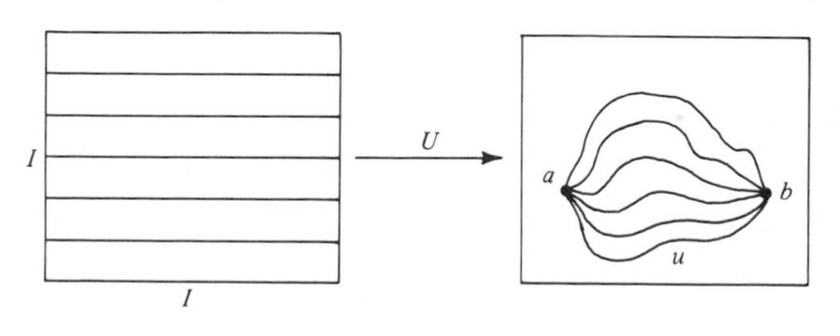

Figure 1

The method of proof is adequately described by considering the case of just one coordinate q^1 in configuration space (i.e., for $n = 1$).

First assume that Lagrange's equations hold. We wish to show that $J(u)$ is stationary; that is, that the derivative $dJ/d\epsilon = 0$ for every family U. For sufficiently smooth functions we may calculate this derivative of the integral J by differentiation under the integral sign, as in

$$\frac{dJ}{d\epsilon} = \int_0^1 \frac{dL}{d\epsilon} dt = \int_0^1 \left[\frac{\partial L}{\partial \dot{q}} \frac{d\dot{q}}{d\epsilon} + \frac{\partial L}{\partial q} \frac{dq}{d\epsilon} \right] dt.$$

Now along the lifted path in $B.C$ one has

$$\frac{d\dot{q}}{d\epsilon} = \frac{d}{dt} \left[\frac{dq}{d\epsilon} \right].$$

With this substitution, the first term in the integral above can be integrated by parts according to the familiar formula for two functions v and w of t;

$$\int_0^1 v \frac{dw}{dt} dt = - \int_0^1 \frac{dv}{dt} w dt + (vw)_{t=1} - (vw)_{t=0}. \tag{2}$$

In the present case $v = \partial L / \partial \dot{q}$ and $w = dq/d\epsilon$; also all the varied paths have the same endpoints, so that $w = dq/d\epsilon$ is zero at $t = 0$ and at $t = 1$. Thus we get

$$\frac{dJ}{d\epsilon} = \int_0^1 \left[-\frac{d}{dt} \left[\frac{\partial L}{\partial \dot{q}} \right] + \frac{\partial L}{\partial q} \right] \frac{dq}{d\epsilon} dt. \tag{3}$$

Now set $\epsilon = 0$ in L. The term in the large brackets is then exactly the left-hand side of Lagrange's equation for the function L, so its vanishing (for all t along the path) does imply that $dJ/d\epsilon = 0$ and hence that J is stationary, as desired.

For the converse, assume that J is stationary for every family U, pick a smooth function $\eta: I \to \mathbf{R}$ with $\eta(0) = \eta(1) = 0$ and use these functions to construct a *variation* of the path as

$$q(t,\epsilon) = q(t) + \epsilon\eta(t).$$

For this special family U the formula (3) still holds, this time with $dq/d\epsilon = \eta$, so

$$0 = \int_0^1 \left[-\frac{d}{dt} \left[\frac{\partial L}{\partial \dot{q}} \right] + \frac{\partial L}{\partial q} \right] \eta(t) dt. \tag{4}$$

This holds for every such smooth function η. The following lemma then shows that the term in the large brackets must vanish–in other words, Lagrange's equation must hold.

Lemma. *If M: $I \to \mathbf{R}$ is a smooth function of t for which*

$$\int_0^1 M\eta dt = 0 \tag{5}$$

for every smooth η: $I \to \mathbf{R}$ with $\eta(0) = \eta(1) = 0$, then M is identically zero.

PROOF. Suppose not, so that $M(t_3) \neq 0$–say $M(t_3) > 0$–for some t_3 in the interval I. We can assume that $t_3 \neq 0$ and $t_3 \neq 1$. Since M is continuous, one must then have $M(t) > 0$ in some small interval about t_3. Now choose b to be a bump function b: $I \to \mathbf{R}$: smooth, zero outside this small interval, positive inside the interval and equal to 1 at t_3. Next choose the variation to be $\eta = bM$. Then

$$\int_0^1 M\eta dt = \int_0^1 bM^2 dt > 0,$$

in contradiction to the hypotheses (5).

The methods used in this argument are not limited to mechanics; they also apply when the Lagrangian is replaced by other smooth functions K. In the plane with coordinates x and y, consider all smooth curves $y = y(x)$ from a point (x_0, y_0) to a point (x_1, y_1). One wishes a curve for which an integral

$$\int_{x_0}^{x_1} K(y, y', x) dx \tag{6}$$

is stationary, where $y' = dy/dx$. The proof above shows that it is necessary and sufficient that the curve satisfy Euler's equation

$$\frac{d}{dx}\frac{\partial K}{\partial y'} - \frac{\partial K}{\partial y} = 0, \tag{7}$$

which for $K = L$ is Lagrange's equation. Such problems arise early in many connections–for example, in the problem of finding the path of quickest descent in a vertical plane (the brachistocron) from the point (x_0, y_0) to the point (x_1, y_1). For this purpose, one really wishes not just a curve where the integral (6) is stationary, but one where it is actually a minimum (or a maximum) among suitable comparison curves. This study gave rise to a remarkable array of rigorous methods in the Calculus of Variations–including methods which would apply when the curves used are allowed to have corners or when the minimizing curve is subjected to various kinds of "side conditions". On the other hand, in mechanics the idea of characterizing trajectories as those paths which minimize a suitable integral has several different forms, with different choices of integral and "side conditions". Several of these forms go by the name "Principle of least action". For our form, the "action" can be defined as the integral over the trajectory of the difference between kinetic and potential energy (the Lagrangian $L = T - V$). This is in contract to Newton's second law, which describes the same trajectory in terms of a local property (a second order differential equation). For an eloquent description of the physics of the principle of least action we refer to Feynmann, Leighton, and Sands, vol. II, lecture 19.

To summarize: The explanation of the invariant form of Lagrange's equations depends on minimizing a suitable integral over families of curves, and is intimately tied to the development of the Calculus of Variations–which in recent times has reappeared as the theory of optimal control.

7. Hamilton's Equations

Another, more invariant form of the equations of motion is provided by Hamilton's equations. They use as coordinates not positions and velocities but positions and the corresponding momenta; they assume that the forces are conservative and so are derived from a potential function V. In place of the Lagrangian, they use the total energy H

$$H = T + V$$

(the *hamiltonian*), considered as a function of position and momentum.

Consider first the case of N particles, with position determined by $3N = k$ coordinates x^i (or q^i) and the trajectories given by Newton's laws in the form (3.9)

$$m_i \frac{d^2x^i}{dt^2} = F_i, \qquad i = 1, \ldots, k. \tag{1}$$

Since the forces are conservative, $F_i = -\partial V/\partial x^i = -\partial H/\partial x^i$, while the ith component of momentum is $p_i = m_i dx^i/dt$. These equations are thus rewritten as

$$\frac{dp_i}{dt} = -\frac{\partial H}{\partial q^i}, \qquad i = 1, \ldots, k, \quad (q^i = x^i). \tag{2}$$

On the other hand $\dfrac{dq^i}{dt} = p_i/m_i$ can be found from the kinetic energy

$$T = \frac{1}{2} \sum m_i \left(\frac{dx^i}{dt} \right)^2 = \frac{1}{2} \sum \frac{p_i^2}{m_i}.$$

as $p_i/m_i = \partial T/\partial p_i$, so that, since V does not depend on the momenta,

$$\frac{dq^i}{dt} = \frac{\partial H}{\partial p_i}. \tag{3}$$

In other words, in the $2k$ dimensional momentum phase space, with $k = n$ and with cartesian coordinates $q^1, \ldots, q^n$ (for position) and $p_1, \ldots, p_n$ (for momenta) the trajectories are the solutions of the $2n$ first order differential equations

$$\frac{dp_i}{dt} = -\frac{\partial H}{\partial q^i}, \qquad \frac{dq^i}{dt} = \frac{\partial H}{\partial p^i}, \qquad i = 1, \ldots, n. \tag{4}$$

These are called *Hamilton's equations.*

These equations hold in much greater generality, for any constrained motion subject to conservative forces. To show this, we will deduce them from the $2n$ first order Lagrange equations

$$\frac{d}{dt}\left[\frac{\partial L}{\partial \dot{q}^i}\right] - \frac{\partial L}{\partial q^i} = 0, \qquad \frac{d}{dt}q_i = \dot{q}_i, \qquad i = 1,2, \ldots, n, \tag{5}$$

in the $2n$ coordinates $q^1, \ldots, q^n, \dot{q}^1, \ldots, \dot{q}^n$. When the first n coordinates $q^1, \ldots, q^n$ are fixed, they determine a point x in the configuration space C, the second n coordinates $\dot{q}^1, \ldots, \dot{q}^n$ are then those for a vector in the tangent space B_xC to C at x. We propose to simplify these equations by replacing these latter n coordinates by the momenta $p_1, \ldots, p_n$, defined as functions of the q's and $\dot{q}$'s by

$$p_i = \frac{\partial L}{\partial \dot{q}^i}, \qquad i = 1, \ldots, n. \tag{6}$$

In the cartesian coordinate case this formula does give exactly the usual components of momentum. Moreover, as in that case, we will assume that the matrix with entries

$$\frac{\partial^2 L}{\partial \dot{q}^i\,\partial \dot{q}^j}, \qquad i,j = 1, \ldots, n.$$

is always non-singular, so that we can solve the n equations (6) for the $\dot{q}^j$ as functions

$$\dot{q}^j = \dot{q}^j(q^1, \ldots, q^n, p_1, \ldots, p_n) \tag{7}$$

(we will also use the partial derivatives of these functions). Thus L, originally given as a function on the velocity phase space,

$$L = L(q^1, \ldots, q^n, \dot{q}^1, \ldots, \dot{q}^n) \tag{8}$$

can then also be expressed as a function of the p's and q's on the momentum phase space. (Mind your p's and q's!)

Now define the Hamiltonian as the quantity

$$H = \sum_{j=1}^{n} p_j \dot{q}^j - L, \tag{9}$$

to be considered via (7) as a function of the p's and q's. Its partial derivatives relative to the p's are then found by the usual chain rule for the derivative of a composite as

$$\frac{\partial H}{\partial p_i} = \dot{q}^i + \sum_{j=1}^{n} p_j \frac{\partial \dot{q}^j}{\partial p_i} - \sum_{j=1}^{n} \frac{\partial L}{\partial \dot{q}^j} \frac{\partial \dot{q}^j}{\partial p_i};$$

in the partial derivatives on the right, the other variables to be held constant are those indicated in the functional dependencies (7) and (8). But because of the definition (6) of the p_i, the two sums here cancel; with the second set of Lagrange's equations (5) the result becomes just

$$\frac{\partial H}{\partial p_i} = \frac{dq_i}{dt}, \qquad i = 1, \ldots, n. \tag{4a}$$

For the remaining coordinates $q^1, \ldots, q^n$ the partials of H, again with dependencies (7) and (8), are

$$\frac{\partial H}{\partial q^i} = \sum_{j=1}^{n} p_j \frac{\partial \dot{q}^j}{\partial q^i} - \frac{\partial L}{\partial q^i} - \sum_{j=1}^{n} \frac{\partial L}{\partial \dot{q}^j} \frac{\partial \dot{q}^j}{\partial q^i}.$$

By the definition of the p_i's, the two sums again cancel; with the first of Lagrange, it becomes

$$\frac{\partial H}{\partial q^i} = -\frac{d}{dt}\left(\frac{\partial L}{\partial \dot{q}^i}\right) = -\frac{dp_i}{dt}, \qquad i = 1, \ldots, n. \tag{4b}$$

We have thus deduced Hamilton's equations (4) from Lagrange's equations; we have presented this as a sort of trick which makes a transformation of variables so as to simplify the form of the Lagrange equations. The trick seems to consist in looking at the Lagrange equations, putting p_i for the terms $\partial L / \partial \dot{q}^i$ so that the Lagrange equations are automatically in the solved form for dp_i/dt—and then pulling H out of the air by the definition (9). Indeed, many standard treatments derive the equations in just this way, usually without bothering to explain (as we did in (7) and (8)) which other variables are to be held constant in which partial derivatives.

But what appears as a trick is in fact an idea—an idea which must have been clear to Hamilton when he did it. But we claim that in general most of the formal tricks appearing in Mathematics are really ideas in disguise–ideas presented as manipulations because the manipulations can be made explicit while the ideas are a bit nebulous. So let us get behind this one trick.

Fix the coordinates $q^1, \ldots, q^n$; they determine a point x in the configuration space C. The velocity is then the vector with components $\dot{q}^1, \ldots, \dot{q}^n$ in the tangent space $B_x C$ to C at the point x; call this vector space W, as in the diagram (10):

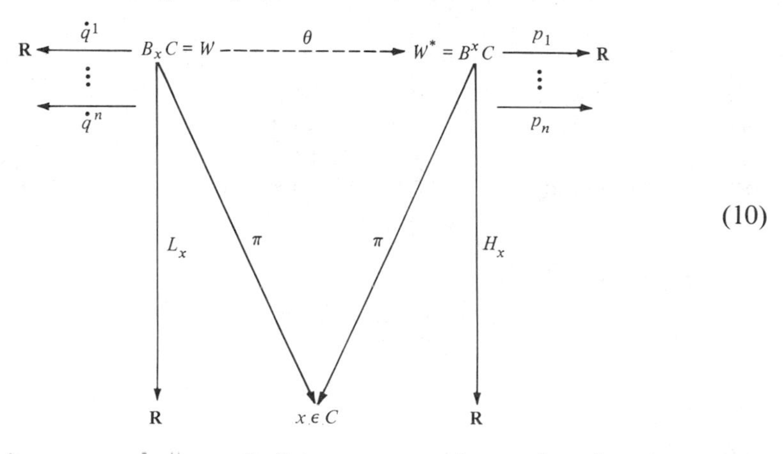

in this diagram each "quantity" is represented as a function (arrow) on the appropriate space to the reals **R**. In particular, everything starts with the Lagrangian L, which for our fixed point x is a quantity L_x on $B_x C = W$; this is a vector space on which the n quantities $\dot{q}^1, \ldots, \dot{q}^n$ are the coordinates. Now the main terms of the Lagrange equations have the n partial derivatives $\partial L_x / \partial \dot{q}^i$. They depend on the particular choice of coordinates $\dot{q}^1, \ldots, \dot{q}^n$, but they are exactly the coefficients (for these coordinates) of the total differential dL_x of the function L_x:

$$dL_x = \sum_{j=1}^{n} \frac{\partial L_x}{\partial \dot{q}^i} d\dot{q}^i. \tag{11}$$

Now a differential is just a vector in the cotangent space $W^* = B^x C$, the space dual to W–hence our picture also includes the cotangent space, projected down by π to $x \in C$. This should actually be said more carefully. The differential dL_x is taken at some point w of W–the point with coordinates $\dot{q}^1, \ldots, \dot{q}^n$; hence the differential is a (cotangent) vector in the cotangent space to W at this point w. But the tangent space to W at any one point w may be identified with W–just translate each vector from the origin in W to a vector starting at w. By the same device (a "canonical" isomorphism) the cotangent space to W at this point may be identified with the dual space W^*. Now we have the idea behind the trick change of variables: It is to transform each point w of the tangent (or velocity) space into that point of the dual space W^* given by the differential dL_x of L_x at that point w. The usual dual coordinates $p_1, \ldots, p_n$ of this point in the dual space are then, as in (11), the quantities

$$p_i = \frac{\partial L_x}{\partial \dot{q}^i}$$

exactly as in (6). The transformation now appears in the figure (10) as the arrow θ: $W \to W^*$; it is classically called the *Legendre transformation* for the quantity L.

Now that we have the transformation θ we would like it to have an inverse θ^{-1}—and it would be handy to have that inverse given as the Legendre transformation for some suitable quantity K on the cotangent bundle $B^{\cdot}C$. This means that we would like to have $\dot{q}^i = \partial K/\partial p_i$, so as a first guess we might set $K = \Sigma p_j \dot{q}^j$. But this won't be quite right, because the $\dot{q}^j$, via θ^{-1}, will be functions of the p's and so the partial derivatives will be

$$\frac{\partial K}{\partial p_i} = \dot{q}^i + \sum_{j=1}^{n} p_j \frac{\partial \dot{q}^j}{\partial p_i} = \dot{q}^i + \sum_j \frac{\partial L}{\partial \dot{q}^j} \frac{\partial \dot{q}^j}{\partial p_i} = \dot{q}^i + \frac{\partial L}{\partial p_i} .$$

This means that we will get the right quantity H if we set $H = K - L$. Then $\dfrac{\partial H}{\partial p_i}$ is $\dot{q}^i$, and

$$H = \sum p_j \dot{q}^j - L.$$

This is the formula for the Hamiltonian, previously pulled out of the air at (9). Now it is no wonder that the differential equations become simpler. The full picture (for all the points x of the configuration space) is

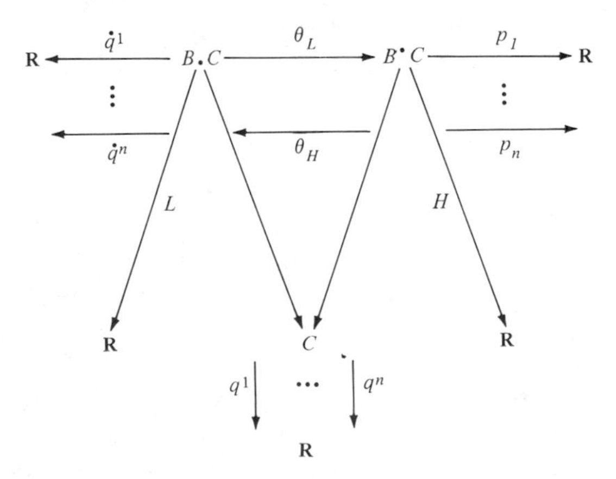

There is still another twist: In the standard case, the kinetic energy is a Riemann metric on the configuration space. In more detail, the kinetic energy in the N particle case is a positive definite quadratic form

$(1/2)\Sigma m_i(\dot{q}^i)^2$ in the velocity coordinates; hence in cases of constraints it will still be such a form, perhaps $T = \Sigma\, g_{ij}\dot{q}^i\dot{q}^j$, with coefficients g_{ij} in the coordinates $\dot{q}^i$ of the tangent space $W = B_x C$. Such a form is at once an inner product on that space W; we have already observed (in Chapter VIII) that a Riemann metric on a manifold C may be described as a smooth way of specifying inner products on the tangent spaces to the manifold. As a homogeneous quadratic function, T can be expressed by the formula

$$2T = \sum_i \dot{q}^i \frac{\partial T}{\partial \dot{q}^i} = \sum_i \dot{q}^i p_i ,$$

where we have used the definition of the momenta p_i.

Therefore the Hamiltonian H defined in (9) can be written

$$H = \sum p_i \dot{q}^i - L = 2T - (T - V) = T + V. \tag{12}$$

In other words, in these cases the Hamiltonian is exactly the total energy $T + V$.

Moreover an inner product $<-, ->$ in a finite dimensional vector space W determines a canonical isomorphism ϕ of W to its dual; indeed, as in §VII.5, ϕ sends each vector v in W to the linear function $\phi v = <v, ->: W \to \mathbf{R}$. In the present case the inner product of vectors $v = (v^i)$ and $w = (w^i)$ is written as $\Sigma\, g_{ij} v^i w^j$, where the entries g_{ij} make up a positive definite symmetric matrix. Thus the linear function $<v, ->$, inner product with v, has as its jth coordinate $\Sigma\, g_{ij}\, v^i$. But, except for a factor 2 and a change of notation (v^i for q^i) this is just the quantity

$$p_i = \frac{\partial T}{\partial \dot{q}^i} = \frac{\partial}{\partial \dot{q}^i} \sum g_{ij} \dot{q}^i \dot{q}^j$$

used to define the ith component of momentum. In other words, the Legendre transformation θ replacing the $\dot{q}^i$ by the p_i is, on each tangent space $B_x C = W$, just the canonical isomorphism (§VII.5) from the inner product space W to its dual W^*. In this way, the Riemann metric given by the kinetic energy determines, via duality, the Legendre transformation.

8. Tricks versus Ideas

Hamilton's elegantly symmetrical equations may be deduced from the Lagrange equations–either by a trick or by conceptual use of various underlying ideas. For the trick we merely plug in new coordinates p_i for the old $\dot{q}^i$, put down a formula for H and perform a few partial differentiations and lo–Hamilton's equations. The conceptual analysis is longer but more revealing. First, the new coordinates p_i are physically the components of momentum, defined by $\partial L / \partial \dot{q}^i$. Taken together, they are the components of a differential, the differential dL_x of the Lagrangian along the tangent space at a point x. (It is sometimes therefore called the "fiber derivative" of L.) The change from the $\dot{q}^i$ to the p_i is not just a juggling of variables, it is thus a transformation θ from the tangent space to the cotangent space, as determined by L. The inverse transformation is then given by a different function on the cotangent space—and this function is the Hamiltonian. Therefore the equation $\partial H / \partial p_i = dq^i/dt$ (from Hamilton's equations) just is a statement that H does give the inverse transformation. Moreover, the kinetic energy is a quadratic form and therefore an inner product on the tangent space W—and the transformation θ is really just the standard isomorphism of such a space to its dual, the cotangent space.

These remarks help to understand what is happening and they serve to connect this development with all sorts of other mathematical ideas, in particular, ideas from linear algebra and manifold theory. They surely would not have been formulated, at least in this language, at the time Hamilton first set up his equations. However, many of the ideas involved, though not these technical formulations, might well be in the backs of the minds of people who present the calculation simply as a quick trick.

This case is of considerable interest because of the wide further development of Hamiltonian dynamics, hinted at below. It also implicitly raises the question of the possible conceptual background of other tricks. Analysis is full of ingenious changes of coordinates, clever substitutions, and astute manipulations. In some of these cases, one can find a conceptual background. When so, the ideas so revealed help us to understand what's what. We submit that this aim of understanding is a vital aspect of Mathematics.

Understanding is not easy, as I may perhaps indicate by anecdote. It has taken me over fifty years to understand the derivation of Hamilton's equations. I first saw them in 1929 in a book on *Theoretical Mechanics* by Sir James Jeans, a noted British applied mathematician. That came in a course taught by Professor E. W. Brown, an expert on celestial dynamics. He must have felt that the formal presentation by Jeans was inadequate, for he turned away from his previous steady reliance on Jean's text and lectured to us from his own handwritten notes. I don't really recall understanding the lectures, because I was more impressed with the yellowed and dog-eared condition of those notes. The very next year I heard the subject again, this time from Leigh Page, a professor of physics. I can tell now what he said, because it was all much as in his book *Introduction to Theoretical Physics*. He plays the tricks and pulls the Hamiltonian H out of the air (and a touch of the Calculus of Variations); there was of course no mention of tangent or cotangent spaces. Nearly 40 years later, I took to lecturing myself on the subject. My then students took and published careful notes of my lectures. In these notes the trick is duly decked out with tangent and cotangent spaces. Evidently that didn't satisfy, because there is then the sentence "We want to understand better how the scheme produced these equations". There follow two pages of attempts at understanding ending "The Hamiltonian function arises from asking the question: When is θ invertible? (This is probably not the way Hamilton found it.)". Two years later I found this measure of understanding inadaquate, so tried again in an article "Hamiltonian Mechanics and Geometry" in the American Mathematical Monthly Vol. 77 (1970), pp. 570–586. Then after a twelve year respite the understanding there purveyed seemed formal (e.g., in the identification of the tangent spaces to W with W itself) so I tried again . . . to get the presentation above.

The point of this cautionary tale is not the individual events, but the difficulty of getting to the bottom of it all. Effective or tricky formal manipulations are introduced by Mathematicians who doubtless have a guiding idea–but it is easier to state the manipulations than it is to formulate the idea in words. Just as the same idea can be realized in different forms, so can the same formal success be understood by a variety of ideas. A perspicuous exposition of a piece of Mathematics would let the ideas shine through the display of manipulations.

9. The Principal Function

There is another, remarkably fruitful, method for integrating the differential equations of motion for a conservative system. This method, suggested by the study of waves in geometrical optics, leads, as does the theory of waves, to a *partial* differential equation (P.D.E.) which is useful in classical mechanics and also in quantum mechanics.

Start with Hamilton's principle, which asserts that the trajectories $q^i = q^i(t)$ of the motion in the configuration space C are those paths $q = (q^1, \ldots, q^n)$ which render the integral

$$\int_0^{t_1} L(q,\dot{q},t)dt$$

of the (time-dependent) Lagrangian L stationary–and this in comparison with other smooth paths starting (at $t = 0$) at the same point $q^i = a^i$ and arriving at the same point $q^i = q_1^i$ at the same time t_i; here we use the subscript 1 to indicate time and position of arrival. The solutions for the ($2n$, first order) Euler equations for this variational problem depend on the initial values $q^i = a^i$ and $\dot{q}^i = b^i$ of the position and velocity vectors q and $\dot{q}$. These solutions will usually be smooth functions

$$q^i = u^i(a,b,t), \qquad i = 1, \ldots, n. \tag{1}$$

Now a trajectory of a particle in mechanics is much like a ray of light in optics, and the latter subject uses a "principal function" W which is essentially the value of the integral above, taken along the actual trajectories, for all values of the initial conditions a^i, b^i. In other words the principal function for the conservative mechanical system is

$$W(a,b,t_1) = \int_0^{t_1} L(q,\dot{q},t)dt. \tag{2}$$

Now Hamilton's principle compares paths with the same starting and arrival points (not starting velocities b). Hence we wish in (2) to replace the initial velocities b by the final positions q_1. Indeed, the equations (1) for the trajectories do determined (for $t = t_1$) the coordinates q_1^i of the point of arrival. The implicit function theorem, under suitable conditions, states that these equations (with $t = t_1$) can be solved for the initial velocities b^i as functions

$$b^i = f^i(a,q_1,t_1), \qquad i = 1, \ldots, n \tag{3}$$

of initial position, final position, and final time (implicitly, also functions of initial time). Because of the form of Hamilton's principle, we wish to express "everything" in terms of these quantities a^i, q_1^i, and t_1, regarded as $2n + 1$ coordinates on the manifold $C \times C \times I$. For example, by substituting the solution (3) in (1), the trajectories themselves have coordinates given as functions

$$q^i = g^i(a,q_1,t_1,t) \tag{4}$$

of these quantities and time t. Here t is the time "along" the trajectory, so that the components of velocity are

$$\dot{q}^i = \frac{\partial g^i}{\partial t}(a,q_1,t_1,t). \tag{5}$$

Similarly, the principal function W expressed in terms of these variables is called S; thus

$$\begin{aligned} S(a,q_1,t_1) &= W(a,f(a,q_1,t_1),t_1) \\ &= \int_0^{t_1} L(g,\dot{g},t)dt, \end{aligned} \tag{6}$$

with g the vector with components g^i as in (4) because the integral is taken *along the trajectory*.

Now we calculate some of the partial derivatives of S, considered as a function of a, q_1 and t_1, so that each partial derivative with "respect" to one of these quantities is taken while holding the rest of them fixed. By differentiating (6) under the integral and using (4) and (5) we get

$$\frac{\partial S}{\partial q_1^j} = \int_0^{t_1} \sum_i \left[\frac{\partial L}{\partial \dot{q}^i} \frac{\partial}{\partial t} \frac{\partial g^i}{\partial q_1^j} + \frac{\partial L}{\partial q^i} \frac{\partial g^i}{\partial q_1^j} \right] dt.$$

Integrating the first terms by parts yields

$$\frac{\partial S}{\partial q_1^j} = \sum_i \left[\frac{\partial L}{\partial \dot{q}^i} \frac{\partial g^i}{\partial q_1^j} \right]_{t=0}^{t=t_1} + \int_0^{t_1} \sum_i \left[-\frac{\partial}{\partial t} \frac{\partial L}{\partial \dot{q}^i} + \frac{\partial L}{\partial q^i} \right] \frac{\partial g^i}{\partial q_1^j} dt.$$

In the remaining integral the term in brackets is just the ith term of the Lagrange equations; hence the integral vanishes. At the start $(t = 0)$, $g^i = a^i$ is constant, so the partial $\partial g^i / \partial g_i^j$ vanishes; at the end $(t = t_1)$, $g^i = q_1^i$, so $\partial g^i / \partial g_1^j = \delta_j^i$ and the final result is simply

$$\frac{\partial S}{\partial q_1^j} = p_j, \qquad j = 1, \ldots, n, \tag{7}$$

because the momenta are defined by $p_j = \partial L / \partial \dot{q}^j$! The result is simple, and so is the proof, provided one pays careful attention to which variable is where and a function of what–which is why we have here departed from convention by writing g for the functions representing q along the trajectory.

Next we consider the partial derivative of S with respect to t_1. By (6) and (1) we may express the function W in terms of S as

$$W(a,b,t_1) = S(a,u(a,b,t_1),t_1). \tag{8}$$

Hence, holding the other variables in W constant, the chain rule gives

$$\frac{\partial W}{\partial t_1} = \frac{\partial S}{\partial t_1} + \sum_{i=1}^{n} \frac{\partial S}{\partial q_1^i} \frac{\partial q_1^i}{\partial t_1}.$$

But the partial derivative on the left is the derivative of the integral of L with respect to its upper limit, so is just L. Also $\partial S / \partial q_1^i$ is p_i by (7). Hence, solving for $\partial S / \partial t_1$,

$$\frac{\partial S}{\partial t_1} = L(q_1, \dot{q}_1, t_1) - \sum p_i \dot{q}_1^i.$$

The quantity on the right is by definition the negative of the value of the hamiltonian H at the endpoint $t = t_1$. Thus

$$\frac{\partial S}{\partial t} + H(q,p,t) = 0$$

holds at the endpoint t_1 of each trajectory. Since any point on the trajectory is the endpoint (of a shorter trajectory) it holds for all t. Moreover, $p_i = \partial S / \partial q^i$ by (7), so the last equation reads

$$\frac{\partial S}{\partial t} + H(q^1, \ldots, q^n, \frac{\partial S}{\partial q^1}, \ldots, \frac{\partial S}{\partial q^n}, t) = 0. \tag{9}$$

This is the Hamilton–Jacobi partial differential equation for the principal function S. It is just $\partial S/\partial t$ plus the Hamiltonian H, with each argument p_i in H replaced by the corresponding partial derivative $\partial S/\partial q^i$. This is a conceptually appropriate replacement, since at each point of the configuration space C the principal function S, considered just as a function of the coordinates q of the point, has a differential involving just these partial derivatives

$$dS = \frac{\partial S}{\partial q^1} dq^1 + \cdots + \frac{\partial S}{\partial q^n} dq^n,$$

and hence determines a cross section $q \mapsto (q, dS)$ of the cotangent bundle (phase space) $B^\bullet C$, where the coordinates are the q's and p's. The Hamilton–Jacobi P.D.E. (9) simply uses the value of the Hamiltonian H on this cross section.

To summarize: A conservative mechanical system with Legrangian L determines a "principal function" S as the integral of L along the trajectories. As a function of time and the end-point of the trajectory, S satisfies the Hamilton–Jacobi (HJ) partial differential equation (9). Conversely, as the next section will indicate, suitable solutions of the HJPDE give the trajectories, often called the *characteristics* of the P.D.E. This is a first indication of the importance of P.D.E. in mechanics (and also in quantum mechanics) and a starting point of the subject "Differential equations of Mathematical Physics"–see Frank–v. Mises [1930].

This also reveals the connection between mechanics and geometric optics. In the latter subject, one wishes to determine the path of a light ray through a specified medium from a point A to a point B. It is given by Fermat's principle: The actual path from A to B is the one which makes the time from A to B (as a function of possible paths) a minimum, more exactly, stationary. On the other hand the wave fronts, for rays emanating from a point A are the loci in coordinates q where the corresponding principal function $S(a,q)$ is a constant. Both mechanics and optics thus involve a "least action" principle, which therefore plays a major role in the calculus of variations (see Caratheodory [1965]). This calculus is a striking example of a single Mathematical form fitting a variety of different facts.

10. The Hamilton–Jacobi Equation

Now consider solutions of this P.D.E.; one other partial differential equation–the wave equation

$$\frac{\partial^2 u}{\partial t^2} = c^2 \frac{\partial^2 u}{\partial x^2} \tag{1}$$

–has already appeared, in §VI.11. It has many solutions–for any smooth function $k(x)$ a wave $u = k(x - ct)$, travelling forward, and for any $g(x)$ a corresponding wave $u = g(x + ct)$ moving backwards. Typically, P.D.E. have solutions depending in this way on arbitrary smooth functions such as g and k.

For similar reasons the Hamilton–Jacobi P.D.E.

$$\frac{\partial S}{\partial t} + H(q, \frac{\partial S}{\partial q}, t) = 0 \tag{2}$$

will have many solutions for S as a function of $q^1, \ldots, q^n$, and t. We may not be able to identify which one is *the* principal function of our mechanical system. Instead we search for a solution S which is like the principal function, in that it depends on n parameters a^i, as a function $S(a^1, \ldots, a^n, q^1, \ldots, q^n, t)$, and this in such a way that the $n \times n$ matrix with the entries

$$\frac{\partial^2 S}{\partial a^i \partial q^j}, \qquad i,j = 1, \ldots, n, \tag{3}$$

is non-singular for all relevant values of the arguments. Such a family of solutions S is known as a *complete solution* of the P.D.E. (2). It may not be the intended principal function for the conservative mechanical system determined by the Hamiltonian function H. Nevertheless, the Hamilton–Jacobi theorem asserts that any such complete solution does provide trajectories, by a family of solutions of Hamilton's equations depending on $2n$ parameters a^i and c^i, and that this family is constructed in just the way sketched in §9 for the actual principal function. Specifically we introduce n new parameters c^j, to be considered as the initial values of the momenta, and use them to set up n equations

$$\frac{\partial S}{\partial a^i} = -c^i, \qquad i = 1, \ldots, n \tag{4}$$

analogous to the equations (9.7). These n equations involve the q^i in $S(a,q,t)$; by the hypothesis on the matrix (3), they can be solved (locally) for the q's as functions of a, c, and t. The momenta are then defined as in (9.7) by the n equations

$$\frac{\partial S}{\partial q^i} = p_i, \qquad i = 1, \ldots, n. \tag{5}$$

On replacing here the arguments q^i in S by the functions $q^i(a,c,t)$ just determined, we obtain the p's expressed also as functions $p_i(a,c,t)$. The

Hamilton–Jacobi theorem asserts that the q^i and p_i so expressed as functions of t will satisfy Hamilton's equations and so will provide trajectories for the mechanical system specified by the Hamiltonian function H.

We now sketch the proof of this theorem, omitting such details as the careful specification of the domains of a's and c's where the hypotheses are to hold. First imagine that the proposed solutions $q(a,c,t)$ are substituted for the variables q in the H–J P.D.E., and then differentiate (2) partially with respect to (say) a^1. By the usual chain rule for the partial derivatives of a composite, the result is

$$\frac{\partial^2 S}{\partial a^1 \partial t} + \sum_{i=1}^{n} \frac{\partial^2 S}{\partial a^1 \partial q^i} \frac{\partial H}{\partial p^i}\left(q, \frac{\partial S}{\partial q}, t\right) = 0.$$

The same second partial of S (except for the order of the two differentiations) can also be found by differentiating (4) with respect to t

$$\frac{\partial^2 S}{\partial t \partial a^1} + \sum_{i=1}^{n} \frac{\partial^2 S}{\partial a^1 \partial q^i} \frac{\partial q^i}{\partial t} = 0.$$

We intend to assume S smooth, so that the order of partial differentiations does not matter. Then taking the difference of these two equations (and similarly for a^1 replaced by a^j) we get

$$\sum_{i=1}^{n} \frac{\partial^2 S}{\partial a^j \partial q^i} \left[\frac{\partial H}{\partial p_i} - \frac{\partial q^i}{\partial t} \right] = 0, \qquad j = 1, \ldots, n.$$

But we assumed above that the matrix (3) of coefficients here is nonsingular. Hence, for each i,

$$\frac{\partial H}{\partial p_i} - \frac{\partial q^i}{\partial t} = 0, \qquad i = 1, \ldots, n. \tag{6}$$

This is the first n of Hamilton's equations.

Now return to H–J and differentiate with respect to q^i, to get

$$\frac{\partial S}{\partial q^i \partial t} + \frac{\partial H}{\partial q^i} + \sum_{j=1}^{n} \frac{\partial H}{\partial p_j} \frac{\partial^2 S}{\partial q^i \partial q^j} = 0.$$

We can obtain the same second partial of S, up to the order of differentiation, by differentiating the definition (5) of the momenta p_i relative to t:

$$\frac{\partial^2 S}{\partial t \partial q^i} + \sum_{j=1}^{n} \frac{\partial^2 S}{\partial q^j \partial q^i} \frac{\partial q^j}{\partial t} - \frac{\partial p_i}{\partial t} = 0, \qquad i = 1, \ldots, n.$$

In view of (6), the first of Hamilton's equations, the summation here agrees with that in the previous equation. Hence, subtracting, we have

$$\frac{\partial H}{\partial q^i} + \frac{\partial p_i}{\partial t} = 0, \qquad i = 1, \ldots, n. \tag{7}$$

This is, as intended, the second set of Hamilton's equations.

Thus we conclude that each complete solution $S(a,q,t)$ of the H–J Partial differential equation does yield a family of solutions $q^i(t)$, $p_i(t)$ of Hamilton's equations (6) and (7), and that this family depends on the $2n$ parameters $a^1, \ldots, a^n, c^1, \ldots, c^n$. There is of course no assurance that the a^i and c^i represent the initial values of coordinates q^i and momenta p_i at $t = 0$. We know only that they provide the necessary number $(2n)$ of constants of integration.

The case when the Lagrangian function L and hence the Hamiltonian H is independent of time has convenient special properties. The H–J P.D.E. then has the form

$$\frac{\partial S}{\partial t} + H\left[q, \frac{\partial S}{\partial q}\right] = 0. \tag{8}$$

For solutions $q(t)$, $p(t)$ of Hamilton's equations one then has

$$\frac{d}{dt} H(q,p) - \sum_{i=1}^{n} \left[\frac{\partial H}{\partial q^i}\frac{\partial q^i}{dt} + \frac{\partial H}{\partial p_i}\frac{dp_i}{dt}\right] - 0,$$

zero because the two terms in the summation cancel by Hamilton's equations. This proves that $H(p,q)$ is a constant k along each trajectory. Since H is the total energy, this conclusion is just a formulation of the principle of the conservation of energy. Along each trajectory, it follows that $\dfrac{\partial S}{\partial t} = -k$. Hence one may replace S by a function K,

$$K = K(q^1, \ldots, q^n) = S + kt,$$

which is independent of t. This function K satisfies a somewhat simpler (and quite useful) version of the Hamilton–Jacobi P.D.E.

11. The Spinning Top

Mechanics developed by the treatment of many specific problems. That notable child's toy, the spinning top, presents a striking problem in mechanics which is essentially that of the gyroscope, a grown-up's instrument. As in Osgood's excellent presentation (W. F. Osgood [1937] cf. also Pars [1965] p. 113) consider only a heavy top symmetrical about an axis, which is spinning (and tipping) about this axis, but so that the point of the top at the end of the axis remains fixed. Such a top appears in Figure 1. The configuration space now consists of all the possible positions of the

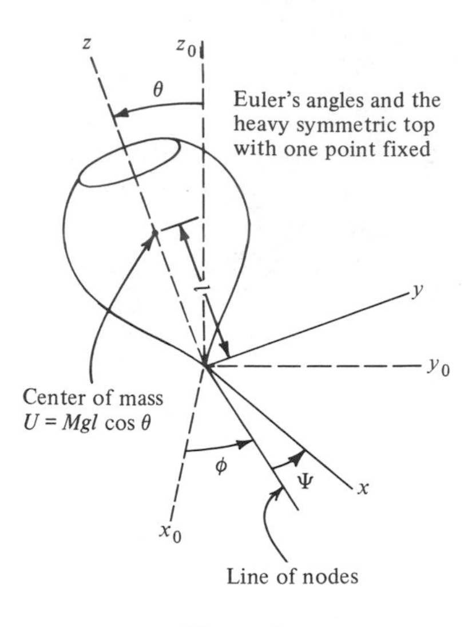

Figure 1

top (with its point fixed). These positions can be described in terms of the angles needed to move from rectangular axes x_0, y_0, z_0 fixed in space (origin 0 at the point of the top) to axes x, y, and z fixed in the top, with the same origin. To describe the position of these axes, think of the point c at which the axis of the top meets a large sphere centered at the origin 0. Let θ be the latitude (measured from the north pole) and ϕ the longitude of this point. These two angles θ and ϕ suffice to describe the positions of the top axis; we need a third angle Ψ to specify how much it has rotated about this axis. Indeed, the top can be brought from an original upright position to its final position in the following three successive steps (Figure 1):

First rotate the top by the longitude ϕ about the vertical axis $0z_0$ in space;
Next rotate the top by the latitude θ about the top's new $0x$ axis;
Finally rotate the top about its new $0z$ axis by an angle Ψ.

These three angles

$$0 \leqslant \phi \leqslant 2\pi, \qquad 0 \leqslant \theta \leqslant \pi \quad \text{and} \quad 0 \leqslant \Psi \leqslant 2\pi$$

suffice to describe completely the position of the top—or the rotation of any rigid body about a fixed point. They are called *Euler's angles*. They are the local coordinates in a configuration space C—but they are not global coordinates. (For example, when $\theta = 0$ or π, the coordinate Ψ is not needed!) This case illustrates well the idea that mechanics often requires that rectangular coordinates be replaced by other coordinates suited to the problems at hand–in the present case, coordinates appropriate to the group of all rotations in 3-space. (This group is also a three dimensional manifold; it is an example of a Lie group.)

In order to formulate the Hamiltonian for this problem, we need elementary facts about the kinetic energy of a rotating body. If a solid flat plate (of any shape) rotates about an axis perpendicular to the plate the only coordinate needed is the angle θ of rotation and the corresponding angular velocity $\omega = d\theta/dt$. A point in the plate with a mass m has coordinates $x = r\cos\theta$, $y = r\sin\theta$ and therefore kinetic energy

$$2T = m(\dot{x}^2 + \dot{y}^2) = mr^2(\sin^2\theta + \cos^2\theta)\omega^2 = mr^2\omega^2. \qquad (1)$$

The total kinetic energy of the plate is then a sum (better an integral) of all these contributions $mr^2\omega^2$. The appropriate integral of the mr^2 is known as the *moment of inertia* I of the plate, and the kinetic energy is

$$T = (1/2)I\omega^2$$

(by the way, this provides another occasion for the use of integrals). In the case of the symmetrical top there are three such moments of inertia, I_x, I_y and I_z, about the respective axes fixed in the top. Also at any one instant the top has instantaneous angular velocities ω_x, ω_y and ω_z about these three axes. Because of the symmetry of the top, these quantities suffice to determine the whole kinetic energy of the top as

$$T = (1/2)\left[I_x\omega_x^2 + I_y\omega_y^2 + I_z\omega_z^2\right] \qquad (2)$$

(without symmetry, there could be cross terms $I_{xy}\omega_x\omega_y$, etc.) In any case, symmetry also gives $I_x = I_y$.

Angular velocities may be represented by vectors. Specifically, a rotation about an axis A with angular velocity ω may be represented by a vector along the line of A, suitably directed, and with magnitude ω. This representation is used because it is effective in combining two angular velocities; by calculating the composite of two rotations one can prove that the effect of combining two such angular velocities is represented by the sum of the two corresponding vectors. In particular, if the angular velocities $\dot{\phi}$, $\dot{\theta}$, and $\dot{\Psi}$ for the Euler angles are represented by vectors, one can add these three vectors by taking their components along the axes x, y, and z fixed in the top to get (use Figure 1)

$$\omega_x = \dot{\phi}\sin\theta\cos\Psi - \dot{\theta}\sin\Psi,$$

$$\omega_y = \dot{\phi}\sin\theta\sin\Psi + \dot{\theta}\cos\Psi,$$

$$\omega_z = \dot{\phi}\cos\theta + \dot{\Psi}.$$

Therefore the kinetic energy T of (2) has the expression

$$T = (1/2)[I_x(\dot{\phi}^2\sin^2\theta + \dot{\theta}^2) + I_z(\dot{\phi}\cos\theta + \dot{\Psi}] \tag{3}$$

in terms of the Euler angles. The corresponding momentum coordinates $p_i = \partial T/\partial \dot{q}^i$ are thus, in an evident notation,

$$\begin{aligned} p_\theta &= I_x\dot{\theta}, \\ p_\phi &= (I_x\sin^2\theta + I_z\cos^2\theta)\dot{\phi} + I_z\cos\theta\,\dot{\Psi}, \\ p_\Psi &= I_z(\dot{\phi}\cos\theta + \dot{\Psi}). \end{aligned} \tag{4}$$

The Hamiltonian $H = T + V$, as a function of the p's and the q's, then is

$$H = (1/2)\left[\frac{p_\theta^2}{I_x} + \frac{p_\Psi^2}{I_z} + \frac{1}{I_x}\left(\frac{p_\phi - p_\Psi\cos\theta}{\sin\theta}\right)^2\right] + Mgl\cos\theta. \tag{5}$$

Here $V = Mgl\cos\theta$, where M is the mass of the top and l is the distance from the point of the top along its axis to the center of mass, while g is the acceleration of gravity. The Hamilton–Jacobi P.D.E. then has the form

$$\begin{aligned} \frac{1}{I_x}\left(\frac{\partial S}{\partial\theta}\right)^2 + \frac{1}{I_z}\left(\frac{\partial S}{\partial\Psi}\right)^2 + \frac{1}{I_x\sin^2\theta}\left(\frac{\partial S}{\partial\phi} - \frac{\partial S}{\partial\Psi}\cos\theta\right)^2 \\ = -2\left(\frac{\partial S}{\partial t} + Mgl\cos\theta\right). \end{aligned} \tag{6}$$

To get a complete solution of this P.D.E. for $S(t,\theta,\phi,\Psi)$, we use two devices. The first is "separation of variables": We try to find a solution which is the sum of four separate functions of one variable (t,θ,ϕ,Ψ) each. Second, we note that of these variables only θ appears explicitly (as $\cos\theta$) in the P.D.E. (6), while the others (t,ϕ and Ψ) turn up only in the denominators of partial derivatives. Such variables are said to be "ignorable"; the corresponding trick is to make these partial derivatives constant. Thus, all told, we try a solution of the form

$$S = a_1 t + a_2\phi + a_3\Psi + R(\theta) \tag{7}$$

involving three constants a_i and a function R. It will satisfy the P.D.E. (6) if

$$\left(\frac{dR}{d\theta}\right)^2 = -2I_x(a_1 + Mgl\cos\theta) - \frac{I_x}{I_z}a_3^2 - (a_2 - a_3\cos\theta)^2/\sin^2\theta. \tag{8}$$

Thus the separation of variables has replaced the P.D.E. (6) by a first order ordinary differential equation in three parameters a_i for the function $R(\theta)$. In this equation the denominator $\sin^2\theta$ on the right can be rewritten as $1 - \cos^2\theta$. The θ appears only in terms of $u = \cos\theta$, so it is natural to use u as a new variable. Now this equation (8) has the general form

$$(dR/d\theta)^2 = F(u)/(1 - u^2)$$

where $F(u) = F(u,a_1,a_2,a_3)$ is an expression independent of ϕ and Ψ; in fact it is a polynomial in its four arguments, cubic in u, quadratic in a_2 and a_3 and linear in a_1, with $\partial F/\partial a_1 = -2I_x(1 - u^2)$. Thus the function R is given by an integral

$$R = \int_0^\theta \left[\frac{F(u)}{1-u^2}\right]^{1/2} d\theta = -\int_u^\theta \frac{(F(u))^{1/2}}{1-u^2} du, \tag{9}$$

where we have entered the suggested change $du = \sqrt{(1-u^2)}d\theta$ in the variable of integration. This does yield a formula of sorts for S. However, we do not need S, but only its partial derivatives. Since $\partial F/\partial a_1 = -2I_x(1-u^2)$, they are

$$\frac{\partial S}{\partial a_1} = t + I_x\int_0^\theta \frac{du}{\sqrt{Fu}}, \quad \frac{\partial S}{\partial a_2} = \phi + \frac{\partial R}{\partial a_2}, \quad \frac{\partial S}{\partial a_3} = \Psi + \frac{\partial R}{\partial a_3}. \tag{10}$$

In the first equation, the integral is now that of the inverse of a square root of a cubic polynomial in u. It is a so-called *elliptic integral*–not expressible in the usual tables of integrals in terms of the elementary functions, but much studied in classical analysis, in particular in complex variable theory (Chapter X). Since R does not depend on ϕ and Ψ, the matrix of second derivatives described in (10.3) is essentially

$$\begin{bmatrix} \dfrac{\partial^2 S}{\partial a_2\,\partial\phi} & \dfrac{\partial^2 S}{\partial a_2\,\partial\Psi} \\[2ex] \dfrac{\partial^2 S}{\partial a_3\,\partial\phi} & \dfrac{\partial^2 S}{\partial a_3\,\partial\Psi} \end{bmatrix} = \begin{bmatrix} 1 & 0 \\ 0 & 1 \end{bmatrix};$$

it thus follows that we have a complete solution in the three parameters a_i.

We are then instructed to find the trajectories (for all initial conditions) by taking three new constants c^i and solving the equations $\partial S/\partial a_i = -c^i$ for the coordinates θ, Ψ and ϕ as functions of time. The first of these equations reads

$$t = -c^1 - I_x\int_0^\theta \frac{du}{\sqrt{Fu}}. \tag{11}$$

Knowing the elliptic integral, this will determine θ in terms of time t, the constants a_1, a_2 and a_3 and c^1; indeed $-c^1$ can be read as the initial value of t when $\theta = 0$. The second of these equations will then determine ϕ in terms of t, while the third gives Ψ in terms of θ, ϕ and t and thus ultimately in terms of t. The first two equations are of most interest; for the axis of the top they provide the latitude θ and the longitude ϕ in terms of time. A full discussion of the consequences has been presented by Klein and Sommerfield [1965] in a famous four-volume book [1897–1910].

One may briefly examine some of the qualitative properties of this solution. Since $F(u) = ku^3 + \cdots$ is a cubic polynomial, with a positive leading coefficient k, the values of $F(u)$ range from $-\infty$ to $+\infty$; moreover, $F(u) = 0$ generally has three roots, with two roots u_1 and u_2 ($u_1 < u_2$) between -1 and $+1$ (Figure 2). These are the only roots of physical interest, since the substitution $u = \cos\theta$ makes $-1 \leqslant u \leqslant 1$. Now $dt/du = I_x(F(u))^{-1/2}$ as above in (11) makes

$$\left[\frac{du}{dt}\right]^2 = F(u)/I_x^2,$$

so the zeros of $F(u)$ are the points where $du/dt = 0$ (i.e., where $\dot{\theta} = 0$). If one uses the positive square root of $F(u)$ in this (ordinary) differential equation, one gets a solution for u increasing from u_1 to u_2 in some time interval. Using the negative square root of F extends this solution, by reflection, to one decreasing from u_2 to u_1. All told, this gives a solution for u of the general form indicated in Figure 3. The function u (and hence the angle θ) is periodic in t. This is the often observed situation where the axis of the top bobs up and down in latitude θ as the top spins (i.e., as ϕ increases).

For the other two angular coordinates ϕ and Ψ the momenta p_ϕ and p_Ψ are given by Hamilton's equations as

$$\frac{dp_\phi}{dt} = -\frac{\partial H}{\partial \phi}, \qquad \frac{dp_\Psi}{dt} = -\frac{\partial H}{\partial \Psi}.$$

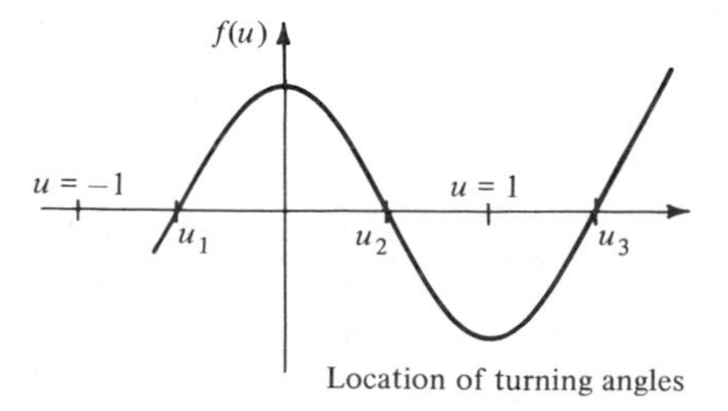

Figure 2

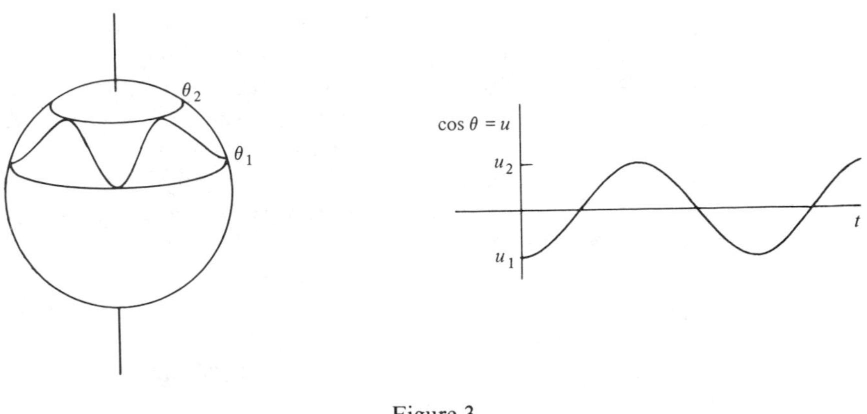

Figure 3

But the Hamiltonian H does not depend on ϕ or on Ψ, so the right hand sides here are zero. Therefore the (angular) momenta p_ϕ and p_Ψ are constants m_ϕ and m_Ψ. In other words, without friction at the point of the top (and our formulation has indeed neglected such friction) the top, once well started, will spin on forever.

The constant momenta m_ϕ and m_Ψ depend on the a's and the c's. However, without studying this dependence one can use these momenta to solve the equation (4) above for the angular velocities $\dot{\phi}$ and $\dot{\Psi}$. One finds, for example, that

$$\dot{\phi} = (m_\phi - m_\Psi u)/I_x(1-u^2).$$

This gives the speed of *precession* of the axis of the top around the vertical axis: As the axis of the top bobs up and down (i.e., as $\cos\theta = u$ varies between limits) the speed of precession varies according to this formula. The magic of mechanics is that it works! Necessarily complex calculations yield realistic results.

12. The Form of Mechanics

Our summary of theoretical mechanics serves to illustrate some aspects of formalism in Mathematics. One starts from qualitative and then quantitative descriptions of motion, such as the motion of planets and of projectiles. The quantitative observations can be brought into some order by empirical laws, such as Kepler's laws. Mathematics proper enters when one finds out how to deduce the empirical regularities from formal rules–in this case, from the laws of motion, the inverse square law, and the formal rules for the calculus and for the manipulation of differential equations. Sometimes this manipulation is combined with knowledgable approximations–neglecting terms empirically known to be negligable. But in principle, these deductions do have a properly formal character–given the initial data and the rules, they proceed to "mechanically" apply these rules without further attention to the phenomena themselves–and yet the result of this formal manipulation successfully predicts the phenomena. While the use of these formal rules can be purely "mechanical" the rules themselves have a conceptual background; they arise from "ideas" such as those of rates of change and initial conditions for paths in a vector field. These ideas, in turn, have arisen from experience. Moreover, the formal rules and the underlying ideas expand outward into other aspects of Mathematics, as we have noted for tangent bundles in §4 and for characteristics of P.D.E. in §9.

Newton's laws are just the beginning. For many other problems of mechanics, additional formal developments are needed. As we have briefly illustrated, the formalism can get to be quite complicated–and this is even more the case in important aspects of mechanics which we have not treated, such as continuum mechanics or the motions of rigid bodies (for rotations, use Euler's angles!) Also many of the necessary formal developments seem at first just to be tricks–devices such as the solution of partial differential equations by "separation of variables". This apparently ad hoc character is emphasized by the conventional presentations, which hurry to get on with the formal developments and so often issue in mystery or ambiguity. Thus one and the same symbol $\dot{q}$ can represent the time rate of change of a coordinate q, or a coordinate in its own right (on a phase space), or a function of some other quantities or parameters. Indeed, when "parameters" appear it is often not clear whether they are constants, variables, or functions. Moreover, the usual use of a symbol for the partial derivative of one variable with respect to another one often does not specify how the first "variable" is a function and so leaves the reader uncertain as to which other variables are to be held constant. The original authors surely know, but this knowledge is often not well transmitted through a sequence of formulas copied from book to book.

Nevertheless, behind every elaborate formal manipulation there is, we claim, a conceptual background which could serve to provide a better understanding. This we have illustrated in §7 and §8 for the derivation of Hamilton's equations from Lagrange's equations. In this case, the relatively quick formal change of variables has a deeper meaning, expressed in the Legendre transformation from the tangent bundle to the cotangent bundle. The exposition of this meaning, though it takes more space, does provide a more secure understanding as well as a remarkable connection to geometric concepts arising in other parts of Mathematics.

For the development of the Hamilton–Jacobi equation we have followed more the quick formal exposition, with little attention to the underlying ideas. They are there and they are well worth the space they would require. We did note briefly that the use of partial differential equations is suggested by an analogy with optics, where the light rays are also grouped into wave fronts.

We have introduced the phase space coordinates p's and q's without much minding their meaning. This meaning centers on the understanding of the formal term $\Sigma\, p_i\dot{q}^i$ which cropped up in the definition of the Hamiltonian. This term turns out to represent a differential form $\theta = \Sigma\, p_i dq^i$; a form such as this is present on any even-dimensional manifold which, like the phase space, is a cotangent bundle—and one can give a coordinate-free description of this form. For subsequent purposes, it is better to use the exterior derivative $\omega = d\theta = \Sigma\, dp_i dq^i$ of this 1-form θ. This ω is a 2-form (a differential form of degree 2) with a number of useful properties. A manifold with such a form is called a symplectic manifold. Much of Hamiltonian mechanics can be developed most clearly on such a manifold, with a greater freedom in the choice of coordinates and with considerable use of certain Poisson brackets of functions defined on the manifold. This "simplectic" approach frees mechanics from its apparent dependence on particular coordinates p and q for momenta and position and explains a classical process of "canonical transformation" by which one may replace the original p's and q's by new coordinates P and Q which no longer need represent momenta and position, but which may be better suited to the problem at hand. This in turn is connected to the study of differential forms and of Lie groups acting on manifolds. Here a Lie group means a "continuous" group such as the group of all rotations about a point in 3-space, as parametrized, say, by those three Euler angles. Formally, a Lie group G is a set which is both a group and a manifold, in such a way that the group operations (product and inverse) are smooth. These operations are therefore differentiable, and this leads from each Lie group to its associated Lie algebra. This is but one of the many connections of mechanics with "abstract" mathematics; it illustrates the way in which abstraction is tied into application. It is to be regretted that there is no really satisfactory modern exposition of the conceptual development of the ideas of classical mechanics. A splendid traditional presentation is given in L.A. Pars [1965].

13. Quantum Mechanics

The Hamiltonian H and the associated Hamilton–Jacobi equation play an essential role in the formulation of quantum mechanics; we summarize this role, but very briefly. The essential observation is that certain microscopic systems, such as the hydrogen atom, have discrete characteristic energy levels represented by the discrete frequencies at which radiation is emitted. These energy levels are explained as the eigenvalues of suitable energy operators (like H) on an appropriate linear space. Such a space W may be obtained from the classical configuration space C with coordinates say q^1, q^2, q^3 by taking W to be the space of all Lebesgue square-integrable complex valued functions Ψ on C; this W becomes a vector space (over the complex numbers) with an inner product

$$<\phi, \Psi> = \int_C \phi\, \overline{\Psi} dq^1 dq^2 dq^3. \qquad (1)$$

The one-dimensional subspaces of W are called "pure" states of the system, while the observables (such as position or energy) are interpreted as self-adjoint operators on W.

Now consider the Hamiltonian H for the classical configuration space, C, expressed as a function of the (rectangular) coordinates q^i and the corresponding momenta p_i. Next introduce Planck's constant h; then in the function H replace each p_i by the differential operators $(h/2\pi i)\partial/\partial q^i$ and each coordinate q^i by the operator "multiply by q^i". This is to mean that functions of q's are to be replaced by the operator "multiply Ψ by that functions". This process turns the function H into an operator–perhaps not everywhere defined–on the Hilbert space W. The general plan is that the eigenvalues of this operator will be the energy levels of the system.

For example, for an electron of mass m and charge e, subject to the usual inverse square law of attraction by a point at the origin, the classical Hamiltonian has the form

$$H = (1/2m)(p_x^2 + p_y^2 + p_z^2) - e^2(x^2 + y^2 + z^2)^{-1/2}. \qquad (2)$$

By the above convention, this turns into an operator which sends each function Ψ into

$$(h^2/4\pi^2 m)\left[\frac{\partial^2\Psi}{\partial x^2} + \frac{\partial^2\Psi}{\partial y^2} + \frac{\partial^2\Psi}{\partial z^2}\right] - e^2\Psi(x^2 + y^2 + z^2)^{-1/2}. \qquad (3)$$

An eigenvector of this operator is then a function Ψ with an eigenvalue E such that

$$\frac{h^2}{4\pi^2 m}\left[\frac{\partial^2\Psi}{\partial x^2} + \frac{\partial^2\Psi}{\partial y^2} + \frac{\partial^2\Psi}{\partial z^2}\right] - \frac{e^2\Psi}{\sqrt{(x^2+y^2+z^2)}} = E\Psi. \qquad (4)$$

It turns out that these eigenvalues agree with the observed values of the spectrum of hydrogen.

This equation is clearly connected to the classical wave equation; this is one of the reasons that some early versions of quantum mechanics were known as wave mechanics. The relationship is actually deeper, and is subtly connected with the classical wave equations at the origin of the Hamilton–Jacobi theorem. Indeed, if one replaces the function Ψ in (4) by

$$\Psi = e^{iS/h}$$

(for S as in the principal function above), substitutes and takes the limit as $h \to 0$, the result is essentially the Hamilton–Jacobi partial differential equation (try it with just one space coordinate). This hints at the precise sense in which quantum mechanics has classical mechanics as a limit (as $h \to 0$).

For more details on this very summary sketch refer to Mackey [1978].

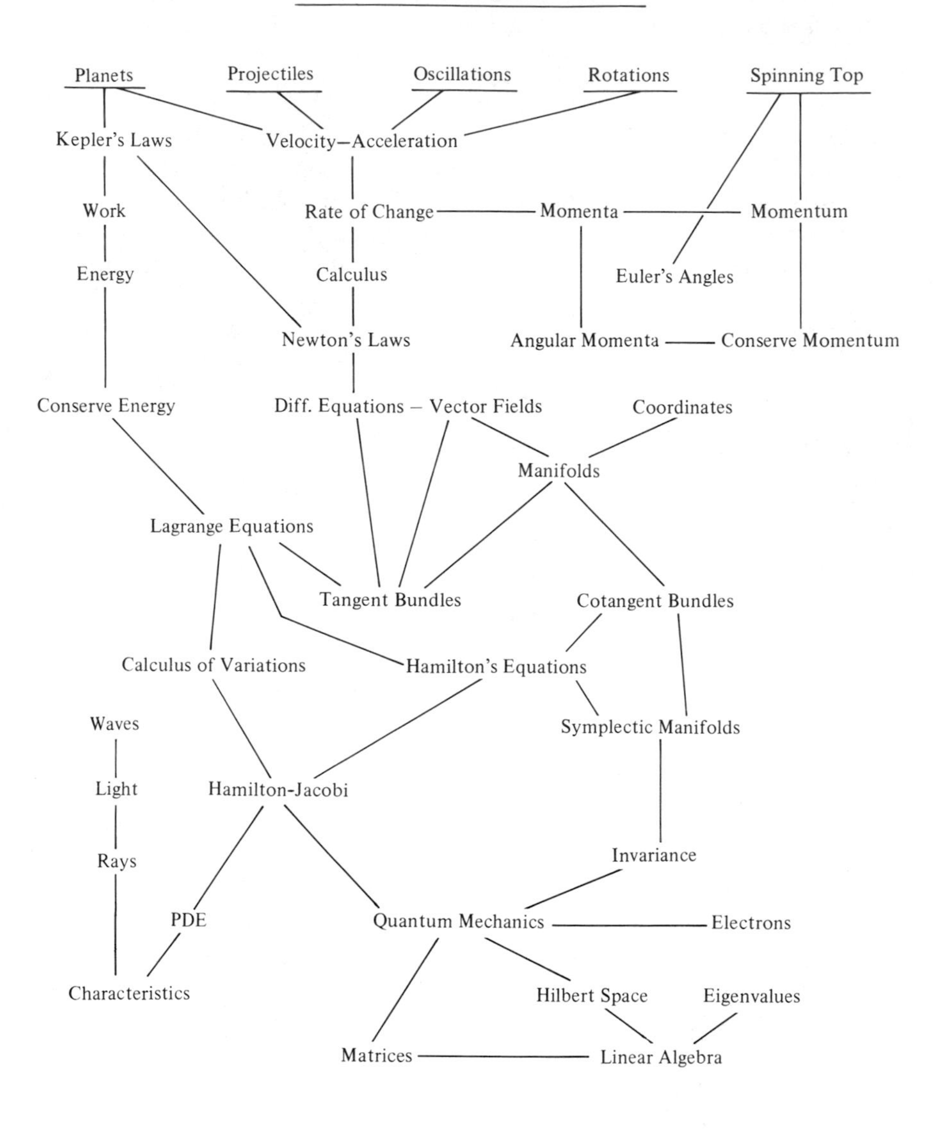
Interconnections Mathematics—Mechanics
Planets
Projectiles
Oscillations
Rotations
Spinning Top
Kepler's Laws
Velocity—Acceleration
Work
Rate of Change
Momenta
Momentum
Energy
Calculus
Euler's Angles
Newton's Laws
Angular Momenta
Conserve Momentum
Conserve Energy
Diff. Equations – Vector Fields
Coordinates
Manifolds
Lagrange Equations
Tangent Bundles
Cotangent Bundles
Calculus of Variations
Hamilton's Equations
Waves
Symplectic Manifolds
Light
Hamilton-Jacobi
Rays
Invariance
PDE
Quantum Mechanics
Electrons
Characteristics
Hilbert Space
Eigenvalues
Matrices
Linear Algebra

CHAPTER X

Complex Analysis and Topology

Since the square of a non-zero real number is always positive, there can be no real square root of -1. Inventing such a square root i and adjoining it to the real numbers, as in §IV.10, leads to extensive and important developments. On the one hand, the resulting complex numbers $x + iy$ represent well the properties of the Euclidean $x-y$ plane and derive part of their "reality" from the geometric reality of the plane. On the other hand, well behaved functions f of such a complex number $z = x + iy$ are those functions f which have a complex derivative, and the properties of these functions are truly remarkable. The resulting study of "complex variables", that is, of differentiable functions of a complex number z, leads to deep mathematical theorems with unexpected practical connections, for example to electrostatic potential and to the steady flow of fluids as well as to aerodynamics. This chapter will introduce these concepts of differentiation and the corresponding integrals and will indicate some of these connections, all with a view to seeing how the apparently simple algebraic device of inventing a "number" i with $i^2 = -1$ has both geometric and analytic consequences–all a striking instance of the remarkable interconnections of formal ideas. Just as many geometric ideas first become apparent in plane geometry (Chapter III), so it is that basic aspects of differentiation and integration are best exemplified by the complex numbers represented in the plane.

1. Functions of a Complex Variable

All the usual elementary functions of real numbers x work equally well (or even better) when x is replaced by a complex number (a "complex variable" z). For example, since the complex numbers form a field, any real polynomial function is also defined for complex arguments z; indeed, for any complex numbers $a_0, \ldots, a_n$ as coefficients one has the polynomial function

$$w = f(z) = a_0 + a_1 z + \cdots + a_n z^n. \tag{1}$$

Geometrically, it is a mapping sending each point of the complex z-plane into a point $w = f(z)$ in the same plane (or sometimes, into the point w in a second complex plane). Just as graphs represent functions of a real variable, so such maps of planes visualize functions of a complex variable z in a geometric way. For example, $w = z + c$ translates the z-plane by the vector c, $w = cz$ with $|c| = 1$ rotates the plane through the angle $\arg c$, and $w = z^2$ sends each point into another with distance from the origin squared and argument $\arg z$ doubled.

Rational functions of z formed from polynomials f and g also yield mappings

$$w = f(z)/g(z) \tag{2}$$

defined for all complex z except for the (finitely many) zeros of the denominator g. For example, if a, b, c, and d are complex constants with $ad - bc \neq 0$, then

$$w = (az + b)/(cz + d) \tag{3}$$

is a *fractional linear transformation* of the z-plane to itself, defined, when $c \neq 0$, at every point except for the zero $z = -d/c$ of the denominator. It is also tempting to say that it sends this point to ∞ and sends ∞ to a/c. (Picture this on a sphere by stereographic projection!)

The exponential function also works for complex arguments. Recall that remarkable real number e and the corresponding function e^x characterized by $e^1 = e$, $e^0 = 1$ and $de^x/dx = e^x$. This function also turns addition into multiplication, according to the formula

$$e^{x+y} = e^x e^y. \tag{4}$$

In fact, this is why the natural logarithm, $\log_e$, defined as the inverse of the exponential function, turns multiplication into addition, as in the use of logarithms for the calculation of products. This formula (4) suggests that a complex exponential should have $e^z = e^{x+iy} = e^x e^{iy}$. Since we already know the real exponential e^x, this leads us to find the imaginary exponential e^{iy}; in other words, to find a complex-valued function of a real argument y which turns addition in y into multiplication. But this is precisely what the two addition formulas for $\cos y$ and $\sin y$ will yield, as in

$$(\cos y + i \sin y)(\cos u + i \sin u) = \cos(y+u) + i \sin(y+u).$$

Therefore we are led to define the complex exponential by

$$e^{x+iy} = e^x(\cos y + i \sin y). \tag{5}$$

This has the desired property that $e^z e^w = e^{z+w}$; it turns out to have all the other desirable properties; in particular, e^z is its own (complex) derivative.

By this definition, e^z is never zero. (How could it be; were $e^z = 0$, then $e^{z+w} = 0e^w = 0$ would be always zero!). Also $z \mapsto e^z$ maps the infinite horizontal strip $\{y \mid 0 \leqslant y < 2\pi\}$ of width 2π in the z-plane onto the whole of the w-plane, omitting only the point $w = 0$. Each horizontal line ($y = $ const.) becomes a ray ($\theta = $ const.) from the origin in the w-plane; each vertical interval ($x = $ const.) becomes a circle. All told, each point $w \neq 0$ is covered infinitely often (once from each horizontal strip of width 2π) by points from the z-plane. Therefore the logarithm function, defined as the inverse to the exponential, must be "many-valued". Explicitly, write $w \neq 0$ in the polar form $w = s(\cos \varphi + i \sin \varphi)$ and observe that the positive real number s has a real logarithm $\log_e s$, while $\cos \varphi$ and $\sin \varphi$ have period 2π. Therefore, for any integer k the definition (5) gives

$$e^{\log_e s + i(\varphi + 2k\pi)} = s(\cos \varphi + i \sin \varphi) = w.$$

In other words, for each choice of k and for $w \neq 0$ the (single valued) function of w,

$$\log_e w = \log_e s + i(\varphi + 2k\pi), \qquad 0 \leqslant \varphi < 2\pi \tag{6}$$

has $e^{\log_e w} = w$, so is an inverse of the exponential. It is called (in an old-fashioned terminology) a *branch* of the "many-valued" function $\log_e$. Under our formal definition of a function, there is of course no such thing as a "many-valued" function. However, with Riemann surfaces one can (geometrically!) paste all these branches together, much as in §VIII.4, so as to get an honest single-valued function defined not on the plane but on a Riemann surface.

The definition (5) of the exponential allows us to write the polar form of any complex number $z \neq 0$ as

$$z = re^{i\theta}, \qquad r = |z|, \qquad \theta = \arg z. \tag{7}$$

This formula includes the remarkable special case $1 = e^{2\pi i}$ —a formula which is really just a consequence of the fact that $\cos \theta$ and $\sin \theta$ have period 2π! Using the logarithm, one can define a "many-valued" exponential to any complex base $c \neq 0$ as $c^z = e^{z \log_e c}$.

Next, how can one define $\sin z$ and $\cos z$ for complex z?

Start by noting that the definition (5) of the complex exponential for $x = 0$ gives

$$e^{i\theta} = \cos \theta + i \sin \theta, \qquad e^{-i\theta} = \cos \theta - i \sin \theta \tag{8}$$

and hence

$$\cos\theta = (e^{i\theta} + e^{-i\theta})/2, \qquad \sin\theta = (e^{i\theta} - e^{-i\theta})/2i.$$

But here the right-hand sides have meaning when $\theta = z$ is taken to be complex; this result we adopt as the definition of $\cos z$ and $\sin z$, so that

$$\cos z = (e^{iz} + e^{-iz})/2, \qquad \sin z = (e^{iz} - e^{-iz})/2i. \tag{9}$$

By their derivation, the functions so defined agree with the usual ones when z is real. Moreover, one may verify that $\cos z$ and $\sin z$ so constructed have all the basic properties of the real cosine and sine functions (period 2π, addition formulas, $\cos^2 + \sin^2 = 1$, etc). We shall presently see, using the Taylor series, that these are the only definitions which extend the real functions $\sin x$ and $\cos x$ to "good" functions of a complex argument.

It remains to explicate which functions of a complex variable are "good"–and to show that such "good" functions have convergent Taylor series. Indeed, the familiar Maclaurin series for e^x

$$e^x = 1 + x + x^2/2 + x^3/3! + \cdots$$

will also work when x is a complex number z. The new definition fits!

For the moment, we do have the remarkable observation that the geometric representation of complex numbers *and* the basic facts of trigonometry combine to replace the real variable x by a complex variable in the functions e^x, $\cos x$, $\sin x$, and $\log x$. Analysis, geometry, and trigonometry have intertwined in a fashion which becomes clear only when we use complex numbers.

There are many other interesting functions of complex z. For example, the infinite series $\Sigma\, 1/n$ is known to diverge. However, if $s = \sigma + it$ is a complex number with real part $\sigma > 1$, then the related series

$$\zeta(s) = \sum_{n=1}^{\infty} \frac{1}{n^s}, \qquad n^s = e^{s \log_e n}, \qquad \log_e n \text{ real} \tag{10}$$

does converge, and so defines a function in the half-plane $\sigma > 1$. It is the Riemann zeta function, famous for its use in number theory.

2. Pathological Functions

Formally, a function f on the reals is defined (Chapter V) to be any set F of ordered pairs (x,y) of real numbers containing exactly one pair (x,y) for each real x. This definition does provide a precise way of saying that "y depends on x" without using any vague notion of "dependence". However, it allows for many more functions than might have been intended–including some which are truly bizarre. For example, the definition

$$f(x) = 0 \qquad \text{when } x \text{ is rational,}$$
$$\quad = 1 \qquad \text{when } x \text{ is irrational}$$

does produce a legal function f. But this function f jumps incessantly from value 0 to value 1 and back. One cannot "draw" a graph for it; it is nowhere continuous. Next, the definition

$$g(x) = x \sin 2\pi/x \qquad \text{when } x \neq 0,$$
$$\quad = 0 \qquad \text{when } x = 0$$

also yields a function g which oscillates faster and faster as x approaches zero. (Draw a graph!) This function is continuous everywhere, but has no derivative at $x = 0$. A more elaborate construction yields the function

$$h(x) = \sum_{n=1}^{\infty} 4^{-n} \sin 110^n x$$

which has no first derivative at any point of **R**. Such nowhere differentiable continuous functions do seem to violate natural intuitions!

In the calculus, Taylor's formula (§VI.8) suggests that "good" functions ought all to be expressible by power series. This does work for e^x, $\sin x$ and many other familiar functions—but then there are functions like

$$f(0) = 0, \qquad f(x) = e^{-1/x^2}, \qquad x \neq 0$$

given by an explicit formula which have *all* derivatives zero at the origin—and so cannot have a Taylor series expansion in powers of x (Wilson [1911], p. 66 #9, p. 438 #7). There are other good functions like $1/(1 - x)$ defined everywhere except at $x = 1$ but with an evident power series (long division!)

$$1/(1-x) = 1 + x + x^2 + x^3 + \cdots \tag{1}$$

which converges only when $|x| < 1$.

There are other more geometrical pathologies. For example, we have defined a (parametrized) "curve" in the plane as a continuous function $I \to \mathbf{R} \times \mathbf{R}$ defined (say) on the unit interval I. Intuitively, we think of a curve as something "one-dimensional"—but Peano succeeded in constructing a curve in the sense of this definition which covers *every* point in a square—and which thus is hardly one-dimensional.

This example indicates that all these "correct" formal definitions which make calculus work rigorously and well also reveal all manner of pathological phenomena. In general, a formal definition of an intuitive idea such as the idea of "continuity" may easily lead to counter-intuitive results. The formal is only imperfectly linked to the intuitive!

Now the familiar formal definitions of continuity and derivative apply directly to functions of a complex variable z. Indeed, the complex plane is evidently a metric space; the familiar Euclidean distance between two points z and z' is given (in complex terms) as the absolute value $|z - z'|$. As for any metric space, the complex plane is then also a topological space in which the open sets are arbitrary unions of open discs $D(c,\epsilon)$. Here $D(c,\epsilon)$ denote the *disc* with complex center c and positive real radius $\epsilon > 0$; it consists of all complex numbers z with $|z - c| < \epsilon$. With this topology (or, equivalently, with this metric) the standard definitions describe which functions $w = f(z)$ of a complex variable z are continuous (and where). It then turns out that e^z, $\sin z$, $\cos z$ are all continuous, and so is each branch of $\log_e z$ except when the angle ϕ of (1.6) is 0. In the same way, a straightforward use of the standard definition of the first derivative will apply to functions $w = f(z)$ of a complex z. But here the remarkable fact is that those functions $f(z)$ which do have a first derivative in the standard sense avoid almost all of the pathologies we have mentioned! The use of complex derivatives, as we will see in this chapter, has extensive consequences.

3. Complex Derivatives

As in the calculus, the derivative of a function $w = f(z)$ of a complex variable z at a point z_0 should be the limit

$$f'(z_0) = \lim_{h \to 0} \frac{f(z_0 + h) - f(z_0)}{h} \tag{1}$$

as the complex number h approaches 0. For this definition to make sense at z_0 both $f(z_0)$ and $f(z_0 + h)$ must be defined, the latter for all sufficiently small (complex) values of the increment h. This means that the set where the function f is defined must include with each point z_0 an open disc about z_0; this set is thus an open set in the topology of the plane. Hence we define: A function $f \colon U \to \mathbf{C}$ defined in an open set U of the complex plane is *holomorphic* in U (one also says *regular analytic* in U) if and only if the limit (1), for h complex, exists at each point z_0 in U. The value of this limit is then the *complex derivative* f' of f in U. Here "limit" has its usual formal definition in terms of absolute values (i.e., in terms of the metric of the plane): For each real $\epsilon > 0$ there is a real $\delta > 0$ such that for all complex h with $|h| < \delta$ on has

$$\left| \frac{f(z_0 + h) - f(z_0)}{h} - f'(z_0) \right| < \epsilon. \tag{2}$$

The idea behind this definition is the familiar one: the derivative $f'(z_0)$

measures the instantaneous "rate of change" of f for a change h in the complex variable z. But there is more to the definition than first meets the eye, because that change h in z_0, with $|h| < \delta$, may be real, purely imaginary, or a combination of these. In order to see what is going on, one may write $z = x + iy$ and then $w = f(z)$ in terms of its real and imaginary parts as

$$f(z) = u(x,y) + i\, v(x,y), \tag{3}$$

where u and v are now two real-valued functions defined for points (x,y) in the given open set U. If now we first take the change h in z to be real, the existence of the limit (1) means that the expression

$$\frac{u(x_0+h,y_0) - u(x_0,y_0)}{h} + i\frac{v(x_0+h,y_0) - v(x_0,y_0)}{h}$$

must have a limit as the real h approaches 0. By the definition of the real *partial* derivatives of the functions u and v, this limit must be

$$\frac{\partial u}{\partial x}(x_0,y_0) + i\frac{\partial v}{\partial x}(x_0,y_0); \tag{4}$$

in particular, both these partial derivatives must exist.

On the other hand, if we take the change h in z to be purely imaginary, say as $h = ik$, and if we use the fact that $i^{-1} = -i$, we get for $f'(z_0)$ the limit

$$-i\frac{\partial u}{\partial y}(x_0,y_0) + \frac{\partial v}{\partial y}(x_0,y_0). \tag{5}$$

Since the desired complex derivative is given by *either* of the expressions (4) or (5), these two expressions must agree. In other words, when $f = u + iv$ is holomorphic in an open set U, the real and imaginary parts of f must both have partial derivatives in U which satisfy the equations

$$\frac{\partial u}{\partial x} = \frac{\partial v}{\partial y}, \qquad \frac{\partial u}{\partial y} = -\frac{\partial v}{\partial x}, \qquad (x,y)\in U. \tag{6}$$

These are the *Cauchy–Riemann equations*. They are necessary if $f = u + iv$ is to be holomorphic in U.

This can be put differently. If we simply take two well-defined real valued functions $u(x,y)$ and $v(x,y)$ with partial derivatives, the combination $u(x,y) + iv(x,y)$ *is* a function of the complex number z, but it is not likely to be a holomorphic function. It will be so only if the partial derivatives of u and v satisfy these Cauchy–Riemann equations. But one may verify that these equations do hold for the exponential function

$e^z = e^x(\cos y + i \sin y)$ and also for the functions $\sin z$ and $\cos z$ as we have defined them. Hence these functions are indeed holomorphic (in the whole z-plane).

These equations have further consequences, both in physics and in geometry. It will presently appear that the existence of the first derivative $f'(z)$ for a holomorphic function $f(z)$ necessarily implies the existence of all higher complex derivatives $f^{(n)}(z)$ as well. This in turn means that the real and imaginary parts u and v will have continuous partial derivatives of all orders. Hence also $\partial^2 v/\partial x \partial y = \partial^2 v/\partial y \partial x$ and so, differentiating both sides of the Cauchy–Riemann equations (6), one has

$$\frac{\partial^2 u}{\partial x^2} + \frac{\partial^2 u}{\partial y^2} = 0, \qquad \frac{\partial^2 v}{\partial x^2} + \frac{\partial^2 v}{\partial y^2} = 0 \tag{7}$$

everywhere in U. In other words, both u and v satisfy Laplace's equation $\Delta u = 0$, where $\Delta = \partial^2/\partial x^2 + \partial^2/\partial y^2$ is the Laplace operator in the plane. This equation is a two-dimensional version of the Laplace P.D.E. (3) of §VI.11. Generally, a twice-differentiable function $u(x,y)$ with $\Delta u = 0$ is called a *harmonic* function (cf. §VI.11).

Such functions arise in theoretical physics as expressions of both gravitational and electromagnetic potentials. Consider for example the electrostatic field arising from electric charges distributed uniformly along one or several very long cylinders perpendicular to the (x,y)-plane. The resulting potential u is then effectively constant in the coordinate perpendicular to this plane, so it can be considered just as a function of x and y, which then must satisfy the Laplace equation (7). In other words, every holomorphic function f yields such a potential u as its real part. The level curves $u(x,y) =$ constant are then the "equipotential" curves, while the imaginary part v of the holomorphic function f gives curves $v(x,y) =$ constant which represent the "lines of force". At all points $z = x + iy$ where $f'(z) \neq 0$ these lines of force are orthogonal to the equipotentials. For a given electrostatic potential of this type, finding the right potential function means finding a holomorphic function of z for which the real part $u(x,y)$ has the desired boundary values.

Next consider the geometry, by considering a function $w = f(z)$ holomorphic in an open set U as a mapping of the set U by $z \mapsto f(z)$ into the w-plane. Since f, as noted above, has all higher derivatives, this will be a smooth mapping and so must carry smooth curves and their tangents in U into smooth curves and their tangents in the w-plane. We assert that this mapping, at any point $z_0 \in U$ where $f'(z_0) \neq 0$, is *conformal* in the sense that it preserves angles between tangents to curves at z_0. For, consider some smooth curve in the z-plane given in terms of a parameter t as $z = g(t) = x(t) + iy(t)$ and passing through the point $z_0 \in U$ when $t = t_0$, so that $z_0 = g(t_0)$. The first derivative has $g'(t) = x'(t) + iy'(t)$, so when $g'(t_0) \neq 0$ the tangent line to the curve at z_0

makes the angle $\arg(g'(t_0))$ with the x-axis. The image curve in the w-plane is given by the composite function $w = h(t) = f(g(t))$. By the Chain Rule for the differentiation of composite functions (which can be seen to hold here)

$$h'(t) = f'(z)g'(t).$$

Thus if both $f'(z_0) \neq 0$ and $g'(t_0) \neq 0$, the arguments are both defined and the argument of the product $f'g'$ is the sum

$$\arg h'(t_0) = \arg f'(z_0) + \arg g'(t_0).$$

In other words, under the mapping f the tangent line to the curve has been rotated counterclockwise by the angle $\arg f'(z_0)$. Hence, given two parametrized curves meeting at z_0, each tangent line is rotated by the same angle $\arg f'(z_0)$. Therefore the angle between the curves is preserved, so the map $z \mapsto f(z)$ is conformal, as claimed.

However, if $f'(z_0) = 0$ then this deduction gives $h'(t_0) = 0$, so the first derivative f' is not enough to determine the slope of the tangent line to the image curve. At such a point z_0 the mapping $z \mapsto f(z)$ need not be a conformal one. For example, the map $w = z^2$ doubles all angles between curves meeting at the origin!

One may also show that the stereographic projection from the (Riemann) sphere to the complex plane is conformal, so that the geometric viewpoint also works well on the sphere.

The preservation of angles at points where $f'(z_0) \neq 0$ should "really" be viewed as a consequence of the Cauchy–Riemann equations. Indeed the mapping $z \mapsto f(z)$, when written in real coordinates as $(x,y) \mapsto (u(x,y),v(x,y))$, induces at each point (x,y) of the plane a linear mapping of the tangent space with matrix the Jacobian

$$\begin{bmatrix} \dfrac{\partial u}{\partial x} & \dfrac{\partial u}{\partial y} \\ \dfrac{\partial v}{\partial x} & \dfrac{\partial v}{\partial y} \end{bmatrix}.$$

The Cauchy–Riemann equations imply that the columns of this matrix are orthogonal–or, equivalently, that the rows are orthogonal. The matrix itself is thus just an orthogonal matrix times a real scalar, the square root of its determinant, which by Cauchy–Riemann again is

$$\frac{\partial u}{\partial x}\frac{\partial v}{\partial y} - \frac{\partial u}{\partial y}\frac{\partial v}{\partial x} = \left[\frac{\partial u}{\partial x}\right]^2 + \left[\frac{\partial u}{\partial y}\right]^2 = |f'(z_0)|^2.$$

In other words, the linear transformation induced by f on the tangent spaces preserves angles and multiples lengths by $|f'(z_0)|$.

This discussion indicates that the use of the idea of tangent spaces, coming from differential geometry, is a helpful adjunct to complex variable theory–even though many texts on complex variables do not make this connection.

Holomorphic functions also apply to fluid flow. Consider the flow of a fluid in the plane (i.e., a "laminar" flow) which is steady in the sense that the fluid velocity at each point (x,y) is independent of time. Write the components of this velocity as $P(x,y)$ and $Q(x,y)$. The component of the velocity along the "infinitesimal" vector (dx,dy) is just the inner product $Pdx + Qdy$. Hence the line integral

$$\int_C P(x,y)dx + Q(x,y)dy$$

around a closed curve C represents a physical quantity called the "rotation" of the fluid around that curve. By the Gauss lemma (VI.10.3) this equals the integral of $\dfrac{\partial Q}{\partial x} - \dfrac{\partial P}{\partial y}$ over the interior of C. We assume that this integral is zero for all curves C, meaning that the flow is *irrotational* (intuitively, there are no vortices). This assumption implies that $\partial Q/\partial x = \partial P/\partial y$. Hence, under suitable continuity assumptions, there is a function $u(x,y)$ with the partial derivatives

$$\frac{\partial u}{\partial x} = P(x,y), \qquad \frac{\partial u}{\partial y} = Q(x,y). \tag{8}$$

Also, as in §VI.10, the flow across a curve L is just the line integral of $Qdx - Pdy$ along L. In particular, the total flow across a closed curve C in the plane is

$$\int_C Q(x,y)dx - P(x,y)dy.$$

To assume that the fluid is *incompressible* is to assume that this integral is zero for all smooth closed curves C, and hence that $\partial Q/\partial y = -\partial P/\partial x$. In terms of the function u of (8) above, this states that $\dfrac{\partial^2 u}{\partial y^2} + \dfrac{\partial^2 u}{\partial x^2} = 0$, so that u is again a harmonic function. It might thus be regarded as the real part of a holomorphic function $w = f(z)$. In this way holomorphic functions represent the steady plane irrotational flows of incompressible fluids.

All told, we have thus some remarkable connections, all starting from the requirement that a function $z \mapsto f(z)$ of a complex variable z have a complex derivative. This requirement leads to the Cauchy–Riemann equations which describe maps which are conformal and which in turn lead to harmonic functions, as they appear both in electrostatic potential and in fluid flow. The deeper ideas of Mathematics do have varied connections!

So much follows from the easy derivation of the Cauchy–Riemann equations that one might be tempted to try to get even more from the assumption that $z \mapsto f(z)$ has a complex derivative—for example, more by differentiating along some slant lines. There is no more. A theorem of Looman and Menchoff states that if two real valued functions $u(x,y)$ and $v(x,y)$ are defined and continuous for (x,y) in an open set U and have partial derivatives which satisfy the Cauchy–Riemann equations (6) everywhere in U, then the function $f(x) = u + iy$ is indeed holomorphic in U.

All this is just the beginning of the remarkable properties of holomorphic functions f. The definition requires just the existence of a first derivative, but this will imply the existence (and continuity) of derivatives of all orders, as well as the presence of suitably convergent Taylor series. In other words, holomorphic functions of a complex variable escape all the pathologies attending ordinary functions f of a real variable. The proof of these remarkable results depends essentially upon the use of integration (a process already strongly suggested by the applications we have noted). The integral of $f(z)$ around a closed path in the complex plane is zero, provided that f is holomorphic in an open set containing the path *and* its "interior". This Cauchy integral theorem plays a central role in complex analysis. To formulate it, we next consider integration and then properties of the paths of integration (§5).

4. Complex Integration

The complex integral

$$\int_h f(z)dz = \int_h f\,dz \tag{1}$$

of a holomorphic function f along some path h in the complex plane should be, as the notation suggests, the infinite sum of values $f(z)$ of f at points z on the path multiplied by "infinitesimal" increments $dz = z' - z$ of z along the path. This idea is turned into a precise definition of the integral as the limit of a suitable finite sum. Let f be holomorphic in an open set U, while $h: I \to U$ is a path in U, parametrized by t, $0 \leqslant t \leqslant 1$ and smooth (i.e., $h'(t)$ is continuous). So divide the unit interval I at $n+1$ points $0 = t_0 < \cdots < t_n = 1$. Note that the increment in $z = h(t)$

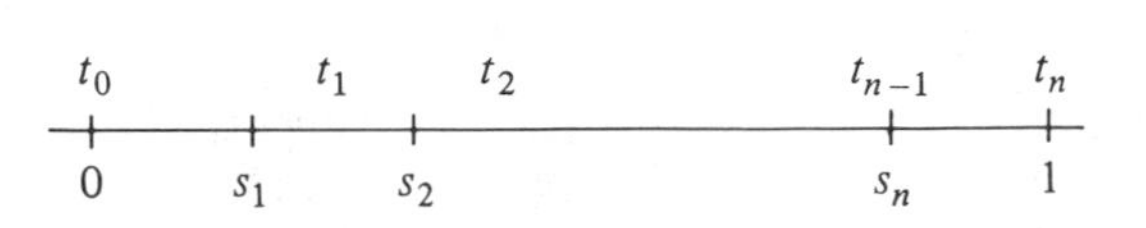

from t_{i-1} to t_i is $h(t_i) - h(t_{i-1})$, choose a point s_i in each interval and evaluate the holomorphic function f at that point. This produces the sum

$$\sum_{i=1}^{n} f(h(s_i))(h(t_i) - h(t_{i-1})). \tag{2}$$

The integral (1) is then defined to be the limit of this sum as n approaches infinity and as the maximum of the $t_i - t_{i-1}$ approaches zero. This limit (and hence this integral) exists if h has a continuous first derivative $h'(t)$ for each t in I (where the derivative at the endpoints of I is to be defined by a one-sided limit). The choice of the parameter t for the path is not important; all that really matters is that the path is smooth and is "directed". This can be formulated by proving that the integral of f is independent of the choice of parameter. A new choice would consist in taking a smooth function $k:I \to I$ which is monotonic increasing ($t < t'$ implies $k(t) < k(t')$) and has $k(0) = 0$, $k(1) = 1$; then the integral of f over the path h is identical to that over the path $h \cdot k:I \to U$. (One must here use the fact that k, continuous on the compact set I, is uniformly continuous.)

As we will soon see, it is also convenient to be able to integrate over the path given by the boundary of a rectangle and over other paths with corners. Such paths are said to be *piecewise differentiable*—they are described exactly as the paths obtained by piecing together ("composing") a finite number of differentiable (i.e., smooth) paths. For such paths h the limit above and hence the integral exists.

Except for the condition that the function $f(z)$ be holomorphic, this integral is just the line integral introduced in §VI.10 and motivated there by various physical applications (for example, to work). Explicitly, if we write $z = x + iy$, $dz = dx + idy$, and $f(z) = u(x,y) + iv(x,y)$ in terms of real and imaginary parts, the integral just defined *is* the line integral

$$\int_h f(z)\, dz = \int_h [u(x,y)dx - v(x,y)dy] + i \int_h [v(x,y)dx + u(x,y)dy]. \tag{3}$$

Going further, one can substitute the function $h(t)$ describing the path to reduce this integral to ordinary real integrals in t, from 0 to 1. All these formulas express the age old idea: The whole is the sum of its parts!

The size of the parts yields an evident upper bound for the size of the integral. Specifically, the function $t \mapsto |f(h(t))|$ is continuous on a compact set (to wit, the interval I), hence has a maximum value M there. On the other hand, the path of integration has a length L_h, which can be defined geometrically as the limit of the sum of the lengths of small inscribed chords, or equivalently as the integral $\int_0^1 |h'(t)|\, dt$. Then, looking at the sums (2) used to define the integral (1), one may prove, much as in (VI. 4.3), that

$$\left| \int_h f(z)dz \right| \leqslant ML_h. \tag{4}$$

This may seem to be just a rough upper estimate of size, but it turns out to be exceedingly handy. We will see one example of its use (in §6) but there are many more throughout analysis, which often requires judicious estimates of the size of all sorts of integrals.

The integral also has useful algebraic properties. It is an additive operation on holomorphic functions, in the sense that

$$\int_h (f(z) + g(z))dz = \int_h f(z)dz + \int g(z)dz, \tag{5}$$

it is also additive on paths: If one forms a "composite" path $h_2 \cdot h_1$—first the path h_1, then followed by the path h_2, one has

$$\int_{h_2 \cdot h_1} f(z)dz = \int_{h_2} f(z)dz + \int_{h_1} f(z)dz. \tag{6}$$

The direction of a path matters; if one integrates backwards over a path h (literally, using the path h^{-1} given by $h^{-1}(t) = h(1 - t)$) one changes the signs of all of the terms $h(t_i) - h(t_{i-1})$ in the sum (2) and hence the sign of the whole integral

$$\int_{h^{-1}} f(z)dz = \quad \int_h f(z)dz. \tag{7}$$

Since the complex integral can be reduced to a real integral, we can also use the fundamental theorem of the integral calculus. Thus, for example, if h is a path from the point $c_0 = h(0)$ to the point $c_1 - h(1)$, while $f(z) = z^3$ then

$$\int_h z^3 dz = z^4 \Big]_{c_0}^{c_1} = c_1^4 - c_0^4.$$

In particular, if h is a closed path ($h(0) = h(1)$) the result is zero: The integral of a polynomial in z around a closed piecewise smooth path is zero. That need not be true for other functions such as $w = 1/z$. In this case the integral around a circle, center at the origin with $z = re^{i\theta}$, is

$$\int \frac{dz}{z} = \int \frac{d(re^{i\theta})}{re^{i\theta}} = \int i\,d\theta = 2\pi i. \tag{8}$$

This is surely not zero. Here $dz/z = d(\log_e z)$, while the logarithm, as we saw, is not a single-valued function. Indeed, if we follow a branch of $\log_e z$ continuously around that (counterclockwise) circle, it changes to the "next" branch by exactly $2\pi i$. Moreover, $f(z) = 1/z$ is not defined at $z = 0$; so is *not* a holomorphic function inside the circle.

These calculations and many like them, suggest that the integral of a holomorphic function f around a closed path h *ought* to be zero when

$h = \partial A$ is the boundary of some region A in which f is holomorphic, as in Figure 1. In this case the Gauss lemma of §VI.10 states that a line integral such as $\int \omega$, for ω a differential, may be replaced by a double integral

$$\int_{\partial A} P\,dx + Q\,dy = \iint_A \left[\frac{\partial Q}{\partial x} - \frac{\partial P}{\partial y} \right] dxdy. \tag{9}$$

In particular, the line integrals appearing in the formula (3) for $\int f dz$ become double integrals

$$\begin{aligned} \int_h (u\,dx - v\,dy) &= -\iint_A \left[\frac{\partial v}{\partial x} + \frac{\partial u}{\partial y} \right] dxdy, \\ \int_h (v\,dx + u\,dy) &= \iint_A \left[\frac{\partial u}{\partial x} - \frac{\partial v}{\partial y} \right] dxdy. \end{aligned} \tag{10}$$

If the whole area A is contained in the open set U where f is holomorphic, the Cauchy–Riemann equations state that both the integrands on the right are zero, and hence so is $\int f\,dz$. This gives a first form of Cauchy's integral theorem:

Theorem. *If $w = f(z)$ is holomorphic with a continuous first derivative $f'(z)$ in an open set U containing a closed piecewise differentiable path h and all of its "interior" area A, then the integral of $f(z)$ around h vanishes*:

$$\int_h f(z)dz = 0. \tag{11}$$

This can also be viewed as a statement that the integral is "independent of path": If we consider two smooth paths h_1 and h_2 both joining a point c to a point d, and if h_1 can be "deformed" into h_2 through an area within which f is holomorphic, as in Figure 2, then

$$\int_{h_1} f(z)dz = \int_{h_2} f(z)dz. \tag{12}$$

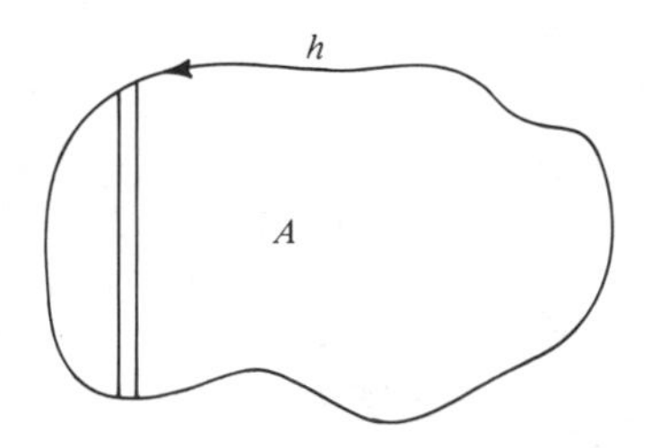

Figure 1

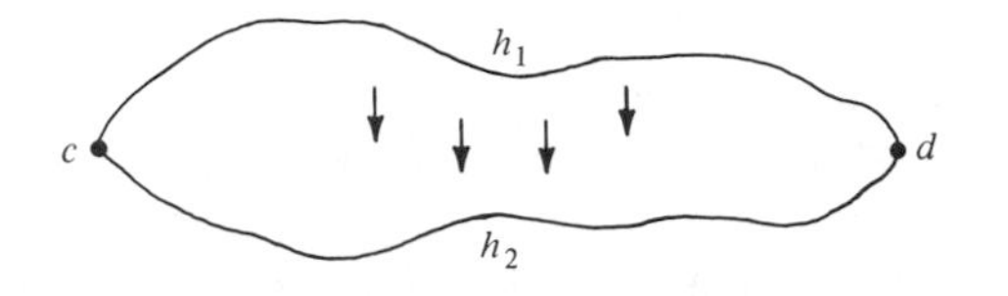

Figure 2. Deformed path.

This form of Cauchy's theorem is not very satisfactory, on several counts. First, the Gauss lemma involving those double integrals normally assumes that the partial derivatives $\partial Q/\partial x$ and $\partial P/\partial y$ are continuous everywhere in the area A; for this reason our statement of the theorem assumes not just that f is holomorphic, but also that its first derivative f' is continuous. By other methods, as we will see, it is possible to avoid this added hypothesis on f. Next, the standard proof of the Gauss lemma (it was merely indicated in (VI.10.5)) involves slicing the area A vertically and perhaps horizontally; in case the contour h is at all convoluted, some fussy adjustments are needed to make this slicing process really work—things may not be as easy as Figure 1 may suggest. Finally, the theorem describes A as the "interior" of the contour h. Now a smooth oval curve, such as that of Figure (1), has an easily recognized interior, but this might not be so with a highly convoluted curve (Figure 3) such as this outline of a maze. More formally, a path $h{:}I \to P$ is *closed* if $h(0) = h(1)$ and *simple* if it does not cross itself; that is, if $t_1 \neq t_2$ implies $h(t_1) \neq h(t_2)$ except for $h(0) = h(1)$; then the image set $h(I)$ is called a *simple closed curve* in the plane $\mathbf{C}$. The *Jordan curve theorem* asserts that every simple closed curve divides the plane into two parts: The plane $\mathbf{C}$ with $h(I)$ removed is not connected, but is the union of two connected sets—an "inside" and an "outside". This theorem is true but is by no means easy to prove. Even for a polygonal curve it is hard. In the simplest such case, when the curve is the boundary of a triangle, the Jordan curve theorem rests on geometric facts such as Pasch's axiom (§III.2). The search for higher dimensional analogs of the Jordan curve theorem has been a major motive in algebraic topology.

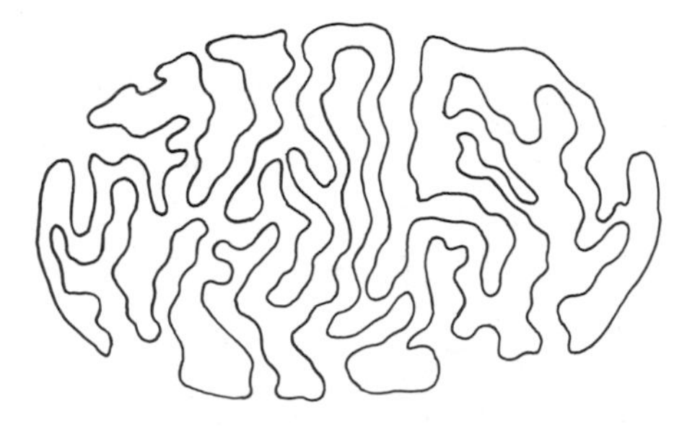

Figure 3. Inside or outside?

To summarize: Cauchy's theorem that an integral $\int f dz$ of a holomorphic function is independent of path is a direct extension of the fundamental theorem of the integral calculus, and so is another representation of the idea that the whole is the sum of its parts. The proof of the theorem by the Gauss lemma is an illuminating indication of the reasons behind this theorem, but in any generality it runs into a morass of difficult geometric and topological considerations. This is one of the reasons that complex analysis leads directly to general questions about the topology of the plane–and of other spaces. Nevertheless, we will soon show how a different approach will give a more direct proof of the Cauchy theorem.

5. Paths in the Plane

The properties of complex integration just presented directly involve several operations on paths in the plane: The composition of two paths (one followed by another), the inverse of a path (the path run backwards), and the deformation of a path, as in the independence of an integral from the choice of path in the Cauchy integral theorem. This situation has made it gradually clear that it is useful to consider these operations on paths in their own right, both for the study of higher-dimensional integrals and for the analysis of the connectivity of a space (such as a Riemann surface). For this latter purpose, it is useful to consider all continuous paths, not just the smooth or piecewise differentiable ones.

A *path* in the subset S of the complex plane $\mathbf{C}$ is thus a continuous map $h: I \to S$ of the real unit interval $I = \{t \mid 0 \leqslant t \leqslant 1\}$ into S. Its *starting* point is the point $h(0)$, it *ends* at $h(1)$. If k is a second path starting at $h(1)$, then $k \cdot h$ denotes the *composite path*: First follow h; then follow k; it may be described formally as the path $k \cdot h: I \to \mathbf{C}$ specified by the equations

$$\begin{aligned} (k \cdot h)(t) &= h(2t), \qquad 0 \leqslant t \leqslant 1/2, \\ &= k(2t-1), \qquad 1/2 \leqslant t \leqslant 1. \end{aligned} \tag{1}$$

The resulting function $k \cdot h$ is indeed continuous on I because $h(2(1/2)) = h(1) = k(0) = k(2(1/2) - 1)$; in other words, the two paths fit together in the middle. As in (4.6), a line integral of a differentiable form ω along the composite path $k \cdot h$ will be the sum of the separate integrals along k and along h.

The *inverse* path h^{-1} (go backwards along h) is given by the equation

$$h^{-1}(t) = h(1 - t), \qquad 0 \leqslant t \leqslant 1. \tag{2}$$

This is continuous, but it isn't quite an inverse for the composite defined in (1), because the composite $h^{-1} \cdot h$ is a path that goes over from $h(0)$ to

$h(1)$ and back again at the same speed; it is not the identity path, which ought to be the path staying put at $h(0)$. At best, one can deform $h^{-1}\cdot h$ into this identity. Also the composite $k\cdot h$ is not really associative; if m is a third path starting at the end $k(1)$ of k, then the composite $m\cdot(k\cdot h)$ follows along m for $1/2 \leqslant t \leqslant 1$, while the associated composite $(m\cdot k)\cdot h$ follows m only for parameter values $3/4 \leqslant t \leqslant 1$.

What is needed is a process of deforming $m\cdot(k\cdot h)$ to $(m\cdot k)\cdot h$. This is provided by the formal concept of a homotopy. If $h_0 h_1 : I \to S$ are two paths in the subset $S \subset \mathbf{C}$ with the same endpoints $h_0(0) = h_1(0)$, $h_0(1) = h_1(1)$, then a *homotopy* holding endpoints fixed (a continuous *deformation*) from h_0 to h_1 is a continuous map $H: I \times I \to S$ with

$$H(t,0) = h_0(t), \qquad 0 \leqslant t \leqslant 1, \tag{3}$$

$$H(t,1) = h_1(t), \qquad 0 \leqslant t \leqslant 1, \tag{4}$$

$$H(0,s) = h_0(0),\ H(1,s) = h_1(1), \qquad 0 \leqslant s \leqslant 1. \tag{5}$$

In other words, this homotopy H maps the unit square with coordinates s and t into S so as to give a "continuous family" of paths h_s, each starting at $h_0(0)$ and ending at $h_0(1)$. The paths are parameterized by s, starting from $s = 0$ with the first path h_0 and ending for $s = 1$ with h_1, as in Figure 1. This figure pictures the deformation of the initial path h_0 into the final one, h_1. For example, one may deform the closed path given by the circle of radius 1 about the origin into another closed path given by a circle of radius 2 by using the homotopy $H(s,t) = (1 + s)e^{2\pi i t}$, as in Figure 2.

A homotopy from h_0 to h_1 and a subsequent homotopy from h_1 to h_2 can clearly be "composed" to give a homotopy from h_0 to h_2, while the "inverse" (replace s by $1-s$) of a homotopy $h_0 \sim h_1$ from h_0 to h_1 is a homotopy from h_1 back to h_0. Therefore the relation "h_0 is homotopic to h_1" is reflexive, symmetric, and transitive. Then taking the equivalence classes for this relation produces for each path h_0 the homotopy class, call it $[h_0]$, of all paths h_1 homotopic to h_0 with the same end points. Since homotopies $h_0 \sim h_1$ and $k_0 \sim k_1$ will yield a homotopy $k_0 \cdot h_0 \sim k_1 \cdot h_1$,

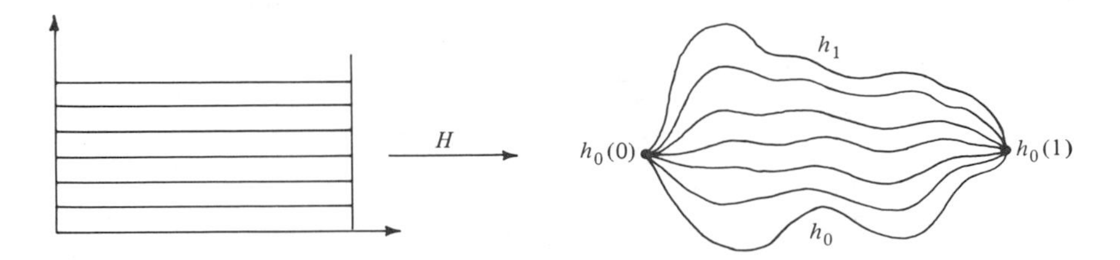

Figure 1. A homotopy.

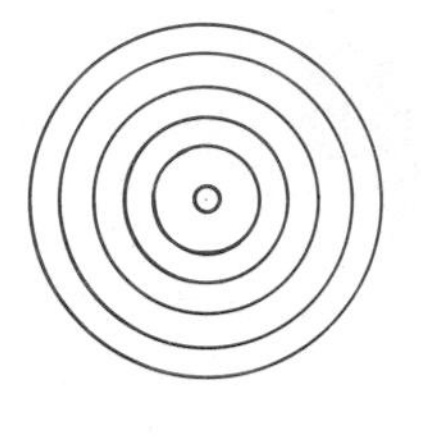

Figure 2

one can define the composite of two homotopy classes as the class $[k]\cdot[h] = [k\cdot h]$. This composition of classes is now associative, in view of a homotopy $m\cdot(k\cdot h) \sim (m\cdot k)\cdot h$ which we can picture as in Figure 3. At each horizontal level of the deformation, the paths follow h from $t = 0$ to $t = 1/4 + s/4$, then k, then m from $t = 1/2 + s/4$ to $t = 1$.

If we consider just the *closed* paths (the loops) starting and ending at one and the same point z_0 these considerations will prove the

Theorem. *For any subset S of $\mathbf{C}$ and for any point $z_0 \in S$, the homotopy classes of closed paths in S from z_0 to z_0 form a group under composition.*

This group is called the Poincaré *fundamental group* $\pi_1(S,z_0)$. Its identity element is the class of the trivial path, constant at z_0, while the inverse of any homotopy class $[h]$ of a closed path h is the class $[h^{-1}]$, given by h run backwards.

This fundamental group construction assigns an algebraic object—the group $\pi_1(S)$—to a geometric object, the subset $S \subset \mathbf{C}$. For example, if S is the unit circle S^1 in $\mathbf{C}$ while z_0 is the point 1 on that circle, the corresponding fundamental group $\pi_1(S^1,1)$ turns out to be the infinite cyclic group $\mathbf{Z}$. A generator of this infinite cyclic group is the path p going once around the circle, in a counterclockwise direction. With some care, we can prove that any closed path h from 1 to 1 for the circle S^1 is homotopic to p^n, where the integer n is the net number of times h winds around the circle in a counterclockwise direction. (This number is also called the *degree* of h.) In other words, the algebraic isomorphism $\pi_1(S^1) \cong \mathbf{Z}$ expresses a geometric fact: That a continuous map $S^1 \to S^1$ is determined,

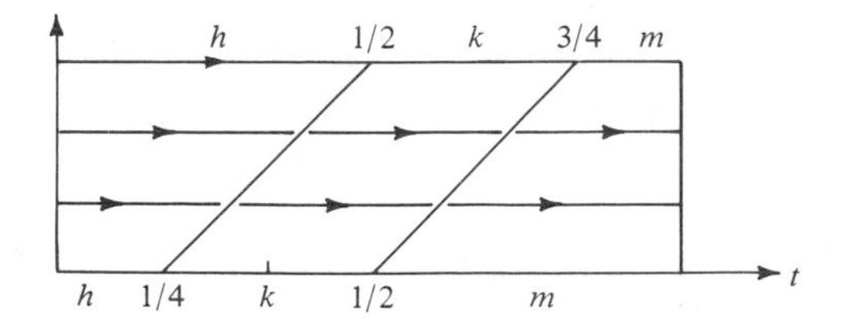

Figure 3. Homotopy associativity.

up to homotopy, by the net number of times the map winds the first circle about the second one; i.e., by the "winding number". (Observe that a closed path $h: I \to S^1$ really amounts to a map $S^1 \to S^1$, where the first circle S^1 is obtained by identifying the ends of the unit interval I.)

If $\mathbf{C} - \{0\}$ is the entire complex plane with only the origin 0 deleted, its fundamental group $\pi_1(\mathbf{C} - \{0\}, 1)$ is still the infinite cyclic group. Indeed, an isomorphism $\pi_1(\mathbf{C} - \{0\}, 1) \to \pi_1(S^1, 1)$ can be constructed by taking any closed path h in $\mathbf{C}$ which misses the origin and deforming it by radial projection from the origin into a path running along the unit circle S^1. This same fundamental group also appears in the behavior of the many-valued function $\log_e z = \log_e r + (\theta + 2\pi k)i$, which is defined for all $z = re^{i\theta}$; i.e., defined in the set $\mathbf{C} - \{0\}$. Specifically, starting with one branch k of this logarithm and following a closed path with winding number n brings one to the branch $k + n$.

On the other hand, consider the set B in $\mathbf{C}$ consisting of two circles tangent at one point (say the point 1). On B there are two closed paths h and k running once smoothly in a counterclockwise direction around each of these two circles (see Figure 4); they give elements $x = [h]$ and $y = [k]$ in the fundamental group of B. A plausible process will deform any closed path in B into a path running at a unform rate first around one circle and again around either the same circle or the other circle and so on. In this way any element in the fundamental group can be written as a composite $x^{m_1} y^{n_1} x^{m_2} y^{n_2} \cdots$ of products of integral powers of x and y. Any group containing two elements x and y must contain all such products; in the present case, two such products are equal only when they become equal by formal cancellations; for this reason this fundamental group is known as the *free group* on two generators x and y. The complex plane $\mathbf{C} - \{0,1\}$ with two points deleted has the same fundamental group, generated by paths x and y going once counterclockwise around 0 or 1, respectively. Then for a function $f(z)$ such as $1/z(z-1)$ holomorphic in the set $\mathbf{C} - \{0,1\}$, the Cauchy integral theorem (in one form to be discussed) will say for two closed paths h and k that

$$\int_h f(z)dz = \int_k f(z)dz \tag{6}$$

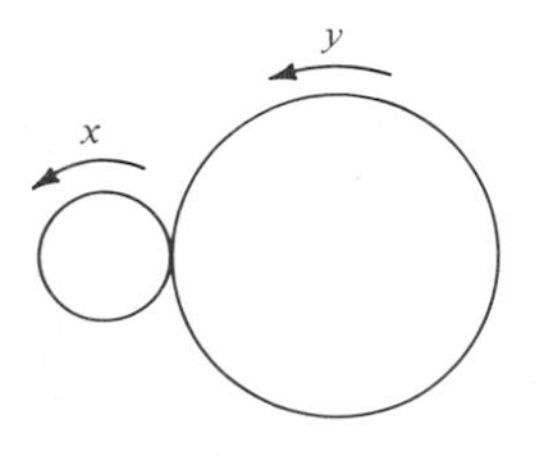

Figure 4

whenever h and k determine the same element in this fundamental group.

In differential geometry (Chapter VIII) smooth functions and smooth paths at a point give rise to tangent and cotangent vectors at that point, the two being paired by the combination "directional derivative of the function along the path at that point". What we now have is somewhat analogous: holomorphic functions f (or integrands) and paths h in a set $S \subset \mathbf{C}$, paired by the integral $\int_h f\,dz$.

In defining the fundamental group $\pi_1(S,z_0)$ we have used only the closed paths starting and ending at some arbitarily chosen base point z_0 in S. As long as S is pathwise connected, however, this homotopy group is independent of the choice of the base point z_0, up to an isomorphism. This can be seen most readily by considering the algebraic system formed by the homotopy classes $[h]$ of *all* paths in S, closed or not. Under the operation of composition these classes do not form a group, because the composite class $[k]\cdot[h]$ is not always defined. However it is defined whenever k starts where h ends, and the triple composite $[m]\cdot([k]\cdot[h])$ is associative whenever it is defined. Moreover, each class $[h]$ has a two sided inverse. With these properties, the homotopy classes under composition form an algebraic system called a *groupoid*, which we will meet again (under more general auspices) in the next chapter. Using this algebraic system, one can readily produce the desired isomorphisms $\pi_1(S,z_0) \cong \pi_1(S,z_1)$ for any path-connected set S. Just choose some path k starting at the first base point z_0 and ending at the second z_1. Thus for any closed path h at z_0 the correspondence

$$[h] \mapsto [k]\cdot[h]\cdot[k^{-1}] \tag{7}$$

does give the desired isomorphism, by the algebraic properties of groupoids. Indeed this correspondence (7) is just conjugation in the groupoid playing much the same role as conjugation in other geometric examples of transformation groups (§V.6), where conjugation by k carries the subgroup fixing a point x to the subgroup fixing the point kx.

A holomorphic function $f(z)$ is defined in some open set U of $\mathbf{C}$, usually in a connected open set. Since U is open, the requirement that it be connected (not the union of two disjoint non-empty open subsets) is equivalent to the requirement that it be path connected. To make the integral $\int f(z)dz$ independent of path in U, by Cauchy's theorem, one may require that U be *simply connected* in the sense that its fundamental group reduces to the identity element 1. This amounts to the requirement that every closed path in U is homotopic (with ends fixed) to the constant path or equivalently that every continuous map $h:S^1 \to U$ of the unit circle S^1 into U can be extended to a continuous map $H:\{z \mid |z| \leqslant 1\} \to U$ of the unit disc into U: Any such extension clearly produces a homotopy from h to the constant path, as in Figure 5.

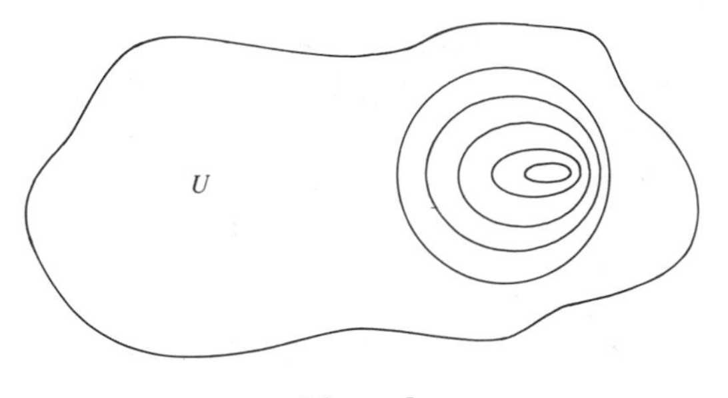

Figure 5

The whole complex plane **C** is simply connected. This result, though it may seem related to the Jordan curve theorem, is much easier to prove: Given a closed path in **C**, simply deform the path by that motion which takes each point at a constant velocity on a straight line to the base point.

Now observe that our definition of the fundamental group $\pi_1(S,z_0)$ of a subset S of the complex plane **C** really used nothing about the complex numbers except that they form a topological space under the standard distance; this implies at once that any subset S is also a topological space with the same metric (or, equivalently, with all the intersections $S \cap U$ as open sets, for any U open in **C**). Therefore, the same definitions will provide a fundamental group $\pi_1(X,x_0)$ for any topological space X, which will be, for example, independent of the basepoint x_0 as soon as the space X is path-connected. For instance, such groups can be defined for surfaces; in particular for Riemann surfaces, and this is one of the reasons they are useful in complex analysis. For example, the fundamental group of the torus $S^1 \times S^1$ (an example of a Riemann surface) is just the group $\mathbf{Z} \times \mathbf{Z}$–the free abelian group on two generators–one path each around each of the circles S^1 in $S^1 \times S^1$. Fundamental groups are also valuable for all other manifolds and other spaces. (For example, the fundamental group of the projective plane described in §VIII.4 is the cyclic group of order 2).

The fundamental group is just a first step in algebraic topology–the analysis by algebra of connectivity and homotopy properties of geometric figures which arise in topological spaces. In this section, our attention has been diverted from complex analysis to topology because complex analysis was historically one of the sources of algebraic topology. Another important source of topology was the qualitative study of trajectories in mechanics. This diversion will thus supplement our too-abrupt introduction in Chapter I of the notion of a topological space. In every such space X, it is convenient to call a set C *closed* when its complement $X - C$ (all points of X not in C) is open. Closed sets (e.g. circles) abound in the complex plane. The axioms for a topological space X, stated in §I.10 in terms of open sets, may be equally well formulated just in terms of closed sets; in particular, the intersection of any family of closed sets is closed. This makes it possible to define for each subset A of a space X a *closure* $\bar{A}$ (the smallest closed set containing A) as the intersection of *all* the closed sets of X which contain A. Then $\bar{A}$ is closed. The axioms for a topological space can also be stated in terms of the unary operations $A \mapsto \bar{A}$, "take the closure", on all subsets of X. In any topological space X it is also convenient to say that a *neighborhood* of a point $x \in X$ is any open set U of X which contains x. Then continuity of a function $f: X \to Y$ can be described naturally in terms of neighborhoods: f is continuous at $x \in X$ if to every neighborhood N of $f(y)$ there is a neighborhood M of x with $f(M) \subset N$—note that we do not have to say "arbitrarily small" neighborhood; the requirement for *every* N suffices!

Thus "topological space" may be defined equally well by way of open sets, closed sets, closure, or neighborhoods. This again illustrates the general thesis that one idea has many formalizations. In this case, the formalizations provide a handy language for the study of the many remarkable and deep properties of topological spaces, in particular of the topology of the plane.

6. The Cauchy Theorem

We return to the problem of a proof of the Cauchy integral theorem.

A *primitive* of a function f holomorphic in an open set U of $\mathbf{C}$ is defined much as in the calculus: it is a function F holomorphic in U with first derivative $F' = f$ in U. When there is such a primitive F, the Cauchy integral theorem does hold for the integral of f around any closed piecewise differentiable path h in U:

$$\int_h f(z)dz = 0. \tag{1}$$

Indeed, introducing the real parameter t for the path h and using the chain rule for differentiation gives

$$\begin{aligned} \int_h f(z)dz &= \int_0^1 f(h(t))h'(t)dt = \int_0^1 F'(h(t))h'(t)dt \\ &= \int_0^1 \frac{d}{dt}F(h(t))dt = F(h(1)) - F(h(0)) = 0, \end{aligned}$$

where zero results because the path is closed ($h(1) = h(0)$).

In particular, z^n has a primitive $z^{n+1}/(n+1)$ for each positive n, so the Cauchy theorem does hold when $f(z)$ is a polynomial. However, this result still does not tell us how to construct a primitive for more general holomorphic functions f. In order to do this, we will consider first very special closed curves: The boundaries of rectangles, taken counterclockwise. This will altogether avoid the problems presented by convoluted boundary curves.

The Cauchy–Goursat Theorem. *Let R be a rectangle in* **C**, *with sides parallel to the real or the imaginary axes, for which the interior of R and the boundary ∂R are both contained in the open set U where f is holomorphic. Then*

$$\int_{\partial R} f(z)dz = 0. \tag{2}$$

The proof will illustrate the use of $\epsilon-\delta$ methods in approximating integrals and will also introduce certain algebraic classes of boundaries which later develop into homology, a part of algebraic topology. First record the perimeter p and the diagonal diameter d

$$p = \text{perimeter } R, \qquad d = \text{diagonal } R \tag{3}$$

of the rectangle. If the intended integral is not zero, contemplate its absolute value

$$a = \left| \int_{\partial R} f(z)dz \right| . \tag{4}$$

We intend to successively subdivide till we find a very small rectangle where the corresponding integral is too big (for f to be holomorphic there). Specifically, divide the given rectangle into four smaller congruent rectangles, as in Figure 1. Now integrate over the boundaries of all four smaller rectangles, always in a counterclockwise direction; then the integrals over the "inside" boundaries cancel, as shown on the right of the figure. Therefore, by the additivity of the integral under composition of paths, the integral of $f(z)$ over ∂R is the sum of the four integrals over the boundaries of the four smaller rectangles. There must therefore be at least one of these four rectangles, call it R_1, with an integral of size $A/4$ or more

$$\left| \int_{\partial R_1} f(z)dz \right| \geqslant \frac{1}{4}A. \tag{5}$$

Pick such an R_1, subdivide it into four subrectangles, choose an offending one of these as before, and continue indefinitely. The result is a nested sequence of rectangles

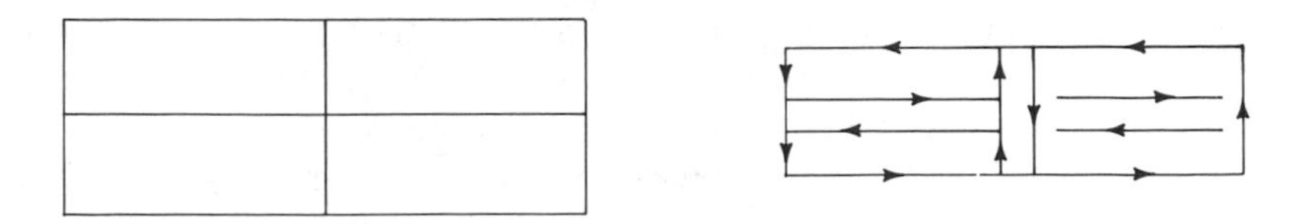

Figure 1

$$R \supset R_1 \supset R_2 \supset \cdots \supset R_k \supset \cdots$$

with

$$\left| \int_{\partial R_k} f(z)dz \right| \geqslant \frac{1}{4^k}A, \text{ perimeter } R_k = \frac{1}{2^k}p, \text{ diameter } R_k = \frac{1}{2^k}d. \quad (6)$$

There is exactly one point c in the complex plane $\mathbf{C}$ common to all these rectangles (which form a nested sequence of closed sets); for example, one can describe the x coordinate of this point c as the least upper bound of the x-coordinates of the lower left corners of the successive rectangles R_k. Since f is holomorphic in U, it has a first derivative $f'(c)$ at the point c. This means that for each positive $\delta > 0$ there is a neighborhood of c in which

$$\left| \frac{f(z) - f(c)}{z - c} - f'(c) \right| < \delta.$$

Since the size of the rectangles approaches zero with k, there is an index k such that all of the rectangle R_k is in the neighborhood; thus on ∂R_k we have the estimate

$$f(z) - f(c) - f'(c)(z - c) = \epsilon(z), \qquad |\epsilon(z)| < \delta |z - c|. \quad (7)$$

Therefore the integral is

$$\int_{\partial R_k} f(z)dz = \int_{\partial R_k} f(c)dz + \int_{\partial R_k} f'(c)(z - c)dz + \int_{\partial R_k} \epsilon(z)dz. \quad (8)$$

On the right, the first two integrals are integrals of polynomials around closed curves, hence are zero. The remaining integral of $\epsilon(z)$ can then be bounded by the general formula (4.4) for the size of an integral. By (7), $|\epsilon(z)| \leqslant \delta$ diameter R_k, so with (6) we have

$$\frac{1}{4^k}A \leqslant \delta \text{ (diameter } R_k)(\text{perimeter } R_k) = \delta \frac{1}{4^k}pd.$$

In other words, $A \leqslant \delta\, pd$. Since all of this holds for each positive δ, one must have $A = 0$. This proves the theorem.

The point of this proof is that the nested rectangles close down on one point c. At this point one can then use the basic idea of the differential calculus–that the first derivative gives a *linear* approximation to the function (here a complex linear approximation (7)). For this reason the proof works assuming just the existence of the derivative $f'(z)$ and not its continuity (as was needed in §4 in the argument by the Gauss lemma).

The proof also illustrates several central ideas of topology. Thus the existence of that point c comes from the fact that in the plane any nested sequence $C_1 \supset C_2 \supset C_3 \supset \cdots$ of closed sets has a non-trivial intersection. This basic fact, about the point-set topology of the plane, is a statement that the points are all there (the plane is complete); it is comparable to the Dedekind cut axiom for the reals.

On the other hand, we have integrated over a sum $h_1 + h_2 + h_3 + h_4$ of the four boundaries $h_i = \partial R_i$ of the four smaller rectangles of Figure 1. Such sums form an abelian group, the group of "chains"; similarly the Figure 1 suggests a 2-dimensional "chain" $R_1 + R_2 + R_3 + R_4$. Since h_1 is the boundary $h_1 = \partial R_1$ one says that h_1 is "homologous" to zero. Also we have subdivided each edge e (of the rectangle) into two pieces e' and e''; one says that there is a "singular" homology $e \sim e' + e''$ because $e - e' - e''$ is a boundary; indeed, a triangle can be collapsed (vertically in Figure 2) so that its boundary is $e - e' - e''$. This is the start of an idea of "singular" homology and of simplicial (i.e., triangular) approximation. These ideas of homology and chains are effective in all dimensions in handling geometric problems of algebraic topology.

We return to exploit the Cauchy–Goursat theorem. First let $f(z)$ be a function holomorphic in an open disc

$$D(c,r) = \{z \mid |z - c| < r\} \tag{9}$$

with radius r and center the complex number c. Then the Cauchy theorem

$$\int_h f(z)dz = 0 \tag{10}$$

holds for any closed piecewise differentiable path h in the disc (9), because we can construct a primitive F for f as the integral, for $w \in D$:

$$F(w) = \int_c^w f(z)dz$$

taken along a rectilinear axis-parallel path in the disk from c to w (see Figure 3). Indeed, by the Cauchy–Goursat theorem, any other axis-parallel rectilinear path from c to w will give the same integral $F(w)$; from this one may readily show that $F(w)$ *is* a primitive of f, so that (1) applies.

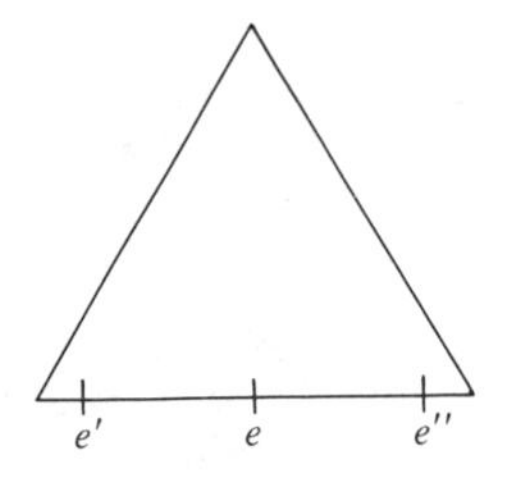

Figure 2

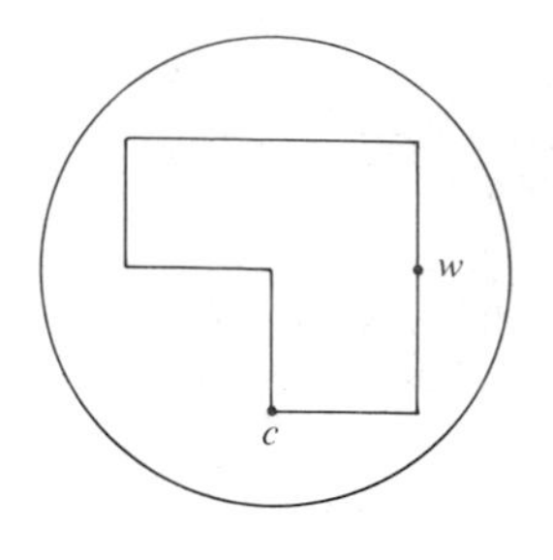

Figure 3

The result is actually more general, since the same argument evidently works if the disc is replaced by an open set which is *convex* (i.e., which contains the line segment joining any two of its points). Indeed, the same argument can be made to work if the disc D is replaced by any simply connected and connected open set U. One again wishes to compare two rectilinear, axis parallel paths in U from c to w, much as in Figure 3. Since U is simply connected, these paths are homotopic; the trick is to use suitable approximations to make the homotopy take place through rectangular changes in path.

Another, explicitly homotopy-theoretic version of the Cauchy theorem asserts that the integrals

$$\int_h f(z)dz = \int_k f(z)dz$$

over the paths h and k with the same endpoints are equal when h is homotopic to k (fixing the endpoints) in the open set U where f is holomorphic. There is still another, more subtle version (harder to prove): If h is a simple closed piecewise differentiable curve in $\mathbf{C}$, while f is holomorphic in the interior of k and continuous on h, then $\int_h f(z)dz = 0$.

All of these somewhat different formal statements are realizations of the one idea underlying the Cauchy integral theorem. The care required for the proof is one of the reasons why complex analysis has served as an excellent training ground for mathematical rigor. Indeed, the proofs *can* all be made rigorous (and most of them perspicuous as well). Historically, this developed gradually, with Cauchy's foundations for the calculus, Riemann's emphasis on geometrical insights, and the careful $\epsilon-\delta$ methods of Weierstrass in complex analysis, calculus of variations, and elsewhere. The better understanding and analysis of these methods has in its turn led to extensive developments in topology.

Though this book does not present many complete proofs, we emphasize that there is revealed here an objective notion of a *rigorous* proof: Each step in the proof is a logical consequence of prior steps or theorems or axioms (for the real numbers, or for sets, as the case may be);

moreover, consequence means consequence according to the explicit rules of inference from mathematical logic (Chapter XI). To be sure, practicing mathematicians often learn these rules just 'intuitively'—but one should not take too seriously the sometime response of a student of a noted and careful professor W, a specialist in complex variables. Asked to tell when a proof is rigorous, the student responded: "When it is acceptable to Professor W".

7. Uniform Convergence

The use of infinite series is the quintessence of the notion of infinitary Mathematical operations. Thus

$$\pi/4 = 1 - 1/3 + 1/5 - 1/7 + 1/9 - \cdots, \tag{1}$$

$$\log_e 2 = 1 - 1/2 + 1/3 - 1/4 + 1/5 - \cdots, \tag{2}$$

while for any real $r < 1$ an "infinite" long division yields the *geometric series* which *converges* to $1/(1-r)$,

$$\frac{1}{1-r} = 1 + r + r^2 + r^3 + r^4 + \cdots. \tag{3}$$

To be sure, sometimes an infinite sum gets out of hand, as in the case of the geometric series (3) for $r = 1$ or $r > 1$ and also with the harmonic series

$$1 + 1/2 + 1/3 + 1/4 + 1/5 + 1/6 + \cdots. \tag{4}$$

Indeed, adding up this series in groups of successive lengths 1,1,2,4, . . . , one finds that the sum grows faster than 1, $1\frac{1}{2}$, 2, $2\frac{1}{2}$, ... It is thus a *divergent* series (but it does suggest the Riemann zeta function; see (11.7) below).

Now, in point of fact, there is no such thing as actually summing an infinite series; nobody has every actually added up an infinite number of terms, no matter how conveniently convergent. Here as elsewhere in Mathematics, the use of the infinite is an intuitively suggestive description of what is in fact a finite formal process of successive approximations. This process can be described, in a completely rigorous way, in set-theoretic and logical terms. For example, an "infinite series"

$$a_0 + a_1 + a_2 + a_3 + \cdots \tag{5}$$

of real or complex numbers is determined by giving a function $n \mapsto a_n$ on the set **N** of natural numbers to the reals **R** (or to **C**). This function also determines the finite *partial sums*

$$s_n = a_0 + a_1 + \cdots + a_n ,$$

one for each n. The series (5) is then said to *converge* (§IV.4) to a number s as sum when the sequence s_n has a limit $\lim_{n\to\infty} s_n = s$. As in §IV.4, this statement about a limit means that for each real $\epsilon > 0$ there exists a natural number N such that, for all $n \geqslant N$, $|s_n - s| < \epsilon$; here the absolute value is that of the real numbers or the complex numbers, as the case may be. As in other cases, the idea that the limit s is something approximated (better and better as N increases) by finite sums gets expressed formally in terms of the logical quantifiers "for all ϵ" and "there exists N". In other words, convergence involves repeated *finite* approximations (up to ϵ) to the non-existent infinite sum.

The series (5) is said to *converge absolutely* when the corresponding series of absolute values,

$$|a_0| + |a_1| + |a_2| + \cdots$$

is convergent. Then absolute convergence implies convergence, as one sees by applying the triangle law $|z_1 + z_1| \leqslant |z_1| + |z_2|$ for absolute values to show that the sequence s_n satisfies the Cauchy condition for convergence. On the other hand, there are many convergent series such as the alternating series

$$1 - 1/2 + 1/3 - 1/4 + \cdots$$

which are not absolutely convergent.

In the geometric series (3) the successive terms are functions of a variable r (there r is real, but the same series converges when $r = z$ is a complex number with $|z| < 1$). More generally, consider a series of functions f_n of one or several parameters (here written z and w) in the "infinite series" form

$$f_0(z,w) + f_1(z,w) + f_2(z,w) + \cdots . \tag{6}$$

Objectively, this means that we are given some function $(n,z,w) \mapsto f_n(z,w)$ to (say) the complex numbers from the product $\mathbf{N} \times D \times E$, where D and E are sets of complex numbers, with $z \in D$ and $w \in E$. The partial sums $s_n(z,w)$ of such a series are then also functions $s{:}\mathbf{N} \times D \times E \to \mathbf{C}$, and the series may converge for some z and w and diverge for other values. Now in such a series the "speed" of the convergence may very much depend on the values of the parameters z and w. For example, in the geometric series (3) the partial sums are $s_n = (1 - r^{n+1})/(1-r)$; they converse to $1/(1-r)$ more and more slowly as r approaches 1. More explicitly, given a level $\epsilon > 0$ of approximation desired, one must choose an N which depends on r (and which necessarily gets larger as r approaches the number 1) to get $|s_n - 1/(1-r)| = |r^{n+1}/(1-r)|$ less than ϵ for $n \geqslant N$; one cannot to a given ϵ choose any one N which works for all r. One says that N cannot be chosen "uniformly" for all $r < 1$.

This suggests that we describe formally when convergence is uniform; this notion will turn out to matter. The definition is straightforward: The series (6) of functions converges *uniformly* on $D \times E$ to a function $s: D \times E \rightarrow \mathbf{C}$ when to each $\epsilon > 0$ there is some $N \epsilon \mathbf{N}$ such that, for all $n \geqslant N$ and all $\langle z,w \rangle \epsilon D \times E$, one has

$$|s_n(z,w) - s(z,w)| < \epsilon. \tag{7}$$

This is just the formal assertion that the same N works for all z and w at issue. If we put it in logistic notation, where $\forall$ is "for all" and $\exists$ stands for "there exists", it reads

$$\forall \ \epsilon > 0 \ \exists \ N \epsilon \mathbf{N} \quad \forall n \geqslant N \quad \forall \langle z,w \rangle \ \epsilon \ D \times E \qquad \text{(7) holds.}$$

Here again what is crucial is the careful use of the logistic quantifiers $\forall$ and $\exists$, written in the right order (!). For later purposes (in the discussion of quantifiers in the next chapter) we observe that in each quantifier the variable at issue ranges over a definite set—the variable ϵ is a positive real, n is a natural number greater than or equal to N, z is in D and so on. In the applications of this definition, the set D (and the set E) is usually a closed set in the topology of $\mathbf{C}$—because convergence over an open set is less likely to be uniform (as with the set of r with $r < 1$ for the harmonic series).

The idea of "uniformity" which is expressed here is really the *same* idea as that which appeared in the foundations of the calculus in §VI.7 with the notion of uniform continuity of a function. There, as here, the formal expressions of the idea of "uniform approximation" depend on the use of the quantifiers. This idea of uniformity can be further developed; there is for example a notion of a "uniform space" (specialized from topological space). The idea of uniform limits was once (early 1900's) regarded as one of the most difficult in Mathematical pedagogy. The difficulty now seems much less.

One point of uniform convergence is that it provides for smooth operations with infinite series. For example, if each function $f_n(z)$ is continuous for z in some open set U and if $\Sigma f_n(z)$ converges uniformly in every closed subset of U, then the sum $\Sigma f_n(z)$ is a function of z continuous in U. Similarly, if in the uniformly convergent series (6) each term $f_n(z,w)$ is complex differentiable in $z \epsilon U$ for all w, so is the sum $s(z,w)$; moreover, this sum is the limit of the series of derivatives $\partial f_n / \partial z$. In brief, a uniformly convergent series of holomorphic functions may be differentiated "term-by-term". Similarly, uniformly convergent series may be integrated term-by-term; we do not enter into the detailed statement.

There are many tests of convergence. For example, uniform and absolute convergence can be verified by comparison with a known convergent series of positive terms, such as the geometric series (3). Specifically, in (6), if there is some positive real $r < 1$ and some fixed M such that

$$|f_n(z,w)| \leqslant Mr^n \qquad \text{for all } n,z \text{ and } w.$$

then (6) converges and the convergence is uniform and absolute.

The properties of infinite series enter vitally in complex analysis by way of the expression of holomorphic functions in terms of Taylor series. They are also vital for calculations (for example, of tables of logarithms and of trigonometric functions).

8. Power Series

The whole subject of complex analysis can now be started over again from the idea that a "good" function of a complex variable z should be a function defined by a power series. Historically, this was the point of view advocated by Weierstrass in his influential lectures in Berlin. One may still see the contrast between the Weierstrass viewpoint and the more geometric approach (Riemann and Klein) in the book *Funktionen Theorie* by A. Hurwitz and R. Courant. First consider the elementary functions. For real x, $\sin x$, e^x, and $\cos x$ can be calculated from their Taylor series, and the same series converge (uniformly in compact sets) to provide good definitions of these functions for any complex argument z:

$$e^z = 1 + z + z^2/2 + z^3/3! + \cdots,$$

$$\sin z = z - z^3/3! + z^5/5! - \cdots.$$

So consider more generally series in powers of $z-c$, such as

$$a_0 + a_1(z-c) + a_2(z-c)^2 + a_3(z-c)^3 + \cdots; \qquad (1)$$

what about convergence? The essential fact is that for each such series (1) there is a real number R, with $0 \leqslant R \leqslant \infty$, such that (1) converges for all z with $|z-c| < R$ and diverges for $|z-c| > R$. (Of course if $R = \infty$, R is not a real number, but the intent is still clear!) This number R is called the *radius of convergence*.

To show this, it suffices to consider the case when $c = 0$. So call a real number $r \geqslant 0$ "good" if all the terms $|a_n| r^n$ have a common bound; that is, if there is some real number M_r with $|a_n| r^n \leqslant M_r$ for all natural numbers n. Then take R to be the least upper bound of all the "good" r (this of course might give an $R = \infty$). If $|z| < R$, then there is a good r with $|z| \leqslant r < R$, so $|a_n z^n| \leqslant |a_n| r^n$; there is even a good s with $r < s < R$. Then for $|z| \leqslant r$

$$|a_n z^n| \leqslant |a_n| r^n \leqslant |a_n| s^n \left(\frac{r}{s}\right)^n \leqslant M_s \left(\frac{r}{s}\right)^n .$$

Hence, by comparison with the convergent geometric series in r/s, the power series (1) converges for $|z| < R$. Indeed, the proof shows that it converges uniformly in every smaller closed circular disc $|z-c| \leqslant r$ for any $r < R$. On the other hand, the series cannot converge for $|z| > R$–for if it did so converge, its terms $a_n z^n$ must all have a common bound, contrary to the definition of R.

The circle $|z - c| = R$ is the *circle of convergence*. At points on this circle, the series may or may not converge. For example, $\Sigma\, z^n/n$ has radius of convergence $R = 1$; for $z = 1$ it becomes the harmonic series (7.4), so diverges, but it converges at every other point with $|z| = 1$!

Now we can restart the development of complex analysis. Consider a function $f: U \to \mathbf{C}$ defined on an open subset U of the complex plane. Call it *regular analytic* in U if, for each $c \in U$ the function $f(z)$ is given in some nontrivial circle about c by a power series (1), convergent at least in the circle. As shown above, the convergence is then uniform in any smaller concentric circle, so the power series (1) can be differentiated term-by-term to get as derivative a convergent series

$$f'(z) = a_1 + 2a_2(z-c)^1 + 3a_3(z-c)^2 + \cdots . \tag{2}$$

Therefore, every function f regular analytic in U is holomorphic in U; moreover the first derivative of f at $z = c$ is $f'(c) = a_1$. By taking higher derivatives we get $f^{(n)}(c) = n!a_n$. This determines the coefficients in the series (1), so the series must have the form

$$f(z) = f(c) + f'(c)(z-c) + \cdots + (1/n!)f^{(n)}(c)(z-c)^n + \cdots . \tag{3}$$

This is just the Taylor series for $f(z)$ expanded about the point $z = c$. This argument shows that the Taylor series (3) is indeed the *only* possible convergent power series at $z = c$ for a regular analytic function. It also shows that regular analytic functions are holomorphic. Soon we will show (by the Cauchy integral) that holomorphic functions are necessarily regular analytic.

The strict power series point of view can be carried out elegantly and systematically (but not in this book, which argues rather for a variety of approaches). From the intended perspective, every good function begins life as a power series, and so is initially defined only inside its circle of convergence. Thus, we might begin with the function g defined by the geometric series

$$g(z) = 1 + z + z^2 + z^3 + \cdots ,$$

but only for $|z| < 1$. How does one use series to recover all the rest of the intended function $1/(1-z)$? We will see in §11 that this question leads to some fundamental ideas about "analytic continuation".

9. The Cauchy Integral Formula

The Cauchy theorem yields a surprising way of representing any function f holomorphic in U as an integral. For each point w in a simply connected open set U, take some simple closed piecewise differentiable path h in U which goes once around w in a counterclockwise direction; for example, take h to be a small circle about w. Then $f(w)$ is represented as

$$f(w) = \frac{1}{2\pi i}\int_h \frac{f(z)}{z-w}dz. \tag{1}$$

Indeed, the right-hand side can be written trivially as the sum of two integrals,

$$\frac{1}{2\pi i}\int_h \frac{f(z)}{z-w}dz = \frac{1}{2\pi i}\int_h \frac{f(w)}{z-w}dz + \frac{1}{2\pi i}\int_h \frac{f(z)-f(w)}{z-w}dz.$$

In the first integral, $f(w)$ is constant along the path h while h can be deformed into a small circle about w; for this case we have already calculated the integral $\int dz/(z-w)$ to be $2\pi i$, as in (4.8). Hence the first term is just $f(w)$. In the second integral we may again deform the path of integration to a very small circle, of circumference $2\pi r$, about the point w. The integrand $(f(z) - f(w))/(z-w)$ is just the difference quotient for f: by definition it approaches the value $f'(w)$ of the derivative of f at w, as $z \to w$. This means that the integrand is bounded by some constant on (and inside) the small circle, while the length of the circle can be made as small as desired. Hence, by the standard maximum and length estimate (4.4) for a complex integral, this second integral can be made as small as desired–hence zero. This proves (2)–without the explicit display of ϵ's which the reader can readily add in.

This *Cauchy Integral Formula* (1) expresses a remarkable fact about a holomorphic function: Its values everywhere *inside* a circle are completely determined by its values *on* that circle–and similarly for any other simple closed boundary curve. This is a tight control on the behavior of holomorphic functions which has no analog for functions of a single real variable. It follows also that the values of each derivative of a holomorphic function f are determined just by the values of f on the boundary–as we will soon see explicitly in equation (5) below.

Cauchy's formula will also produce a power series for a holomorphic function f. For example, suppose that f is holomorphic in a simply connected open set containing the origin $w = 0$. In Cauchy's formula (1), take h to be a circle with center 0 of some radius r and lying inside U. In the integrand the quotient $1/(z-w)$ can be expanded as a geometric series

$$\frac{1}{z-w} = \frac{1}{z}\frac{z}{(z-w)} = \frac{1}{z}\left[\frac{1}{1-\frac{w}{z}}\right]$$
$$= \frac{1}{z}\left[1 + \frac{w}{z} + \left[\frac{w}{z}\right]^2 + \cdots + \left[\frac{w}{z}\right]^n + \cdots\right], \tag{2}$$

convergent for $|w| < |z| \neq 0$. So choose some r' with $0 < r' < r$, where r is the radius of the circle h; then for all z on this circle and all w with $|w| \leqslant r'$ this series is uniformly convergent because it converges for all these w at least as fast as the geometric series $\Sigma\,(r'/r)^n$. Thus it can be integrated over z term-by-term to give a convergent series in w

$$f(w) = a_0 + a_1 w + a_2 w^2 + \cdots + a_n w^n + \cdots, \tag{3}$$

where the nth coefficient a_n is an integral

$$a_n = \frac{1}{2\pi i}\int \frac{f(z)}{z^{n+1}}dz. \tag{4}$$

Moreover, the argument shows that the series (3) converges in any circle, with center at w, inside the given set U where f is holomorphic.

We have thus shown that a function f holomorphic in U is necessarily regular analytic in U! In other words, the existence of a complex derivative for f at each point of U is equivalent to the existence of a convergent power series at each point. Henceforth we thus can (and will) drop the term "regular analytic" (which is anyhow old fashioned).

Since the convergent series (3) is the unique power series for f around $w = 0$, the nth coefficient a_n in (4) must be exactly the nth coefficient in the Taylor series (8.3) for f at $w = 0$ (i.e., the Maclaurin series). Thus, in (4), $a_n = f^{(n)}(0)/n!$. The corresponding result holds at every point c in U, so that

$$f^{(n)}(c) = \frac{n!}{2\pi i}\int_h \frac{f(z)}{(z-c)^{n+1}}dz, \tag{5}$$

where the path of integration is any sufficiently small circle about $z = c$. This is Cauchy's integral formula for the derivatives of $f(z)$. There is an easy mnemonic for this formula: Take the original Cauchy integral formula (1), differentiate (under the integral sign) n times with respect to w, and then change w to c.

The proof has also shown that the Taylor series for a holomorphic function f at a point $z = c$ will converge inside any circle about c which is contained in an open set U where f is holomorphic. This leads to the slogan: The Taylor series for a holomorphic function f converges in the circle out to the nearest singularity (i.e., the nearest point where f is undefined or is defined but is not holomorphic). As stated, this is a slogan and not a theorem, because the function might have been defined in a different way, for example as the function f with $f(z) = z^2$ when $|z| < 1$ and $f(z) = 1$ when $|z| \geqslant 1$. This function is holomorphic in $|z| < 1$ but not in any larger circle about the origin (and not in any open set containing $z = 1$). Nevertheless, the Taylor series of the function at the origin is just z^2, so converges everywhere; the function "should have been" z^2 to begin with. The slogan can (with a little trouble) be reformulated to be rigorous, but the interest is already clear. For example, the geometric series for $1/(1 - z)$ converges for $|z| < 1$ and for no bigger circle about the origin precisely because $z = 1$ is the nearest singularity.

There are many familiar functions holomorphic in the whole plane–for example polynomials, e^z, $\sin z$, and $\cos z$. Such functions are commonly called *entire* functions. In the cases just listed, however, each such entire function (except the constant polynomials) is unbounded in the whole plane. This is no accident; Liouville's theorem asserts that any function $f(z)$ holomorphic in the whole plane and bounded there is necessarily a constant. For, consider the Taylor expansion (3) for such an f around the origin; it converges in the whole plane. Its coefficients are given by the formula (4). If M is a bound for $f(z)$ in the plane, so that $|f(z)| \leqslant M$ for all z, the formula (4) with the path h a circle of radius r and circumference $2\pi r$ yields the estimate

$$|a_n| \leqslant Mr^{-n}.$$

For $n \geqslant 1$ we can let r approach infinity, to get $|a_n| = 0$ for all such n. Thus $f(w) = a_0$ is indeed constant.

This proof illustrates again the remarkable utility of the basic estimate (4.4) for the size of an integral.

From Liouville's theorem one readily derives a proof of the fundamental theorem of algebra (§IV.10). For let

$$g(z) = b_0 + b_1 z + \cdots + b_n z^n, \qquad b_n \neq 0,$$

be a polynomial of degree $n > 0$ with complex coefficients b_i. If g has no complex zeroes at all, then $f(z) = 1/g(z)$ is defined and holomorphic in the whole complex plane. One readily shows also that f is bounded there (because, for large z, the term $|b_n z^n|$ dominates $|g(z)|$. Hence, by Liouville, $f(z)$ is constant–nonsense.

This proof, by an indirect argument, is clearly not constructive, since it gives no indication of where to find a complex zero of $g(z)$. It is a splendid example of an elegant but circuitous proof, since a really elementary fact about complex polynomials is here proved expeditiously–once one has in hand the substance of power series and complex integration and their application to holomorphic functions. There are other, more topological, proofs of the fundamental theorem and still others in a more algebraic style–including some which are constructive.

The Cauchy integral formula has many other striking and useful consequences. For example, any holomorphic function $f: U \to \mathbf{C}$ is an *open mapping* in the sense that the image $f(V)$ of each open subset V of U is an open subset of $\mathbf{C}$. With this one may also establish a *maximum principle*: If f is defined in a bounded connected open set U and on its "boundary" ∂U so as to be holomorphic in U and continuous on the union of U and ∂U, while M is the maximum value of $|f(z)|$ on the boundary ∂U, then f is either constant or, for z in U, $|f(z)| < M$. In other words, f does not take on any maximum value in the "interior" of U. In this formulation, the *boundary* ∂U of U is defined to be the complement of U in the closure $\overline{U}$ of U in $\mathbf{C}$; this insures that the closure $\overline{U}$ is the union of U and ∂U.

There are a number of other forms of this *maximum principle*, as well as corresponding maximum principles for harmonic functions $u(x, y)$.

10. Singularities

Consider a function f with an isolated singularity; it will suffice to consider such a singularity at the origin $z = 0$. This means that the function f is holomorphic at least in an open set D of all complex z with $0 < |z| < R$; such a set is a *punctured disc* of radius R. The Laurent theorem for such a function states that there are then *two* convergent series, one in powers of z and the other in powers of $1/z$, so that $f(z)$, for any z in the punctured disc, is given as a sum

$$f(z) = a_0 + \sum_{n=1}^{\infty} a_n z^n + \sum_{m=1}^{\infty} a_{-m} z^{-m}. \tag{1}$$

Moreover, the coefficients a_i in this "Laurent series" are uniquely determined by f. The coefficient a_{-1} of z^{-1} is called the *residue* of f at $z = 0$.

The proof (which we do not give), starts with the concentric circles $|z| = r$ and $|z| = s$ within the disc and with $r < s$. (See Figure 1.) By an argument like that for Cauchy's integral formula one can prove, for each complex number w between these two circles, that

$$f(w) = \frac{1}{2\pi i} \int_{|z|=s} \frac{f(z)}{z - w} dz - \frac{1}{2\pi i} \int_{|z|=r} \frac{f(z)}{z - w} dz.$$

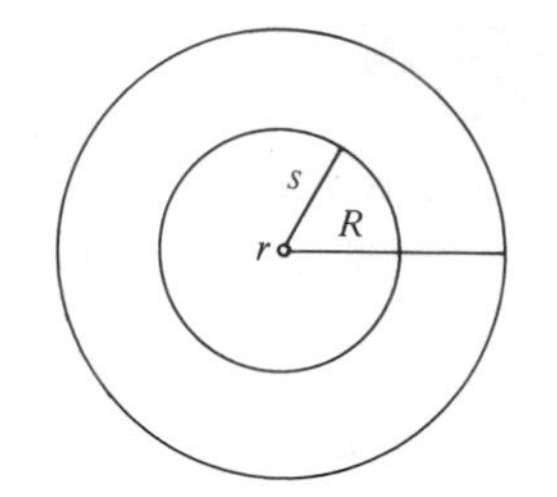

Figure 1. Punctured disk.

For the first integral, $|w/z| < 1$, so we can get from it a power series in w, much as in (2) and (3) of §9. For the second integral, $|z/w| < 1$, so this yields similarly a power series in w^{-1}; together these give (1), with z there replaced by w. We can think of the formula (1) as a representation of $f(z)$ as the sum of two functions, the first holomorphic inside $|z| < R$, the second holomorphic for $|z| > 0$.

An example is the function $e^{1/z}$. From the usual power series for e^z one has

$$e^{1/z} = 1 + \frac{1}{z} + \frac{1}{2!z^2} + \cdots + \frac{1}{n!z^n} + \cdots. \tag{2}$$

This is a series like (1) with infinitely many non-zero coefficients a_{-m} with negative index. Such a function is said to have an *essential singularity* at the origin $z = 0$. When there are only a finite number of coefficients $a_{-m} \neq 0$, the function f is said to have a *pole*. It is convenient to write $V_0(f) = -m$ for the negative of the *order* of the pole at 0.

When there are no negative powers of z present in the Laurent expansion the function f (if not already defined at $z = 0$) may be defined there by setting $f(0) = a_0$; it then becomes holomorphic in $|z| < R$; it is said to have had a *removable singularity* at 0. If f is defined (and holomorphic) near $z = 0$, it has a *zero* there if $f(0) = a_0 = 0$. The smallest n with coefficient $a_n \neq 0$ is then the order of the zero at the origin; one writes $V_0(f) = n$ for this order. The operator V_0 (order at $z = 0$) has for functions f and g the formal properties

$$V_0(fg) = V_0(f) + V_0(g), \qquad V_0(f+g) \geqslant \mathrm{Min}(V_0 f, V_0 g). \tag{3}$$

These same properties will reappear in number theory (Chapter XII). Such an operator V_0 may be called a *valuation*: if one defines $\|f\| = 2^{-V_0 f}$, the properties (3) of such a valuation take the suggestive form

$$\|fg\| = \|f\| \cdot \|g\|, \qquad \|f + g\| \leqslant \mathrm{Max}(\|f\|, \|g\|);$$

the "absolute value" $\|f\|$ is "large" when the function f has a pole of high order at 0, and "small" when f has a high order zero there.

One may also consider singularities of a function "at ∞", where the point at ∞ corresponds under stereographic projection to the north pole of the Riemann sphere. More explicitly, the point at ∞ can be brought to the origin by the conformal transformation $z \mapsto 1/z$ which carries functions holomorphic near ∞ (i.e., outside some large circle) into functions holomorphic in a circle about 0. Hence, by so transforming the Taylor series at the origin, one sees, as in (2), that a function f holomorphic for $|z| > R$ and at ∞ has a (unique) expansion

$$f(z) = a_0 + \sum_{m=1}^{\infty} a_{-m} z^{-m} \tag{4}$$

convergent for $|z| > R$, in powers of $1/z$. The coefficient a_0 is its value $f(\infty)$; when $a_0 = 0$, it has a zero at infinity. The behavior of a holomorphic function is largely controlled by the location and orders of its singularities. For example, a function which is holomorphic in the extended complex plane (including ∞), except for a finite number of poles, is necessarily a rational function.

The function e^z has an essential singularity at ∞; notice, however, that it takes on all complex values except 0 in any neighborhood of ∞ (a neighborhood of infinity contains the outside of a circle). This behavior is typical of essential singularities; the so-called big Picard theorem states that a holomorphic function with an essential singularity at a point c will take on every complex value, with at most one exception, in every neighborhood of c.

For a function such as a_{-1}/z with a pole of order 1 at 0, we know by (4.8) that the (counterclockwise) integral along a small circle h around this pole is

$$\int_h \frac{a_{-1}}{z} dz = (2\pi i) a_{-1}\,; \tag{5}$$

it is completely determined by the residue a_{-1} at the pole. On the other hand $\int a_{-m} z^{-m} = 0$ when $m > 1$. For integrals around several poles the contributions add. This observation can be made into a proof of the

Residue Theorem. *If $f(z)$ is holomorphic in an open set U except for a finite number of poles at points $c_1, \ldots, c_k$ in U, while h is a simple closed piecewise differentiable curve homotopic to 0 in U and going once (counterclockwise) around all these poles, then*

$$\int_h f(z) dz = 2\pi i \sum_{j=1}^{k} \operatorname{Res}_{c_j} f. \tag{6}$$

In brief, the integral is $2\pi i$ times the sum of the residues at all poles inside h. This famous result provides a convenient way to evaluate certain definite integrals without finding antiderivatives. Often an integral of a (real valued) function along some interval of the real axis can be made part of a "contour integral" which can be evaluated by (6), usually so that the integral along other (nonreal) parts of the contour h approach zero. The technique can be learned from the appropriate texts; it is another illustration of the power of the Cauchy integral theorem.

11. Riemann Surfaces

Now we return to the geometric properties of holomorphic functions. They involve topological ideas, such as that of homeomorphism: The idea is that two spaces are homeomorphic (topologically the same) if the first can be deformed into the second without tearing and without identifying points. Formally, as in §VIII.6, a continuous map $f:X \to Y$ of topological spaces is a *homeomorphism* if it has a two-sided continuous inverse $g:Y \to X$. Another idea is that of a "covering" map. A simple example is the function $\theta \mapsto 5\theta$ for $0 \leqslant \theta < 2\pi$; it wraps the circle S^1 with coordinate θ five times around itself, covering the circle evenly, as suggested by Figure 1. This means that S^1 viewed just locally from below looks like a stack of separate slices; for each point y of the second circle there is a small neighborhood U, illustrated in Figure 1, such that the inverse image of U under p consists of disjoint pieces (slices), each one homeomorphic under p to U. (Here the "inverse image" of U means the set of all points x with $px \in U$.) There are many such "covering" maps; thus trigonometry rests on the covering map $x \mapsto (\cos x, \sin x)$ (the line covers the circle by the wrapping function of §IV.2).

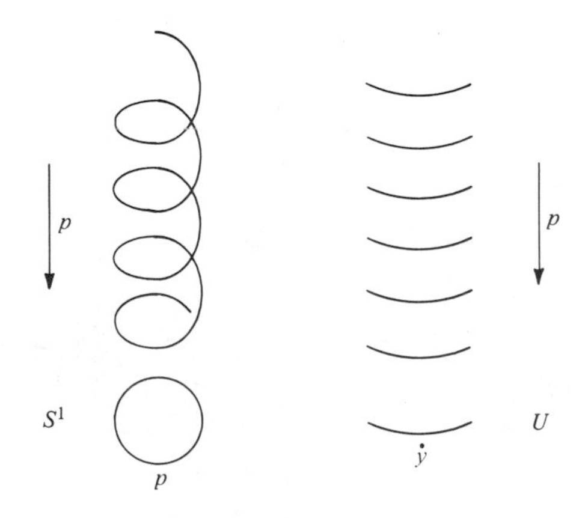

Figure 1. Covering space.

Another covering map is $(x,y) \mapsto (e^{2\pi ix}, e^{2\pi iy})$, which covers the torus $S^1 \times S^1$ by the plane $\mathbf{R}^2$. Each point on the torus has a small neighborhood which is the image of (denumerably many) small disjoint neighborhoods from the plane, as indicated in Figure 2. This suggests the general definition of a covering space Y: A *covering* map $p: Y \to X$ between topological spaces is a continuous map such that each $x \in X$ has a neighborhood U for which the inverse image $p^{-1}U$ is the union of disjoint open sets U_j, on each of which p restricts to a homeomorphism $U_j \to U$.

Covering maps appear prominently with Riemann surfaces. The example of Figure 1 suggests the holomorphic function $w \mapsto w^5 = z$, which maps the w-plane $\mathbf{C}_w$ onto the z-plane $\mathbf{C}_z$, multiplying the argument of each complex number $w \neq 0$ by 5, and so covering the z-plane (except for the origin) five times. If $\mathbf{C} - \{0\}$ is the complex plane with the origin removed, this map $(\mathbf{C}_w - \{0\}) \to (\mathbf{C}_z - \{0\})$ is a covering map in the sense of our definition. Moreover, $w^5 = z$ means that $w = \sqrt[5]{z}$, so each of the five fifth roots of $z \neq 0$ appear. Thus $\mathbf{C}_w - \{0\}$ is the Riemann surface on which the many-valued function $\sqrt[5]{z}$ becomes single-valued—like the Riemann surface for $\sqrt{z}$ discussed in §VIII.4.

The *punctured plane* $\mathbf{C}_z - \{0\}$ has other coverings, an n-fold covering for $w^n = z$ and one with infinitely many sheets. Consider for instance the exponential function

$$w = u + iv \mapsto z = e^w = e^u(\cos v + i \sin v). \tag{1}$$

By the definition (1) of the exponential, the horizontal strip with $0 \leqslant v < 2\pi$ of the w-plane is mapped onto the whole z plane (omitting 0), the segment $u = 0$ along the imaginary v-axis of the strip is mapped onto the unit circle in the z-plane, while the horizontal lines through B, C, and D in the w-plane become rays from the origin in the z-plane (see Figure 3). Each parallel horizontal strip $2\pi k \leqslant r < 2\pi(k+1)$ simi-

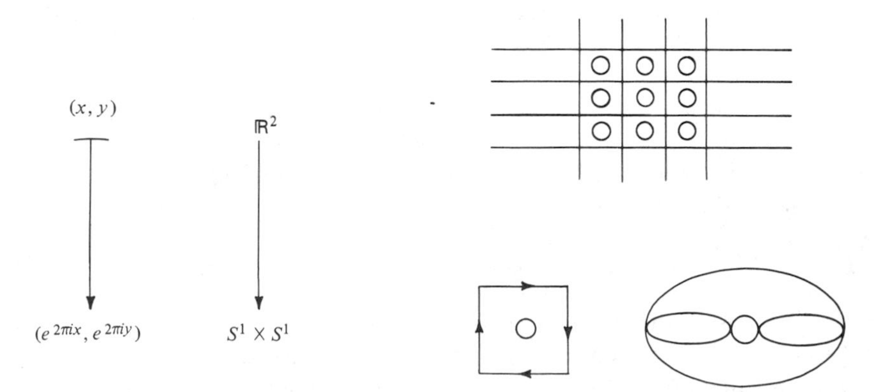

Figure 2

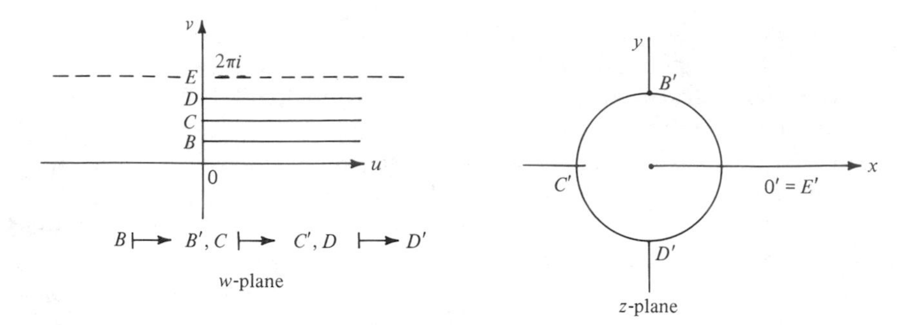

Figure 3. The exponential function.

larly covers the whole z-plane, omitting 0. This exponential map is a covering map $\mathbf{C}_w \to (\mathbf{C}_z - \{0\})$, since each open disc about a point $z \neq 0$ and not containing the origin is covered by a denumerable number of disjoint open sets from the w-plane, one from each strip. Actually, this covering space $\mathbf{C}_w$ is both connected and simply connected. One can prove that any connected and simply connected covering space $p: Y \to (\mathbf{C}_z - \{0\})$ of the punctured plane is necessarily homeomorphic (preserving the projection p) to this covering space $\mathbf{C}_w$; for this reason $\mathbf{C}_w$ is called the *universal covering space* of the punctured plane $\mathbf{C}_z - \{0\}$. Moreover, the vertical translations of the w-plane by $2\pi ik$ (sheets to sheets) are denumerably many, and correspond to the elements of the (infinite cyclic) fundamental group of the punctured plane. Other coverings of the punctured plane correspond to subgroups of the fundamental group, so group theory enters here also.

The plane $\mathbf{R}^2$ is also, as above, the universal covering space of the torus.

The inverse of the function $w \mapsto e^w$ of (1) is the many-valued logarithm function $w = \log_e z$; it is single-valued on the surface $\mathbf{C}_w$, which is thus the Riemann surface of $\log_e z$; it can be defined as the manifold of all pairs (z,w) of complex numbers which satisfy $z = e^w$, so it is just the manifold $\mathbf{C}_w$ of all complex numbers w.

This leads directly to the general definition of a Riemann surface—as a special kind of surface—one with a complex structure. Recall from §VIII.7 that a 2-dimensional manifold (a surface) S is a topological space which is the union of open sets U which form an atlas of charts, which are homeomorphisms

$$\phi : U \to V, \qquad U \text{ open in } S, \qquad V \text{ open in } \mathbf{R}^2 \tag{2}$$

into open sets of the plane $\mathbf{R}^2$ such that each overlap of domains U and U' yields a homeomorphism. Specifically, the charts ϕ and ϕ' overlap if the intersection $U \cap U'$ of their domains is not empty; then restriction of ϕ^{-1} and ϕ' to maps ϕ_1^{-1} on $\phi(U \cap U') = W$ and ϕ'_1 on $U \cap U'$ compose to give the "overlap map"

$$\phi'_1 \cdot \phi_1^{-1}: W \to W' \tag{3}$$

between two open subsets W and W' of the Euclidean plane $\mathbf{R}^2$. Since this plane can be identified with the complex plane, we define: A Riemann surface is just a surface with an atlas in which all the overlap maps are holomorphic (i.e., each $\phi'_1 \cdot \phi_1^{-1}$ of (3) is a holomorphic function on the open set $W \subset \mathbf{C}$). In other words, to get a Riemann surface S take a bunch U_i of topological spaces, each identified by a bijection ϕ_i as in (2) with some open set V_i of $\mathbf{C}$, and paste the U_i together by suitable holomorphic identifications (3). Since the plane $\mathbf{C}$ had *one* complex dimension (and real dimension 2), one may also say that a Riemann surface is just a one-dimensional "complex manifold". Higher dimensional such manifolds also occur, in algebraic geometry and analysis.

Two different such atlases A and A' on the same space S describe the same Riemann surface just when each overlap map (from a chart of A to one of A') is holomorphic. Much as for smooth surfaces, one can then describe the Riemann surface intrinsically by using the maximal such atlas. Also, any open subset S_0 of a Riemann surface is itself a Riemann surface (just restrict each chart to $U \cap S_0$). Often Riemann surfaces (in particular S_0) are required to be connected.

The charts serve to replace open sets of the surface S by open sets of $\mathbf{C}$. For example, they can be used to define when a complex valued function $f: S \to \mathbf{C}$ on the surface is holomorphic on S: It is so when for each chart ϕ of S the composite

$$\mathbf{C} \supset V \xrightarrow{\phi^{-1}} U \xrightarrow{f \mid U} \mathbf{C} \tag{4}$$

is holomorphic on the codomain $V \subset \mathbf{C}$ of the chart; here $f \mid U$ is the restriction of the given function f to U. In particular, by the overlap condition, each chart ϕ is itself a holomorphic function $s: U \to V \subset \mathbf{C}$ on the domain U of the chart; since the function s is a homeomorphism to V, it is often called a *uniformizing parameter* at each point $q \in U$; by translating V in $\mathbf{C}$, one can arrange that the parameter value $s(q)$ is 0 at the point q. Then the definition (4) reads: f is holomorphic in a neighborhood U of q if it is a holomorphic function g of the uniformizing parameter s there; in symbols

$$f(q) = g(sq); \tag{5}$$

here $g = (f \mid U)\phi^{-1}$ is the composite function of (4). This is the classical language; it may reflect the classical view that functions f of abstract objects (points q) ought to be replaced by functions of "real" objects (numbers). This is no longer our view; in particular, it is not invariant, because it depends on a choice of the uniformizing parameter. At each point, there many such choices.

The definition of holomorphic functions on a Riemann surface by charts is just like the definition by charts of smooth functions on a differentiable manifold, as in (VIII.8.3). The same idea has still other formulations, as for example in manifolds modelled on Banach spaces.

The complex plane, an open disc, any open set in the complex plane, and the Riemann sphere (with charts given by stereographic projections) are all Riemann surfaces. So is the torus–with suitable charts. A function holomorphic on a surface S has a first derivative at each point with respect to any uniformizing parameter at that point–although the derivative clearly depends on the choice of uniformizing parameter.

Riemann constructed his surfaces for particular functions f because these surfaces could display the way such functions (even when "many valued") depend essentially just on the location and character of the singularities. To illustrate, we may, with Weierstrass, start with a function $f(z)$ defined just in some circle $|z| < R$ by a power series $\Sigma a_n z^n$ convergent there. At each point c' in the circle the power series determines all the derivatives of f, and so gives the Taylor series $\Sigma f^{(n)}(c')(z-c')^n/n!$ which is convergent in some circle $|z-c'| < R'$ about c'; in part, the circle may extend beyond the initial disc. Continuing from the second circle one gets a sequence of Taylor series in circular discs which can be pasted together wherever they match. This gives a connected Riemann surface S on which the evident extension of the original power series is defined and holomorphic. Also each point on this surface S comes from a point (say the point c') in the complex plane, so the surface is equipped with a holomorphic function $p: S \to \mathbf{C}$; this is usually a covering map (over part of $\mathbf{C}$). For example, the Riemann surface for $\sqrt{z}$, constructed as in §VIII.4, covers $\mathbf{C} - \{0\}$ twice.

This general process of extension, suitably formulated, is called *analytic continuation*. For example, one might start with the power series

$$\frac{1}{2-z} = \frac{1}{2}\left[1 + \frac{z}{2} + \frac{z^2}{4} + \cdots + \frac{z^n}{2^n} + \cdots\right] \tag{6}$$

convergent in the disc $|z| < 2$. At the point $z = i$ the resulting new power series will converge in a circle of radius $\sqrt{5}$–out to the nearest singularity, the pole at 2. Using the values in this circle one can form the Taylor series about the point $z = 2 + i$; it converges in a circle of radius 1. This process, as suggested in Figure 4, will continue until the whole z-plane except $z = 2$ is covered. It really yields the intended functions $1/(2 - z)$ and not something else because of the following readily established general theorem:

The Principle of Analytic Continuation. If two functions f and g are holomorphic in a connected open set U of $\mathbf{C}$ and if $f = g$ on some non-empty open subset $U_0 \subset U$, then $f = g$ throughout U.

For a proof, consider the difference $h = f - g$ and the subset $E \subset U$ where *all* the derivatives $h^{(n)}$ are zero. This set E includes U_0, because any

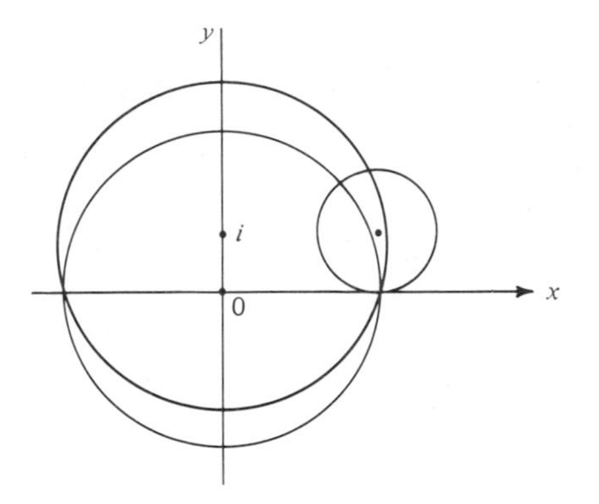

Figure 4. Analytic continuation.

derivative there can be calculated by taking limits of difference quotients within U_0. E is open, because the Taylor series for h at each point of E is zero. On the other hand, each $h^{(n)}$ is continuous, so the set of its zeros and hence their intersection, the set E, is closed. Since E is non-empty and clopen (closed and open, quickly spoken) in a connected set U, it must be all of U.

This is a typical use of the definition of "connected" (§VIII.6): a set which is not the union of two disjoint non-empty open sets.

This process of analytic continuation will readily produce Riemann surfaces which are not part of the complex plane. Thus if we start at $z = 1$ with the binomial series (i.e., the Taylor series) for $\sqrt{z} = (1 - (1 - z))^{1/2}$ and continue as above around the origin, we return to $z = 1$ after one circuit with the opposite value of the square root. Thus the family of all such power series continuations of $\sqrt{z}$ produces the Riemann surface of this function, just as it has been described in §VIII.4.

These examples of analytic continuation via power series may not be persuasive, because they simply reproduce an already known function $1/(2 - z)$ or $\sqrt{z}$ out of one of its Taylor series. But the process can be used to better effect. One example is the Riemann zeta function. This starts from the (divergent) harmonic series $1 + \frac{1}{2} + \frac{1}{3} + \cdots$. For s complex with real part $s > 1$ the corresponding series

$$\zeta(s) = 1 + \frac{1}{2^s} + \frac{1}{3^s} + \cdots + \frac{1}{n^s} + \cdots \tag{7}$$

does converge, where n^s means $e^{s \log n}$ with real logarithm $\log n$. If one writes the complex number s as $\sigma + it$ for σ and t real (this is the standard notation in connection with ζ), the series converges absolutely and uniformly in any closed half-plane $\sigma \geqslant c > 1$; hence ζ is holomorphic in the whole half-plane $\sigma > 1$. It can be continued into the whole complex plane with the exception of a pole (with residue 1) at the point $s = 1$;

however, the analytic continuation is best accomplished via suitable integrals and functional equations and not just by power series. This result is a striking example of the power of analytic continuation, which here has produced from the initial definition (7) a holomorphic function defined in a much larger domain!

The fundamental theorem of arithmetic states that each rational number n can be written, and uniquely, as a product of prime numbers. Hence the infinite series on the right of (7) is formally the product of all the series

$$\frac{1}{1-\frac{1}{p^s}} = 1 + \frac{1}{p^s} + \frac{1}{p^{2s}} + \cdots + \frac{1}{p^{ks}} + \cdots$$

for all the primes p. This suggests Euler's formula for $\zeta(s)$:

$$\zeta(s) = \prod_p \frac{1}{\left(1-\frac{1}{p^s}\right)} \tag{8}$$

as an infinite product. There is a straightforward way to justify such hypothetical "infinite" products as limits of actual finite products. Thus formula (8), which is really an analytic restatement of the fundamental theorem of arithmetic, is just the first indication of the numerous important connections of the ζ-function with number theory.

The zeta function has no zeros for $\sigma > 1$, but has zeros at $s = -2, -4, \cdots$. Riemann conjectured that all of the other zeros of $\zeta(s)$ are on the "critical" line $\sigma = 1/2$. This conjecture, if proven, would have many remarkable number-theoretic consequences.

Here's to the zeros of ζ of s
G. Bernard Riemann has made a good guess
They're all on the critical line, said he
And their density's one over 2π log t.

This conjecture of Riemann has got them all started
But many a good man from this life has departed
Without ascertaining with suitable rigor
What happens to zeta when log t gets bigger.

This ditty (to the tune of "Sweet Betsy from Pike") goes on to record that Hardy has at least proved that there are infinitely many zeros on the critical line. More seriously there are many notable applications of complex analysis to number theory; they constitute the subject known as *analytic number theory*. For instance, this theory gives the most perspicuous (but not the most elementary) proof of the *prime number theorem*: If $\pi(x)$ is the number of prime numbers less than or equal to x, then

$$\lim_{x\to\infty} \frac{\pi(x)\log x}{x} = 1. \tag{9}$$

Covering spaces arise with Riemann surfaces, but their utility extends well beyond complex analysis. Any "well-behaved" connected space X (in particular, any connected manifold) has a covering space $p:\tilde{X} \to X$ which is connected and simply connected, while the elements h of the fundamental group of the base space X act as continuous transformations $h:\tilde{X} \to \tilde{X}$ with $ph = p$; they are then called *covering transformations*. This covering space is *universal*, in the sense that for every other covering space $q: Y \to X$ there is a covering space $p':\tilde{X} \to Y$ with $p = qp'$, as in

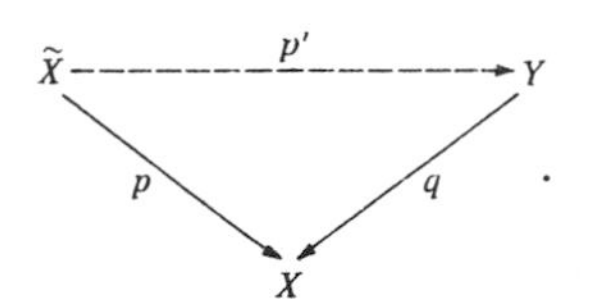

The universal covering space $\tilde{X}$ can be constructed by taking its points to be homotopy classes of paths from a fixed base point in X, with endpoints of the path fixed under the homotopy. Both Massey [1967] and Pontryagin [1939[give clear expositions of covering spaces.

The covering space serves to visualize the way the fundamental group of a space "acts" on other objects associated with the space. It is just one example of the interaction between the quite explicit constructions of complex analysis and the general development of topology.

12. Germs and Sheaves

The process of analytic continuation by pasting together circles of convergence can be stated more conceptually in terms of germs and sheaves of germs. Consider two functions f and g holomorphic in neighborhoods U and V of a point c in $\mathbf{C}$. They are said to have the same *germ* at c if there is an open neighborhood W of c with $W \subset U \cap V$ and $f = g$ in W. (With the same words, we can define when two continuous functions have the same germ at a point in any topological space.) The binary relation "f and g have the same germ at c" is reflexive, symmetric and transitive; hence we can form for each f its equivalence class $[f]_c$, the germ of f at c, consisting of all g holomorphic in some neighborhood of c and having there the same germ as f.

By Taylor's theorem, f and g have the same germ at c if and only if they have the same Taylor series at c. Therefore a germ at c can be identified (and so made more "concrete") with a power series in $z - c$, convergent in some open circle about c. No such easy identification is possible for germs of real-valued functions which are merely continuous.

Let A_c denote the set of all germs of holomorphic functions at the point c, while the union A of all the A_c is the set of all germs at all points of $\mathbf{C}$. It is in a natural way a topological space, called the *sheaf* of *germs* of *holomorphic* functions on $\mathbf{C}$. To describe this topology, take for each holomorphic $f: U \to \mathbf{C}$ the set of germs

$$W_{f,U} = \{\text{all germs of } f \text{ at points of } U\} = \{[f]_c \mid c \in U\}$$

and call this set a neighborhood of the germ $[f]_c$ in A. In brief, neighboring germs are those which come from the same holomorphic function. This informal description of "neighborhoods" can be made formal, by stating that the sets $W_{f,U}$ are a *base* for the open sets of the intended topological space; this means that an open set of A is any union (possibly an empty or an infinite union) of sets $W_{f,U}$. To show that the axioms for a topological space hold, it is then enough to show that the intersection of two such open sets is again open, and for this it is enough to show that any intersection $W_{f,U} \cap W_{g,V}$ is either empty or a neighborhood. This is easy: If $[f]_c = [g]_c$ at some point $c \in U \cap V$, then $f = g$ in an open set U' containing c; the desired intersection $W_{f,U} \cap W_{g,V}$ is therefore the union of such sets $W_{f,U'}$. This proof that A is a topological space is a typical use of the idea of neighborhood and the notion of a base for a topology. In this particular case, A is actually a Hausdorff space; this would not be the case for sheaves of germs of simply continuous functions.

There is a natural projection $p:A \to \mathbf{C}$ which sends each germ $[f]_c$ to the indicated point c. This projection is continuous; in fact, it has a much stronger property: It is a *local homeomorphism*. To each point a in A there is a neighborhood W such that $p(W)$ is open in $\mathbf{C}$ while p, when restricted to W, is a homeomorphism $W \to pW$. In other words, each germ a has a neighborhood which, under p, behaves exactly like a neighborhood of pa in the complex plane; the sets $W_{f,U}$ are such neighborhoods. Observe that the projection $p:\tilde{X} \to X$ of any covering space is a local homeomorphism in the sense of this definition. However, the projection $p:A \to \mathbf{C}$ for germs is by no means a covering map. About each point c in $\mathbf{C}$ one can find holomorphic functions with smaller and smaller radii of convergence. Hence there is no one neighborhood of c which reappears around every germ about c. In short, the space A is very large; it can evidently be described as the space of all possible convergent power series $\Sigma\, a_n(z-c)^n$, for all c, where each power series is regarded not as a function of z, but simply as a thing—a point of A. In general topology, "points" can be almost anything, and "spaces" can be very large and still conceptually useful. There is also a "very large" continuous function $F:A \to \mathbf{C}$, which sends each germ $[f]_c$ to the value $f(c)$ of the original f. In symbols, $f([f]_c) = f(c)$.

This one space A with the function F "represents" all possible holomorphic functions by way of "cross sections", as follows:

Theorem. *For any holomorphic function $g: V \to \mathbf{C}$ on an open subset V of the complex plane $\mathbf{C}$ there is a unique continuous cross section $s: V \to A$ of p with $F \cdot s = g$. Conversely, each continuous cross section $s: V \to A$ of p over an open set V of $\mathbf{C}$ yields a holomorphic function g on V as the composite $g = F \cdot s$.*

Here a *cross section* s of p means a map $s: V \to A$ such that the composite $p \cdot s$ is the "identity"; that is, the inclusion of V in $\mathbf{C}$ as in the diagram below.

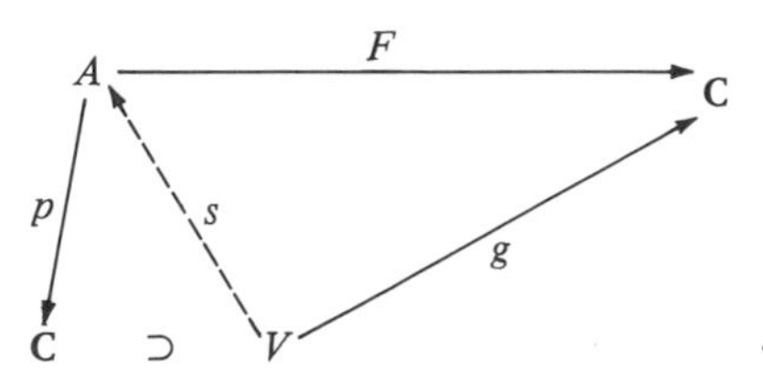

Proof. Given g, the only choice for s is the mapping with $s(c) = [g]_c$ for each c in V. This definition provides a continuous cross section; moreover for F as above the composite $F \cdot s$ is, as in the diagram, just the given holomorphic function g.

Conversely, let s be any cross section of p over an open set V in $\mathbf{C}$. It must send each point $c \in V$ to the germ $[f]_c$ at c of some function $f = f_{s,c}$ holomorphic in some open set U containing c. The composite $g = F \cdot s$ then sends c to $F(sc) = f(c)$; it remains to show g holomorphic. But the continuity of s at c means that some neighborhood V_0 of c in $\mathbf{C}$ is sent by s into the open set $W_{f,U}$ of germs of f. Hence, for c' in V_0, $sc' = [f]_{c'}$ so that $gc' = fc'$. Therefore g, like f, is holomorphic.

It may at first seem strange that the mere requirement of continuity of a cross section over V produces a function g on V which is not just continuous, but even holomorphic. This results because the holomorphic character is built into the construction of A: Each point of A is the germ of some function which *is* holomorphic.

Holomorphic functions on $\mathbf{C}$ yield a presheaf H by

$$H(V) = \{g \mid g: V \to \mathbf{C} \text{ is holomorphic}\}.$$

Indeed, if $V_1 \subset V$, each holomorphic function on V restricts to one on V_1, and these restriction maps make $H(V)$ a presheaf, as defined in §VIII.11. Moreover, this presheaf H is a sheaf. If $V = \cup V_j$ is any covering of $V \subset \mathbf{C}$ by open sets V_j, while there are corresponding holomorphic functions $g_j: V_j \to \mathbf{C}$ such that each pair g_j and g_k agree on the intersection $V_j \cap V_k$, then there is a function $g \in H(V)$ which restricts to g_j on each V_j; indeed, g is defined at each point $c \in V$ as the common value of all the $g_j(c)$ for all indices j with $c \in V_j$; this g is clearly holomorphic.

By the theorem, this sheaf H may also be described by taking each $H(V)$ to be the set of cross sections over V of the projection $p: A \to \mathbf{C}$. In other words, the idea of a holomorphic function can be formulated by a sheaf or as cross sections of the big space A.

The space A is actually a 2-dimensional manifold *and* a Riemann surface (but *not* connected), because each point $[f]_c$ of A has a neighborhood

$W_{f,U}$ which is homeomorphic to the open set $U \subset \mathbf{C}$, and these homeomorphisms serve as the charts for the desired Riemann surface structure. However, A is far from being connected. It does consist of connected pieces, according to the following very general decomposition theorem.

Theorem. *Any n-dimensional manifold M is the union of disjoint connected subsets M_j, each of which is itself a manifold of dimension n and an open subset of M.*

The subsets M_j are called the connected *components* of M; this result is true for much more general topological spaces; we need it only for manifolds, where "connected" is equivalent to "path connected" (see §VIII.6).

To prove the theorem, take any point q in M. The component M_j containing q is then defined to be the set of all of those points of M which can be joined to q by a (continuous) path in M. Since each point q' has an open neighborhood homeomorphic to an open set of $\mathbf{R}^n$, every path from q to q' can be prolonged (e.g., by straight segments in $\mathbf{R}^n$) to any point in a suitable open neighborhood of q'. Therefore, the set M_j is open. It can be described as the equivalence class of a point q under the reflexive, symmetric, and transitive relation "can be connected by a continuous path".

In the space A each connected component A_j is just a connected Riemann surface. Indeed, a point q of A is a point c in $\mathbf{C}$ together with an open disc with center c and a power series $f(z) = \Sigma\, a_n(z-c)^n$ convergent there. A path from q in A then looks like a path from c in $\mathbf{C}$ with a disc and series attached at each point, so that they overlap, as in Figure 1. The figure shows only a finite number of discs, which suffice to cover because the interval is compact. It thus follows that the connected component of the point q in A consists of all the power series obtained by analytic continuation of the original series f; hence it is exactly the Riemann surface of f. This conceptual description of the Riemann surface

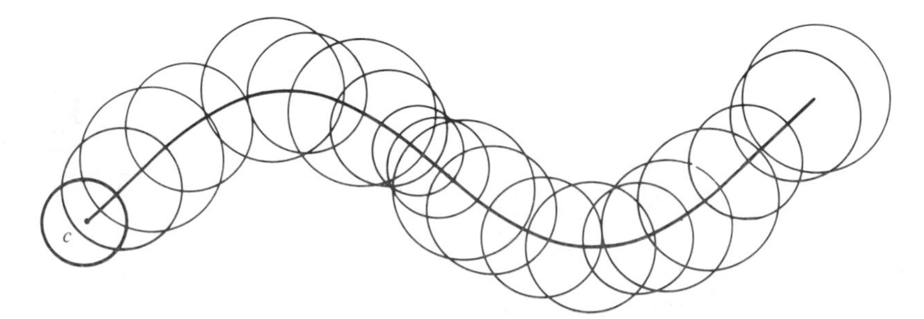

Figure 1. Continuation along a curve.

as a connected component of the big space A helps to understand examples like the analytic continuation of $1/(z-2)$ displayed in Figure 11.2.

This example illustrates how very general abstract concepts can give a better understanding of concrete examples. Once we have constructed the big space A and applied to it the easy geometric idea of breaking it up into components we have *all* the Riemann surfaces, including for instance the Riemann surface so meticulously constructed in §VIII.4 for the function $\sqrt{(z^2-1)(z^2-4)}$.

This cheerful statement is not quite true, but it can be easily repaired. First, one should enlarge A to include also germs at ∞, giving a local homeomorphism $p:A_\infty \to S^2$ to the Riemann sphere. Second, to each component of A_∞ one should adjoin "branch points", like the point 0 for $\sqrt{z}$ and the four branch points $z = \pm 1$, $z = \pm 2$ for $\sqrt{(z^2-1)(z^2-4)}$. These examples are all branch points of "order" 2. Generally, a branch point has some positive integer k as order and can be considered as the "germ" of a convergent power series in a uniformizing parameter $t = (z-c)^{1/k}$. Third, when a function f has a pole the "germ" of the corresponding Laurent series should be adjoined to its Riemann surface: This is to include poles at branch points. This will insure that the Riemann surface for a rational function such as $1/(z-1)$ is the whole Riemann sphere, including the pole at $z = 1$ and the zero at ∞. In particular this Riemann surface is compact.

There is an important *monodromy* theorem which specifies when analytic continuation along two different paths to a point c' from a germ at c yields the same germ at c'. The proof depends on an important topological process of "lifting" a homotopy from $\mathbf{C}$ to A.

Our conceptual definition of a Riemann surface is just the starting point for many beautiful and subtle developments. For example, an analytic isomorphism $f:S \to S'$ of two Riemann surfaces is a holomorphic function f together with a two-sided holomorphic inverse $g:S' \to S$. The Riemann mapping theorem asserts that any connected and simply connected open set U of $\mathbf{C}$ which is not all of $\mathbf{C}$ is analytically isomorphic to the unit disc. (One can also describe all the analytic isomorphisms of the disc to itself.) More generally, the *uniformization theorem* asserts that any connected and simply connected Riemann surface is analytically isomorphic to one of the following: The unit disc, the complex plane, or the Riemann sphere. The proof of this theorem requires analytic methods much more subtle than any we have summarized.

Such analytic isomorphisms also have practical uses. Thus the exterior of the unit circle in $\mathbf{C}$ can be mapped conformally upon the exterior of the profile of an airplane wing, as in Figure 2. A steady flow of air around the wing can then be replaced by a flow around the circle (see §3 above). There are very useful explicit formulas for constructing and approximating such conformal maps.

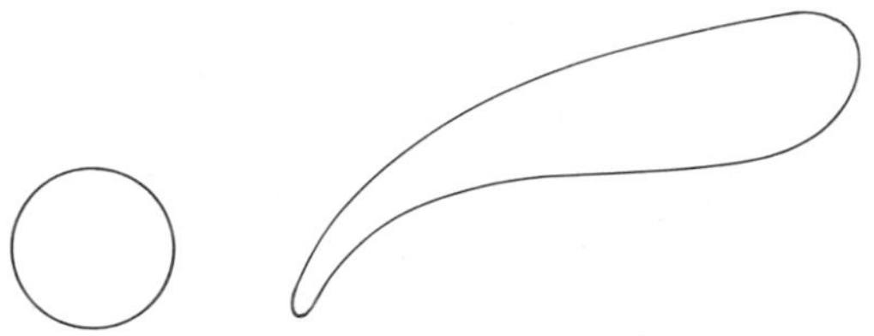

Figure 2. Airfoil.

13. Analysis, Geometry, and Topology

This chapter has not been limited to complex analysis. It has been arranged to illustrate the remarkable interaction between analysis and geometry, in which various immediate and practical problems lead to new and general concepts. These concepts in turn will help to understand the solutions of the original problems and turn out to have specific applications elsewhere; moreover they help develop subtle general theorems like the big Picard theorem and the uniformization theorem.

Complex analysis starts from the notion of differentiation, as it was needed to formulate velocity and acceleration in mechanics. Complex differentiation yields the class of holomorphic functions and their remarkable properties. At the same time, these functions provide conformal maps and have application to electrostatic potential and fluid flows. The construction of a variety of holomorphic functions suggests the process of analytic continuation along a curve, which can then be codified and understood as the choice of a connected component in the big space A of germs of holomorphic functions. The explicit Riemann surfaces, used to handle many-valued functions by way of slitting and pasting planes together, all fit in the big topological space A.

This is typical of the development of abstract concepts. The original and tangible notion that a space is something physically visible, consisting just of points with no extension, gets gradually enlarged, until one arrives at wildly big topological spaces such as the space A of germs in which a point is "really" a whole disc carrying a power series. This same general process–topologize the "manifold" of all the objects under study–has many other uses in Mathematics.

The notion of function also changes. The holomorphic functions (on open sets U of $\mathbf{C}$) form a sheaf H which can be represented as the sheaf of cross sections of a local homeomorphism $p{:}A \to \mathbf{C}$. This was historically the first example of a sheaf–a notion which now crops up for all sorts of manifolds, complex, differentiable, smooth, or algebraic. More generally, give a sheaf S on any topological space X, one may construct a space A, whose points are the germs of S, together with a local homeomorphism $p{:}A \to X$. The given sheaf S is then the sheaf of cross sections of p, so that

$S(U)$ is the set of cross sections over U. In other words, any good class of functions can be represented simply as continuous cross sections. This idea has applications even in set theory, as we will note in the next chapter.

CHAPTER XI

Sets, Logic, and Categories

The rich multiplicity of Mathematical objects and the proofs of theorems about them can be set out formally with absolute precision on a remarkably parsimonious base. Thus almost all the objects of Mathematics can be described as sets: A natural number is a set of sets (a cardinal), a rational number is a set of pairs (an equivalence class), a real number is a set of rationals (a Dedekind cut), and a function is a set of ordered pairs (a table of values). Similarly the theorems of Mathematics can all be written as formulas in a very parsimonious formal language which uses only set-membership, the basic connectives of logic (or, not, there exists) and the needed primitive terms of each subject (thus "point" and "line" for incidence geometry). Finally, most of the proofs of Mathematical theorems can be stated with absolute rigor as a sequence of inferences, each an instance of a finite number of basic schemes of inference.

A "foundation" for Mathematics is provided by this remarkable reduction of Mathematics to sets, well formed formulas, and formal inference. It does miss a little of Mathematics–formally undecidable theorems and constructions which need the "universe" of all sets. Since we hold that Mathematics is based on human activities and scientific problems we have delayed the discussion of the foundations to this point, when we have at hand a considerable sample of the things to be "founded".

Almost all Mathematicians now use the convenient language of set theory to describe Mathematical structures, but very few Mathematicians can recite the axioms which found set theory, while almost none of them (except for the logicians) can specify the exact rules of logical inference. Hence in this chapter we start with the standard Zermelo–Fraenkel (ZF) axioms of set theory and then the calculi of propositions and of predicates, complete with *modus ponens* and the other needed rules of inference. This allows a formal definition of "proof" and a description of the Gödel incompleteness theorem (§7). (In a recursive formal system strong enough to code arithmetic, there are formal statements which can be neither proved nor disproved.) Moreover, the Zermelo–Fraenkel axioms do not settle the famous continuum hypothesis–there are models of ZF for which the hypothesis holds and others for which it fails. The demonstration of the latter fact by Cohen's method of forcing (§8) suggests other "independence" results and leads to the construction of many alternative models of set theory. Another result is the introduction of a considerable variety of axioms meant to supplement ZF.

For these reasons "set" turns out to have many meanings, so that the purported foundation of all of Mathematics upon set theory totters. But there are other possibilities. For example, the membership relation for sets can often be replaced by the composition operation for functions. This leads to an alternative foundation for Mathematics upon categories–specifically, on the category of all functions. Now much of Mathematics is dynamic, in that it deals with morphisms of an object L into another object of the same kind. Such morphisms (like functions) form categories, and so the approach via categories fits well with the objective of organizing and understanding Mathematics. That, in truth, should be the goal of a proper philosophy of Mathematics.

1. The Hierarchy of Sets

The explicit recognition of sets as objects of serious Mathematical study came about in the 19th century in at least two ways. On the one hand, Dedekind used sets to explain Kummer's ideal numbers as suitable sets of ordinary numbers (§XII.3) and to explain real numbers as cuts–that is, as sets of rational numbers. On the other hand, questions about Fourier series forced the consideration of quite general (and bizarre) sets of real numbers. Fourier, in studying the conduction of heat (1822) was led to represent periodic functions, of period say 2π, as sums of infinite series of sines and cosines,

$$\begin{aligned} f(x) \sim a_0/2 + a_1 \cos x + b_1 \sin x + \cdots \\ + a_n \cos nx + b_n \sin nx + \cdots. \end{aligned} \tag{1}$$

It was remarkable how many such functions could be so represented (§VI.11) but Fourier's methods of showing this were intuitive and unreliable. Because of their evident utility not just for the heat equation but for other partial differential equations, Mathematicians turned to the careful examination of such series. This led to the lively development of the subject now known as *harmonic analysis*, not just for the rotation group of the circle (which is present but hidden in (1) by the use of the angular coordinates x, $0 \leqslant x < 2\pi$, for the circle) but for many other groups.

One problem about such trigonometric series, as posed by Riemann, is the problem of *uniqueness*: When a function f is the limit of such a series

(1) for all x, does f determine the coefficients a_n and b_n uniquely? George Cantor found that the answer was "yes"; indeed, he found that he did not need to assume that the series (1) converged to $f(x)$ for all x, but only for a great many x. So he was led to the question: For what sets S of reals x will

$$\sum_n a_n \cos nx + b_n \sin nx = 0$$

for all $x \in S$ imply that all the coefficients a_n and b_n are zero? It turned out that some of these sets S required elaborate descriptions. This started him thinking about general sets; so he became the founder of set theory.

Cantor developed set theory extensively, introducing the cardinal number, card A, of a set A (§II.8) and the ordinal number of a well-ordered set (§II.9). Two cardinal numbers are compared by the rule that Card $A \leqslant$ Card B if there is an injection $A \to B$. The Schroeder–Bernstein theorem asserts that Card $A \leqslant$ Card B and Card $B \leqslant$ Card A together imply Card $A =$ Card B. (For a proof, see *Survey*, Chapter XIII.) From this and the axiom of choice it follows that the cardinal numbers are linearly ordered. Using a "diagonal" argument, Cantor proved that the set $\mathbf{R}$ of real numbers is not denumerable. This means that Card $\mathbf{N} <$ Card $\mathbf{R}$. He then raised the question: Is there any cardinal number properly between these two? The statement that there is no such is Cantor's *continuum hypothesis*. ($\mathbf{R}$ is commonly called "the continuum".)

Bigger and bigger sets can be constructed by the *power set* operation P, which assigns to each set A its power set $P(A)$, consisting of all subsets of A. For example, $P(\mathbf{N})$ has the same cardinal as $\mathbf{R}$. The diagonal argument mentioned above shows that the cardinal of $P(A)$ exceeds that of A. For if $f\colon A \to P(A)$ were a bijection, one could define a subset $S \subset A$ as

$$S = \{x \mid x \in A \text{ but not } x \in f(x)\}; \tag{2}$$

here $x \in A$ means "X is an element of A", as usual. Since $S \in P(A)$ and f is a bijection, there must be some a in A with $f(a) = S$. Then $a \in f(a)$ would, by the definition (2), give $a \notin f(a) = S$; while $a \notin S$ means by (2) again that $a \in f(a)$–a contradiction. This is called a "diagonal" argument because the set S is the complement in A of the "diagonal" set consisting of all those elements x with $x \in f(x)$. Recall that the Russell paradox (II.8.5) also involves a diagonal construction

$$R = \{S \mid \text{not } S \in S\}. \tag{3}$$

Inevitably we come to the question: What is a set? Initially, we described sets as (well defined) collections of Mathematical objects, without really specifying what a "collection" is or what an "object" is (a number? a point? a series?). Now note that many constructions produce sets all of whose elements are themselves sets: The set $P(A)$ of all subsets of A; the set $\mathbf{Z}_m$ of all congruence classes of integers, modulo m, or the cardinal number of A. This raises the possibility of considering *only* those sets whose elements x are themselves sets, the elements t of those elements x again being themselves sets, and so on. This "and so on" will be more persuasive if we start from the bottom with the empty set, written as $\varnothing$. It has only one subset, itself, so the set $P(\varnothing)$ of all subsets of $\varnothing$ is just the set $\{\varnothing\}$ whose only element is the empty set. Write these two sets as

$$R_0 = \varnothing, \qquad R_1 = P(R_0) = \{\varnothing\}. \tag{4}$$

Now the one-element set $\{\varnothing\}$ has two subsets, itself and the empty set, so its power set is a two-element set

$$R_2 = P(R_1) = \{\varnothing, \{\varnothing\}\}. \tag{5}$$

This set in its turn has four different subsets, so its power set is

$$R_3 = P(R_2) = \{\varnothing, \{\varnothing\}, \{\{\varnothing\}\}, \{\varnothing, \{\varnothing\}\}\}; \tag{6}$$

here $\{\{\varnothing\}\}$ of course denotes the set whose only element is the set whose only element is the empty set $\varnothing$. Iterating this process produces sets R_n with 2^n elements. The union of all these sets R_n is a denumerable set R_ω; here ω designates the first infinite ordinal number.

The construction started in (4), (5), and (6) can be formalized as a recursion over the ordinal numbers. Here an ordinal $\gamma \neq 0$ which is not the successor $\gamma = \alpha + 1$ of some ordinal α is called a *limit ordinal*; ω is the first such. The definition of the sets R_α by recursion on the ordinal number α then reads

$$R_0 = \varnothing, \qquad R_{\alpha+1} = P(R_\alpha), \qquad R_\gamma = \bigcup_{\beta<\gamma} R_\beta, \tag{7}$$

where the last clause applies when γ is a limit ordinal. Since the ordinal numbers α are well-ordered (i.e., each non-empty set of ordinals has a first element), one can define the *rank* of any set x in this hierarchy (7) to be the first ordinal α with $x \in R_\alpha$. All the elements of x then have smaller rank, so that no set x can be a member of itself. Each set x in the hierarchy has only sets as elements, and can be pictured as a "tree" which presents the elements of its elements. . . . For example, the set (6) above is the following tree

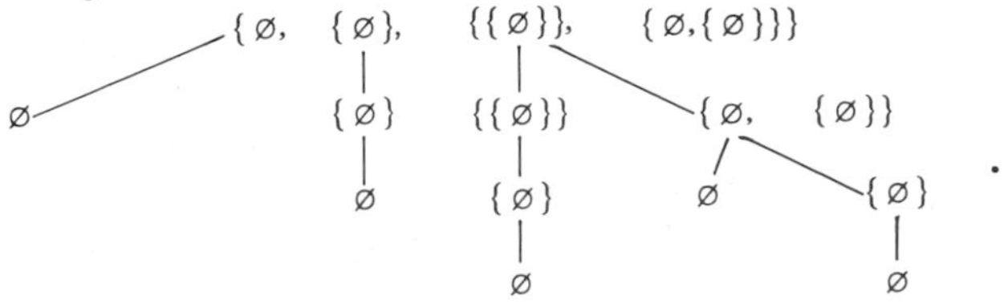

```
              {∅,   {∅},   {{∅}},      {∅,{∅}}}
             /        |       |    \
            ∅        {∅}    {{∅}}   {∅,    {∅}}
                      |       |     /  \
                      ∅      {∅}   ∅    {∅}          .     (8)
                              |          |
                              ∅          ∅
```

The whole *cumulative hierarchy* (7) is intended (or imaged) to produce all the sets that we need for Mathematics. Though it does not produce any "universal set" of all sets, it does produce sets of all finite sizes as well as many infinite sets. Note that the description of this hierarchy uses the (informal) ideas of empty set, union of a set of sets, subset, and power set, and it rests also on the intuitive idea of the ordinals (or some equivalent notion of "stages" of construction). On this basis, the hierarchy provides a systematic "picture" of all needed sets. It is speculative and not formal (see also §4 below).

2. Axiomatic Set Theory

The standard foundation for Mathematics intends to build up all Mathematical concepts out of sets. We have already indicated some of the steps: Functions and then bijections are defined as sets of ordered pairs of sets; cardinal numbers are equivalence classes of sets, under bijections; natural numbers are finite cardinals; integers use pairs of natural numbers; rational numbers are equivalence classes of pairs of integers; real numbers are sets (Dedekind cuts) of rationals; functions on the reals are suitable sets of pairs of reals; complex numbers are pairs of reals; functions of complex numbers are sets of pairs of complex numbers. Riemann surfaces are sets of charts where each chart is a suitable function which again is a set of pairs. And so on. With some care and some pedantry, (almost) all Mathematical concepts can in this way be reduced to constructions on sets. Of course a class–as in the case of equivalence classes–is here taken to be a set, while an ordered pair of sets is to be defined as a certain set, as indicated in §V.4.

This approach to Mathematics has the advantage that every concept can be made absolutely clear and explicit. Thus a Riemann surface *is* just a set of charts and not some mysterious configuration formed by ingeniously pasting together sheets cut from the complex plane. But if this approach is to be dependable, it must avoid the Russell Paradox and its ilk. That paradox seems to spring from the naive conception of a set as any collection of identifiable things. To avoid this, one may adopt axioms for sets which appear to hold in the cumulative hierarchy we have described. For this purpose, sets are not described as "collections"; one simply posits that there are sets $x, y, z, \ldots$ which may stand in the relation $x \in y$ of membership (x is an element of y) and that these sets satisfy certain axioms formulated strictly in terms of this relation $\in$. Everything else about sets and the Mathematical objects constructed from them is to be defined (ultimately) in terms of this membership relation and all theorems are to be proved from these axioms.

The Zermelo–Fraenkel axioms system (ZF) is standard. One first defines the inclusion relation $\subset$ between sets x and y by

$$x \subset y \iff \text{For all } t, \quad (t \in x \Rightarrow t \in y). \tag{1}$$

where $\iff$ is short for "if and only if" while $\Rightarrow$ is short for "implies". Then the equality of sets is defined by

$$x = y \iff x \subset y \ \& \ y \subset x. \tag{2}$$

With (1), this amounts to stating that two sets are equal if and only if they have the same elements–a view basic to the idea of a set.

The Zermelo axioms now come in the following list:

Extensionality, Null set, Pairing, Power set, Union, Infinity, Comprehension, and Regularity. (3)

In detail, these axioms read:

Extensionality.

$$x = y \ \& \ y \in z \quad \text{imply} \quad x \in z. \tag{4}$$

This states that equals may be substituted for equals on the left in any statement $y \in z$. The definition (2) of equality has already provided for substitution on the right in $t \in y$:

$$x = y \quad \text{and} \quad t \in y \Rightarrow t \in x.$$

Sometimes both $\in$ and $=$ are taken as (independent) primitive notions for sets. In this case (4) is just a statement of a basic property of $=$ (an axiom), while the definition (2) becomes in this version the axiom of extensionality.

Null set. There is a set $\varnothing$ with no elements.

By the definition (2), any two such sets are equal.

Pairing. For any sets x and y there is a set u so that, for all t,

$$t \in u \iff (t = x \quad \text{or} \quad t = y). \tag{5}$$

By (2) again, any two such sets u are equal, so one writes $u = \{x,y\}$. This axiom provides also for a *singleton*: A set $\{x\} = \{x,x\}$ with just one element x. It also produces ordered pairs $\langle x,y\rangle = \{\{x\},\{x,y\}\}$ as explained in (V.4.4).

Power set. For any set x there is a set u with, for all s,

$$s \in u \iff (s \subset x). \tag{6}$$

This set u, again unique by (2), is the *power set* $u = Px$.

Union. For any set x there is a set u with, for all t,

$$t \in u \iff \text{There is an } s \quad \text{with} \quad t \in s \in x .$$

This set u, again unique, is the union of all the members of x, usually written $\bigcup_x$:

$$t \in \bigcup_{s \in x} s \iff (\exists s) \qquad (t \in s \,\&\, s \in x).$$

In particular, if $x = \{y,z\}$, this set is the union $y \cup z$.

Before formulating the next axiom (that of infinity), notice that we can now construct for each set x the singleton set $\{x\}$, the doubleton $\{x,\{x\}\}$ and hence the union of the latter set, which is called the *successor* of x

$$s(x) = x \cup \{x\}. \tag{7}$$

In particular, this will yield all the finite ordinal numbers, as in the definition suggested in (II.9.5):

$$0 = \varnothing, \qquad 1 = s(\varnothing) = \{0\}, \qquad 2 = s(1) = \{0,\{0\}\} = \{0,1\} \cdots . \tag{8}$$

Each ordinal is thus the set of all smaller ordinals, so the first infinite ordinal ω should be the set of all finite ordinals. Hence the existence of at least one infinite set can be insured by requiring that there be a set containing this ω; that is, a set containing this $\varnothing$ and closed under the operator s of (7).

Infinity. There exists a set w with $\varnothing \in w$ and such that $x \in w$ implies $s(x) \in w$.

Comprehension. For any set u and any property P of elements of u there is a set s with

$$x \in s \iff x \in u \quad \text{and} \quad x \text{ has } P. \tag{9}$$

That set s, again unique, is usually written as

$$s = \{x \mid x \in u \text{ and } x \text{ has } P\}. \tag{10}$$

We have used this construction repeatedly; for example, it provides for the intersection $u \cap v$ of two sets. It also allows one to "cut out" or separate *from* a *given set* u, the subset s of all those elements x of u which do have the specified property P; hence it is sometimes called the *axiom of separation*. This formulation of the axiom does not allow an immediate construction of Russell's paradoxical set R of (1.3), because it provides only for the construction of subsets of a *given set u*, and not, like R, for subsets of the universe of all possible sets. This formulation is also in accord with the way sets are constructed in the cumulative hierarchy of §1 above.

Actually, this axiom (9) is really an infinite family of axioms, one for each property P; it is thus called an *axiom scheme*.

These axioms are essentially those formulated by Zermelo in 1908. The last axiom (9) is not in satisfactory form, because it refers not just to the one primitive notion $\in$, but also to a "property". With Skolem and Fraenkel, one must then append here the explanation that a "property" P of x means something specified by an explicit set-theoretic formula; in other words, something formulated only in terms of x, membership and the usual logical connectives. Thus for example, "For all t and s, $t \in x$ and $s \in x$ imply $t \subset s$" is a property of x.

Regularity. If a set x is not empty it has an element w which has no elements in common with x.

In other words

$$x \neq \varnothing \Rightarrow (\exists w)\, w \in x \quad \text{and} \quad x \cap w = \varnothing. \tag{11}$$

To see what this axiom means, observe first that it shows that there can be no set x with $x \in x$ (no set is a member of itself). For if this were so, the singleton set $\{x\}$ is non-empty, but its only element, to wit x, has an element (x again) in common with x. Similarly, the axiom insures that never $x \in y \in x$. For if this were so, $\{x,y\}$ would be a non-empty set which has elements (x or y) in common with each of its elements. Also, once we have defined sequences y_n of sets, the axiom of regularity will insure that there is no infinite "descending" sequence with

$$\cdots y_n \in y_{n-1} \in y_{n-2} \in \cdots \in y_1 \in y_0, \tag{12}$$

for then the set $x = \{y_0,y_1,\ldots\}$ violates this axiom. For this reason, the axiom of regularity is also called the axiom of *foundation*: Given a set y_0, there is no infinite regress of elements of its elements.

The intuitive intent is that one should be able to "see" that all these axioms appear to be valid for the cumulative hierarchy of sets R_α described in §1. For example, in the case of the regularity axiom, each non-empty set x has some ordinal rank α, while all the elements of x have smaller rank. Among the ordinal ranks of elements $t \in x$ there is (by well-ordering of ordinals) a first ordinal β. Any element $w \in x$ of this rank β has for x the property required in the regularity axiom (11). Put differently, an infinite descending sequence of sets like (12) would produce an infinite descending sequence of ordinals. This intuitive argument depends on a prior intuitive notion of ordinal.

These Zermelo axioms are usually supplemented by the *axiom of choice*. After functions have been defined, this axiom reads

Choice. For every set x with no non-empty members y there is a function f with domain x such that $f(y) \in y$ for each $y \in x$.

In other words, f "chooses" for each $y \neq \varnothing$ in x a particular element $f(y) \in y$; thus f is a "choice function" for x. It is needed only when x is an infinite set, and many "infinite" arguments in Mathematics use this axiom, often in one of the equivalent forms (a Zorn lemma). Zermelo proved that the axiom of choice implies that every set can be well-ordered.

Just as in the axiomatization of geometry, the operational meaning of the (Zermelo) axioms for sets is that "everything" about sets should be formally demonstrable from the axioms without using the intuitive idea that a set is some sort of collection of things. We now illustrate this process briefly, but only for the ordinal and cardinal numbers.

First the intuition. The basic idea, as already used, is that of a well-ordered set. In particular, the ordinals themselves are well-ordered:

$$0 < 1 < 2 < 3 < \cdots < \omega < \omega + 1 < \cdots .$$

Generally, an ordered set is well-ordered when every non-empty subset has a first element in the order or, equivalently, when there are in the order no infinite descending sequences (compare the sequence (12)). Now the axioms easily produce sets which are "models" for the finite ordinals: $0 = \varnothing$, while the successor function s of (7) produces finite ordinals such as

$$1 = \{0\}, \qquad 2 = \{0,1\}, \ldots, n = \{0,1,2, \ldots, n-1\}.$$

The axiom of infinity suggests that the first infinite ordinal is

$$\omega = \{0,1,2,3,4, \ldots\}.$$

On inspection, each of these sets is "transitive". Here a set t is called *transitive* when $x \in y \in t$ implies $x \in t$; in other words, every element y of t is also a subset of t.

Next the formal development. With von Neumann, one can now define an *ordinal* to be a transitive set α such that every element $t \in \alpha$ is also transitive. (This is the case for ω and the finite ordinals n.) It then follows from the axioms that the elements of an ordinal α are linearly ordered by the membership relation $\in$. With the regularity axiom one can then prove that this linearly ordered set α is well-ordered–i.e., that each non-empty subset $x \subset \alpha$ has a first element in the order of elements of α. For, by regularity, there is some element $y \in x$ with no element in common with x. In particular, $z < y$ for some $z \in \alpha$ means that $z \in y$ and so by transitivity of x that $z \in x$, a contradiction to the fact that x and y have no elements in common. Hence y is the required first element of x.

This proof is one of the critical uses of the regularity axiom.

This formal definition of the von Neumann ordinals α will lead to all the needed properties of ordinals; in particular one can prove that every well-ordered set P is order-isomorphic to a von Neumann ordinal. This in turn means that the von Neumann ordinals "represent" all the ordinals given by the older definition (§II.9): An ordinal number is the order type ord P of some well-ordered set, where

$$\text{ord } P = \{P' \mid P' \in V \text{ and } P \sim P'\}, \tag{13}$$

while $P \sim P'$ means that P and P' are ordinally isomorphic and V is some "type" or "universe". (Some sort of universe is needed to make the comprehension axiom apply here; we cannot simply take *all* sets P' with $P \sim P'$.) Whatever the universe, if it is big enough it will contain a von Neumann ordinal α with $P \sim \alpha$, so the older ordinal, or P, *is* realized or "represented" by α.

(One can describe a universe as a set V such that the elements of V by themselves satisfy the Zermelo axioms, but that requires an added axiom–there exists a universe–and might (and has!) led to the use of a whole succession of universes!)

The naive definition of a cardinal number (§II.8) in terms of the relation of cardinal equivalence reads

$$\text{card } S = \{S' \mid S \equiv S'\}; \tag{14}$$

again an illegal use of comprehension. This difficulty can be avoided by using a universe, much as in (13). More elegantly, one can first show from the axioms that one can carry out definitions by recursion over the ordinals, as in the definition (1.7) of the hierarchy R_α. This also allows the definition from the axioms of the ordinal *rank* $\rho(T)$ of each set T, as the least α with $T \in R_\alpha$. Then one can describe the cardinal number of a set S by collecting only the equivalent sets of smallest rank. Formally, one can legitimately define

$$\text{card } S = \{S' \mid S \equiv S' \text{ and } \forall T,\ T \equiv S \Rightarrow \rho(S') \leqslant \rho(T)\}. \tag{15}$$

Alternatively, the troubles with comprehension in (14) can now be avoided by defining a cardinal number to be an ordinal which is not cardinally equivalent to any smaller ordinal. Thus ω is a cardinal, but $\omega + 1$ is not. Again, the cardinal numbers so defined provide representatives for each of the (perhaps non-existent?) equivalence classes (14)–because every set can be well-ordered.

This treatment of the basic Mathematical idea of number–ordinal and cardinal–illustrates the clarification of ideas possible by a suitable axiomatization. Here the initial (and central) idea of a well-ordered set has come into clearer focus by way of the models provided by the regularity axiom and the notion of a transitive set.

From the infinite set ω and the iterated power set axiom one obtains a tower

$$\omega, P\omega, P^2\omega, \ldots, P^n\omega, \ldots$$

of sets which, by the diagonal argument, get larger and larger. To get a still larger set one would need to form the union of all these sets $P^n\omega$ for all natural numbers n, and then apply P again. But one cannot form this union by the axioms unless one knows that there is a set consisting of all the sets $P^n\omega$; in other words, that the image of ω under the function $n \mapsto P^n\omega$, for $n \in \omega$, is a set. The final axiom (replacement) does insure that the image of any set under a function is again a set; here a "function" $x \to y$ is to be described by a property $R(x,y)$ specified by a formula–much as in the case of the comprehension axiom. This again is an axiom scheme (one axiom for each property)!

Replacement. If $R(x,y)$ is a formula stated in terms of the sets x,y and the membership relation while u is a set such that for each $x \in u$ there is exactly one y with $R(x,y)$, then there is a set consisting of exactly all these y.

This set is of course the image of u under the function defined by R.

The replacement axiom scheme implies the comprehension axiom scheme. When replacement is adjoined to the Zermelo axioms, one gets the Zermelo–Fraenkel (ZF) axioms for set theory. Normally, one also includes the axiom of choice, to get ZFC.

3. The Propositional Calculus

To achieve complete clarity, Mathematical statements can be formulated in a specific symbolic language. This was already needed, for example, to explain in principle what is meant by a "property" in the comprehension and replacement axioms of set theory. The use of a language plays a central role in logic.

As a start, we consider a language limited to the standard logical connectives "and", "or", and "not". (In symbols, $\wedge$, $\vee$, and $\neg$.) They are used to connect statements p, q, and r. For example, these might be statements of arithmetic such as

$$x < 2, \qquad x + 3 = 7, \qquad 3 < 4, \qquad 3 < 2.$$

Consider a statement p, like the first two here, which involves just one "variable" x. It corresponds to a set P; namely, to the set P of all those numbers x for which the statement holds true. Then a conjunction of statements such as $p \wedge q$ (for p & q) corresponds to be intersection $P \cap Q$ of the corresponding sets of numbers. In this spirit we have the following table of connectives.

Name	Form	Reading	Boolean Form	Name
Conjunction	$p \wedge q$	p and q	$P \cap Q$	intersection
disjunction	$p \vee q$	p or q	$P \cup Q$	union
negation	$\neg p$	not p	P'	complement
implication	$p \Rightarrow q$	p implies q	$P' \cup Q$	–

The last connective, implication, is sometimes called *material implication* and sometimes written as $p \supset q$ or $p \to q$; it may also be read as "If p, then q". Other connectives can be defined from these; thus "p if and only if q" is defined as $(p \Rightarrow q) \wedge (q \Rightarrow p)$ and is written $p \Leftrightarrow q$. However, we cannot rely upon the verbal reading "if and only if" to explain what we mean. The whole formal use of these connectives can be given by *truth tables* which specify the *truth* value ($+$ for true and $-$ for false) of the result when the truth values of the arguments p and q are given. These tables are

p	q	$p \wedge q$	$p \vee q$	$\neg p$	$p \Rightarrow q$
+	+	+	+	−	+
+	−	−	+	−	−
−	+	−	+	+	+
−	−	−	−	+	+

These tables do reflect the intended meanings. From the third column we can read off the formula $p \wedge q$ as a function

$$\wedge\colon \{\pm\} \times \{\pm\} \to \{\pm\}, \qquad \{\pm\} = \{+, -\}$$

on the product with itself of the two element set $\{\pm\}$ of truth values; $\Rightarrow$ and $\vee$ are also such functions, while $\neg$ is a function of one variable.

There are also composite functions, represented by "formulas". Specifically, certain strings of letters p, q, r, connectives, and parentheses are called *formulas* (sometimes *well-formed formulas*). They are defined by recursion as follows

(i) A letter $p, q, r, \ldots$ is a formula F;
(ii) If F and G are formulas, so are $F \wedge G$, $F \vee G$ and $F \Rightarrow G$;
(iii) If F is a formula, so is $\neg F$.

Formulas arise only in these ways. (Of course, we also insert parentheses, as in $F \Rightarrow (G \wedge H)$, to make a formula unambiguously readable.) By using the truth tables, any formula F involving n letters $p_1, \ldots, p_n$ is represented by a composite function $\{\pm\}^n \to \{\pm\}$. Then a formula F is called a *tautology* when *all* the values of the corresponding function are $+$. For example, the formula $(p \wedge q) \Rightarrow (p \vee q)$ is a tautology.

Since $p \Leftrightarrow q$ can be defined in terms of the other connectives, we can also treat formulas involving " $\Leftrightarrow$ ". Some useful tautologies are

$$(p \Rightarrow q) \Leftrightarrow (\neg p \vee q), \tag{1}$$

$$(p \wedge q) \Leftrightarrow \neg(\neg p \vee \neg q). \tag{2}$$

The first is the famous definition of material implication: "p implies q" means "either not p or q". In particular, a false statement implies any other statement! Philosophers who find this conclusion uncomfortable have introduced other versions of propositional logic, such as strict implication or relevance logic. As yet, these variants have not proved helpful in the formulation of Mathematics.

Since $r \Leftrightarrow s$ is true precisely when r and s are both true or when r and s are both false, the tautology (1) states that $p \Rightarrow q$, as a function of truth values, is identical to the function $\neg p \vee q$; in other words, this provides a possible definition of $\Rightarrow$ in terms of $\neg$ and $\vee$. Similarly (2), which is one of the de Morgan laws, can be used to define $\wedge$ in terms of $\neg$ and $\vee$. These two latter connectives thus would suffice for this calculus. (It is even possible to replace these two by a single binary connective, but the result is not illuminating.)

The propositional connectives correspond (according to the first table) to Boolean operations on sets; more exactly, on subsets of some fixed set V. In this correspondence, a tautology becomes a Boolean function with value V for all arguments, while an equivalence (a tautology $F \Leftrightarrow G$) becomes an identity in Boolean algebra. Any complete system of axioms (there are many) for Boolean algebra can be turned into a system of axioms (with rules of inference) for the propositional calculus. We do not need such a system, since the consequences of the axioms can be described directly and simply as "all tautologies".

Thus everything about propositions p turns on truth—but this is not yet a complete explication of that weighty word!

4. First Order Language

To describe Mathematics objectively, we must describe a "language" in which all Mathematics can be written down. The quantifiers "there exist" and "for all", as we have seen, are necessary in any language formulating Mathematical concepts. Each language is intended to discuss some domain of individuals using variables $x, y, z, \ldots$, in adequate supply, each ranging over the domain. Formulas of the language are built up from variables and constants (0,1, . . .) using some primitive predicates and primitive function symbols; we consider as an example the case of a language $L(B, +)$ with one binary predicate $B(x,y)$ and one binary function symbol $+$. On this basis, one first defines *terms*:

(i) Each variable and each constant is a term;
(ii) If $+$ is a binary function symbol and s and t are terms, then $s + t$ is a term.

Terms are obtained only in this way. From the terms one builds up *formulas* by a similar recursion:

(iii) If B is a binary predicate symbol while s and t are terms, then $B(s,t)$ is a formula;
(iv) If F is a formula, so is $\neg F$;
(v) If F and G are formulas, so is $F \vee G$;
(vi) If x is a variable while F is a formula, so are $(\forall x)F$ and $(\exists x)F$.

Formulas are obtained only in these ways. The last clause (vi) is usually employed when F is a formula involving the variable x. This variable is said to be *bound* in the formulas $(\forall x)F$ and $(\exists x)F$ and in any further formula built up from these. Variables which are not bound are said to be *free*. The intent is that a formula F with free variables $y,z,w, \ldots$ is a statement of some property of the elements of the domain represented by these free variables, while the property involves other "bound" variables ranging over the whole domain. For example in a formula such as $[(\forall x)B(x,y)] \vee B(x,z)$ the first x is bound while the last is free; it is usually good practise to rename the latter (say as w).

The formulas as defined above use only the two propositional connectives $\neg$ and $\vee$; others can be introduced by their definitions, as indicated in the last section. We have described a language $L(B, +)$ with just one binary predicate B and one binary operation $+$; if B were $<$, it could be called a "language of arithmetic". There are languages with several predicates, unary, binary or n-ary; there can also be more function symbols, unary, binary, ternary, . . . or even none. It is a striking observation about actual Mathematical systems that the primitive functions involved are almost all unary or binary (successor or product), while the primitive predicates B are usually unary or binary. There are occasional exceptions, such as the ternary betweenness relation in the foundations of geometry (Chapter III); even then one gets rid of "betweenness" as quickly as possible by defining it in terms of the "less than" relation for real number coordinates. On the other hand, there is clearly no way in which binary relations or functions could all be replaced by unary ones. In philosophical terms, everything cannot be reduced to properties (unary predicates) of things, as was the apparent intent of Aristotelian logic.

The neglect of ternary and higher operations is curious. Does it lie in the nature of Mathematics, or is it just a failure of imagination? For example, affine geometry is intrinsically based on a ternary operation $(p,q,w) \mapsto (1-w)p+wq$ which assigns to two points p and q and a real number w the *weighted average*—but one usually quickly chooses an origin (which is not invariant) and so express the average in vector space terms, where there are only unary and binary operations. There have been occasional attempts to study objects called n-groups, for example a 3-group is a set with a ternary operation such as $(x,y,z) \mapsto (x-y+z)$. These n-groups have not prospered, although it is no trick to define, as above, a language to fit this case.

A language L, as we have defined it above, is called a one-sorted *first order* predicate language, essentially because the quantifiers $\forall x$ and $\exists x$ involve variables x ranging over only one domain. A *two-sorted* language would have two types of variables, to range over two different domains (for the case of vector spaces, domains of scalars and of vectors). A *second order* language would be one with additional quantifiers on variables which range over predicates or over sets. The use of this word "variable" is doubtless a carry-over from the use of this word in calculus, which was intended to deal with quantities (times and distances) which really "vary". Here a variable is just a symbol; its chief role is that one can substitute other symbols (terms) for a free variable.

The important language $L(\in)$ of set theory is a first order language with one sort (sets), no primitive function symbols, and just one binary predicate $\in$, that of membership. In this language, one frequently uses *bounded quantifiers*, in which the variable x in question is restricted to range only over the elements of some specified set. Such quantifiers may be written

$$\begin{aligned} (\forall x \in u)F \quad &\text{for} \quad (\forall x)\,(x \in u \Rightarrow F), \\ (\exists x \in u)F \quad &\text{for} \quad (\exists x)\,(x \in u \wedge F). \end{aligned} \tag{1}$$

In most Mathematical discussions (as the reader of this book has been asked to note) the quantifiers used are all bounded ones. Quantifiers without a bound crop up chiefly in the higher reaches of set theory (e.g., quantifiers over all ordinals).

With the exact description of the set-theoretic language $L(\in)$, we now have an explicit description of the idea of a "property", as this arose in the Comprehension and Replacement axiom schemes of ZF. A property of x and y means a formula of the language $L(\in)$ with just two free variables x and y. In using the comprehension axiom for ordinary parts of Mathematics it usually suffices to consider only formulas with bounded quantifiers. This may be stated as follows:

Bounded Comprehension (BQ). For any set u and any formula $F(x)$ of the language of set theory in which all quantifiers are bounded, there is a set s with

$$x \in s \iff x \in u \wedge F(x). \tag{2}$$

For most Mathematics, the appropriate axioms for set theory seem to be ZBQC: The Zermelo axioms with comprehension replaced by bounded comprehension and with choice added. This approach (which has nowhere been developed in detail) is a pragmatic choice of axioms: They suffice (but not for definitions like (2.11)).

Some set theorists assert that they can "see" that the ZFC axioms are "true" for the sets described in the cumulative hierarchy of §1. Note that in this hierarchy at stage $\alpha + 1$ each new set $s \in P(R_\alpha)$ is a subset of R_α, so consists of elements already present in R_α. But in the description of the hierarchy there is no real explanation of what is meant by a "subset". If each subset of R_α is to be described by giving a property of its elements, it seems plausible that the first order formula expressing this property should refer only to those sets already at hand and hence should involve only quantifiers over R_α. On this ground, bounded comprehension is much easier to "see" in the hierarchy than unbounded comprehension. (Much the same argument would apply for "bounded" replacement.) This again illustrates the observation that an intuitive idea (here, the cumulative hierarchy) has more than one formal realization.

Within ZFC set theory, the cumulative hierarchy can be formally constructed. Before axioms are at hand, the hierarchy is a Platonic myth, clearly visible only to those with a sixth sense for sets.

5. The Predicate Calculus

Complete rules of proof can be formulated in each first order language. For this, we need substitution (make the variables "vary"): If F or $F(x)$ is a formula in a predicate language which involves a free variable x while t is a term of the language we write (as usual) $F(t)$ for the formula obtained from F by substituting t for each free appearance of x.

A *tautology* of a predicate language is a formula obtained from a propositional tautology T (§3) by replacing each propositional letter p of T by a formula of the predicate language. Thus

$$(F(x) \Rightarrow G(x)) \Rightarrow (\neg G(x) \Rightarrow \neg F(x)), \tag{1}$$

$$F(x) \iff \neg\neg F(x), \tag{2}$$

are tautologies of the predicate language, which are obtained from $(p \Rightarrow q) \Rightarrow (\neg q \Rightarrow \neg p)$ and from $p \iff \neg\neg p$, respectively.

Now we can define proof. A *formal proof* in a predicate language is a finite sequence of formulas of that language in which each formula is either an axiom or results from prior formulas of the sequence by one of the rules of inference. Here the axioms (actually, they are axiom schemes) are

(i) All tautologies of the language.
(ii) For each term t and for each formula $F(x)$ with a free variable, all formulas of either of the forms

$$((\forall x)F(x)) \Rightarrow F(t), \qquad F(t) \Rightarrow (\exists x)F(x). \tag{3}$$

The rules of inference are

(iii) Modus Ponens: From F and $F \Rightarrow G$ infer G.
(iv) Generalization: If the variable x does not occur free in F, then
 (iva) From $F \Rightarrow G(x)$ infer $F \Rightarrow (\forall x)G(x)$,
 (ivb) From $G(x) \Rightarrow F$ infer $((\exists x)G(x)) \Rightarrow F$.

Conventionally, these rules of inference are presented as two-line figures, with premises above the line and conclusions below. Thus modus ponens (think of an Aristotelean syllogism) is the figure

$$\frac{F \quad F \Rightarrow G}{G}, \tag{4}$$

while the generalization rules are the figures (x not free in F)

$$\frac{F \Rightarrow G(x)}{F \Rightarrow (\forall x)Gx}, \qquad \frac{G(x) \Rightarrow F}{(\exists x)G(x) \Rightarrow F}. \tag{5}$$

These rules do correspond to ordinary principles of argumentation: What holds for an unspecified x holds for all x; what follows from the property G of an unspecified x follows from the existence of any x whatever with this property G. By (3) and modus ponens, what holds for all x holds for any term.

An essential point is that a formal definition of proof requires both some syntax (how to build formulas) and some inference (how to deduce new formulas from old). From tautologies and the specified rules of inference one can derive others; for example a second form of modus ponens: From $F \Rightarrow G$ and $G \Rightarrow H$ infer $F \Rightarrow H$.

To see explicitly how these rules work, let us prove

$$(\forall x)F(x) \Leftrightarrow \neg(\exists x)\neg F(x), \tag{6}$$

a result which could be used to define the universal quantifier in terms of the existential quantifier. In one direction the classical tautology $p \Leftrightarrow \neg\neg p$ yields the following proof:

$\neg F(x) \Rightarrow (\exists x)\neg F(x)$,	Axiom scheme (ii),
$\neg(\exists x)\neg F(x) \Rightarrow \neg\neg F(x)$,	Tautology (1) and Modus Ponens,
$\neg(\exists x)\neg F(x) \Rightarrow F(x)$,	Tautology (2) and Modus Ponens (Second form),
$\neg(\exists x)\neg F(x) \Rightarrow (\forall x)F(x)$,	Generalization rule (iva).

The reader may wish to construct his own formal proof of the reverse implication, using (ivb) and the other axiom scheme of (ii). These illustrations may support the conviction that such formal proofs are always possible and always pedantic. There have been elaborate demonstrations (for example, *Principia Mathematica* in three volumes by Whitehead and Russell) which show that the usual proofs of Mathematics can be forced into this mold.

A predicate calculus with equality is often used. This means that there is a binary predicate "=" together with appropriate axioms: The reflexive, symmetric, and transitive laws for equality, plus the axiom stating that "equals may be substituted for equals anywhere". It suffices to require the latter for the primitive predicates and function symbols, as in

$$(x = x' \wedge y = y') \Rightarrow (B(x,y) \Leftrightarrow B(x',y')), \tag{7}$$

$$(x = x' \wedge y = y') \Rightarrow x + y = x' + y', \tag{8}$$

for the language $L(B, +)$ of §4. Our surprisingly short list of axioms and rules turns out to be adequate, both practically and theoretically. Practical adequacy means that all Mathematical proofs can be formulated in this way–in the appropriate language, with the appropriate special axioms adjoined. Thus "all" proofs in set theory can be formulated with these rules and the ZFC axioms in the language $L(\in)$. Theoretical adequacy is provided by Gödels completeness theorem, which refers to *sentences* (formulas with no free variables): A sentence S in a first order language L can be proved from a list A of axioms if and only if S holds in every set-theoretical model of these axioms. Here a *model* of a first order language L such as $L(B, +)$ is (roughly) a set M to represent the intended domain of the variables, a subset $B_0 \subset M \times M$ to represent the binary predicate symbol B and a function $f: M \times M \to M$ to represent the binary function symbol $+$ of the language, all such that the axioms hold for these sets.

The important *compactness theorem* asserts that if T is a set of sentences of a first order language L such that every finite subset of T has a model, then T has a model. It is called the "compactness" theorem because it can be interpreted as asserting that a certain topological space is compact (a remarkable example of the utility of "geometric" ideas such as compactness in this part of Mathematics).

Peano arithmetic (§II.2) can be readily formulated in a first order predicate language with equality, one constant 0 and one unary function symbol s (for successor). These specifications already take care of the first two

Peano axioms (i) and (ii) of §II.2, stating that 0 is a number and that the successor of a number is a number. Next come the axioms

$$(\forall n) \neg (sn = 0), \tag{9}$$

$$(\forall n)(\forall m)(sn = sm \Rightarrow n = m), \tag{10}$$

and the axiom scheme for induction: For every formula $F(n)$ of the language, with one free variable n,

$$(F(0) \wedge (\forall n)[F(n) \Rightarrow F(sn)]) \Rightarrow (\forall m)F(m). \tag{11}$$

For an adequate arithmetic one needs addition and multiplication of natural numbers. The definition of these binary operations by recursion (as in §II.2) requires a recursion theorem involving set theory. Hence an adequate Peano arithmetic independent of set theory requires an enlarged language with binary operators $+$ and $\times$ together with, as axioms, the recursion equations (II.2.1) and (II.2.2) describing these operators. The statement (11) provides for induction for all those properties of a natural number n which can be expressed by a formula of the language. (This is analogous to the use of formulas in the comprehension axiom.) Since a formula is just a finite well-formed sequence of symbols, there is only a denumerable number of formulas in this language. Therefore, the presentation of Peano arithmetic provides only for a denumerable number of proofs by induction!

This observation has the strange result that these Peano axioms have a "non-standard" model in ZFC. For consider a language with one additional constant δ and the following denumerable list of potential additional axioms:

$$0 < \delta, \qquad s0 < \delta, \qquad ss(0) < \delta, \qquad sss(0) < \delta, \quad \cdots.$$

Any finite subset of this list has a model. Therefore, by compactness, the whole list has a model–and in that model of the Peano axioms there is a "natural number" δ larger than any $s^n(0)$!

An alternative view embeds Peano arithmetic in set theory. A model is then a set $\mathbf{N}$ with an element $0 \in \mathbf{N}$ and a function $s\colon \mathbf{N} \to \mathbf{N}$ which together satisfy (9) and (10) and which have the induction property: For every subset $T \subset \mathbf{N}$,

$$0 \in T \wedge (\forall n)(n \in T \Rightarrow s(n) \in T) \Rightarrow T = \mathbf{N}. \tag{12}$$

In this case, the strength of the induction axiom depends on the set theory–on how many subsets T are there provided. With ZF or ZBQ set theory one can then prove the recursion theorem of §II.2 and from this show that any two such sets $\mathbf{N}$ are isomorphic. Located in *this* way within a set theory, Peano arithmetic has no non-standard models.

Some other parts of Mathematics can be readily formulated in the first order predicate calculus. This is the case, in geometry, for the first four groups of axioms (Incidence, Order, Congruence, and the Parallel Axiom) in Hilbert's foundation of plane Euclidean geometry (§III.2). However, the formulation of the fifth group (The continuity axiom) uses sets, either in Hilbert's completeness form (can be embedded in no larger systems; i.e., no larger set) or in the Dedekind cut form (the cut is a set). Hence it is convenient to consider the Hilbert plane within set theory as a *set* of points and a *set* of lines, with their various predicates having the properties stated in the axioms of Chapter III.

The conventional axioms for the real numbers are also formulated within a set theory; specifically the Dedekind completeness axiom requires that every bounded *set* of real numbers has a least upper bound. The alternative but equivalent Cauchy or Weierstrass completeness axioms (§IV.5) also involve sets in order to describe what is meant by a sequence.

For group theory there is also an evident first order language with equality, with variables ranging over the group and with one binary operation, that of multiplication. This language suffices to formulate the axioms for a group, but the subsequent development of group theory soon goes beyond the language. It uses integers (the orders of group elements) and sets, both cosets and sets upon which the group operates. For groups and for the other parts of abstract algebra and topology the free use of sets is a major convenience.

6. Precision and Understanding

Proof in Mathematics is both a means to understand why some result holds and a way to achieve precision. As to precision, we have now stated an absolute standard of rigor: A Mathematical proof is rigorous when it is (or could be) written out in the first order predicate language $L(\in)$ as a sequence of inferences from the axioms ZFC, each inference made according to one of the stated rules. Here "written out" need not mean only in the parsimonious language $L(\in)$; all sorts of symbols for derived concepts may appear, since each occurrence of such a symbol can be replaced by its (suitably precise) definition. To be sure, practically no one actually bothers to write out such formal proofs. In practice, a proof is a sketch, in sufficient detail to make possible a routine translation of this sketch into a formal proof. When a proof is in doubt, its repair is usually just a partial approximation of the fully formal version. What is at hand is not the *practice* of absolute rigor, but a *standard* of absolute rigor.

Moreover there are different versions of the standard. There are alternative axiomatizations of set theory: ZF, ZFC, Z, ZBQC, or others. The first order predicate calculus might be expanded by using quantification over properties to give a second order calculus. An advocate of strict implication might prefer one of the many systems of model logic. An intuitionist could assert that an existence statement such as $(\exists x)F(x)$ means that one has exhibited such an x; hence this existence statement is not logically equivalent (as would follow from (5.5)) to $\neg(\forall x)\neg F(x)$: The latter statement would give the existence of x only as the result of a contradiction, using $p \Leftrightarrow \neg\neg p$. Hence the intuitionist gives up the tautology $p \Leftrightarrow \neg\neg p$, but the standard of rigor remains. There is a fully formal intuitionist logic, complete with appropriate axioms and rules of inference. The idea of rigor remains applicable, here and in all the other variants, except for certain advocates of "constructive" Mathematics who claim that their views cannot possibly be reduced to *any* formalism. One might also except those geometers who hold that future geometric insights will transcend any present logic. In any other variant of logic, the test for the correctness of a proposed proof is by formal criteria and not by reference to the subject matter at issue.

Historically, much of this concept of rigor was already present in Greek geometry. The full development of rigor came in successive steps in modern times, with the 19th century rigorous understanding of calculus, the development of complex analysis, the careful study of the calculus of variations, the Hilbert foundation of plane geometry, the subsequent extensive use of axiomatic methods, the formulation of symbolic logic (Frege, Whitehead–Russell) and the axiomatization of set theory by Zermelo. Note that Whitehead and Russell used a theory of "types" (classes, then classes of classes, and so on) rather than a set theory. Russell's first publication of his theory of types came in the same year, 1908, as Zermelo's publication of his axioms for sets.

For the concept of rigor we make a historical claim: That rigor is absolute and here to stay. The future may see additional axioms for sets or alternatives to set theory or perhaps new more efficient ways of recording (or discovering) proofs, but the notion of a rigorous proof as a series of formal steps in accordance with prescribed rules of inference will remain.

Moreover, there are good reasons why Mathematicians do not usually present their proofs in fully formal style. It is because proofs are not only a means to certainty, but also a means to understanding. Behind each substantial formal proof there lies an idea, or perhaps several ideas. The idea, initially perhaps tenuous, explains why the result holds. The idea becomes Mathematics only when it can be formally expressed, but that expression must be so couched as to reveal the idea; it will not do to bury the idea under the formalism. For example, proofs of the Pythogorean theorem (§III.1) require formal concepts (area or similarity) but they also serve to explain why the square on the hypotenuse is a sum. . . .

The proof of Cauchy's integral theorem by rectangular paths (§X.6) is an example. On the one hand it is rigorous; it replaces a vague slicing up of areas (Stokes' lemma) by a precise use of rectangular axis-parallel paths. To be sure the use of rectangles is accidental; much the same proof could be done, and has been done, by using triangles instead. However, the rectangles are a clear vehicle for exposing the underlying reason: that a holomorphic function has at each point the same derivative in every direction; in particular, in the two directions exhibited along axis-parallel rectangular paths used to construct a primitive for the given function. This is the underlying idea–but it is not enough to wave one's hands about the idea. Reduced to formal statement, it becomes a proof. Proofs serve both to convince and to explain–and they should be so presented. Rigorous proofs serve to avoid errors–an important function.

But proofs have their limitations, as we will now see.

7. Gödel's Incompleteness Theorems

If naive set theory were so formalized as to allow the construction of Russell's paradoxical set R, it would allow the proof of both $R \in R$ and $\neg(R \in R)$; this is a contradiction. In general, a *contradiction* in a formal system arises when there is a sentence p of the system and proofs of both p and $\neg p$. In view of the tautology $p \wedge \neg p \Rightarrow q$, the presence of such a contradiction allows the proof in the system of any sentence q. Hilbert aspired to obtain a secure foundation for Mathematics by proving that a suitable formal system (including at least arithmetic and analysis) is *consistent* in the sense that it has no contradictions. Such a result would be a theorem about proofs in the system; that is, a Mathematical theorem in which the objects of study are not sets or points or numbers but (formal) proofs. Such a study is called *proof theory* or *metamathematics* (The Mathematical study of Mathematics). It is possible only in the presence of a fully formal notion of inference and proof. Hilbert intended that this metamathematics itself should use only "finite" methods, which therefore would be wholly secure.

In 1931 Kurt Gödel, using a subtle diagonal argument, proved an "incompleteness" theorem which showed that Hilbert's objective can not be reached, except perhaps by some novel extension of the Hilbert idea of finite methods. Gödel considered a formal theory T which contained ordinary arithmetic, which is consistent (in a strong sense to be explained) and in which the axioms and rules of inference are either finite in number or are specified in a recursive way; the latter is the case for all the systems we have considered. In this system T he showed how to construct a sentence G such that neither G nor $\neg G$ could be proved within the system. Such a G is then undecideable. Its existence is Gödel's first incompleteness theorem.

Since the rules of T are recursive, Gödel also showed that one can formulate within the system T a sentence con_T which, when interpreted, states that "T is consistent", he then showed that this sentence could not be proved within the system. In other words, no such system T is strong enough to establish its own consistency. This would apply in particular to ZFC set theory, and shows that the (hoped for) consistency of ZFC cannot be proved with the means of ZFC. It is this second Gödel incompleteness theorem which blocks the Hilbert program. (All "finite" methods can presumably be formulated within ZFC.)

We will now give a very informal sketch of the proofs of these two Gödel incompleteness theorems. They depend on a coding by "Gödel numbers" which translate statements about proofs of T to statements about numbers within T. Since T contains ordinary arithmetic, it has constants such as 0, $0' = 1$, $0'' = 2$ for numerals, variables $x, y, z, \ldots$ for numbers, equality, a successor function, and symbols $+$ and $\times$ for addition and multiplication, as well as the apparatus of the first order predicate calculus. By a subtle use of the Chinese remainder theorem, in T one can define "prime number" and prove in T the fundamental theory or arithmetic: Every natural number is uniquely (except for order) a product of primes.

Metamathematics must discuss formulas and proofs in T. These formulas (and their proofs) are all finite formal expressions in a denumerable list of basic symbols, so the collection of all formulas and all proofs is denumerable. By explicitly enumerating them in some code, we can replace statements about proofs by statements about the corresponding numbers, their codes. Here is a code. First label the basic symbols (in some order) by the primes:

$$\begin{array}{ccccccccccc} \neg, & \vee, & \exists, & 0, & ', & =, & +, & \times, & x, & y, & z \\ 2, & 3, & 5, & 7, & 11, & 13, & 17, & 19, & 23, & 29, & 31. \end{array}$$

Each formula is a finite list of symbols, so is coded by a list of numbers $n_1, n_2, \ldots$; this can be replaced by a single number $m = 2^{n_1} 3^{n_2} 5^{n_3} \ldots$. Conversely, give a natural number m, one can factor it into primes and so read off the corresponding list of basic symbols. Usually this list will be meaningless, but sometimes it will represent a well-formed formula. When this is the case, m has a certain property which we write, in informal arithmetic, as

$$\text{form}(m) = \text{“}m \text{ is the code of a formula of } T\text{”}.$$

This property is a (complicated) arithmetic statement: "Factor m and ...", so there is also within the system T a formula $F(x)$ in a free variable x which expresses "exactly" this property. (We return below to the meaning of "exactly".) (Note, however that in the formula $F(x)$ each quantifier $\forall\, k$ or $\exists\, l$ has a numerical bound, as all $k < m$ or there exists $l < m$. For example, to test that m is prime we consider only possible factors $l < m$.)

We also consider proofs in T. Each formal proof is a finite sequence of formulas, coded by a sequence of numbers $m_1, m_2, \ldots$. We turn this sequence into a single number $n = 2^{m_1}3^{m_2}5^{m_3}\cdots$ called the code of the proof. Conversely, given n one can factor it to get the exponents m_i and the corresponding expressions, and then test whether these expressions are well formed formulas which in that order constitute a proof. When this is so n has a (complicated) informal property $\mathrm{Prf}(n)$ which is recursive, since the system T was so given. Hence $\mathrm{Prf}(n)$ can also be translated into the formal system as a statement $P(y)$. Now select a variable w of the formal system and a corresponding informal variable h. The translation process is continued as in the following table:

About T		Within T	
$\mathrm{Form}(m)$ =	"m is the code of a formula",	$F(x)$	
$\mathrm{Prf}(n)$ =	"n is the code of a proof",	$P(y)$	
$\mathrm{Dem}(m,n)$ =	$\mathrm{Form}(m)$ and $\mathrm{Prf}(n)$ and the proof ends at the formula with code m.	$D(x,y)$	
$\mathrm{Sub}(m,k)$ =	The code of the following formula "Take the formula (if any) with code m; in it replace the variable w by the kth numeral".	$S(x,y)$	(a term)
$\mathrm{Arg}(m,n)$ =	$\mathrm{Dem}(\mathrm{Sub}(m,m),x)$	$A(x,y) = D(S(x,x),y)$	
For all n, not $\mathrm{Dem}(\mathrm{Sub}(h,h),n)$		$(\forall y)\neg A(w,y)$	code p

Now the last formula in the right-hand column, like every formula of the system, has a code p; let $\bar{p}$ be the corresponding numeral. Plug in p for h and $\bar{p}$ for w to continue the table as follows:

For all n, not $\mathrm{Dem}(\mathrm{Sub}(p,p),n)$	$(\forall y)\neg A(\bar{p},y) = G$.

Call the last formula on the right G. It is obtained by taking the formula just above it in the table, with code p, and substituting therein the numeral $\bar{p}$ for the variable w. Hence by the definition of Sub above, G *is* the formula with code $\mathrm{Sub}(p,p)$. However, its translation on the left states "There is no proof for $\mathrm{Sub}(p,p)$"; that is, there is no proof for G. In other words, G when translated back to the left says "There is no proof for me" (in the system).

Our further examination of this startling situation will use the fact that the translations above (left to right) are "good" in the following sense:

Translation Principle. When a statement such as $\mathrm{Arg}(m,n)$ holds in informal number theory for numbers m and n, then for the corresponding numbers $\bar{m}$ and $\bar{n}$ there is in T a formal proof for the translation $A(\bar{m},\bar{n})$. Moreover, if $\mathrm{Arg}(m,n)$ fails, then there is a formal proof for $\neg A(\bar{m},\bar{n})$. This is to apply only to statements such as $\mathrm{Arg}(m,n)$ in which quantifiers like $(\forall k)$ or $(\exists k)$ have numerical bounds; say all $k < m$.

Lemma. *For every number k the informal statement*

$$\mathrm{Arg}(p,k) = \mathrm{Dem}(\mathrm{Sub}(p,p),k)$$

does not hold.

For suppose it did. This displayed informal statement says that the number k is the code for a proof of G, whose code is $\mathrm{Sub}(p,p)$. In other words, there is a formal proof (code k) for G. But by the translation principle applied to our informal statement there is also a formal proof for $A(\bar{p},\bar{k})$ and hence a proof for $(\exists y)A(\bar{p},y)$—namely, $\bar{k}$ is the y claimed to exist. But $(\exists y)A(\bar{p},y)$ is (classically) equivalent to $\neg(\forall y)\neg A(\bar{p},y)$, which is $\neg G$. So there is also a proof of $\neg G$. Contradiction. Since the system is consistent, there can be no such contradiction.

Given this lemma, suppose there is a proof for G. That proof has some code k. Since G has the code $\mathrm{Sub}(p,p)$ this means that $\mathrm{Dem}(\mathrm{Sub}(p,p),k)$. By the lemma, this is not so.

On the other hand, suppose there were a proof for $\neg G$; that is, for $\neg(\forall y)\neg A(\bar{p},y)$ and hence for its equivalent $(\exists y)A(\bar{p},y)$. Now according to the lemma, for each number k, $\mathrm{Dem}(\mathrm{Sub}(p,p),k)$ does not hold. Hence, by the second half of the translation principle, there is a proof for the negation of the translation; that is, for $\neg A(\bar{p},\bar{k})$ where $\bar{k}$ is any numeral. In other words, for the numerals 0, $0'$, $0''$, and so on we have proofs of

$$\neg A(\bar{p},0), \qquad \neg A(\bar{p},0'), \qquad \neg A(\bar{p},0''), \ldots,$$

while at the same time we have a proof for $\neg G$; that is of

$$(\exists y)A(\bar{p},y).$$

In other words, this y, proved to exist, can be none of the numerals $0,0',0'',\ldots$ of the formal system. This would be a bizarre situation. It is excluded if we assume that the formal system T is ω *consistent*: That there is no formula $H(y)$ in one free variable and proof of all of

$$\neg(\forall y)H(y); \qquad H(0), \quad H(0'), \quad H(0''), \ldots.$$

(The above case is that where $H(y) = \neg A(\bar{p},y)$.) The "strong consistency" mentioned above in our first discussion of Gödel's first incompleteness theorem is precisely this notion of ω-consistency. (There is a subtle refinement of the construction of G which renders this stronger consistency assumption unnecessary.) This completes our sketch of Gödel's first incompleteness theorem.

For Gödel's second incompleteness theorem one must formalize the notion of consistency. As we have already noted, any statement q (for example, $0 = 1$) can be proved in a system which is inconsistent. Hence T is consistent if there is no proof of $0 = 1$, and this becomes a statement Con_T in informal arithmetic which has a translation into a sentence C_T of the formal theory. Now our proof of the first theorem showed that if T is consistent, there is no proof of G. But "T is consistent" translates to C_T, while "there is no proof of G" translates to G. Hence, translating the proof of the first Gödel theorem gives in the system T a proof of

$$C_T \Rightarrow G.$$

If T could prove its own consistency there would also be a proof in T of C_T. These two and modus ponens yield in T a proof of G–and this we know is impossible. Therefore a consistent and recursive formal system T cannot prove its own consistency. This is Gödel's second theorem.

This proof heavily uses $\mathrm{Sub}(p,p)$; it is thus a "diagonal" argument.

Our discussion above is a simplified summary of the presentation of Gödel's theorems on pages 204–223 of S. C. Kleene's "Introduction of Metamathematics". That source gives more detail, including a proof of the translation principle (there called the property "numeralwise expressible").

The undecideable sentence G is also "true"–because it says "there is no proof of G". It might thus be added to the system as another (very complicated) axiom–but for this enlarged system T_G one could construct a new undecideable sentence G'. One is thus left with the striking fact that every such formal system can state sentences which it cannot decide.

The instructions for constructing G are quite indirect; if they were carried out explicitly, the result would be a very long formal sentence, involving many intercalated references to prime factorizations. Considered just as a sentence of arithmetic, G is thus not very interesting. Recently J. Paris has constructed a true and relevant sentence of Peano arithmetic which is not provable in that arithmetic; for an exposition, see the article by J. Paris and Leo Harrington on pp. 1133–1142 of the *Handbook of Mathematical Logic* (Barwise [1977]).

The Gödel theorems are closely related to a well-known theorem of Tarski, which asserts that the truth (of sentences of arithmetic) cannot be defined within the language of arithmetic. These theorems provide a profound commentary on the nature of Mathematics and its formalisation.

8. Independence Results

For Zermelo–Fraenkel set theory many sentences–including a number of interesting ones–are independent of the axioms and so remain undecided and undecideable. One such is the continuum hypotheses CH, already mentioned: There is no cardinal number between the cardinal of the integers and that of the reals. It turns out that CH can be neither proved nor disproved in ZFC: It is *independent* of ZFC.

The second half of this result was due to Godel: CH is consistent with ZFC; that is, its negation cannot be proved from ZFC. To show this, he started with a model V of ZFC and built up within it a smaller model, the model L of "constructible" sets; he then showed that L satisfies ZFC and CH.

Gödel's constructible sets [Gödel, 1940] are built up in an ordinal hierarchy like the cumulative hierarchy of §1. In that case, given the set R_α of all sets of ordinal rank α, the next step consists in forming the set $R_{\alpha+1} = P(R_\alpha)$ of "all" subsets of R_α. But here the notion of "all" possible subsets of a given set R_α is not wholly clear. How is subset defined? Perhaps the subsets should be those obtained by some use of the comprehension axiom. The constructible hierarchy is built by taking as subsets only those given by "essential" uses of this axiom in producing subsets of a set B already present. Given B, enlarge the language $L(\in)$ of set theory by adding to the language one new constant symbol b' for each element $b \in B$. Use the evident interpretation of formulas F in the enlarged language and consider only formulas with quantifiers ranging over B. Now call a set S "defineable" from B if S can be described as

$$S = \{x \mid x \in B \wedge F(x)\}$$

for some such formula $F(x)$, with just one free variable x, in the enlarged language. (Each such formula will involve only a finite number of the added constants b'.) The collection $D(B)$ of all these subsets $S \subset B$ is the set of sets *defineable* from B. It is a subset of the power set $P(B)$.

The *constructible hierarchy* L_α is now defined by recursion over the ordinals α as

$$L_0 = \varnothing, \qquad L_{\alpha+1} = D(L_\alpha), \qquad L_\gamma = \bigcup_{\beta<\gamma} L_\beta, \tag{1}$$

where γ denotes a limit ordinal. This is parallel to the definition (1.7), within ZFC, of the cumulative hierarchy R_α; comparison at once shows that $L_\alpha \subset R_\alpha$ for every α. By a careful analysis one can show that the class L of *constructible* sets (all those in some L_α) does satisfy ZFC and CH. Thus CH is consistent with ZFC.

Subsequently Paul Cohen established the other half of the independence of CH. (CH cannot be proved from ZFC.) It was known, via the important Löwenheim–Skolem theorem, that any model M of ZFC must contain a denumerable submodel M_0. Starting from M_0, Cohen adjoined a "generic" set u from M, using it and M_0 to build up a larger model M^1 containing M_0 and u and satisfying ZFC. The assurance that this added set u was indeed "generic" (had no properties other than those specified) was achieved by a method called "forcing". By a subtle argument, it then could be shown that the larger model M^1 did satisfy ZFC but not CH.

CH is thus an example of an "interesting" statement about sets which is independent of the axioms of ZFC. Similarly, the axiom of choice is independent of ZF.

The method of forcing has been used to establish many other independence results. One striking such result concerns Lebesgue measure, as it is used to construct the Lebesgue integral of real analysis (§VI.11). Using the axiom of choice, one can prove that there are sets of real numbers which are not Lebesgue measurable. By using forcing, one can construct a model of ZF (and hence a model of the reals within it) which does not satisfy choice and in which every set of reals *is* Lebesgue measurable. There are also surprising independence results in algebra. One such concerns a problem of J.H.C. Whitehead. Let F be a free abelian group (i.e., a direct sum of copies of the additive group $\mathbf{Z}$ of integers), while g: $A \to F$ is an epimorphism of abelian groups. By the axiom of choice, g must *split* (that is, there is a homomorphism h: $F \to A$ with gh the identity; this h makes F a direct summand of A). To construct h, send each free generator x of F to a chosen element $a \in A$ with $ga = x$. Whitehead's problem now reads: If H is an abelian group such that every epimorphism g: $A \to H$ with kernel Z splits, is H necessarily free? On first sight, this seems to be just a question of algebra. However Shelah, using forcing, has constructed models of set theory which show that it is independent of ZFC. (For an exposition, see Eklof [1976].)

These results, and others too numerous to mention, show that many interesting Mathematical questions cannot be settled on the basis of the Zermelo–Fraenkel axioms for set theory. Various additional axioms have been proposed, including axioms which insure the existence of some very large cardinal numbers and an axiom of determinacy (for certain games) which in its full form contradicts the axiom of choice. This variety and the undecideability results indicate that set theory is indeterminate in principle: There is no unique and definitive list of axioms for sets; the intuitive idea of a set as a collection can lead to wildly different and mutually inconsistent formulations. On the elementary level, there are options such as ZFC, ZC, ZBQC or intuitionistic set theory; on the higher level, the method of forcing provides many alternative models with divergent properties. The platonic notion that there is somewhere *the* ideal realm of sets, not yet fully described, is a glorious illusion.

This situation bears some resemblance to that in geometry after the discovery of consistency proofs for non-Euclidean geometry showed that there was not one geometry, but many. This meant that geometries could be formulated with many different systems of axioms, some of which were relevant to higher analysis and some to physics (as with non-Euclidean metrics in relativity). Similarly, the initial idea of a collection leads to substantially different versions of set theory, some of which (e.g. forcing, see

§14 below) have relevance to other parts of Mathematics, though not yet (?) to Physics.

Moreover, it is thus appropriate to consider alternatives to set theory as a foundation for Mathematics.

9. Categories and Functors

One way to organize algebra is to look not just at the objects at issue (sets, groups, or rings) but also at the mappings between them: Functions between sets, homomorphisms between groups or between rings, and generally "arrows" between "objects". This approach also fits topology, where the arrows are continuous maps of spaces, and geometry, with arrows the smooth maps of manifolds. It leads to the notion of a category.

A *category* **C** consists of *objects* $A,B,C,\ldots$ and *arrows* $f,g,h,\ldots$. Each arrow f has an object A as *domain* and an object B as *codomain*; one then writes

$$f\colon A \to B. \tag{1}$$

To each object A there is an *identity arrow* $1_A\colon A \to A$. If the domain of an arrow $g\colon B \to C$ is the codomain of f there is defined a *composite* arrow

$$(g\cdot f)\colon A \to C. \tag{2}$$

These primitive terms are subjected to two axioms. The *associative law*: Given arrows

$$f\colon A \to B, \qquad g\colon B \to C, \quad \text{and} \qquad h\colon C \to D \tag{3}$$

the two resulting triple composites are equal, as in

$$h\cdot(g\cdot f) = (h\cdot g)\cdot f\colon A \to D. \tag{4}$$

The *identity law* for the identity arrows asserts that for each $f\colon A \to B$ one has

$$f\cdot 1_A = f = 1_B\cdot f\colon A \to B. \tag{5}$$

This whole description could be viewed as a definition within set theory, in which case a category C is to be regarded as a set of objects and a set of arrows such that (1)–(5) hold.

Alternatively, this is an independent set of axioms in a first order language with two sorts, "objects" and "arrows".

There are many models of this simple system of axioms. Here are some categories, in each of which "composition" is the standard composition of maps:

Set: Objects are sets, arrows are functions,
Grp: Objects are groups, arrows, homomorphisms of groups,
Ab: Objects are abelian groups with homomorphisms as arrows,
Vect: Objects are vector spaces (over **R**), arrows linear transformations,
Euclid: Objects are inner product spaces, arrows orthogonal transformations,
Top: Objects are topological spaces, arrows continuous maps,
Man: Objects are C^∞-manifolds, arrows C^∞-maps between them.

All these, and many more, are "big" categories. They do not legally exist within ZFC because the collections of all the objects and all the arrows in question are not sets but are "classes", like the class of all sets or the class of all ordinals. These categories also have "small" versions, such as the category of all finite subsets of **N** or of all manifolds in $\mathbf{R}^n$. But note in particular that in each of these categories each function (or map) is *not* just a set of ordered pairs of elements, but is also equipped with both a domain and a codomain. Thus the identity $\mathbf{N} \to \mathbf{N}$ and the inclusion $\mathbf{N} \to \mathbf{Z}$ count as *different* arrows. Good reasons for this convention, will soon appear, in (7) below.

There are many categories which are *small* (i.e., legal sets). Each group G is a category in which there is only one object, while the arrows are the elements of G with group multiplication as composition. Each partially ordered set S and more generally each preordered set is a category (a *preordered* set S is one with a reflexive and transitive binary relation $s \leqslant t$). This category S has objects the elements of S and one arrow $s \to t$ exactly when $s \leqslant t$: the transitive law for the order then provides uniquely for the composition of arrows, while the reflexive law provides the identity arrows. Conversely, any category in which there is at most one arrow with given domain and codomain is a preordered set. In particular, each ordered set is a category, thus the ordinal number $4 = \{0,1,2,3\}$ is the category with four objects and the following arrows

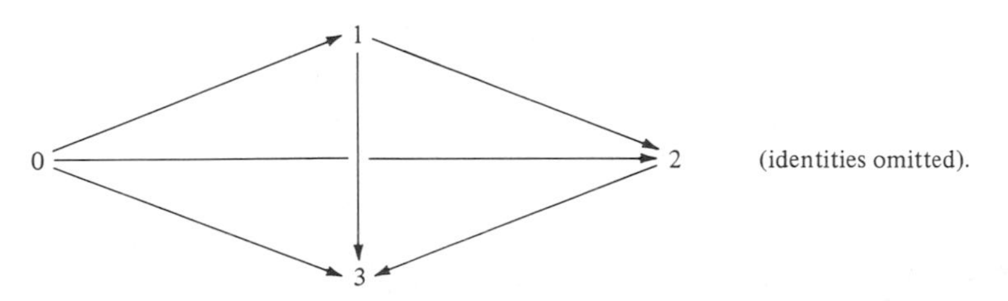

The homotopy classes of paths in a topological space (§X.5) form a category under composition of paths. In this category every arrow is invertible; that is, has a two-sided inverse. A category with this property is called a *groupoid*.

Big categories arise with each new description of Mathematical objects. Given a type of Mathematical structure, what are the corresponding morphisms preserving these structures? They are the arrows of a category. The slogan "What are the morphisms?" applies to categories themselves. A morphism F: $\mathbf{C} \to \mathbf{D}$ of categories (usually called a *functor*) must send objects to objects, and arrows to arrows, preserving domains, codomains, identities, and composites. Thus a functor F: $\mathbf{C} \to \mathbf{D}$ assigns to each object C of $\mathbf{C}$ an object FC of $\mathbf{D}$ and to each arrow f: $C \to C'$ of $\mathbf{C}$ an arrow Ff: $FC \to FC'$ of $\mathbf{D}$ in such a way that $F1_C = 1_{FC}$ and $F(g \cdot f) = (Fg) \cdot (Ff)$ whenever the composite $g \cdot f$ is defined. Thus a functor produces a "picture" of $\mathbf{C}$ in $\mathbf{D}$:

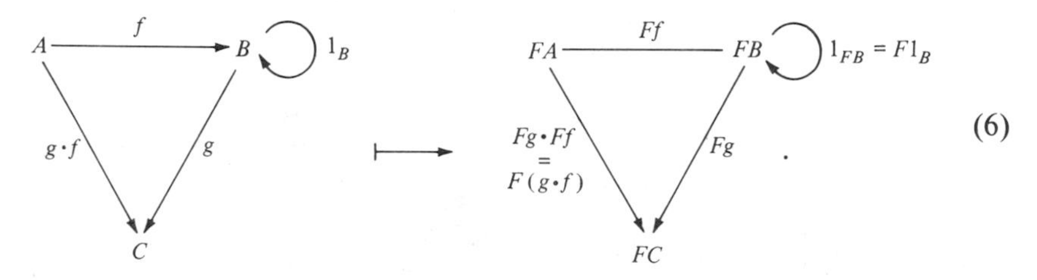

(6)

Such diagrams as these, in which any two of the arrows or composite arrows from one vertex to another one are equal, are called *commutative diagrams*.

Functors typically arise from many constructions of Mathematical objects; one gets a functor when the construction also applies suitably to the morphisms. An example is the tangent bundle T^*M of a smooth manifold M. Here any smooth map f: $M \to M'$ carries tangent vectors at a point p in M to tangent vectors at the point fp in M', thus giving a map f^*: $T^*M \to T^*M'$ which is also smooth. Labeling this latter map T^*f specifies the "tangent bundle" functor T^* on the category of smooth manifolds to itself.

The fundamental group π_1 is another example. A *pointed* topological space X is one with a selected point x_0, called its *base point*. The corresponding morphisms $(X,x_0) \to (Y,y_0)$ are then continuous maps f: $X \to Y$ which preserve base points in the sense that $f(x_0) = y_0$; the resulting category is called Top_*. Now the construction in (§X.5) of the fundamental group $\pi_1(X,x_0)$ at x_0 provides to each such space a group and to each such morphism a morphism $\pi_1(f)$: $\pi_1(X,x_0) \to \pi_1(Y,y_0)$ of groups, which sends each closed path h in X into the closed path fh in Y. Thus π_1 is really a functor from pointed spaces to groups. It provides an algebraic picture, by groups and their homomorphisms, of spaces and their continuous maps. Since topology is concerned not just with the spaces themselves but also with the maps between them it can make effective use of such functors, from topology to algebra. There are many other such functors: homology groups, cohomology groups, K-groups, and the like.

This example also illustrates why it is useful to require that each map or arrow have a specified codomain. Consider a closed circular disc D and its circumference, a circle S^1. Then the identity function 1_{S^1} and the inclusion function j,

$$1_S\colon S^1 \rightarrow S^1, \qquad j\colon S^1 \rightarrow D \tag{7}$$

have the same effect on each individual point of S^1 and so (as functions) are the same sets of ordered pairs. However, if we apply the fundamental group functor π_1 (with any chosen point of S^1 as base point) we get $\pi_1(S^1) = \mathbf{Z}$ and

$$\pi_1(1_S) = 1_{\mathbf{Z}}\colon \pi_1(S^1) \rightarrow \pi_1(S^1) = \mathbf{Z},$$

the identity, while for the inclusion j, $\pi_1(j)$ collapses $\mathbf{Z}$ to the unit element since $\pi_1(D)$ is the trivial one-point group. Thus the functor π_1 has very different effects on the two arrows of (7); these two functions differ only in their codomain, but that makes a difference of substantial topological interest!

Some functors turn the arrows around. For a vector space W over a field K the dual space (§VI.4) W^* consists of all the linear transformations $f\colon W \rightarrow K$. Hence if the arrow $T\colon V \rightarrow W$ is a linear transformation of spaces each vector f in W^* yields by composition with T a vector $f \cdot T\colon V \rightarrow K$ in V^*. If we write T^*f for this vector $f \cdot T$, while if $S\colon U \rightarrow V$ is another linear transformation, then

$$(T \cdot S)^*f = (S^* \cdot T^*)f,$$

and also $1_W{}^*f = f$. Thus $(T \cdot S)^* = S^* \cdot T^*$, so that composition of linear transformations is reversed under the operation "take the dual". This operation, written $*$ or $V \mapsto V^*$, $T \mapsto T^*$, is thus what is called a *contravariant* functor (on the category of vector spaces over K). A formal definition of contravariance can be avoided by introducing to each category $\mathbf{C}$ an *opposite* category $\mathbf{C}^{op}$. This has the same objects as does $\mathbf{C}$ and for each arrow $f\colon A \rightarrow B$ of $\mathbf{C}$ an arrow $f^{op}\colon B \rightarrow A$ in $\mathbf{C}^{op}$, with the same identities and a composition defined (when possible) by $f^{op} \cdot g^{op} = (g \cdot f)^{op}$. A "*contravariant*" functor G on a category $\mathbf{C}$ to a category $\mathbf{D}$ is then just an (ordinary or "covariant") functor $\mathbf{C}^{op} \rightarrow \mathbf{D}$. For example, the cotangent bundle construction T^*M on manifolds M is a contravariant functor, because a smooth mapping $M \rightarrow M'$ takes each cotangent vector on M' (determined by some smooth function on M') *back* into a cotangent vector on M.

In any category $\mathbf{C}$ two objects A and B determine the collection

$$\hom(A,B) = \{f \mid f \text{ is an arrow of } \mathbf{C} \text{ and } f\colon A \rightarrow B\}. \tag{8}$$

It is called a "hom set", though it might be too big (e.g., a class). $\mathbf{C}$ is said to have *small hom sets* if these classes are all sets; this is the case even for all the big categories we have listed above. For each object B in such a category $A \mapsto \hom(A,B)$ provides a contravariant function $\mathbf{C}^{op} \rightarrow$ *Sets*. Similarly, $A \mapsto \hom(B,A)$, often written $\hom(B,-)$, is a covariant function on $\mathbf{C}$ to *Sets*.

The open sets U in a topological space X form a partially ordered set under inclusion and hence a category open(X) with arrows $V \to U$ the inclusions. For example, for U open in the complex plane the set $H(U)$ of all functions holomorphic on U is a contravariant functor on this category; any inclusion $V \subset U$ of open sets yields, by restriction of holomorphic functions, a map $H(U) \to HV)$. The statement that H is a functor means exactly that the composite $H(U) \to H(V) \to H(W)$ is $H(U) \to H(W)$ whenever $W \subset V \subset U$. Similarly, a presheaf P on any topological space (§VIII.11) is just a contravariant functor

$$P\colon (\text{Open } X)^{op} \longrightarrow Sets. \tag{9}$$

Understanding Mathematical operations leads repeatedly to the formation of totalities: The collection of all prime numbers, the set of all points on an ellipse, the manifold of all lines in 3-space, the manifold of all positions and velocities of a mechanical system, the set of all subsets of a set, the set of all power series expansions for a function (its Riemannian surface) or the category of all topological spaces. There are no upper limits; it is useful to consider the "universe" of all sets (a class) or the category *Cat* of all small categories as well as CAT, the category of all big categories. This is the idea of a *totality*, and these are some of its many formulations. After each careful delimitation, bigger totalities appear. No set theory and no category theory can encompass them all–and they are needed to grasp what Mathematics does.

10. Natural Transformations

Our slogan proclaimed: With each type of Mathematical object, consider also the morphisms. So, what is a morphism of functors; that is, a morphism from F to G where both F and G are functors $F,G\colon \mathbf{C} \to \mathbf{D}$ between categories $\mathbf{C}$ and $\mathbf{D}$? An example is already at hand in the category $\mathbf{V}$ of all vector spaces V over a fixed field K. Here the formation of the double dual $V \mapsto V^{**}$ is a (covariant) functor $\mathbf{V} \to \mathbf{V}$. With the evaluation map $e\colon V^* \times V \to K$ each vector $v \in V$ gives a function $e(-,v)\colon V^* \to K$ which is an element of the double dual V^{**}. Hence there is for each V a linear transformation

$$\kappa_V\colon V \longrightarrow V^{**}; \qquad v \mapsto e(-,v), \tag{1}$$

said to be "natural" (§VII.4) because independent of any choice of basis. Now think of κ as a transformation of the identity functor $V \mapsto V$ into the double dual $V \mapsto V^{**}$. For each linear map $T\colon V \to W$ one readily shows from the definitions that the following diagram commutes

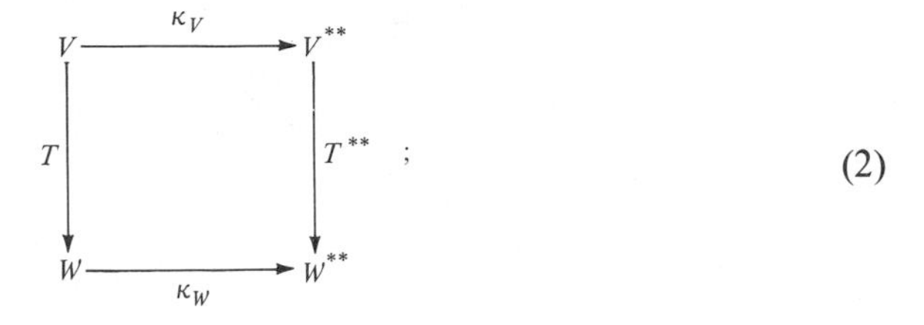

(2)

in other words, we get the same result when κ is applied before or after T. We will adopt this property (2) as the general condition under which a transformation such as κ between functors is a morphism of functors (in the standard terminology, a "natural transformation").

If F, G are two functors $\mathbf{C} \to \mathbf{D}$ a *natural transformation*

$$\tau \colon F \mathrel{\dot{\to}} G$$

is thus taken to be any rule which assigns to each object A of $\mathbf{C}$ an arrow $\tau_A \colon FA \to GA$ of $\mathbf{D}$ in such a way that the diagram

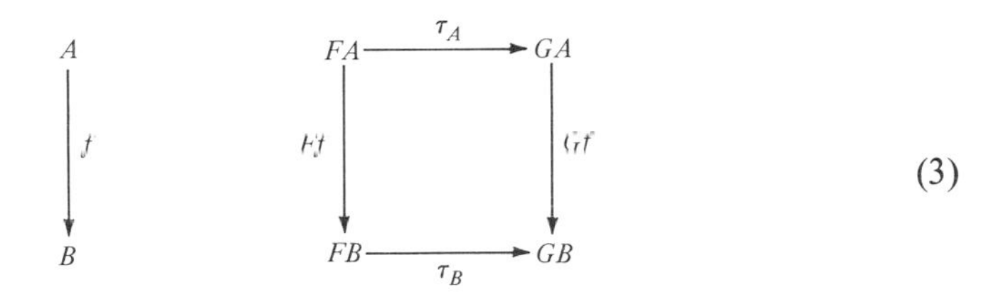

(3)

commutes for every arrow $f \colon A \to B$ of $\mathbf{C}$, in other words $Gf \cdot \tau_A = \tau_B \cdot Ff$ for all f. One can usefully enlarge the picture implicit in this diagram. The given functors F and G carry each commutative diagram $(A \to B \to C \to \cdots)$ in $\mathbf{C}$ into a commutative diagram in $\mathbf{D}$. The natural

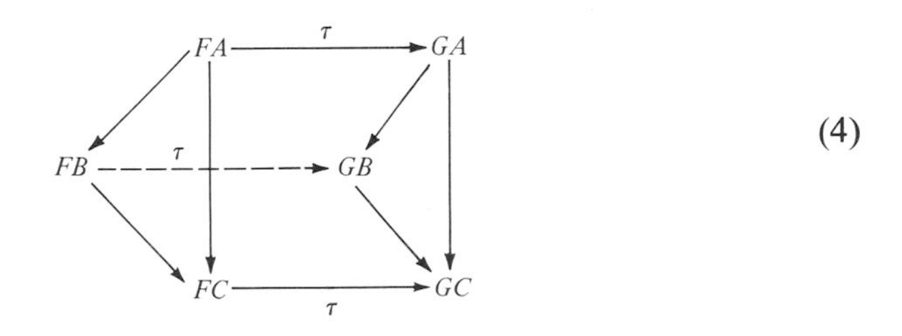

(4)

transformation τ "translates" the first picture on the left into the second, on the right, so that the whole diagram is commutative.

For example, if H is the functor giving the set $H(U)$ of all holomorphic functions g on the open set $U \subset \mathbf{C}$, then the operation $g \mapsto dg/dz$, take the first derivative, is a natural transformation $H \to H$ between functors on the category open($\mathbf{C}$).

Now that we have morphisms of functors, we can form a category of functors, once we note that the composite of two natural transformations $\tau\colon F \to G$ and $\sigma\colon G \to H$ is a natural transformation $\sigma \cdot \tau\colon F \to H$. More explicitly, given categories **C** and **D** the *functor category* $\mathbf{D}^{\mathbf{C}}$ has

$$\text{objects}(\mathbf{D}^{\mathbf{C}}) = \{F \mid F \text{ is a functor } \mathbf{C} \to \mathbf{D}\}, \tag{5}$$

$$\text{arrows}(\mathbf{D}^{\mathbf{C}}) = \{\tau \mid \tau \text{ is natural } \tau\colon F \to G\}. \tag{6}$$

If we regard these as sets, they can be illegitimately big. In particular, for each category **C** we can construct $\text{Set}^{\mathbf{C}}$, the category of all set-valued functors on **C**. It is big (at least as big as *Set*) but not dangerous. It is, rather, useful; mapping **C** into $\text{Set}^{\mathbf{C}}$ by a hom functor embeds **C** in a category with better properties!

11. Universals

The language of categories allows one to give a common description of certain standard constructions, such as products and pullbacks, which are used in many different explicit categories. This common description usually has the form: A certain arrow to or from the constructed object is "universal" among all possible arrows with some desired property.

First consider products. The "cartesian" product of two sets X and Y is normally described as are the cartesian coordinates in the plane: It is the set of all pairs $\langle x,y \rangle$ of elements $x \in X$ and $y \in Y$. This description at once yields two functions, the projections $p\colon \langle x,y \rangle \mapsto x$ and $q\colon \langle x,y \rangle \mapsto y$ on the factors X and Y, as in the usual geometric figure of the product of two intervals:

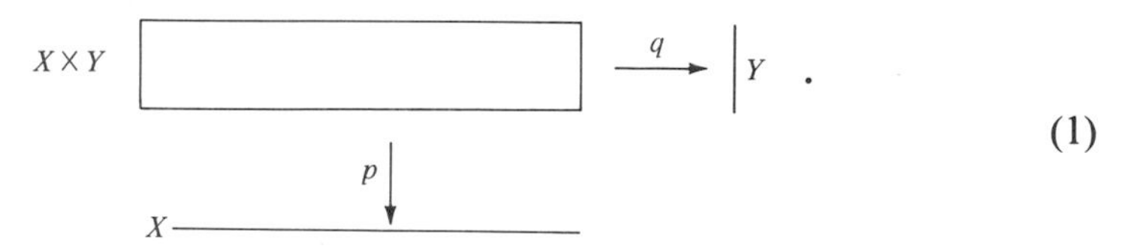

(1)

This also allows the construction of functions into the product $X \times Y$ from separate functions into X and Y. Thus, given two functions

$$f\colon Z \to X, \qquad g\colon Z \to Y \tag{2}$$

from a common domain Z, there is a function $h\colon Z \to X \times Y$ which sends $z \in Z$ into $\langle fz,gz \rangle \in X \times Y$. This function h can also be characterized in terms of f and g, but without mention of elements z, as that unique function $h\colon Z \to X \times Y$ with $p \cdot h = f$ and $q \cdot h = g$.

This last property uses arrows (and their composition) and not elements, so can be stated in any category as follows.

Definition of product. A *product* of two objects X and Y in a category **C** is an object P and two arrows, p and q,

$$X \overset{p}{\leftarrow} P \overset{q}{\rightarrow} Y \tag{3}$$

such that for each pair of arrows $f\colon Z \to X$, $g\colon Z \to Y$ with a common domain Z there is a unique arrow $h\colon Z \to P$ with $p \cdot h = f$ and $q \cdot h = g$. In other words, there is a unique h which makes the following diagram commutative:

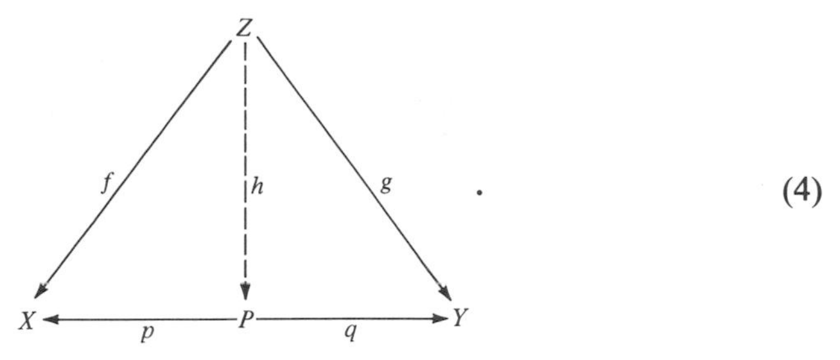

The arrows p and q are the *projections* of the product, while a category is said to have products if there is such a product diagram (3) for every pair of objects of the category.

This definition has the form: Among the diagrams of the shape (3) for given ends X and Y, the product diagram (3) is "universal"–every other such diagram (with middle term Z) maps into the product diagram (3) via a *unique* arrow h. As a result, this property determines the product P and its projections p and q "up to an isomorphism". This means that if there is in **C** another diagram

$$X \overset{p'}{\leftarrow} P' \overset{q'}{\rightarrow} Y \tag{5}$$

which is also a "product" of the given objects X and Y in this sense, then there is an invertible arrow $\theta\colon P' \to P$ with $p\theta = p'$ and $q\theta = q'$. (The universal property of P gives θ; that of P' gives a two sided inverse for θ.)

For the category of groups, we already noted in (V.8.2) that the usual direct product of two groups is a product in this universal sense. The same construction by pairs of elements yields the product for abelian groups. Similarly the (categorical) product $V \times W$ of two vector spaces over the same field consists of pairs $\langle v,w \rangle$ of vectors with termwise operations; this product is commonly called a "direct sum". For two topological spaces X and Y, the set-theoretical product $X \times Y$ has a topology in which the open sets are arbitrary unions of products $U \times V$ with U and V open in X and Y, respectively. This definition of the product topology is calculated precisely to produce a topology in which both of the projections p and q are continuous. Thus *Top* has products.

In all of these cases, the product object consists of pairs of "elements"; in other words, for each type of structure, the product of two sets-with-structure is formed by putting a structure (of the type at issue) on the product of the underlying sets. There is a reason for this, which we skip. (The functor sending set-with-structure to the underlying set has a left adjoint.)

In any category, binary products (those with two factors) yield by iteration products with any finite number of factors (and described by the appropriate universal property). In *Grp*, *Set*, and *Top* there exist also products of any (infinite) set of factors. The interested reader may be able to check that the (Tychonoff) topology on the infinite product of topological spaces is again chosen exactly so as to make all the projections on the factors continuous.

Pullbacks are like products. Let there be given three objects X, Y, and Z and two arrows u, v which form a sort of "corner"

(6)

If the objects are sets, take the set P of all those pairs $<x,y>$ with $x \in X$ and $y \in Y$ for which $u(x) = v(y)$; thus P has the evident projections into X and Y. If the objects are spaces, take the same pairs $<x,y>$; they form a subset of the product space $X \times Y$ and inherit from it the "subset" topology. In either category, P fills out a commutative square

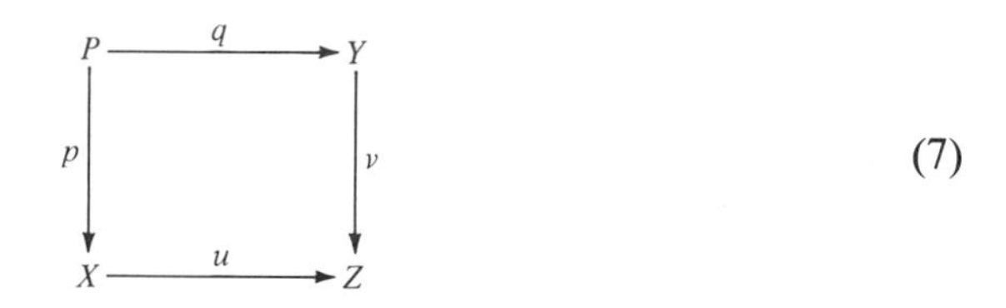

(7)

with the following universal property: For every other commutative square

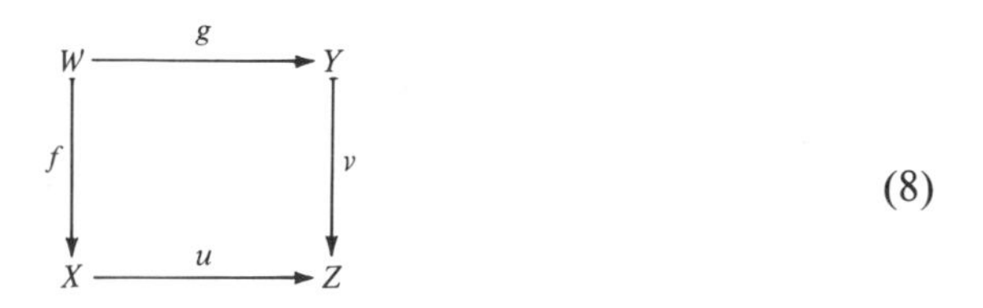

(8)

built on the same corner (6) there is a unique arrow h: $W \to P$ with $p \cdot h = f$ and $p \cdot h = g$; that is, a unique arrow which renders commutative the diagram

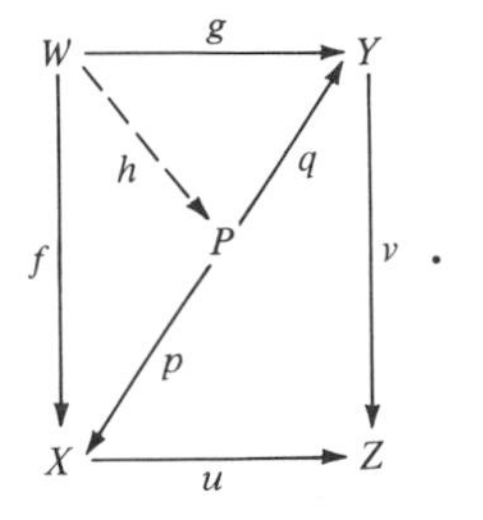

(9)

In other words, the object P is the universal way of completing the corner (6) to a square. In any category, a P with this universal property is called a *pullback* or a *fibered product*, written $P = X \times_Z Y$. (This notation is incomplete, since P is not determined just by the objects X, Y, and Z, but also by the arrows u and v.) One also says that W is the pullback of Y along u, or is obtained by "changing the base" from Z to X.

Pullbacks are omnipresent. If u and v are inclusions of subsets of Z, the pullback P is their intersection. If v: $T^*M \to M$ is a tangent bundle, while u: $N \to M$ is the inclusion of a submanifold, the pullback is the tangent bundle to N.

An object T in a category **C** is called a *terminal object* if to every object X of the category there is a unique arrow $X \to T$. As for products, it follows from this universal property that a terminal object T in **C** is unique, up to an isomorphism. In the category of sets, any one-point set is terminal (and of course any two such sets are isomorphic). Similarly the group with only one element and the topological space with only one point are terminal in their respective categories. Given a terminal object T, products can be constructed from pullbacks, because the product $X \times Y$ is just the pullback $X \times_T Y$ formed from the necessarily unique arrows $X \to T$, $Y \to T$. In fact, given the (binary) pullbacks P and a terminal object, one can construct pullbacks P for all sorts of fancier finite diagrams

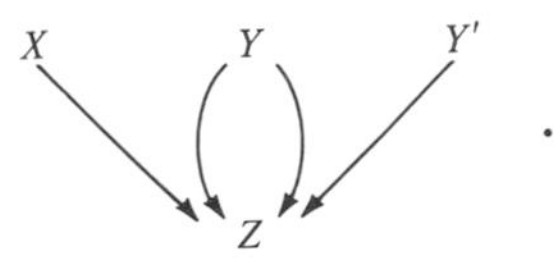

Such a pullback is an object P with an arrow p to each object of the diagram which makes all composites from P equal. Such a pullback is called a *finite limit*.

Each description of a universal object by means of arrows can be "dualized" by writing the same description with all arrows reversed. Thus an *initial* object I (the dual of a terminal one) in a category **C** is an object such that there is a unique arrow $I \to X$ from I *to* every object X. The empty set is an initial object in *Sets*; the one element group is an initial object in groups. A *coproduct* of two objects X and Y is a diagram

$$X \xrightarrow{i} Q \xleftarrow{j} Y \tag{10}$$

such that to every similar diagram $f\colon X \to W \leftarrow Y\colon g$ there is a unique $h\colon Q \to W$ with $f = h \cdot i$ and $g = h \cdot j$, as in the commutative diagram

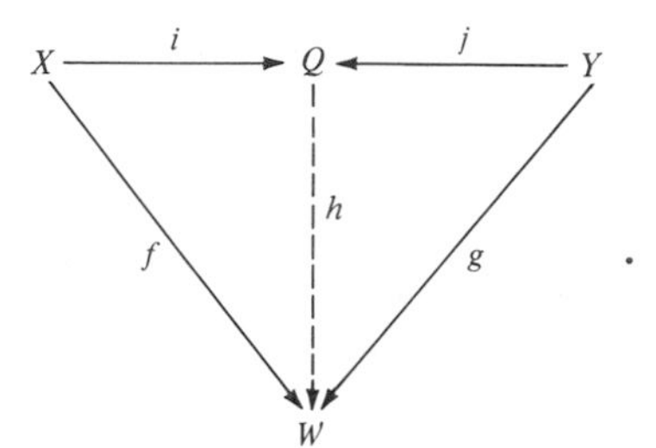

(11)

The arrows i and j are called the *injections* of the coproduct. The coproduct of two sets S and T is their disjoint union, as described in (II.8.1). The coproduct of two vector spaces V and W is their product (a direct sum) with the evident inclusions. The coproduct of two groups is their free product (§V.8). In particular, the free product of two infinite cyclic groups is the free group on two generators, which appears in topology as the fundamental group of the space consisting of two tangent circles (Fig. X.5.4).

The dual of a pullback can also be formed in these categories; it is called a *pushout*—or, in the case of groups, an *amalgamated* product.

We now turn to the construction of "exponentials". If m and n are natural numbers, the power m^n can be described as the number of different functions on a set of n things to a set of m things. Hence we write Z^Y for the *exponential* set or the *function set* consisting of all functions $f\colon Y \to Z$:

$$Z^Y = \{t \mid t\colon Y \to Z\}. \tag{12}$$

This function set can also be characterized in a "universal" way by using a familiar process which turns a function $f(x,y) \in Z$ of two variables $x \in X$ and $y \in Y$ into a function $F(x)$ of one variable x with values $F(x)$ which are functions of the second variable y. The formal definition of this F reads

$$(F(x))(y) = f(x,y), \qquad x \in X,\ y \in Y; \tag{13}$$

here each value $F(x)$ is a function of y, hence an element of the function set Z^Y, so that $F\colon X \to Z^Y$. This is a sort of inference (a bijection) from f to F

$$\frac{f\colon X \times Y \to Z}{F\colon X \to Z^Y} \tag{14}$$

which is reversible: For each F as in the lower line one may construct the corresponding f in the upper line by using the same formula (13), read backwards.

In order to describe the bijection $f \mapsto F$ as a universal process, we need to express formally the left hand side of (13) in which the function $t = F(x)$ is "evaluated" at the argument y. This uses the *evaluation* function e,

$$e\colon Z^Y \times Y \to Z, \qquad (t,y) \mapsto t(y), \qquad t \in Z^Y, \qquad y \in Y. \tag{15}$$

Then $e(F(x),y) = (F(x))(y)$, so the function F described in terms of elements x as in (13) can be described by arrows as follows: To each f there is a unique F which renders the diagram

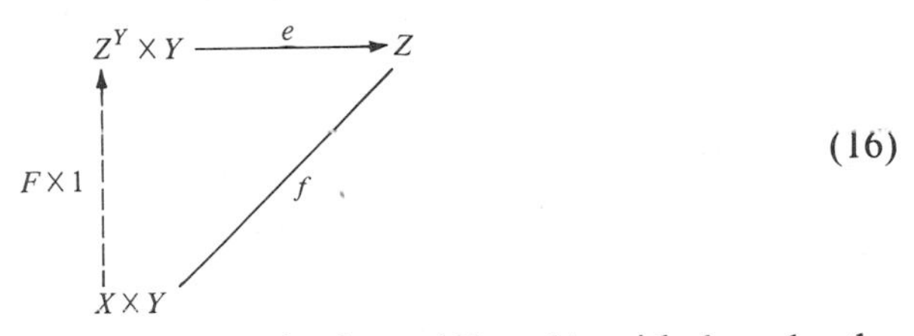

(16)

commutative. As an equation, this reads $f = e(F \times 1)$, with $1 = 1_Y$ the identity arrow of Y. This equation is just a different statement of (13), not using elements.

This form of the description of exponentials applies to any category **C** which has products. The category is said to "have exponents" if to every pair of objects Z, Y there is an object Z^Y and an arrow $e\colon Z^Y \times Y \to Z$ such that, whatever the arrow $f\colon X \times Y \to Z$, there is a unique arrow F which makes the diagram (16) commute. In other words, the evaluation arrow e is universal among arrows f from a product.

Our description has also indicated that the exponential Z^Y, defined in this way, exists for any two objects Z and Y in the category of sets. This holds also in the category of vector spaces (or of abelian groups): Just take Z^Y to be the vector space (or abelian group) of all linear transformations (or of all homomorphisms) $Y \to Z$. However, exponentials do not exist for all spaces in the category *Top*; in other words, there is not always a "good" way to define a topology on the function space Z^Y. This observation has actually led to the suggestion of a suitable restriction (or perhaps an enlargement) of the notion of a topological space, so as to provide a "convenient category" of such modified spaces, in which products and the corresponding exponentials always exist.

The description (16) of the exponential has another formulation, this time in terms of functors. First observe that in a category **C** with products (of objects) one also has products of $f \times g$ of arrows, because given $f\colon X \to X'$ and $g\colon Y \to Y'$ there is by (14) a unique arrow $f \times g$ which makes the following diagram commutative

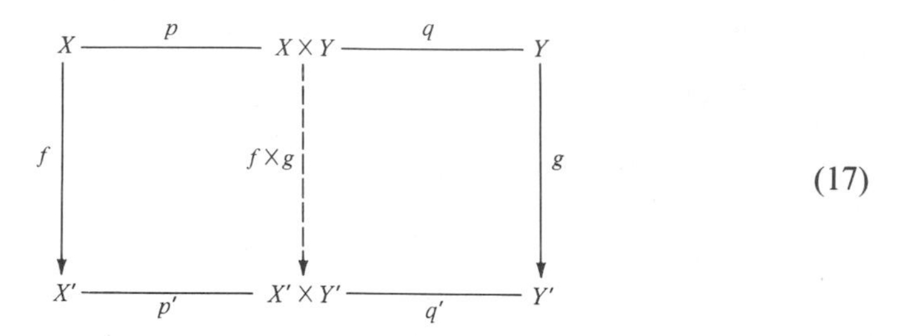

(17)

In this way the product $X \times Y$ is really a functor (of two variables X and Y). In particular, for each fixed Y, the operator $- \times Y$ is a functor of one variable, the "product with Y", sending X to $X \times Y$ and $g: X \to X'$ to $g \times 1: X \times Y \to X' \times Y$. Moreover, Z^Y for this fixed Y is a functor of Z. The correspondence (14) above states that the functor $X \mapsto X \times Y$ uniquely determines the functor $Z \mapsto Z^Y$ by the bijection $f \mapsto F$. Here $F: X \to Z^Y$ is an element of the "hom set" $\hom(X,Z^Y)$, while $f \in \hom(X \times Y,Z)$. Thus the bijection (14), expressing f in terms of F, can be written as a bijection of hom sets (cf. (9.8))

$$\hom(X \times Y,Z) \cong \hom(X,Z^Y). \tag{18}$$

Both sides are functors (of the arguments X, Y and Z) to sets, and this bijection is a natural isomorphism of functors. One says that it makes the functor $(\ )^Y$ the right adjoint of the functor $- \times Y$, while the latter functor is a left adjoint (because it is on the left side of the hom set in (18)).

Generally a functor F: $\mathbf{C} \to \mathbf{X}$ is *left adjoint* to a functor G: $\mathbf{X} \to \mathbf{C}$ when there is a natural bijection

$$\hom(FC,X) \cong \hom(C,GX) \tag{19}$$

between functions $\mathbf{C}^{op} \times X \to Sets$. There are many examples of adjoint functors; each universal construction leads to a pair of adjoint functors, and conversely.

12. Axioms on Functions

The standard "foundation" for Mathematics start with sets and their elements. It is possible to start differently, by axiomatizing not elements of sets but functions between sets. This can be done by using the language of categories and universal constructions. We will now sketch this approach, in order to emphasize again the observation that the leading ideas of Mathematics have a variety of alternative formal expressions.

The intent then is to set up an adequate axiom system for functions and their composition. The functions will be arrows between objects which are (perhaps) sets, but the axioms will not refer to any elements of these sets; they will be formulated strictly in terms of objects, arrows, their domains, codomains and their composites, as well as identity arrows. The axioms will thus be stated in a first order language with equality and with two sorts, objects and arrows.

Axiom I (Category). The objects and arrows form a category.

Axiom II (Finite Limits). In this category there is a terminal object 1 and a pullback for each "corner" $X \to Z \leftarrow Y$.

This axiom provides for the number 1 and the product of two numbers, since pullbacks and a terminal give products, as noted in §11. One could also get zero and addition of numbers by an axiom requiring the existence of an initial object and pushouts. We do not do this because these properties turn out to be consequences of the other axioms to follow.

Axiom III (Exponents). For any two objects there is an exponential.

Next we need to handle subsets. Now a subset S of X can be considered as an inclusion $S \subset X$ and hence as an arrow $m: S \to X$. Such inclusions are injections; they can be characterized as "monic" in the following way: An arrow $m: S \to T$ is *monic* if, for every pair of arrows $f, g: S \to X$, $m \cdot f = m \cdot g$ implies $f = g$; this is just like the description with elements x, y of an injective function m: $mx = my$ implies $x = y$. One might object that the usual inclusion map is not only an injection, but sends each element of the subset S to "itself". This we cannot say. Instead we call two monics $m: S \to X$ and $m': S' \to X$ *equivalent* if and only if there is an invertible arrow $\theta: S \to S'$ with $m' \cdot \theta = m$, as in the diagram

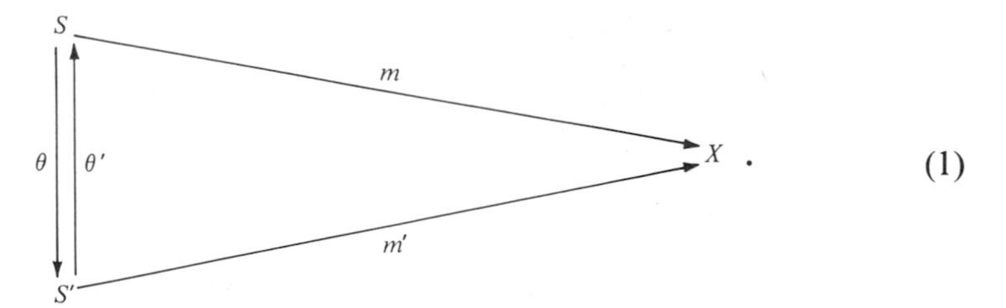

(1)

Equivalent subobjects of X will be represented by the same arrow from X, by taking the arrow to be like a characteristic function ψ. These functions, much used in probability theory, take only the values 0 and 1; ψx is 1 when x is in the subset S at issue and 0 otherwise; it thus expresses a basic fact about a subset S: Either $x \in S$ or not ($x \in S$). Formally, in set theory the characteristic function ψ_S of a subset $S \subset X$ is defined by

$$\begin{aligned} \psi_S(x) &= 1 \qquad x \in S, \\ &= 0 \qquad x \notin S. \end{aligned} \tag{2}$$

Moreover from the characteristic function ψ one can reconstruct the subset S: It consists of exactly those elements $x \in X$ which land on 1 under ψ. This means that S is the pullback of the inclusion i: $\{1\} \subset \{0,1\}$ under ψ:

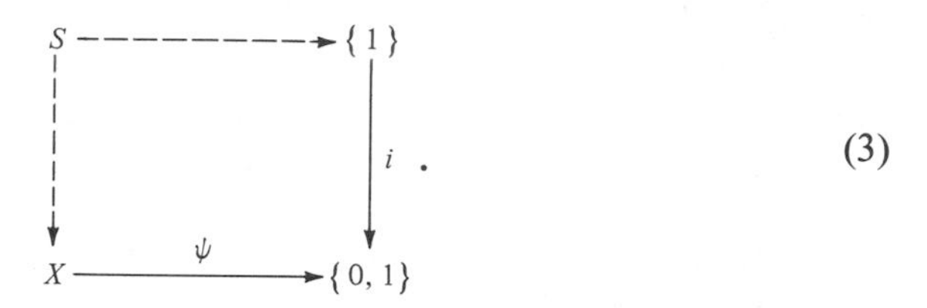

(3)

This description can be purged of all "elements" by using the "universal" description of the pullback and by replacing the traditional set $\{0,1\}$ of two truth values by an arbitrary object Ω, as follows.

Axiom IV (Subobject Classifier). There is an object Ω and an arrow τ: $1 \to \Omega$ from the terminal object 1 such that every monic m: $S \to X$ is the pullback of τ along a unique arrow ψ:

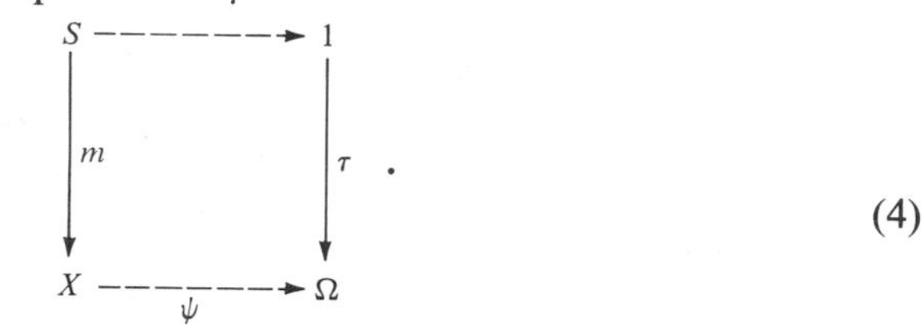

(4)

In this axiom, the arrow $S \to 1$ must be the unique arrow from S to the terminal 1. Also, since 1 is terminal, any arrow (such as τ) from 1 must be monic. Hence the axiom states: There is an arrow τ from 1 which is "universal" among monics, in the sense that every monic arrow is a unique pullback of τ. Observe that axioms II and III are also statements about the existence of universals (i.e., of adjoints).

These four axioms are powerful; a category which satisfies them is called an *elementary topos* (elementary because the axioms are first order; a "topos" because every topological space carries such a topos (of sheaves; as will appear)). For example, arrows $1 \to Y$ act "as if" they were elements of Y (they would be, in the category of sets). Also X is the product $X \times 1$, so there is an isomorphism $X \times 1 \cong X$. Hence, starting with a monic we get successive one-one correspondences to other arrows:

$S \rightarrowtail X$	monic,
$X \to \Omega$	characteristic function,
$1 \times X \to \Omega$	$X \cong 1 \times X$,
$1 \to \Omega^X$	exponential law.

In other words, equivalent monics $S \to X$ correspond to "elements" of Ω^X. Thus Ω^X is an object which is the "set" of all "subsets" of X, so that we have here a categorical description of the power set $\Omega^X = PX$.

As in the axioms for set theory, one needs an axiom to assure the existence of something infinite. In set theory, one axiomatized the properties of the successor function on finite ordinals. Here we axiomatize the fact that the successor function, with 0, provides definitions of other arrows by recursion:

Axiom V (Existence of a "Natural Number Object). There is an object **N** and arrows $0\colon 1 \to \mathbf{N}$, $s\colon \mathbf{N} \to \mathbf{N}$ such that to every diagram $1 \xrightarrow{x} X \overset{h}{\dashrightarrow} X$ there is a unique arrow f which makes the following diagram commute:

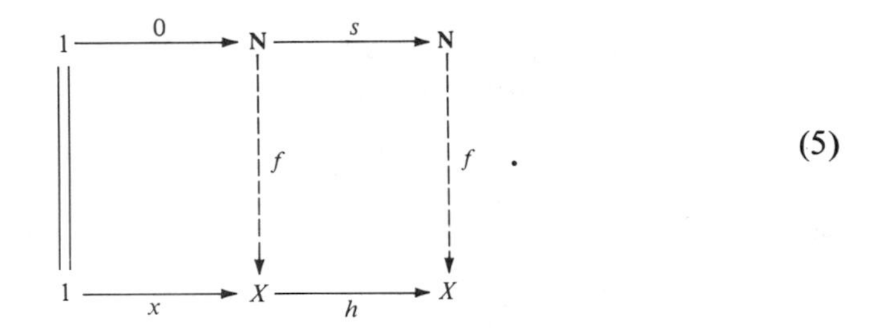

(5)

This diagram (cf. (II.2.5)) is just that defining the function f from h by the recursion

$$f0 = x, \qquad f \cdot s = h \cdot f.$$

Note that this axiom is also a universal construction.

These axioms already allow the development of a good deal of Mathematics, but they are substantially more general than classical set theory, chiefly because they do not require that there are only two truth values, 0 and 1. The logic which they implicitly involve can be intuitionistic, because double negation need not be the identity. This means that the algebra of subobjects of X (i.e., elements of Ω^X) need not be Boolean.

In set theory, the axiom of choice is equivalent to the statement that each epimorphism $e\colon X \to Y$ has a cross section; that is, a function $s\colon Y \to X$ with $e \cdot s = 1$. Indeed, the epimorphism e defines for each $y \in Y$ a non-void subset of X, the "inverse image" $e^{-1}\{y\}$, and the cross section s "chooses" in each such set an element sy. In any category, an arrow $e\colon X \to Y$ is said to be *epic* if, for all arrows $f,g\colon Y \to Z$, $fe = ge$ implies $f = g$. Then we can state choice without elements as follows:

Axiom VI (Axiom of Choice). For every epic $e\colon X \to Y$ there exists an $s\colon Y \to X$ with $e \cdot s = 1$.

This axiom implies that all the algebras of subobjects are Boolean.

In this approach to Mathematics the objects X do not have elements. As noted already, the arrows $1 \to X$ might be regarded as elements of X, but there may not be "enough" of them to make all the usual arguments. In set theory two functions f,g: $X \to Y$ are different only when there is an element x with $fx \neq gx$ where they differ; here there may be two different arrows f,g: $X \to Y$ but no arrow x: $1 \to X$ where they differ; that is, where $f \cdot x \neq g \cdot x$. One can require a topos to be more like classical set theory by asking that it be "well-pointed" in the following sense (where 0 designates an initial object, which can be constructed from the previous axioms).

Axiom VII (Well-Pointed). The objects 0 and 1 differ. Moreover, to any two distinct arrows f,g: $X \to Y$ there is an arrow x: $1 \to X$ with $f \cdot x \neq g \cdot x$.

It is now possible to develop almost all of ordinary Mathematics in a well-pointed topos with choice and a natural number object (Axioms I–VII). The development would seem unfamiliar; it has nowhere been carried out yet in great detail. However, this possibility does demonstrate one point of philosophic interest: The foundation of Mathematics on the basis of set theory (ZFC) is by no means the only possible one!

One can also make a direct comparison between axiom systems. Starting with sets given by ZFC, the conventional description of functions f: $X \to Y$ as sets of ordered pairs does give a category which is a well-pointed elementary topos; indeed, this is the process which we have used to motivate much of the discussion above. Conversely, it is more complex to construct sets from the topos axioms, because one constructs sets as trees. More exactly, one can define when a binary relation on an object of a topos produces a "tree" like the $\in$-tree described in (1.6) above for a set. The resulting trees (sets) satisfy the Zermelo axioms in a form close to ZBQ (bounded comprehension) but do not necessarily satisfy the replacement axiom. This emphasizes again that bounded comprehension suffices for most of Mathematics.

13. Intuitionistic Logic

Examples of toposes include situations with more than the usual two truth values. For consider the category **D** with objects X the functions t: $X_0 \to X_1$ between two sets X_0 and X_1. An arrow in **D** is a pair of functions f_0 and f_1 such that the following diagram commutes:

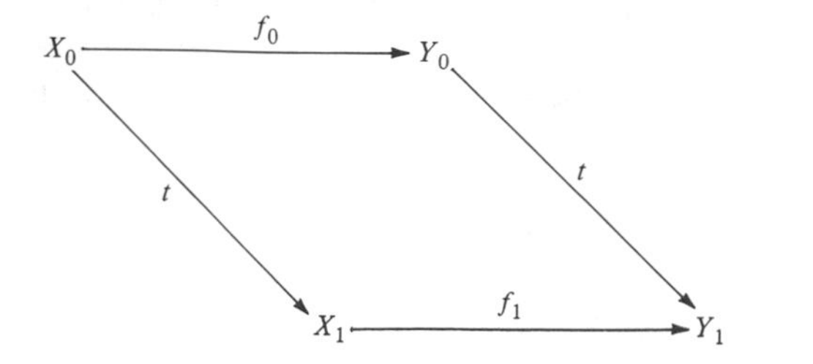

(1)

(we use the same letter t for the functions in the two objects X and Y). Composition of these functions f_i does define composition in **D**. One may think of an object X of **D** as a set "varying with time t"; at the start it is a set X_0 which changes in time to a set X_1; thereby different elements of X_0 may coalesce ($x_0 \neq x'_0$ but $tx_0 = tx'_0$) and new elements may appear.

This category is an elementary topos; that is, it satisfies axioms I–IV above. For example, the product of two objects X and Y is just the object $t \times t$: $X_0 \times Y_0 \to X_1 \times Y_1$, while an arrow $m: S \to X$ is monic in **D** if and only if both $m_0: S_0 \to X_0$ and $m_1: S_1 \to X_1$ are injective, as in the vertical arrows of the figure

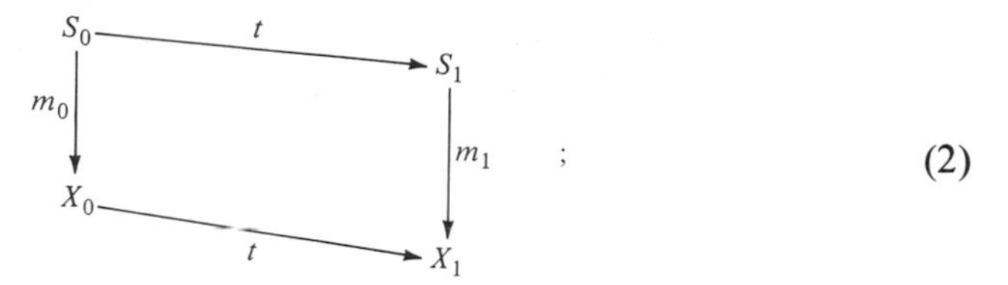

; (2)

this means that $S_0 \subset X_0$ and $S_1 \subset X_1$ are subsets, with m_0 and m_1 the injections. So when does an element $x_0 \in X_0$ belong to the subobject S of X? It may be that $x_0 \in S_0$ (that is, $x_0 = m_0 s_0$ for some element s_0); when that is the case, then also $tx_0 \in S_1$. Or, it may be that x_0 is not in S_0 but that tx_0 is in S_1. Or, it may be that neither $x_0 \in S_0$ nor $tx_0 \in S_1$. Thus there are *three* possible truth values! We can define a "characteristic function" ψ for the subobject S as a function on X_0 to $\{0,1,\infty\}$ for $x_0 \in X_0$ as follows:

$$\begin{aligned} \psi(x_0) &= 0 && \text{if } x_0 \in S_0, \\ \psi(x_0) &= 1 && \text{if } x_0 \notin S_0 \text{ but } tx_0 \in S_1, \\ \psi(x_0) &= \infty && \text{if } x_0 \notin S_0 \text{ and } tx_0 \notin S_1. \end{aligned} \tag{3}$$

To be more complete, we can define an object Ω of **D** as

$$t: \Omega_0 = \{0,1,\infty\} \to \Omega_1 = \{0,\infty\}; \quad t0 = 0, \quad t1 = 0, \quad t\infty = \infty \tag{4}$$

and a function $\psi_1: X_1 \to \Omega_1$ by $\psi_1(x_1) = 0$ if $x_1 \in S_1$ and ∞ if $x_1 \notin S_1$, for all $x_1 \in X_1$. This process defines a diagram of four objects of **D** in the form

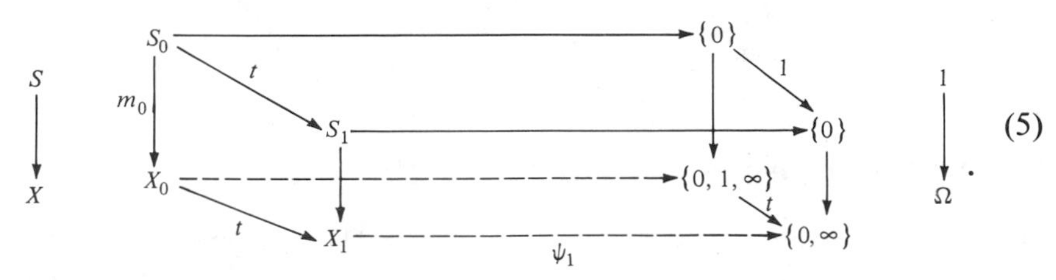

. (5)

Here $\{0\} \to \{0\}$ is a terminal object in **D**. The diagram commutes and is a pullback, so that **D** does satisfy the axiom IV, with the subobject classifier the object Ω described in (4). Thus Ω has three truth values, 0 for "now", 1 for "eventually" and ∞ for "never". By a quite similar construction one can form an elementary topos **E** with objects X the infinite strings of sets and functions

$$X\colon X_0 \xrightarrow{t} X_1 \xrightarrow{t} X_2 \xrightarrow{t} X_3 \to \cdots, \tag{6}$$

while the subobject classifier Ω has infinitely many truth values with

$$\Omega\colon \Omega_0 = \{0,1,2,3, \ldots, \infty\} \to \cdots$$

and the characteristic function $\psi_0\colon X_0 \to \Omega_0$ of a subobject $S \subset X$ is defined by $\psi x = n$ if n is the least integer with $t^n x \in S_n$ and $\psi x = \infty$ if $t^n x$ is never in the subset S_n. Briefly, ψx is the "time till truth".

Classically, the subsets S of a set X form a Boolean algebra under the operations unions, intersection, and complement ($\wedge, \vee, '$); in particular the double complement of a subset S is S itself, $S'' = S$, in parallel to the tautology $\neg\neg p = p$ of the propositional calculus. The simplest Boolean algebra is the algebra $\{0,1\}$ of two truth values, and this algebra is the subobject classifier Ω for the topos of sets. The three Boolean operations, as defined by truth tables, are functions

$$\Omega \times \Omega \xrightarrow{\wedge} \Omega, \qquad \Omega \times \Omega \xrightarrow{\vee} \Omega, \qquad \Omega \xrightarrow{\neg} \Omega.$$

In any elementary topos one can define corresponding arrows on the subobject classifier Ω. They satisfy most but not all of the identities describing Boolean algebra; the resulting algebra is called a *Heyting algebra* because it first arose in Heyting's formal description of Brouwer's intuitionism. In particular, in such an algebra, $\neg\neg$ is not the identity.

We do not give a full description, but we do note a geometric example: The algebra of all the open subsets U of a topological space X is a Heyting algebra. Here the intersection or the union of two open sets is again open, by the axioms of topology, so intersection and union are defined, with the usual properties. However, the complement $\neg U$ of an open set must be defined to be an open set; it is taken to be the largest open set in X disjoint from U. As a result, $\neg\neg U$ may be larger than U. For example, if X is the real line and U the open subset of all positive reals except 1, then $\neg U$ is the set of all negative reals and so $\neg\neg U$, the set of *all* positive reals, is larger than U.

Thus alternative logics are closely related to geometry!

14. Independence by Means of Sheaves

There remains one more surprising connection: The independence proof for the continuum hypothesis, done by the method of forcing, can be reinterpreted as a construction of a suitable elementary topos by the geometric means of sheaf theory.

To begin with, that elementary topos **D** of "sets through time" (§13) has a different description. Take the category **C** with only two objects 0 and 1 and only one non-identity arrow τ: $0 \to 1$. Then a functor F on **C** to *Sets* is a picture of the category **C** in *Sets* as

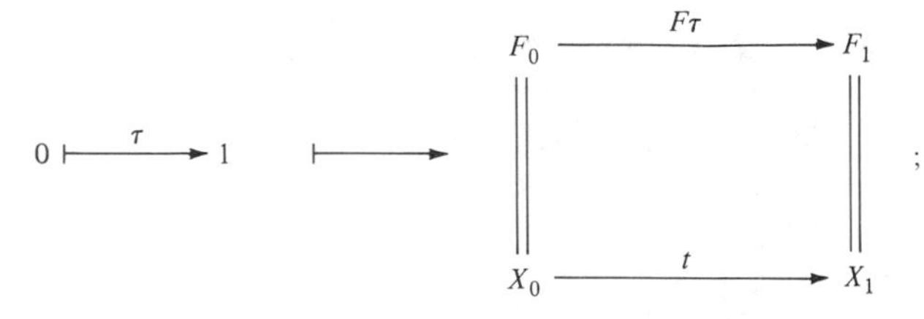

in other words, such a functor is just an object $X_0 \to X_1$ of the category **D**, while a natural transformation f: $F \to G$ between two such functors is just an arrow (f_0, f_1) in **D**, as described in (13.1). Thus **D** is just the functor category $Set^{\mathbf{C}}$, as defined in (10.5) and (10.6).

This is a general situation: If **T** is a topos, while **C** is any category, then the functor category $\mathbf{T}^{\mathbf{C}}$ is a topos. In particular, since the category of sets is a topos, so are all the functor categories $Set^{\mathbf{C}}$. The proof is straightforward, the essential point being the construction of a subobject classifier Ω, much as in the special case of the Ω constructed in §13 for the topos **D** of sets through time.

Next, a presheaf P on a topological space X, as in §VIII.11, is just a contravariant functor

$$P\colon (\mathrm{Open}\, X)^{op} \longrightarrow Sets,$$

as in (9.9), on the category of open sets of X. Hence the category of all presheaves on X is a functor category

$$Set^{(\mathrm{Open}\ X)^{op}}$$

and therefore an elementary topos. In geometry, one passes from the category of all presheaves to the smaller category of sheaves, defined as in §VIII.11. Much the same passage is possible in any topos if the "topology" is replaced by an arrow j: $\Omega \to \Omega$ with suitable properties. The double negation arrow $\neg\neg$: $\Omega \to \Omega$ has these properties. Cohen's proof of the independence of the continuum hypothesis can now be formulated as follows. Start with a model *Set* of sets and within it take a (carefully chosen) partially ordered set P. Regard P as a category **P** and from it form the functor category ("presheaves"):

$$Set^{\mathbf{P}^{op}}.$$

This category is a topos, but its logic is not classical, since its subobject classifier Ω is a Heyting algebra and not Boolean. In this topos, take the "topology" given by the double negation operator and pass to the corresponding category of sheaves

$$Sh_{\neg\neg}(Set^{\mathbf{P}^{op}}).$$

This category is still a topos (this requires a considerable proof) *and* because of the use of double negation in its formation, it is classical. This means that the corresponding subobject classifier Ω is a Boolean algebra–but it may be very large. Choosing in it a maximal prime ideal, one can "divide" it down to be two-valued, like $\{0,1\}$. This division process still produces a topos, which is now a model of set theory. Because of the careful choice of the poset P one can show that CH fails there.

This cursory summary is designed just to indicate how geometric ideas about sheaves, initially introduced with Riemann surfaces, also crop up in foundational studies about sets. In Mathematics, apparently different formal ideas are often closely connected.

Details of this development are presented in Peter Johnstone's *Topos Theory*. For aspects of logic, we refer to Barwise *Handbook of Logic*.

15. Foundation or Organization?

Set theory and logic provide a standard "foundation" for Mathematics: Define all Mathematical objects in the first order language of sets and prove all Mathematical theorems from ZFC and these definitions using the rules of inference specified in logic. We have indicated that there can be alternative foundations, one such being that by elementary topoi. Now in one sense a foundation is a security blanket: If you meticulously follow the rules laid down, no paradoxes or contradictions will arise. In reality there is now no guarantee of this sort of security; we have at hand no proof that the axioms ZFC for set theory will never yield a contradiction, while Gödel's second theorem tells us that such a consistency proof cannot be conducted within ZFC. Similarly, there is no proof of the consistency of the axioms for a well-pointed topos; we know only that they are consistent if ZFC is consistent, because we have provided within ZFC a model for the topos axioms.

Alternatively, set theory and category theory may be viewed as proposals for the organization of Mathematics. The canons of set theory provide guides to the formulation of new concepts and emphasize the extensional character of Mathematics: A "property" is completely determined by knowing all the elements which have that property. Similarly, the canons of category theory emphasize the importance of considering not just the objects but also their morphisms. They also emphasize the use of universal constructions and their associated adjoint functors.

Neither organization is wholly successful. Categories and functors are everywhere in topology and in parts of algebra, but they do not as yet relate very well to most of analysis. Set theory is a handy vehicle, but its constructions are sometimes artificial. Moreover, it is clearly far too general; as Hermann Weyl once remarked, it contains far too much sand. We conclude that there is as yet no simple and adequate way of conceptually organizing all of Mathematics.

Historically, there have been other proposals for the organization of Mathematics or of parts of Mathematics. For the Greeks, Mathematics was geometry, and they formulated real numbers and algebraic operations only in geometric terms. In the 18th century, Mathematics appeared largely in the development of all the aspects of the calculus; this was a natural reflection of the wide opportunities this development offered for formal manipulations and for extensive applications. Subsequently the extraordinary fruitful properties of holomorphic functions made complex variable theory a center about which (much of) Mathematics could revolve. There were competing organizations. In analysis, rigor was enshrined under ϵ and δ. In geometry, Felix Klein proposed that the many varieties of space provided by non-Euclidean and other geometries could be classified and hence organized in terms of their groups of symmetries—the full linear group, the orthogonal group, the projective group, and others. In a way this organization did include complex analysis as the study of groups of conformal transformations; this approach amounts to putting a heavy emphasis on geometric function theory in contrast to analytic function theory and the overuse of $\epsilon-\delta$ methods. Currently the remarkable properties of group representations and their use in arithmetic and physics again suggests that group theory provides an effective organization.

This variety of proposals for organizations reflects the diversity and richness of Mathematics.

Figure 1 may indicate some of the interconnections of the subjects treated in this chapter.

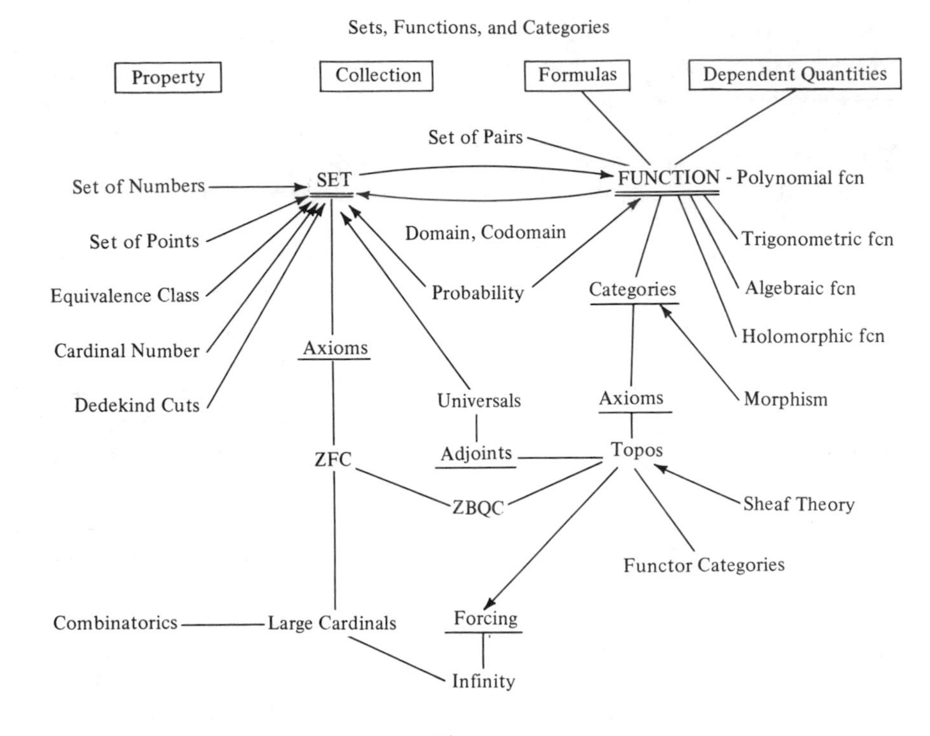
Sets, Functions, and Categories
Property
Collection
Formulas
Dependent Quantities
Set of Pairs
Set of Numbers
SET
FUNCTION - Polynomial fcn
Set of Points
Domain, Codomain
Trigonometric fcn
Equivalence Class
Probability
Categories
Algebraic fcn
Cardinal Number
Axioms
Holomorphic fcn
Dedekind Cuts
Universals
Axioms
Morphism
ZFC
Adjoints
Topos
ZBQC
Sheaf Theory
Functor Categories
Combinatorics
Large Cardinals
Forcing
Infinity

Figure 1

CHAPTER XII

The Mathematical Network

From our examination of a variety of mathematical topics, we now return to the underlying questions: What does this variety say about the nature of Mathematics? How does it illuminate the philosophical questions as to Mathematical truth and beauty and does it help to make judgements about the direction of Mathematical research? In particular, what is the foundation of Mathematics?

Our central observation is this: the development of Mathematics provides a tightly connected network of formal rules, concepts, and systems. Nodes of this network are closely bound to procedures useful in human activities and to questions arising in science. The transition from activities to the formal Mathematical systems is guided by a variety of general insights and ideas. Within the formal network, new developments are stimulated and guided by conjectures, problems, abstractions, and the constant desire to understand more. These observations may be summarized by a diagram:

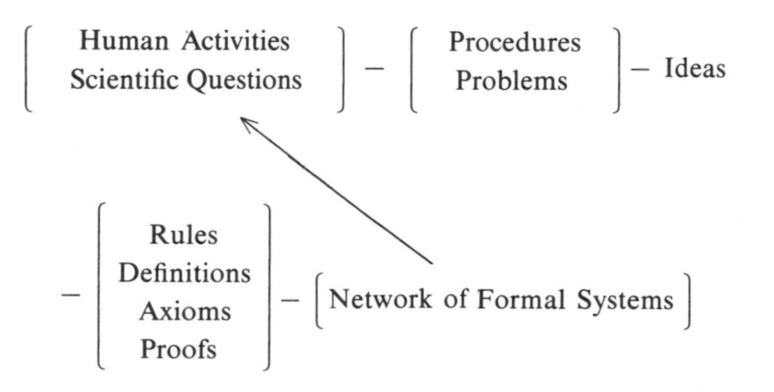

We now continue to explicate the portions of this diagram.

1. The Formal

The presentation of Mathematics is formal: Calculations are done following rules specified in advance; proofs are made from previous axioms and follow predetermined rules of inference; new concepts recognized as relevant are introduced by unambiguous definitions; errors and disagreements are cleared up not by dispute but by appeal to the relevant rules. It is characteristic of any formal procedure that it makes no reference to the meaning or to the applications, but only to the form. The formalism may be imperfect and sketchy, but it carries with it perpetually the possibility of perfection. Because of these characteristics, Mathematics (within its limits) is absolutely precise and independent of persons. The formal can be communicated well, without ambiguity. It develops in many successive stages.

The formal arises first in the rules for arithmetic questions. Given two integers in base ten form, rules tell us how to form a "sum" or a "product". The rules are unambiguous; we can tell when they have been carried out correctly, whether by hand, by abacus, or by computer. Different calculations of the same product, when done carefully, come to the same answer. The rules make no mention of what the numbers might mean, or what decimals are—although in a larger system, still formal, say in Peano arithmetic, such rules can be derived from the Peano axioms, the recursive definitions of sum and product, and the description of what decimals are.

This austerity of the rules of arithmetic is the basis of their applicability. The use of arithmetic is governed by well understood practical operations. The operation of counting uses the successor of each decimal and prescribes how to count, say, the number of matches in a pile. Then the formal process of adding two decimal integers matches the practical process of combining two piles. If the combined count comes out wrong, the failure is never attributed to the rules, but to miscounts or missed matches. The very generality of the rules and their manifold prior uses means that accidental error is never laid at their doorstep. Similarly there are common sense operations to determine areas, say of rectangles: Procedures to recognize when a shape is approximately a rectangle and then ways to measure its length and breadth in feet. The rule for multiplication then provides the area, which again can be checked by a measurement. Thus an area of 96 square feet can be covered (more or less exactly) by 96 square tiles, 12 inches by 12 inches each. If it turns out that the tiles must overlap, then some count or measurement must have slipped, or perhaps the rectangle was really a parallelogram. It is never the rule for multiplication of decimals which is at fault.

In brief, the formal rules of arithmetic are a firm background for the occasionally faulty operation of counting.

The Peano axioms provide a second step in formalization. Now the rules for calculation are deduced from axioms, but again without any attention to the meaning of these natural numbers, as cardinals or perhaps as ordinals. Or, going further, one may adopt the von Neumann definition of ordinal number, as presented in Chapter XI, but then the argument deriving the rules is a formal one from the Zermelo–Fraenkel axioms of set theory and need pay no heed to the idea that an ordinal is the order-type of a well-ordered set. Thus in arithmetic (as elsewhere in Mathematics) there are many stages of formalization, connecting precisely with each other.

Axiom systems provide another prominent example of the formal. The decisive example is the complete axiomatization of plane Euclidean geometry. Intuitively we think that we are dealing with plane figures composed of points, lines, triangles, and circles. But with careful formulation of the (Hilbert) axioms for the plane, all the proofs of theorems can be done strictly from the axioms, as emphasized in §III.2 "proofs without figures". We do not need to refer to the straight line as something without breadth (as it may have been defined by Euclid). We use only the incidence and congruence axioms. This is fortunate; because of it, there is no need to assume that there are real straight lines out there somewhere in the world, nor that there really is a flat plane. Here again the austerity of the Mathematics is coupled with operational rules for its use. Angles can be measured by sighting through the telescope on a surveyor's transit at a fixed vertical rod, then turning the telescope and measuring the "angle" of turn on a protractor built into the transit. Horizontal distances can be measured with chain and plumb–bob. If the distances fail to satisfy (say) Pythagoras, the error is ascribed to the chain man, not to the proof of the Pythagorean theorem. Or, if we carefully set up three points and measure from each the angle subtended by the other two, we might get three angles which do not add to 180°. If so, that is an error of observation or perhaps an indication that some non-Euclidean geometry is in order for this application. Euclidean plane geometry and trigonometry stand firm on the axioms, immune to the deviations of light rays or the vagaries of surveyors. It is firm in its formal standards for deductions; this firmness is known by long experience to be fruitful. Plane geometry is a deductive system, useful in space, but not itself a science of space.

Euclidean geometry also exhibits the rules of inference from logic. A proof in geometry does not (except perhaps verbally) refer to the truth or to successive truths, but only to tautologies and axioms, and to effective applications of rules of inference. And the description of tautologies in terms of truth tables is not a way of appealing to philosophical truth, but only a formal rule for substituting 0's and 1's in a formula with results which have proved to be effective. This observation is most relevant for the case of quantifiers. In the predicate calculus, a statement $(\exists x)F(x)$ does not really mean: I have exhibited a point p with the property $F(p)$. This statement $(\exists x)F(x)$ may be inferred from $F(p)$, or it may be proved by other means; for example, by an indirect proof. So all told,

$(\exists x)F(x)$ is to be viewed as a logistic formula which can be manipulated according to the rules of generalization suggested by experience and specified in §XI.5. This observation does not prevent the construction of other predicate calculi with more restrictive rules for manipulating quantifiers, as they might be suggested by the intuitionist's ideas about existence. But both the classical and some forms of the intuitionist logic can be fully formal; they then simply specify what is meant by a proof. Axioms typify the formal, not just in geometry and arithmetic, but everywhere where they appear in Mathematics. Thus the real numbers began as a source of careful proofs about the reals, giving properties consistent with the operations used in measurement. Similarly, the Zermelo–Fraenkel axioms for sets are not a description of real collections or of totalities of objects; they are formal rules for the manipulation of the binary relation "member of", and long experience has shown them effective in constructing Mathematical concepts. Of course these axioms are suggested by experience, in particular by properties of membership in the intuitive description of the cumulative hierarchy of sets (§XI.1), but these suggestions do not in any Mathematical way determine the axioms. For example, they do not determine whether the comprehension axiom should use bounded or unbounded quantifiers. This is because the hierarchy at each stage forms "all" subsets of the set present at the previous stage, and there is no formal meaning for the notion "subset" here. Also, the hierarchy does not determine how far the ordinals go or what large cardinals may exist. The axioms for set theory have the same status as the axioms of geometry, which are suggested but in no way determined by the practices of surveyors.

Definitions also exhibit the formal. In principle, in every argument the *definiendum* may be replaced by the *definiens*. Or consider particular definitions. A *function* is defined as a certain kind of set of ordered pairs; then properties of functions are those which follow from the definition. The definitions have been chosen so that the properties apply under the standard operations using functions. Thus when x and y represent physical quantities and $y = f(x)$, then the measurements of y depend upon those of x. Or, a function $h\colon U \to \mathbf{C}$ defined on an open set U of the complex plane $\mathbf{C}$ is defined to be holomorphic when it has at each point a of U a complex derivative. Among formal definitions, this one is remarkably well-chosen, as shown by the amazing properties of such functions. However, we know that the properties hold because they are all strictly formal consequences of the definition. To be sure, one might alternatively have followed Weierstrass in defining a holomorphic function $f\colon U \to \mathbf{C}$ as one represented at each point $a \in U$ by a convergent power series in $z-a$. But this, too, is a formal definition, and there is a fully formal proof that it is equivalent to the definition by the existence of the complex derivative. Moreover the theory of holomorphic functions requires formal definitions of other concepts: Integral, piecewise differentiable paths, etc. But all these cases–and their use in training young Mathematicians to understand rigor–exhibit again the clarity which is achieved when all the relevant concepts are defined firmly and formally.

Newtonian mechanics presents another application of the formal. The law $F = ma$ for "force is mass times acceleration" is incomplete without a specification of "force". But when the force is specified by the inverse square law, one can make a completely formal calculation of the orbit of a planet. There are then observational operations (what to examine with a telescope) which tell how to check the result of the formal calculation against results of observation. The case of an earth-bound projectile or rocket is similar: Given the basic equation, one calculates formally *without* looking at the facts what the trajectory should be–and then compares with observations. This comparison is usually not wholly precise, but this does not invalidate the formal calculations; it may only suggest further calculation involving suitable formulas for the forces due to the resistance of the air.

The formal appears clearly in algebra. For finite groups one proves formally (say) that the order of a subgroup always divides that of the group; this can then be checked by numerical examples of particular groups or on figures, using the group-theoretic description of symmetry. In geometry the expert practitioners usually put the pictorial examples and intuition in the foreground, but the formal character is really there, as for instance in the case of the intrinsic geometry defined by a Riemann metric on an abstract surface, while the actual surface sits there as a surface embedded somehow in Euclidean space. In this way, the real generality of geometry depends essentially upon the fully formal definitions of the geometric objects involved.

The ultimate in formality would be the meticulous deduction of all Mathematical theorems from definitions in terms of sets and the Zermelo–Fraenkel axioms. This extreme is never done, but its potential presence serves to define the possible extent of formality. There are indeed many other cases where it was historically necessary to develop a meticulous formalization of parts of Mathematics which had been built up intuitively. One example is the rigorous foundation of the differential and integral calculus in the 19th century–a process which was important not only in itself but also in its contribution to the more conceptual approach to all of Mathematics. Algebraic geometry provides another example. Its development in the late nineteenth century was lively (England, Germany, Italy)–but a bit shaky as reflected in the canard that the highest accomplishment in algebraic geometry consisted in the demonstration of a general theorem and the simultaneous construction of a (special) counterexample to that theorem. The various geometrical insights of this development were subsequently formalized in several successive stages in the period 1925–1975, first by the more systematic use of ideal theory, then by the introduction of general valuation theory, then by the study of specializations, sheaves, and schemes. (This last also illustrates to a remarkable extent the observation that an underlying geometrical idea can have *many* different formalizations.)

Arithmetic, algebraic, or trigonometric calculations, as we have described them, tend to be precise calculations of *the* answer. Many other problems need or allow only a rough answer. In other cases, the principles underlying the calculation may be uncertain and the necessary data is at best approximate. Hence many calculations have the character of estimates. Side effects are neglected, or terms with probable lower order of magnitude are dropped, or unmanageable formulas are replaced by more manageable and (hopefully) adequate alternative formulas. Such procedures can be formal (in part) but are sometimes material, in that the choice of approximations depends on the scientific meaning of the data. Situations such as this prevail in many parts of applied Mathematics. Sometimes estimates in this sense can be fully formalized–notably in the estimates used in analytic number theory and in many basic parts of classical analysis. At other times, estimates in this sense seem to resist reduction to any strictly axiomatic form. Put differently: Some subjects in applied Mathematics do not contain any theorems or proofs of theorems; these subjects thus seem to escape an analysis of Mathematics as formal or formalizable. Thus the boundary lines between formal Mathematics and its applications are vague and moveable. This is a necessary result of the theory that some ideas in Mathematics arise from the applications.

To summarize, the formal in Mathematics appears in many stages:
Rules of calculation (arithmetic calculations, trigonometric identities);
Formulas for differentiation in the calculus;
Rules for estimation and approximation;
Axiom systems for arithmetic and geometry;
Rules for logical inference;
Axioms for abstract algebraic and topological objects;
Axioms for sets and topoi;
The relentlessly full formalization of Mathematics.

Hilbert's doctrine of formalism contends that Mathematics is just the systematic manipulation of formulas. This was a necessary background for his attempt to provide a formal consistency proof for all of Mathematics; it should not be confused with extreme versions of formalism, which regard Mathematics as simply a game with symbols.

We do not take either of these extreme positions but we do assert that the formal is essential to Mathematics. Mathematics is not a scientific study of the facts but a developing analysis of the forms which underly the facts. It is not a science of time and space, but a formulation of the ideas needed to understand time, space, and motion. This understanding depends on ideas. The underlying ideas can be made precise and communicable only by being made formal, as exhibited throughout our chapters.

2. Ideas

Most of the formalizations in Mathematics are based on some underlying idea–an intuitive notion which gives guidance and purpose. It is not easy to give a precise description of the nature of an idea; indeed a deeper idea may be almost impossible to communicate and so may be recognized only after it has been embodied in some formalization. A number of the more general ideas of Mathematics are stimulated more or less directly by human activities. In §I.11, Table 1 we have given a sample of such ideas. It was noted then (and in many subsequent examples) that an idea may have several quite different formal realizations. Thus the idea of "manifold" starts as a geometric locus of all the items or points of some concept. It becomes formal as a topological manifold or as a differentiable manifold (C^∞, C^1 or just smooth in some sense) or perhaps as a complex manifold. In all these cases,, every point of the manifold has a neighborhood described by a "good" chart, but the original idea of manifold should also allow in some way for manifolds with singularities (and I speculate that in the future geometry will see more formal concepts of this sort). In any event, the "idea" of manifold is nebulous enough that it has many formalizations; moreover these different formalizations allow us to study separately different aspects of the idea.

Mathematical ideas arise not just from human activities or scientific questions; they also arise out of the urge to understand prior pieces of Mathematics. A "set" was initially a collection of points in the line or the plane, and then any sort of "collection", considered as one thing and consisting of well-defined Mathematical objects. Subsequently the "idea" of a set was described more specifically in terms of the cumulative hierarchy. Then we have an idea clear enough to be communicated but still fuzzy: At each stage R_α of the hierarchy one next forms the set of "all" subsets of R_α–and there is here no formal description of what subsets are meant. Ideas, as we use the term, are inherently vague. (They are not to be confused with the Platonic idea of the ideal line or the ideal circle or the ideal set.)

Since we are not able to give any precise description of the term "idea", we may simply list examples (see Table 1), this time of ideas which arise from prior parts of Mathematics; again each idea turns out to have several formalizations. We have already seen other examples: The idea of curvature in differential geometry, or the related ideas of eigenvalues and spectrum for self-adjoint operators.

The development of an idea can be long and involved. Thus the study of change and motion suggests the idea that change can be "smooth" rather than sudden. Examples of such smooth changes appear both in mechanics and in geometry. Eventually, this idea "smooth" divides into the separate ideas of "continuous" and "differentiable." The difference between these two, however, is clear only after both have been carefully formulated by means of the descriptions of limits. Differentiability, so

Table 12.1. Some Ideas Arising within Mathematics

Origins	Idea	Formal Versions
Polynomials	Function	Formal expression (in a language)
sin, tan, log		Complete table of values,
Dependent variables		Set of ordered pairs,
		Arrow in a category,
Velocity, acceleration	Rate of change	Derivative
Tangent line		Partial derivative
		Complex derivative
Differential operator	Operation	Linear transformation
Matrix		Binary operation
Arithmetic operations		Unary operations
Linear order	Relation	Binary relation
Partial order		Set of ordered pairs
Congruence (numbers, figures)		Ternary relation
Betweenness		
Prime factorization	Decomposition	Prime ideal factorization
Partial fraction expansion		Product
Components (of a manifold)		Coproduct, etc.

described, then appears in the calculus and in the analysis of the local structure of curves, surfaces, and manifolds; and then eventually (but not until the 1930's) in the precise global definition of a smooth manifold. Continuity, once formalized, is similarly seen to apply to functions of several variables and to functions of curves, thus gradually leading to the 20th century notion of a metric space. Eventually it becomes clear that the metric notion of distance can be replaced by the more qualitative notion of neighborhood, as in the definition of a topological space. Thus it is that the formal properties of open sets eventually serve to codify the very idea of continuity.

There are also many less general ideas: An idea how one might prove some desired theorem. Sometimes the idea fails, and sometimes its successful execution involves complications or additional technical tricks. Again, the proof may be just a routine realization of the idea. In any event, when the formal proof is at hand the original idea (with supporting detail) has been realized. Ideas require formalizations.

The idea of putting together separate pieces of a function leads on the one hand to a concept of a sheaf (as in §VIII.11); on the other hand, where the pieces are power series expansions, it becomes the concept of analytic continuation for holomorphic functions.

Our view that Mathematics is and must be formal is supplemented by the observation that each formalism rests on some underlying or leading idea. The reverse can also happen, as when geometric ideas are formulated so vaguely that they do not constitute a proof of the intended results.

3. The Network

We cannot realistically constrain Mathematics to be a single formal system; instead we view Mathematics as an elaborate tightly connected network of formal systems, axiom systems, rules, and connections. The network is tied to many sources in human activities and scientific questions. We have already sketched some pieces of such a network, in §V.10 with a table of functions, transformations and groups, in §VI.12 with a table of the interactions of the concepts of the calculus, and in §IX.13 with the table of the Interconnections of Mathematics and Mechanics.

In the full, dense network of parts of Mathematics, there are many outside ties. The most basic subjects (the subjects at the "edges" of the network) are tied closely to connections reaching outside of Mathematics. The primary ties are those to various human activities: Counting, measuring, moving, observing, changing, and others as listed in §I.8. In some cases, these ties can also be regarded as connections not to activities, but to phenomena: Multitude (that which can be counted), extent (that which can be measured), motion (which can be observed), change (which again can be observed). By way of these ties and connections, mathematics is grounded in "reality", at least in whatever reality is represented by these activities and these phenomena.

The subjects of Mathematics are also tied or connected to other parts of human knowledge, and most especially to the various sciences. Geometry is tied to mensuration, architecture, surveying, navigation, and, on a more sophisticated level, to space, time, and space-time as these enter into physics. Calculus is tied to mechanics, dynamics, and many other parts of theoretical physics. Differential equations and Fourier analysis are likewise tied to physics–as is also vector analysis, complete with all the sophistication of dual spaces and tensor products. Calculus is also tied to economics, for example, in the use of marginal concepts in Mathematical economics. The number of such connections between Mathematics and science is very great–and often these connections go to subjects which are in the "middle" of the network of Mathematics, and not just to the basic subjects at the edge of the network.

These external connections of Mathematics are numerous and tight, but they do not wholly describe or determine the Mathematical subjects. Basic Mathematical concepts may be derived from human activity, but they are not themselves such activity–nor are they the phenomena involved as the background of such activity. The subjects of applied Mathematics are closely related to branches of sciences–but I argue that they are usually not themselves science. A scientific theory can be falsified by factual data, while a Mathematical theory cannot be so falsified. For example, Euclidean geometry (as we now view it) would not be falsified by measurements (as, for example, measurements of angle sums in some triangle of light rays). Were such a sum to deviate substantially from 180°, it would falsify a scientific theory which asserted that rays of light satisfied the axioms for lines in Euclidean geometry. However, Euclidean geometry itself could be falsified at most by a contradiction derived from its axioms. The axiomatic method is a declaration of independence for Mathematics.

I assert that subjects of Mathematics are *extracted* from the environment; that is, from activities, phenomena, or science–and that they are then later applied to that–or other–environments. Thus number theory is "extracted" from the activity of counting, and geometry is extracted from motion and shaping. The exact mechanism of this "extraction" has not been described in detail here; it will clearly vary considerably from case to case. I have deliberately chosen this work "extraction" to be close to the more familiar word "abstraction"–and with the intent that the Mathematical subject resulting from an extraction is indeed abstract. Mathematics is not "about" human activity, phenomena, or science. It is about the extractions and formalization of ideas–and their manifold consequences.

It is the intimate interconnection of Mathematical ideas which is striking. We have already remarked in Chapter VIII and IX how the geometric notions of manifold, tangent bundle, and cotangent bundle also arise in close parallel in mechanics, with configuration space, velocity phase space, and momentum phase space. In some sense, this connection is a continuation of the way in which calculus and celestial mechanics came out together in the hands of Newton. Similarly it was Riemann who connected the manifold ideas of the differential geometry of surfaces with the complex manifolds now called Riemann surfaces. The full development of these ideas took nearly a century, and has now issued (in part) in the notion of a sheaf (the sheaf of germs of holomorphic functions) and its surprising use in "forcing" methods in set theory.

There are many other deep interconnections not covered in our previous chapters. One important example is a three-way connection between prime ideals in algebraic number theory, points on Riemann surfaces and points on algebraic curves. We sketch these items:

An *algebraic number* y is a root of an (irreducible) polynomial equation

$$y^n + a_{n-1}y^{n-1} + a_{n-2}y^{n-2} + \cdots + a_1 y + a_0 = 0 \qquad (1)$$

with rational coefficients a_i. As in the Galois theory (§V.7) one considers not just y, but all the rational combinations w of y and its powers; they form a field K, called an *algebraic number field*. The degree of the irreducible equation (1) for y is also the dimension of the field K, regarded as a vector space over the ground field $\mathbf{Q}$ of rational numbers; this observation establishes a useful connection with linear algebra. Every number w in the field K satisfies a polynominal equation like (1) of degree at most n (use linear algebra!). When there is such an equation with highest coefficient 1 (the equation is monic) and with all the other coefficients a_i rational integers, the number w is called an *algebraic integer*. The collection of all the algebraic integers w in the field K forms a commutative ring $\mathcal{O}$; indeed this important example (with polynomials) is a basic reason for introducing that concept; see §VI.3. In this way the passage from integers to rational numbers is extended; $\mathcal{O}$ passes to K:

$$\begin{array}{ccc} \mathbf{Z} & \subset & \mathbf{Q} \\ \cap & & \cap \\ \mathcal{O} & \subset & K = \mathbf{Q}(y). \end{array} \tag{2}$$

This is the start of algebraic number theory and its connections with Galois theory.

The quadratic case (degree $n = 2$) appears first. For example, if $y = i$, this ring $\mathcal{O}$ is just the ring of all *complex integers* $m + ni$ with $m,n \in \mathbf{Z}$. In this ring every integer can be factored, and essentially uniquely, into prime (i.e. irreducible) complex integers. But there is no such unique factorization for many other quadratic fields. And when y is a higher root of unity (in which case K is called a *cyclotomic* field) this need not be the case, as Kummer discovered in trying to handle the Fermat problem.

However, unique factorization can be restored, by factoring algebraic integers not into these integers, but into "ideal" factors which are exactly the ideals used in §7.10 as the kernels of ring homomorphisms; this means that one must define a *product* AB of two ideals in $\mathcal{O}$—as the ideal generated by all products ab for $a \in A$, $b \in B$. Also, each element w in $\mathcal{O}$ determines an ideal, the set (w) of all its integral multiples. Now the algebraic number w is prime in $\mathcal{O}$ when it cannot be factored there; that is, when for all u, v in $\mathcal{O}$, $uv \in (w)$ implies $u \in (w)$ or $v \in (w)$. Similarly an ideal P in $\mathcal{O}$ is called *prime* if $uv \in P$ implies $u \in P$ or $v \in P$. Now we can state the first main theorem of algebraic number theory: Every algebraic integer w in $\mathcal{O}$ has a unique decomposition

$$(w) = P_1^{e_1} P_2^{e_2} \cdots P_k^{e_k} \tag{3}$$

as a product of powers of distinct prime ideals (with integral exponents e_i). Moreover, the fact that an ideal is the kernel of a homomorphism $\mathcal{O} \to \mathcal{O}/P$ enters here; the familiar field $\mathbf{Z}/p$ of integers modulo P has been extended to quotient rings (fields) $\mathcal{O}/P$.

Now shift gears and consider an algebraic function w of a complex variable z, as defined by some (irreducible) polynomial equation

$$w^n + a_{n-1}(z)w^{n-1} + a_{n-2}(z)w^{n-2} + \cdots + a_1(z)w + a_0(z) = 0,(4)$$

with coefficients $a_i(z)$ which are rational functions of z. This function w then has a Riemann surface over the z-plane—or better, over the whole Riemann sphere. All the rational combinations of z and w are also analytic (holomorphic except for poles) on this Riemann surface. With the complex numbers $\mathbf{C}$ they form a field $\mathbf{C}(z,w)$, called an *algebraic function field*. Now a rational function of z is determined, up to a constant factor, by giving its zeros and its poles. Similarly, what matters about each algebraic function w are the points P on the Riemann surface where w has a zero or a pole. One may write $V_p w$ for the order of the zero (or, negative, for the order of the pole) of the function w at the point P. Then $V_p w = e_p \neq 0$ for only a finite set of points P. This is vaguely like the factors in the decomposition (3) of ideals. Also, the function w is *entire* (has no poles at points over the finite portion of the Riemann sphere) precisely when the rational coefficients $a_i(z)$ in its irreducible equation (4) are all polynomials. The entire functions in $\mathbf{C}(z,w)$ form a commutative ring $\mathcal{O}$, much like the ring of algebraic integers, and we have an array of four rings

$$\begin{array}{ccc} \mathbf{C}[z] & \subset & \mathbf{C}(z,) \\ \cap & & \cap \\ \mathcal{O} & \subset & \mathbf{C}(z,w); \end{array} \qquad (5)$$

here $\mathbf{C}[z]$ is the ring of all polynomials in z.

This suggests a parallel between algebraic numbers and algebraic functions. The parallel can be carried out by developing each theory by the methods first found for the other. On the one hand, ideals apply to algebraic functions. The ring $\mathcal{O}$ of entire functions on the Riemann surface has unique decomposition of ideals into products of prime ideals P—the prime ideals correspond exactly to the points (on the finite part) of the Riemann surface. Moreover the exponent e_p to which P appears in the decomposition of the ideal (w) of an entire function is precisely the order of the zero of that function at the point P. The remaining points of the Riemann surface (those points over ∞ on the Riemann sphere) can be captured by the conformal transformation $z \mapsto 1/z$; that is, by using prime ideals over the ring $\mathbf{C}[1/z]$. In short, the points on this Riemann surface, which we previously treated geometrically and topologically, can be described algebraically as prime ideals!

The situation can be reversed, using analytic methods for algebra. For each point P on the Riemann surface the function V_p (for order of zero) has the formal properties (X.10.4)

$$V_p(wu) = V_p w + V_p u,\ V_p(w+u) \geqslant \mathrm{Min}(V_p w, V_p u) \qquad (6)$$

and $V_p(c) = 0$ for any complex constant c. Such a function on a field is called a *valuation*, and every valuation of the field $\mathbf{C}(z,w)$ arises from a point P in the Riemann surface, and is V_p, up to a multiple. This is still another description of "point on a Riemann surface".

Now for an algebraic number w, consider the exponent e to which the prime ideal P occurs in the decomposition (3) of w. This exponent V_p *is* a valuation on the algebraic number field $\mathbf{Q}(w)$–and all the prime ideals of $\mathcal{O}$ can be recovered from valuations. Moreover, one can mimic the power series used in analysis. Each valuation V can be turned into a function

$$|w| = 2^{-Vw}$$

like the absolute values used in analysis. A series in powers of $z-a$,

$$c_0 + c_1(z-a) + c_2(z-a)^2 + \cdots$$

with complex coefficients $c_i \in \mathbf{C}[z]/(z-a)$ is replaced by a series

$$c_0 + c_1 p + c_2 p^2 + \cdots \tag{7}$$

in powers of p, with coefficients $0 \leqslant c_i < p$; that is, coefficients in $\mathbf{Z}/(p)$. This series (7) converges when we turn the valuation at the point P into an "absolute value" by setting

$$|w| = 2^{-Vw},$$

because then $|w|$ is small when Vw is large; i.e., when w is divisible by a large power of p. The "p-adic" numbers, imported from analysis, have remarkable algebraic uses. In this way analysis and algebra intersect, to the profit of both.

There is also a connection to plane algebraic curves defined by polynomial equations like (4) in coordinates z and w. This extensive approach eliminates the privileged position of z.

Such ideas arise also with diophantine equations in two variables x and y

$$\sum_{ij=1}^{n} a_{ij} x^i y^j = 0 \tag{8}$$

with integer coefficients. If we replace x and y by z and w this equation becomes complex, and so determines a Riemann surface of some genus g (the genus is the number of handles, so the torus has genus 1). Some time ago, Mordell conjectured that a diophantine equation (8) with genus $g > 1$ would have at most a *finite* number of rational solutions (x,y). Just recently, using a vast extension of the methods we have just sketched, the conjecture has been established by G. Faltings. In other words, the connection between number theory and Riemann surfaces is indeed both tight and effective.

Mathematics is a network . . . and formal concepts abound (ideal, valuation, Riemann surfaces).

4. Subjects, Specialties, and Subdivisions

Since we have described Mathematics as a network, we must specify more the nodes of that network–the various special "subjects" into which Mathematics is (or may be) divided. It will turn out that this notion of a subject–a branch of Mathematics–is both useful and elusive; the meaning of a "branch" will turn out to be various and variable.

Initially, Mathematics may appear to have a simple subdivision, with the main branches arising from basic human activities, perhaps as in Figure 1, where the associated activities are enclosed in boxes.

Here we have the five traditional subject subdivisions (underlined) of Mathematics. *Number Theory* is described by its subject matter: The natural numbers or perhaps the integers; the subject matter is essentially unique, its structure described say by the Peano postulates (considered as a definitive description, despite the nonstandard models and the true sentences which cannot be demonstrated from these axioms). *Algebra* is described by its methods: The manipulation of formulas using letters for variables or unknowns which stand for indeterminate numbers, usually real numbers. *Geometry* is initially a science: The science of space, as it is described by the axioms of Euclidean geometry; there is only one space, and Euclid is its prophet. *Calculus* is the science of variable quantities as they arise in measurement and are represented by (real) numbers, it is also the technique of dealing with such quantities by differentiating and integrating–or, if one presses deeper, by using infinitesimals and infinite operations; more precisely by using limits, $\epsilon - \delta$ and their careful formulation in terms of quantifiers. *Applied Mathematics* is (at first) largely the study of particle and continuum dynamics by the methods of calculus; in return, it raises new problems for the calculus. Thus the study of heat and its differential equations leads to trigonometric series, while the proposals for principles of "least action"–teleology in mechanics–are connected with the calculus of variations.

This first picture of the network is clearly far too simplistic. In the next version (Figure 2) there are added inputs from "ideas" (shown in parentheses).

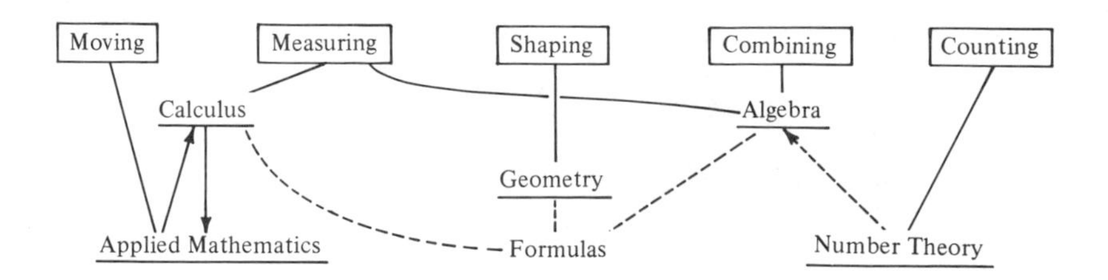

Figure 1

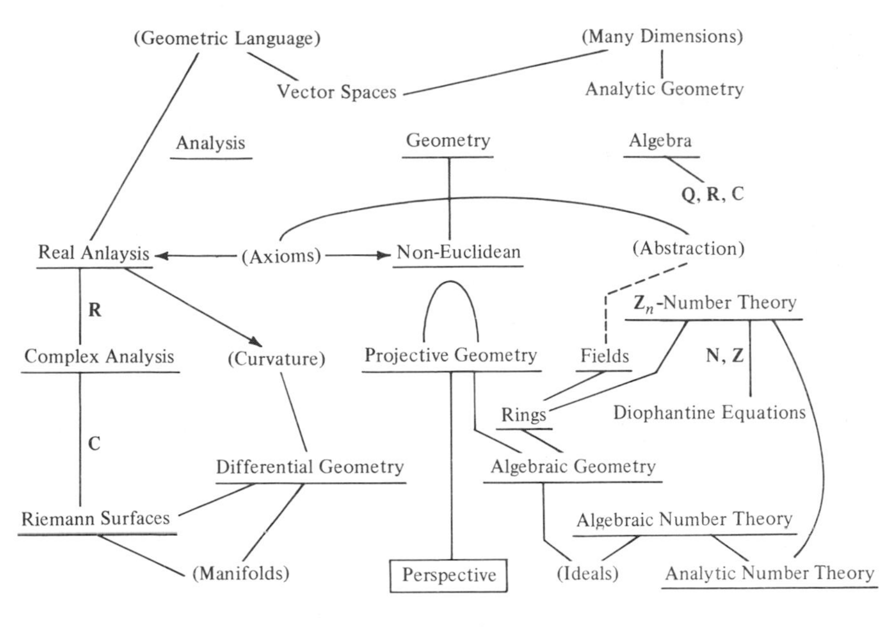

Figure 2

Here the subject "Number Theory" continues, including elementary number theory, the study of Diophantine equations, quadratic forms and the like. But two more subjects now split off. Algebraic number theory is described primarily by its subject matter: Algebraic integers and their properties, but it is closely tied to algebra by its methods. Analytic Number theory is characterized by its methods: The use of analytic techniques, especially complex integration, to tackle the hard questions about the distribution of the primes and the whereabouts of the zeros of zeta of s. (The important line connecting this subject to complex analysis won't fit on the figure.)

"Algebra", once the manipulation of formulas about unknown reals, becomes just the application of the rules for this manipulation to other number systems–integers modulo m or p-adic numbers from algebraic number theory or power series treated "formally", that is, without requiring convergence. Hence algebra splits into several subjects each of which is nominally described as the study of all the set-theoretic models of a list of axioms on the rational operations involved. Thus ring theory is described formally by the ring axioms for addition and multiplication; it subdivides into commutative ring theory, coming from algebraic number theory and from algebraic geometry, and non-commutative ring theory, which arises from matrices, from quaternions, and from the search for more division algebras and which also has connections with algebraic number theory, because the local behavior of primes can be formulated by certain linear algebras. Thus in this case a subject (ring theory) is precisely delimited by a formal list of axioms, while its connections in the Mathematical network exhibit its origins and uses.

Field Theory is again formally described by its axioms (all the good properties of the four rational operations of addition, multiplication, subtraction and division). It arises from the Galois theory–how to understand the solution of polynomial equations in terms of the symmetries of all their roots and of the field of all rational combinations of these roots–and this idea was essential to our discussion in §3 of algebraic number fields and function fields.

Back at the top of the figure (and historically prior to abstract algebra) comes "analytic geometry", a subject which can be described as a method: The remarkable method of so using Cartesian coordinates to reduce problems of plane and solid geometry to routine algebraic manipulation. This method is well known and widespread; it has been claimed that the one notion of modern Mathematics familiar to the man in the street is the idea of coordinates: The tenth floor at 329 West 15th Street. And this reduction of geometry to the manipulation of coordinates inevitably suggests the idea that things dependent upon many inputs (mechanics, statistical mechanics, etc.) can be described in geometric language, as if they were things spread out in a "higher dimensional" physical space. The standard technical formulation of this idea comes in the subject of *vector spaces* (also called *linear algebra* or *matrix theory*). Its substance is determined by its many uses (e.g., eigenvalues): its form is determined by its simple axiomatics: A vector space is a module over some field. It is thus legally subsumed under the subject *Module theory* (not on our figure): A module (§VII.12) is a ring together with an action of that ring on an abelian group, and module theory is a necessary complement to ring theory. In short, the geometric vision is supported by firm and formal algebraic structure.

The discovery of non-Euclidean geometry and its models showed decisively that geometry is not a science of actual space but an axiom-based study of space-like configurations; historically the axioms for geometry are the very model of what an axiomatic approach can be. The various non Euclidean geometries can be subsumed under projective geometry, which has extra-mathematical motivation from the study of perspective; the perhaps parallel subject of "descriptive geometry" has never really been recognized as a branch of mathematics, and projective geometry itself is not now an active field of research. However, the curves, surfaces, and varieties defined in projective spaces as the loci of polynomial equations in the coordinates form the subject matter of the lively field of algebraic geometry, even though these varieties are now described in a context of "schemes"–spaces more abstract than the original projective spaces. The subject is a striking example of the confluence of geometric, algebraic, and topological ideas.

Our figure also exhibits the subjects of real and complex analysis–subjects defined essentially as the study of well behaved functions on specific domains (**R** or **C**). Here the initial ideas of the calculus (rate of change, the whole is the sum of its parts) have an extraordinary and formal development. Through the ideas of tangent lines, curvature and torsion there is a deep connection with geometry: Differential geometry as a subject can be defined either as an application of calculus to geometry or, more intrinsically, as a study of smoothly curved space—which need not have an Euclidean context.

Our Figure 2 does not reflect the study of functions or the pervasive aspects of group theory; this was pictured in part in the Table of §V.10. In principle, group theory can be described formally as the study of all the set-models of the (very simple) group axioms. This does not recognize the variety of interconnections of group theory with other subjects, Figure 3 is a start, beginning of course with the basic examples of symmetry. Group theory as a subject quickly splits into different branches: Finite group theory (which is close to number theoretical ideas) and infinite (combinatorial) group theory. The addition of the commutative law produces the subject of abelian groups, which has a quite different flavor and may not be as "deep" as finite group theory. However abelian groups are extraordinarily useful as topological invariants: they and modules lead to homological algebra (for example, in abelian categories). In a different direction, groups with a smooth structure (Lie groups) or with a topology are closely tied to analysis. Finally the representations of groups by linear transformations ties group theory back to some of its origins in geometry and has remarkable further connections: As carriers of symmetry, such representations can be used to predict the varieties of elementary particles in Physics.

Figure 2 omitted one side: The connections of analysis with classical applied Mathematics. Here the various fields are tightly connected to the

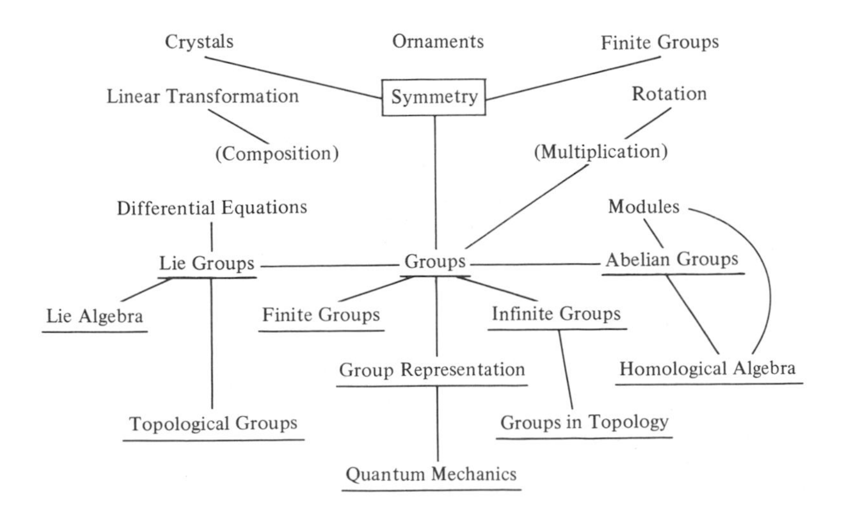

Figure 3. Interconnections for group theory.

associated scientific problems; they might be arranged as in Figure 4, beginning with differential equations, ordinary and partial (ODE to PDE!).

In each of these subjects the physical principles involved provide certain "equations of state", usually partial differential equations. The subsequent development is in part a series of formal deductions in a strictly Mathematical style from these hypothesis (and the appropriate boundary value conditions) and in part a series of approximations and adjustments made with awareness of the intended applications. For that matter the direction taken by the deductions is often guided by the applications. In this sense, our description of Mathematics as the study of formal systems, guided by ideas, must be supplemented by an extensive guidance by scientific questions. The boundary line between science (truths about the physical world) and Mathematics (deductions from well motivated axioms) runs right through the middle of some of these subjects. Classical mechanics, as discussed in Chapter IX, is not quite typical, because it is older, more carefully axiomatized, and (probably as a result) more closely tied to analysis and geometry.

There are also many connections back to core Mathematics, as for example in the use of functional analysis and the various infinite dimensional spaces required for full study of P.D.E. Moreover the quantitative study of the trajectories of dynamics leads to the consideration of dynamical systems, which in its turn played a large role in the growth of topology (qualitative geometry).

There are many other subjects in applied Mathematics which we have hitherto ignored. A small "sample" appears in Figure 5.

Probability as the "frequency" of an event involves serious foundational questions, avoided by the measure-theoretical axiomatization suggested

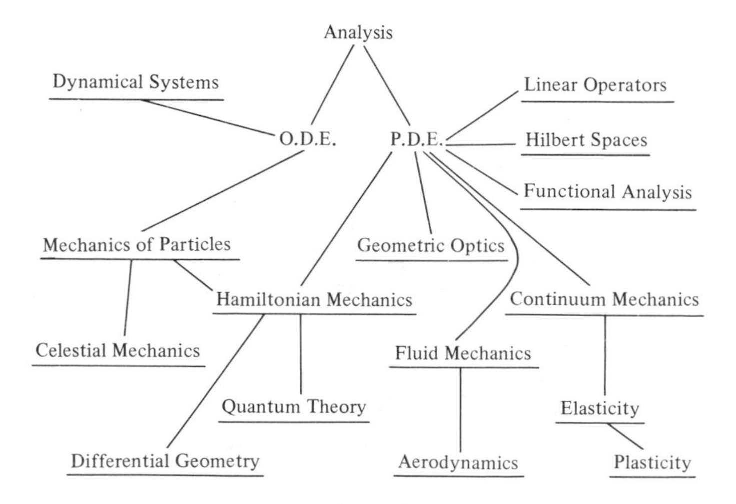

Figure 4

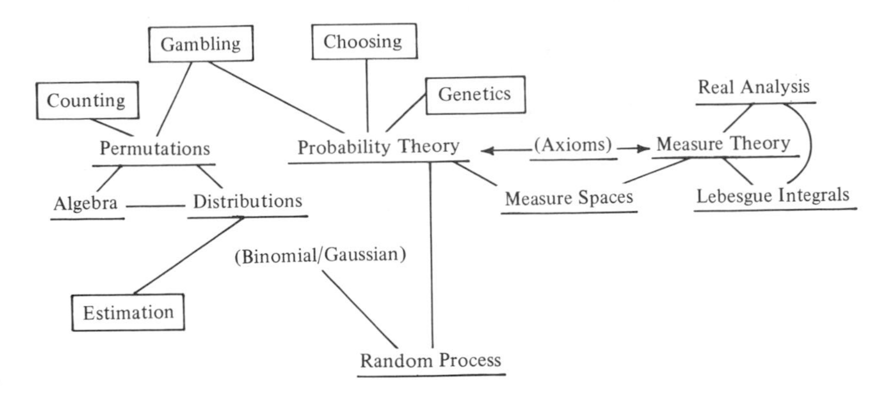

Figure 5. The probable roots of probability.

here. Probability enters vitally into statistical estimates, but in statistics the choice of methods (linear regression is overused) is much dependent upon the application intended, while the foundations are still partially obscure (Bayesian) or at least controversial. For these reasons we cannot count statistics as now a part of Mathematics in our sense.

We have thus exhibited several diagrams of (part of) the network of Mathematics with nodes the various specified "subjects" or "branches". The description of a subject can be quite various. It is sometimes done by a concept (holomorphic function) carefully defined in some context. It is sometimes derived from (our ideas of) a form of experience as in geometry. But then the subject may split into more specific forms, sometimes those described by axioms (topological space) and sometimes by types of formulas (algebraic geometry). Within algebra, different subjects are often delimited by the appropriate axiom systems, but often the direction in which the axioms are exploited is suggested by the applications or by the problems of special interest. When axioms are used, it can easily happen that one subject is "contained" in another in a formal sense. Thus a lattice is a particular kind of partially ordered set (one with greatest lower and least upper bounds of pairs of elements), while a partially ordered set in its turn is a particular kind of category (one in which, given objects A and B, there is at most one arrow from A to B, and this when $A \leqslant B$). Despite these formal inclusions, the subjects in question really count as different ones, because of the very different ways in which the axioms are developed.

Many subjects are subdivided into further specialties, reflecting new techniques, different motivations and sometimes just the pressure of more and more specialists. Thus Mathematical logic was once neatly subdivided into four principal branches as shown just below the four boxes in Figure 6. However, one of the four (Proof Theory) did not have equal status and the other branches proceeded to subdivide again, as indicated in the case

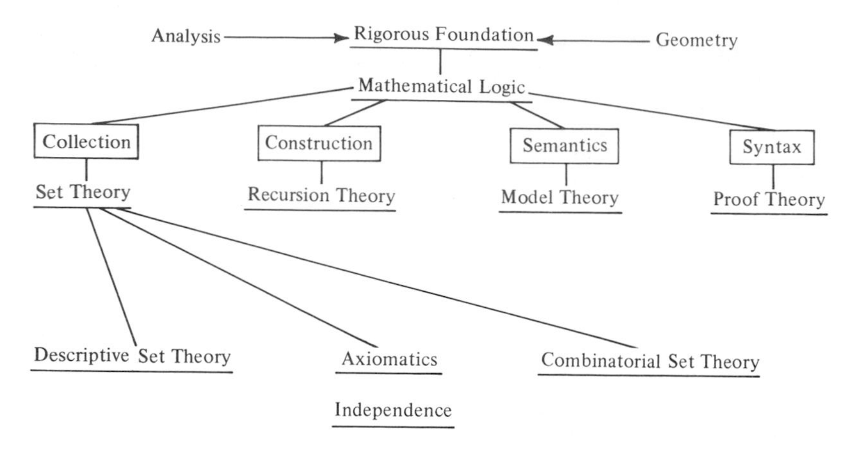

Figure 6

of set theory in the figure. Moreover, this neat four-part subdivision manages to omit some of the smaller activities: Intuitionsistic logic, modal logic, and algebraic logic for example. There are also extensive ties to computer science, especially in recursion theory and the study of algorithms.

This particular figure illustrates well the increasing subdivision of Mathematics attendant upon specialization and a resultant lack of attention to connections (here to the rest of Mathematics) and neglect of some of the original objectives (here questions about the foundations of Mathematics). Many other subdivided subjects show similar effects.

We have thus illustrated many subjects and branches of Mathematics, together with diagrams of the partial networks in which they appear. The full network of Mathematics is suggested thereby, but it is far too extensive and entangled with connections to be captured on any one page of this book.

5. Problems

The development of Mathematics, as reflected in the multiplication of its subfields, is driven by various forces: Ideas arising from human activities, questions posed by science, and problems presented within Mathematics. Many research mathematicians think that the main guide to Mathematical research is the desire to solve specific Mathematical problems—not so much the smaller puzzle-type problems, but the major and famous Mathematical problems. This includes some of the notable problems which have recently been solved, such as Hilbert's fifth problem (continuity assumptions on Lie groups imply analyticity), the negative solution of the decision problem for Diophantine equations (Hilbert's tenth problem), and the positive solution for the Poincaré conjecture in dimension 4 (a simply connected four-dimensional manifold with the homology of a 4-sphere is necessary homeomorphic to a 4-sphere). More especially, one must count here the many famous problems not yet solved: The Poincaré conjecture in dimension 3, the Riemann hypothesis for the zeros of the zeta function, the Goldbach hypothesis (every even number is the sum of two primes), the conjecture that every finite group is the Galois group of a normal extension of the field **Q**, and many more. There are other, less spectacular but equally difficult problems to be found in individual fields of Mathematics. In all these cases, the continued attempt to solve such problems can be a major source of new techniques, new ideas, and even new branches of Mathematics.

Fermat's last theorem ($x^n + y^n = z^n$ has no solution in integers for $n > 2$) is a striking example. For the corresponding equation with $n = 2$ one can find all integral solutions (all integral Pythagorean triangles) by the factorization

$$y^2 = (z^2 - x^2) = (z + x)(z - x)$$

followed by a routine attention to the prime factors of the right hand side. In the nineteenth century, Kummer was tempted to try the same for any n, by the factorization

$$y^n = z^n - x^n = (z-x)(z-\epsilon x)(z-\epsilon^2 x)...(z-\epsilon^{n-1}x), \tag{1}$$

where ϵ is a complex primitive nth root of unity. But the algebraic integers on the right side of (1) do not have a unique factorization into primes (irreducibles), so the suggested attack on Fermat's last theorem fails. Kummer's attempt to understand this gap led to the development of the decomposition into the prime ideals just mentioned in (3.3). This idea led directly to the whole subject of algebraic number theory (see §3), with many further results–none of which yet prove Fermat's conjecture. Algebraic number theory also extended earlier studies by Gauss of the classification of quadratic forms such as

$$ax^2 + 2bxy + cy^2$$

with integral coefficients. Given the analogy of algebraic numbers and algebraic functions, it is likely that algebraic number theory would eventually have developed without the Fermat–Kummer impetus. However, it is still the case that a decisive, well-posed problem–in elementary number theory–led directly to a major development in the higher ranks of number theory.

Another example is Burnside's conjecture on finite simple groups: That every finite simple group which is not cyclic (of prime order) must have even order. The penetrating Feit–Thompson demonstration of this conjecture (§V.9) led to the subsequent elaborate determination of *all* finite simple groups and to an extensive revival of the whole subject of finite group theory. In this case (and in many others) the presence of a pertinent problem is a major dynamic in the advance of Mathematics. However, this is not always so. Hilbert's fifth problem (about continuous groups) was famous until it was solved–and then it dropped out of prominence, though the general subject of Lie groups continued to be active, little influenced by the solution of this problem.

There are also all manner of small problems, some natural, some concocted. Problems are a vehicle of competition and display. In some schools (often in those influenced by Hungarian traditions) the business of mathematics seems to be just the formulation and the eventual solution of hard problems, with perhaps more attention to the difficulty than to the relevance of the problem. Problems in this sense are akin to chess problems–challenging but not necessarily relevant. Planar graphs provide an example. Here a *graph* means an (abstract) finite collection of vertices and edges, with the prescription that each edge joins exactly two (different) vertices. Some graphs cannot be drawn in the plane (or, equivalently, on a sphere) without an unintended crossing of edges. This is the case in a simple conundrum. Tell the gas, electric and piped-heat companies how to run their mains (gas etc.) all at the same level, six feet below ground; to three different houses:

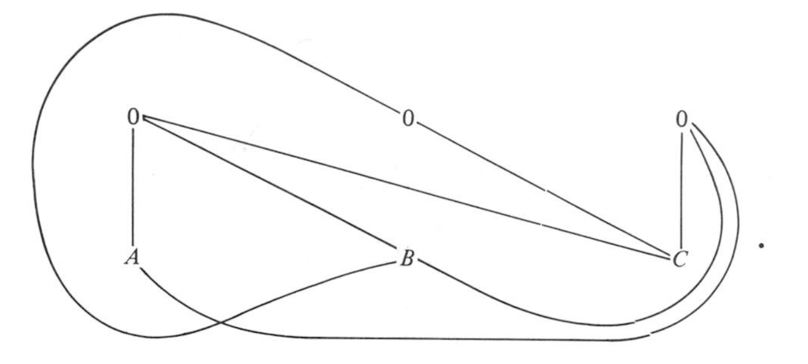

(2)

By simple application of the (deep) Jordan curve theorem, this is impossible, as suggested in the figure. Or try to tell five secretive embasies, located in Bern during a war, how to dig tunnels all six feet under so that they can secretly communicate, each to each:

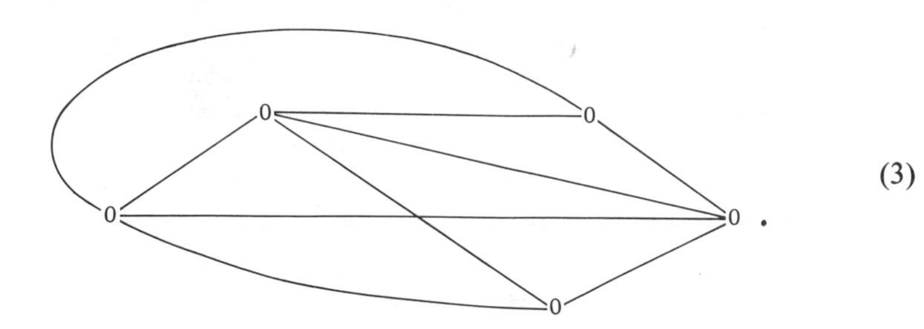

(3)

Jordan again testifies to the inevitable failure of this attempt at secrecy.

Both (2) and (3) are thus non-planar graphs. The story goes that W.L. Ayres tried to prove that every non-planar graph must contain the configuration (3), while C. Kuratowski tried to prove that every non-planar graph must contain (2). Evidently both attempts failed, but the two people met at some international congress and exchanged notes. Then Kuratowski went on to prove the striking theorem: *Every non-planar graph contains either the graph* (2) *or the graph* (3).

Both graphs (2) and (3) *can* be drawn on a torus (or on any surface of higher genus). Soon Paul Erdos proposed a problem:

Find a list of all minimal non-toral graphs (or similarly for other surfaces). What with extensive use of planar electric circuits or computer chips and the related revival of combinatorial questions, someone recently proved: There are exactly 103 different graphs which cannot be drawn on the projective plane and such that at least one of them is contained in any graph not representable on the projective plane. This evidently cumbersome result clearly demonstrates that the Erdos problem is not very fruitful: It has not (at least yet) led to an understanding of why a graph cannot be drawn on the projective plane (or the torus). Mathematics should be concerned with understanding.

We conclude that problems, like theorems, are sometimes fruitful and sometimes not. Solving a supposedly famous problem is not necessarily a famous contribution to Mathematics. However, the emphasis on problems is an effective counterweight to the emphasis on structure and axiomatics in our description of Mathematics. They are an essential component in the dynamic.

6. Understanding Mathematics

A major element in Mathematics is the need to fully understand why the formulas work and why the theorems hold. In lectures, in courses, and in conversations Mathematicians try repeatedly to better understand the received results of each subject. Their methods are various; we now examine some under the headings *analogy, examples, analysis of proofs, shift of attention*, and the search for *invariant form*:

(a) *Analogies.* In many respects, 2-dimensional Euclidean space is much like 3-dimensional space: Points and lines become points, lines, and planes; triangles and circles become tetrahedrons and spheres. These and other analogies strongly suggest a consideration of "spatial" configurations in higher dimensions. Some initial attempts describe 4-dimensional space by points, lines, planes, and hyperplanes; as this becomes cumbersome and as the phenomena requiring still higher dimensions arise there is a search for more effective methods, first by using coordinates and then by vector spaces.

In the beginning uses of measurements, the analogies between measuring distances, measuring angles, and measuring quantities (also weights) leads gradually to the formulation of real numbers as the general means of handling measurements.

The analogy between Boolean operations on sets (intersection, union and complement) and the logistic operations on propositions (and, or, and not) plays a central role in logic.

Or consider linear combinations. Every vector in the plane is the sum of a horizontal and a vertical component. Every complex number is a sum of real and imaginary parts. Every solution of the differential equation $y'' = -y$ is a sum of a multiple of $\cos x$ and a multiple of $\sin x$. All these analogous unique linear decompositions lead to the concept of a basis in a vector space.

Or again, the prime ideals in algebraic number theory are analogous to the points on Riemann surfaces of algebraic functions (§3). Throughout mathematics, analogies are powerful sources of development.

(b) *The study of examples.* The choice of a specific but critical example may well serve to indicate the direction in which a whole theory might develop. Thus the Weierstrassian theory of analytic functions as power series in a complex variable came after his teacher Gudermann had exposed him to a complicated "combinatorial" study of the particular power series for e^x, $\sin x$, $\cos x$, and the like. Hilbert's intensive study of relative quadratic extensions of number fields led him to make a series of conjectures about general abelian extensions of algebraic number fields, and the demonstration of these conjectures occupied the development of class field theory for fifty years. Algebraic geometry started from the careful classification of the quadratic, cubic, and quartic curves in the plane. Riemann surfaces for many-valued functions of a complex variable were first developed by pasting together slit planes carrying branches of algebraic functions, and only later were given a general conceptual formulation.

In a sense, the "general" theories of Mathematics are there in order to conveniently describe the common features of a multitude of special cases—and a crucial special case may be the key to a general theory.

(c) *The analysis of proofs.* There is a continual search to get better proofs of known theorems—not just shorter proofs, but ones which do more to reveal *why* the theorem is true. This study can lead to the discovery of crucial axioms used in the proof (like the discovery that some basic arguments in analysis made hidden use of the axiom of choice). Again, study of a proof may suggest new concepts. For example, the Cauchy integral theorem, initially and perhaps carelessly proved by the use of the Gauss–Green lemma (§X.5), plays a central role in complex analysis. Study of its proof suggested subdivision arguments, for rectangles and triangle, and so suggested homological methods. Other proofs involve deformations and so bring in homotopy, and still other approaches culminate in the Jordan curve theorem and the resulting deeper study of the topology of the plane.

To present a proof, one may often try to put it in a vivid or diagrammatic form. When this done for the Jordan–Holder theorem (§V.7), it naturally leads to the diagrammatic presentation of a partially ordered set and thence to the notion of a lattice. Again, topologists found it suggestive to denote continuous maps from one space X to another space Y by arrows $X \to Y$. This simple notation soon suggested fertile concepts of exact sequences and of categories. In brief, proofs are the meat of mathematics, so chewing them over can produce juicy results!

(d) *Shift of attention.* The understanding of a proof may progress when our attention is shifted from the initially emphasized aspect to some other focus. Thus Galois theory starts as the study of the solutions of a particular polynomial equation; it then becomes the study of the field of all the rational combinations of the roots of that equation. This shift makes it possible to see the Galois group of the equation not as a group of permutations of its roots, but as a group of automorphisms of the corresponding field. In group theory, the cosets of a normal subgroup N of a group G become the elements of the factor group G/N; when, instead, the emphasis is placed on the function sending each element to the coset which it generates, there is revealed a homomorphism of the original group G onto the factor group G/N–and one then sees that the properties of this homomorphism, and not the particular properties of the cosets, provide (§V.6) the more effective description of the factor group. Similarly, shifting from matrices to the corresponding linear transformations substantially changes the emphasis in linear algebra, from calculation to visualization–and lays the groundwork for an effective approach to infinite-dimensional vector spaces.

(e) *Invariant formulation.* The introduction of analytic geometry by Descartes provided for algebraic proofs of geometric facts–but the algebraic proofs depended on a choice of coordinate systems, while the geometric facts–for example, the fact that the medians of a triangle are concurrent–are independent of the choice of coordinates which may be used in the proof. For this reason, the development of analytic methods in geometry has constantly seen parallel attempts to make the results (or even the arguments) independent of such choices. In linear algebra, this appears in formulations independent of the choice of bases. In tensor analysis, a tensor, relative to bases, first appeared as a multiply indexed family t_{ij}^k of scalars t, and the description of a tensor had to be coupled with a complete description of the effect of a change of basis on these t's; alternatively, tensors can be defined (§7.9) in a coordinate-free manner as elements of tensor products of vector spaces. An algebraic variety in affine n-dimensional space is the locus of a finite list of polynomial equations in the coordinates–but it can be described more invariantly in terms of the ideal of *all* polynomials which vanish on the locus; the given equations are then just one set of generators (among the many possible sets) for that ideal.

The search for an "invariant" formulation is not limited to questions of geometry. It can also arise in algebraic contexts. Thus a specific group may originally be described in terms of specified generators and their relations; for example, a dihedral group is generated by a rotation (by $360°/n$) and a reflection. Alternatively, the same group may be described in more invariant terms as the group of all symmetries of an n-sided regular polygon. In universal algebra, a group is defined as an algebraic system with one binary operation (of multiplication), one nullary operation (identity), and one unary operation (of inverse). However, there are other iterated operations intrinsic to a group, such as the ternary operations

$$(x_1,x_2,x_3) \mapsto x_1 x_3^{-1} x_2 x_1^{-1} \quad \text{or} \quad (x_1,x_2,x_3) \mapsto x_3 x_2^2 x_1^{-1} x_2 x_3 .$$

For each n, there are similar n-ary operations defined by "words" in n of the x_i and their inverses. A full and invariant derivation of the "theory" of groups is given by the totality of all such group operations, suitably organized (for example, as a category).

Invariant formulations also appear in mechanics; where the principle of least action presents the laws of mechanics in an essentially coordinate free fashion.

7. Generalization and Abstraction

Another dynamic leading to the extension of Mathematics is the process of generalization or abstraction, which can be described as follows under several headings.

(a) *Generalizations from cases* refers to the way in which several specific prior results may be subsumed under a single more general theorem. Thus the observation of right triangles with given lengths of sides (such as the triangle with sides 3, 4, 5, or sides 5, 12, 13) leads both to the Pythagorean theorem about right triangles and to the algorithm which obtains all solutions in integers of the Pythagorean equation $x^2 + y^2 = z^2$. Similarly, the familiar algebraic expansions of $(x + y)^2$, $(x + y)^3$, and so on are subsumed in the binomial theorem expanding $(x + y)^n$ for any integer n —complete with the formulas for the binomial coefficients, and generalized yet again with infinite binomial series for fractional exponents n.

In these instances, the generalization from cases produced a general theorem subsuming all that was known in the specific cases. Other, more involved instances may see the development of a theory which unites some (but perhaps not all) of the results holding in the given cases. Thus algebraic number theory started with specific properties of quadratic and cyclotomic number fields before going on to arbitrary algebraic number fields, but the deeper results of the class field theory apply only to abelian extensions (those which are normal with an abelian Galois group (§V.7) over some ground field, such as **Q**). Similarly the elaborate study of the various forms of conics, cubic, and quartic curves in the plane finally extracted some systematic facts about intersections; for example, a cubic and a quartic "in general" intersect in $3.4 = 12$ points. Generalization then led to Bezout's theorem: Two plane curves of orders m and n intersect "in general" in mn points. But the validity of this theorem does require some substantial adjustments. Among the intersections we must count possible intersections at "infinity" (in the projective plane) as well as all the intersections at points with complex coordinates–while each intersection must be counted with a suitable "multiplicity". Here special cases, such as tangent lines and osculating circles suggest how to go about defining what is meant by multiplicity; for example, suitable ideals may be used. Similar notions of multiplicity thus develop for the intersections of surfaces and manifolds of higher dimension: They play a major role in algebraic geometry and subsequently in algebraic topology. There they are transmuted into cup products in cohomology.

There are many other examples of generalizations from cases–for instance, various cases in number theory led to the unique decomposition of finite abelian groups into products of cyclic groups (V.8.4).

(b) *Generalization by analogous steps* refers to situations where a more general theory runs in parallel to earlier cases without actually replacing or subsuming those cases. An instance in point is the parallel between real numbers, complex numbers, and quaternions–followed by the search for higher dimensional such algebras over the reals, with the discovery that there were no further finite dimensional division algebras without the loss (say) of the associativity of multiplication. Another striking example is the generalization from real analysis to complex analysis.

(c) *Generalization by modification* refers to instances in which a desirable theorem cannot be generalized in a direct way, but in which suitable concepts are used to modify the theorem so as to make a generalization possible. Thus the unique prime decomposition valid for the ordinary integers fails to hold for the ring of integers in most quadratic and cyclotomic fields of algebraic numbers–but it can be replaced in this and other rings by a unique decomposition of ideals into products of prime ideals. Moreover, the original decomposition of an integer n into its prime factors is reflected in the general theorem by way of the decomposition into prime ideals of the *principal ideal* (n) of all multiples of n. Again, the Jordan curve theorem, valid for simply closed curves in the plane, is not true for all such simple closed curves on the torus–but the classification of the ways in which this theorem fails is a major source of the study of homologous curves and of the introduction of the homology and homotopy groups (which, in effect, measure variously the numbers of ways in which the Jordan curve theorem can fail).

Abstraction. Generalization and abstraction, though closely related, are best distinguished. A "generalization" is intended to subsume all the prior instances under some common view which includes the major properties of all those instances. An "abstraction" is intended to pick out certain central aspects of the prior instances, and to free them from aspects extraneous to the purpose at hand. Thus abstraction is likely to lead to the description and analysis of new and more austere or more "abstract" mathematical concepts. We will describe some types of abstraction under headings "abstraction by deletion," "abstraction by analogy," and "abstraction by shift of attention".

(d) *Abstraction by deletion* is a straightforward process: One carefully omits parts of the data describing the mathematical concept in question to obtain the more "abstract" concept. This often leads to a reverse process, in which it is shown that all (or some) of the abstract objects can have the deleted data restored, perhaps in more than one way. Such a restoration is then called a "representation theorem".

For example, if one starts with the notion of a transformation group, one may delete the elements being transformed but retain the associative, identity, and inverse laws for the composition of transformations. The result is the notion of an "abstract" group. The corresponding representation theorem, due to Cayley, asserts that every abstract group is isomorphic to a group of transformations on some set (§V.6).

The algebra of sets (§I.9) concerns the algebraic operations of intersection, union, and complement on the subsets of some given set. If one deletes all references to the elements of these subsets, but retains the three operations and (a suitable list of) the basic identities which they satisfy, one reaches the notion of a Boolean algebra. Conversely, the Stone representation theorem asserts that every Boolean algebra can be represented as the algebra of some of the subsets of a suitable set.

Sometimes such a deletion is suggested by the sorts of theorems being considered. Thus, in functional analysis, certain theorems about the existence of functions with given properties are like theorems about the existence of vectors in linear spaces. This use of geometric language then suggests an abstraction: The deletion of the numbers on which the functions act, so that the functions (for the intended purpose) become just points or vectors in some Banach or Hilbert space.

(e) *Abstraction by analogy* occurs when a visible and strong parallel between two different theories raises the suggestion that there should be one underlying, less specific theory sufficient to give the common results. Thus systems of numbers or of functions closed under the usual rational operations (addition, subtraction, multiplication, and division by non-zero elements) follow the same algebraic rules and yield the same theorems for the solutions of quadratic and higher equations (Galois theory). These parallels then suggest the abstract notion of a field—which soon turns out to include other instances such as the p-adic numbers and the field of integers modulo p as well as other finite fields. In the same direction, the close parallel between algebraic number fields and fields of algebraic functions suggests the more abstract theory of finite dimensional extensions of any field.

The notion of a modular lattice arose from a similar striking analogy. On the one hand, the Jordan–Hölder theorem for finite groups (V.7.3) asserts that any two composition series for such a group have the same length. On the other hand, for a vector space any two finite bases have the same number of elements–a number which can be described as the maximum length of a descending chain of subspaces (the number is of course the dimension of the total space). Now both subspaces R, S, T of a vector space and normal subgroups of a group form lattices under inclusion–that is, posets with least upper bounds $R \cup S$ and greater lower bounds $R \cap S$, as described in §I.9. Moreover these lattices satisfy an additional condition, the so-called *modular* law:

$$R \supset T \quad \text{implies} \quad R \cap (S \cup T) = (R \cap S) \cup T. \tag{1}$$

One is thus led to formulate and prove the theorem asserting that any two maximal descending chains in a finite modular lattice have the same length. For subspaces of a vector space, the length is the dimension. However, this modular lattice theorem does not contain the original Jordan–Hölder theorem, because the latter concerned not chains of normal subgroups of the whole group but chains of subgroups, each normal in the next larger. The modular lattice theorem shows only that any two *chief* series (maximal chains of normal subgroups) in a group have the same length. However, the modular lattice theorem does apply in other cases, as for example to subspaces of a projective space, and it has infinite-dimensional generalizations, to the so-called *continuous geometries* used by von Neumann to analyse rings of operators on Hilbert spaces. (See Birkhoff [1967].)

Incidentally, the modular law (1) also holds in a Boolean algebra which satisfies the stronger distributive law (for *all* R and T)

$$R \cap (S \cup T) = (R \cap S) \cup (R \cap T).$$

This resembles (1), since $R \supset T$ there means that T on the right of (1) may be replaced by $R \cap T$.

(f) *Abstraction by shift of attention*. Some abstractions arise when the study of a Mathematical situation gradually makes it clear that certain features of the situation–perhaps features which were originally obscure–are really the main carriers of the structure. These features, with their properties, are then suitably abstracted. Thus the Galois theory begins as a theory of the roots of a given polynomial, and the Galois group arises first as a group of permutations of those roots–all permutations leaving invariant all polynomial (and hence, all rational) relations between the roots. Presently, it appears that this group of permutations acts on all the rational expressions in these roots, and that the Galois group can thus be described as a group of automorphisms of the field consisting of all such expressions. With this more abstract view of the Galois group, the whole theory becomes more perspicuous! Abstraction supports understanding!

The development of point set topology exhibits several successive such shifts of attention. Initially, the subject dealt with the properties of sets of points on the line, the plane, or in three space and with the properties of functions continuous on such point-sets. Presently, it became clear that most of the results depended on the internal properties of these point-sets, and not on the way in which the sets were embedded in a Euclidean space. All that mattered was the distance between any two points in the set. This, and the need to handle more general cases arising, say, from the calculus of variations, led to the abstract notion of a metric space, with axioms expressing the properties of distance. However, it then appeared that only "qualitative" properties of the distance were needed in order to define continuity; this led to the more abstract notion of a topological space (and the corresponding representation theorem, which arises here as the problem: When is a topological space metrizable?).

This case illustrates the observation that attention can shift dramatically from one notion to another. Initially, continuity and limits were defined in terms of inequalities and an $\epsilon-\delta$ process. It was then recognized that these numerical equalities amounted to using open intervals (on the line) or open discs (in the plane). Gradually, Mathematicians came to utilize the more general open sets–those which are arbitrary unions of open intervals or open discs–and to the eventual recognition that these open sets are all that is needed for the definition of continuity. In each Mathematical situation the basic notions are disentangled by insight and understanding–which often amounts to a shift of attention, and which may then call for a corresponding abstraction.

There are many other types and examples of abstraction. Thus, the "abstract" description of a Riemann surface as a one-dimensional complex manifold helps understand how concrete Riemann surfaces can be pieced together from sheets with slits. The abstract definition of a symplectic manifold illuminates the role of the p's and q's in the formulation of Hamiltonian mechanics. The abstract definition of a function as a set of ordered pairs explains the functional dependence of one quantity upon another, good whether or not that dependence is expressed by a formula. Much of the dynamics of Mathematics depends on the constant attempt to get better and deeper understanding. Generalization and abstraction are two useful means to such understanding.

8. Novelty

We have seen that problems, generalizations, abstraction and just plain curiosity are some of the forces driving the development of the

Mathematical network. But this network is also tied to a variety of human and scientific activities; thus questions so arising can be important sources of novel Mathematics. Historically, projective geometry was started in part by considerations of practical perspective, while Fourier's study of heat led to his series, so widely useful.

There are many other, current, examples. Dirac, in formulating quantum mechanics, came to consider his well-known delta function $\delta(x)$—defined for all real x, zero when $x \neq 0$, non-zero only for $x = 0$ and such that its integral from $-\infty$ to ∞ is 1. Literally speaking, there is no such function, so that Dirac's maneuver may have seemed Mathematically empty. But V.I. Sobolev and Laurent Schwartz observed that the role of the delta function could be replaced by a suitable operator on functions, say by a "distribution". This impetus led to the development of a considerable branch of Mathematics.

The bizarre phenomenon of a single solitary wave was first observed about 1840, when Scott Russell chased such a wave moving down the Edinburgh–Glasgow canal. There appeared to be no way to derive the existence of such a solitary wave from the accepted partial differential equations for water waves. Finally, several authors did find model equations which would predict such waves; one such is the Korteweg–de Vries equation

$$u_t + u_x + uu_x + u_{xxx} = 0, \qquad \left[u_t = \frac{\partial u}{\partial t} \right], \tag{1}$$

where $u(t,x)$ is the height of the wave above the standard water level in the canal at time t and distance x down the canal. In this equation the basic terms $u_t + u_x$ express standard conditions (§VI.11) for the propagation of waves such as $u = f(x-t)$; they are modified by the effect of dispersion represented by the third partial derivative u_{xxx} and by a term uu_x, non-linear in u. This equation does have explicit travelling wave solutions, appropriate to the observed phenomenon. Moreover, it has surprising applications elsewhere, for instance in magneto-hydrodynamics.

Computer science brings up other new Mathematical ideas. For example there are new algorithms which find the product of two matrices more quickly than the straightforward application of row-by-column multiplication. There are many other new algorithms, plus questions of principle about *computational complexity*: For a computation of specified size, how does one estimate the minimum time really necessary; can it be done in "polynomial time"?

Not all outside influences are really fruitful. For example, one engineer came up with the notion of a *fuzzy set*—a set X where a statement $x \in X$ of membership may be neither true nor false but lies somewhere in between, say between 0 and 1. It was hoped that this ingenious notion would lead to all sorts of fruitful applications, to fuzzy automata, fuzzy decision theory and elsewhere. However, as yet most of the intended applications turn out to be just extensive exercises, not actually applicable; there has been a spate of such exercises. After all, if all Mathematics can be built up from sets, then a whole lot of variant (or, should we say, deviant) Mathematics can be built by fuzzifying these sets.

All these searches for new ideas and for new formalizations are inevitably adventurous and uncertain. It can turn out that an apparently attractive new idea is not really fruitful or relevant, or it can turn out that a proposed new abstract concept or formalization is just too cumbersome or too peripheral to be productive. On the other hand, some new formal concepts can be of central subsequent importance. What really matters is the role of the concepts in the network of Mathematical form.

9. Is Mathematics True?

It is customary to ask of a piece of Mathematics: "Is it true?" For example, one might ask: Is the Jordan curve theorem true of that simple closed curve drawn out there in the middle of my complex plane? The theorem says that the curve separates that plane into an inside and an outside so that no path can connect an inside point to an outside point without meeting the curve itself. That particular curve is quite convoluted (Figure X.4.3), so it's a bit hard to really tell which of the points are inside and which are outside; but is it true?

The whole thrust of our exhibition and analysis of Mathematics indicates that this issue of truth is a mistaken question. There really isn't an absolutely flat Euclidean plane out there, complex or otherwise; there is only the (very slightly) bumpy surface of the blackboard, which furthermore is very far from extending out to "infinity" in any direction whatever. What appears on the blackboard is not really a simple closed curve, but a wavy and somewhat thick line of chalk marks; perhaps the marks even skip a little, so that a careful draughtsman might be able to sneak a path from the "inside" out to some point clearly "outside". Even if the curve were devoid of gaps, I can't really demonstrate that it is continuous; for every ϵ greater than zero I haven't actually chosen a δ such that. . . . Moreover I am not really interested in the inside and outside of this particular Jordan curve; I am rather interested in knowing that my plane is simply connected or that the holomorphic functions I intend to define on it satisfy Cauchy's integral theorem or that the results of this theorem can be used, via some conformal map, to design an airplane wing. For this and all sorts of other reasons the theorems of Mathematics, by Jordan or others, are not simply statements about the behavior of individual objects in the physical world.

This point can be put more explicitly. Our survey has indicated that Mathematics is an extensive network of formal rules, definitions and systems, tightly tied here and there to activities and to science. This description does not in any way provide a physical object for each Mathematical term, or a physical law corresponding to each Mathematical theorem. Instead there are many pieces of Mathematics and a variety of accepted procedures for making practical use of some of these pieces. As a result, it is simply meaningless to ask, given a relation between Mathematical terms, whether it is true of the "corresponding" physical objects. The correspondence is just are not that direct. Instead, our description of Mathematics indicates that the appropriate questions are different ones:

Is this piece of Mathematics *correct*? That is, do the calculations follow the formal rules prescribed, and are the theorems deduced from the stated axioms by rules of inference on which we have agreed?

Is this piece of Mathematics *responsive*? That is, does it settle some problem which had arisen or does it carry further some development which was incomplete?

Is this piece of Mathematics *illuminating*? That is, does it help understand what had gone before, either by further analysis or by abstraction or otherwise?

Is this piece of Mathematics *promising*? That is, though it is a novel departure from precedent or fashion, is there a reasonable chance that it will subsequently fit in the picture?

Is this piece of Mathematics *relevant*? That is, is it tied to something which is tied to human activities or to science?

Many of these questions are questions of degree: Is it more or less relevant? Even that is hard to judge because relevance can be transmitted all along the network of Mathematics. There are many pieces of Mathematics which fit well but are not immediately relevant to any application–and which much later may turn out to be applicable. The promise of a new idea is also difficult to judge; too often the new may be dismissed as "off-beat". But before all comes the question: Is it correct?

Centuries ago, when Mathematics dealt largely with arithmetic and elementary geometry, it was perhaps easier to think of the numbers or the geometric figures as real objects about which one could make true statements. These objects had been long familiar and they were eminently useful, so it was comfortable to consider them as real. Now, looking carefully, we see that this comfort was an illusion. Numbers are the means used in calculating by rules, while figures are the images used to suggest formal geometric proofs; their eminent utility simply suggests that the formal is powerful in practise. By now there are so many and various Mathematical objects that any imputation of their reality offers spectacular cutting opportunities for Occam's razor.

The real world is understood in terms of many different Mathematical forms. For example, the fours group $\mathbf{Z}_2 \times \mathbf{Z}_2$ is not a single object somewhere in the world. It is rather a form, exemplified many times over in the world, by the symmetry present in each rectangular shape. It is also exemplified within Mathematics, say as the Galois group of the field

$\mathbf{Q}(\sqrt{2},i)$ generated by i and $\sqrt{2}$. However, in Mathematics it doesn't exist all by itself, but as a very small part of group theory, illustrating the decomposition theorem for finite abelian groups, but overshadowed by other newer pieces such as the monster group discussed in §V.9. Centuries ago none of these groups were there, because our ancestors had not then recognized *any* groups. In the interim, Mathematics has brought out many new and relevant forms. They arise by contemplating the world, but of themselves they are not in the world. They are forms.

This argument rests not just on the few examples here cited, but on the whole weight of evidence of the character of Mathematics. Our examination has shown that Mathematics consists of formal rules, formal systems, and formal definitions of concepts. The proofs of Mathematics do not depend upon experience, and indeed can often be invented or carried out by young people with little experience. Hence the results cannot be checked by asking how they might refer to real experience. Mathematics is not literally a science because its results cannot be falsified by fact or experiment. Therefore the question "Is Mathematics true?" is out of place.

To be sure, it is easy and common to think that Mathematics is true. This is partly because the formal Mathematics realizes intuitive ideas which themselves can be vivid and forceful. Even more it is because Mathematics, through the decisive notion of formal proof, can appeal to an absolute standard of rigor. Moreover, this standard is external and impersonal: Once the axioms of a system are set, all the statements of the system are either demonstrable, refutable, or (thanks to Gödel) undecidable. No collusion, no political influence, no second thoughts can alter the fact of the matter that the theorem can be proved. Given the straightforward definition of a finite simple group, the existence of the "monster" is ineluctable. Mathematics is not true, but its correct results are certain.

This thesis has consequences. The sets used in Mathematics are not physical totalities or collections; they are instances of an abstract concept defined by careful axioms–and we may even adjust the axioms (it seems that we need to!) These axioms mimic what we might operationally do with a collection of pebbles or of people. But each real collection out there is finite; even the fanciest collectors of coins or of stamps have collections which are finite. Now the axiom of infinity is necessary, both for Zermelo–Fraenkel and for categories. It states a formal property of a collection such that, were that collection a real object, it would be an infinite collection. But there is no such thing. The axiom of infinity is simply a formal statement (with consequences!) inspired or modeled after the practical observation that a collector of stamps or coins can always add one more item to his collection; the axiom imagines that done forever (which has never yet happened). This is the sense of the infinite in Mathematics, but all the actual operations of the Mathematician–his calculations, his definitions, his proofs–are strictly finite. These observations destroy the position of these finitists (like Kronecker) who objected to Cantor's use of infinite sets.

Mathematical existence is not real existence; it is in no way to be compared with the existence of people, political compaigns, or with the non-existence of unicorns. Aspects of such "real" existence did lead to Mathematical (or logical) ideas which in turn were formalized in the axioms for $(\exists x)F(x)$ in the predicate calculus. But in that calculus each theorem $(\exists x)F(x)$ simply means that this formal statement has been deduced according to those formal rules of inference. By virtue of this parallel, those existence theorems are often relevant in the applications of Mathematics. They are also a check on the adequacy of a formalization: When a differential equation is supposed to model some physical phenomenon, an existence theorem for solutions of the equation shows at least that the model is capable of producing *some* result. The fundamental theorem of algebra proclaims that there exists a complex root of any polynomial (with complex coefficients): it doesn't of itself produce the root, though there are some proofs which are constructive in the sense that they prescribe steps which will calculate approximations to the root. A theorem of Zermelo's, with an astute proof, asserts that the axiom of choice implies that the reals can be well-ordered. It does not exhibit a well-ordering. Nobody has exhibited such. In short, Mathematical theorems about existence are those that can be proved from the logical rules about existence, nothing more, nothing less. Indeed, if you want less, it can be done with intuitionistic logic, but this is equally a formal system. Brouwer's own earlier, non-formal intuitionism was based on an assumption of the primacy of the natural numbers in Mathematics—and this flies in the face of the observation that actual Mathematics effectively uses natural numbers and real numbers as equally primary. The pillarstones are geometry *and* arithmetic.

The view that Mathematics is "correct" but not "true" has philosophical consequences. First, it means that Mathematics makes no ontological commitments; in other words, Mathematics makes no statement about the reality of some of its objects. We do not need to know that every Dedekind cut in the rationals determines a "real" real number; we do know only that this description produces forms for real numbers which work with magnificent success. (This success is perhaps *the* essential bond between those pillarstones of geometry and algebra.) Quine has on occasion argued that we make an ontological commitment to the range of a variable when we quantify over that range. This arises, I suspect, from an undue concern with logic, as such. For Mathematics, the "laws" of logic are just those formal rules which it is expedient to adopt in stating Mathematical proofs. They are (happily) parallel to the laws of logic that philosophers or lawyers might use in arguing about reality—but Mathematics itself is concerned not with reality but with rule.

This view means that the philosophy of Mathematics need not involve questions about epistemology or ontology. If Mathematical theorems do not assert truths about the world, we need not inquire as to how we know or would come to know such truths. (We of course *do* need to inquire how we recognize a correct proof, but getting the recognition is a major part of advanced education in Mathematics, and is usually not considered as part of epistemology.) This observation means that the philosophy of Mathematics cannot be much advanced by many of the books entitled "Mathematical Knowledge", in view of the observation that such a title usually covers a book which appears to involve little knowledge of Mathematics and much discussion of how Mathematicians can (or cannot) know the truth. This dismissal applies especially to the later (posthumous) volume of Wittgenstein [1964], where the actual Mathematical content rarely rises above third grade arithmetic, while the actual concern is less with Mathematics than with its use to illustrate some strictly philosophical issue.

But this does not dispose of the hard questions about the philosophy of Mathematics; they are merely displaced. Mathematics *is* a formal network, but the concepts and axioms there are based on "ideas". What are they? We have identified a number of particular such ideas; there are many more. They are not like Platonic ideal forms, because they are intuitive and vague–as we have repeatedly noted, the "same" idea can lead to a number of different formalizations. Indeed, it seems that we can recognize and name an idea only after it has given rise to one or more formal expressions. Before that an idea can be pretty nebulous; it is "up in the air". That may be why it is hard to convey to others some new intuition about Mathematics. Hence:

Question I. What are the characteristics of a Mathematical idea? How can an idea be recognized? described? What is the relation between a (nebulous) idea and a (precise) formulation.

This question may also have historical aspects. Why did the idea of composition (symmetry, groups, categories) arise so late? Was it that no applications were visible? Or were the potential uses in geometry just ignored? Or did it require Galois?

Next comes the conundrum: How do we recognize the presence of formal rules in simple human activities? How do we select the "facts" about numbers from the informal practises of counting? For that matter, how do we recognize that some particular scientific question has a formal or Mathematical component? This is an especially puzzling problem, because it is a common observation that when a scientist comes to a Mathematician with a question, he is very often unable to formulate that question is clear Mathematical terms, so that it may take considerable added inquiry to get to such a formulation. This gives us

Question II. How does a Mathematical form arise from human activity or scientific questions? What is it that makes a Mathematical formulation possible?

This may be a question of epistemology (inverted). How do we know that there is some Mathematics there to be extracted?

After the Mathematical form is extracted its development continues, with the dynamics of problems, generalizations, and abstractions which we have described. What is then mysterious is that these abstractions, inventions, and calculations so often have applications and uses back in the indubitably real world. It's not just that prime numbers, long considered a sort of special and esoteric concern of number theory, now suddenly can be used in cryptology. It is all those differential equations which do produce solutions which fit, all those Hilbert spaces which crop up in the formulation of quantum mechanics, all those tensors in differential geometry which then can be used in relativity, all those problems in the calculus of variations which can be used in studying optimal control, and so on and on. The Mathematician's process of building up his elaborate formal systems so that they fit together also produces models which (partially) fit the real world. This leads to our third and most basic question, like that formulated originally by Eugene Wigner:

Question III. How can one account for the unreasonable effectiveness of Mathematics in providing models for science and knowledge?

Here ontology and reality reappear. And Mathematical knowledge, apparently just about formulas, turns out to sometimes *help* produce knowledge of phenomena.

Is it that the real world is so constituted that parts of it easily fit formal patterns? Or is it that our very notion of formal pattern has been developed over millenia so that it is possible to fit the pattern to the fact?

For a strict formalist, for whom Mathematics is just any manipulation of symbols, this question III cannot be answered.

On our view, the rules, axioms, and definitions of Mathematics are indeed formal, but they were initially chosen in the light of various human activities and scientific phenomena, and then extensively developed to improve understanding. There is therefore reason to expect that the results might be applicable in return, at least to the original phenomena. But the resulting application is often to other and different phenomena!

To fully account for this applicability, the phenomena must in some sense be ready to fit the formulas. This becomes a question about the character of reality. All the experience of Mathematics and of the physical sciences shows that many aspects of reality can be measured, organized, and then understood by way of theories and concepts which have a large formal content. We have agreed that the form is chosen to reflect the facts. It must also be the case that the facts accept the form. We have not explained why this is so.

Put differently, the elaborate formal development of Mathematics yields conclusions which often fit the world. Does this mean that the "real world" is not just a bundle of confusions, but that the variety of events in the world does fall into patterns which are amenable to sophisticated formal descriptions? If there indeed *are* such patterns, how is it that man is able to extract and analyse them so effectively? Numbers in their complex structure, groups in their variety and symmetry, geometry in its many versions, and analysis in its apperception of change all deal with patterns which are highly relevant to the world. This requires explanation on the metaphysical and the epistemological level–and raises thus a variety of questions to which I have no adequate answer.

Here is a sketch of some tentative answers to these questions.

(i) Various phenomena do have underlying similarities and regularities. These regularities are somehow propagated from one situation to another.

(ii) On the basis of millennia of experience, mankind has developed "ideas" about the phenomena which in turn are used to extract from the phenomena a conceptual description of some of these similarities and regularities. Some of the concepts involved can be made strictly formal, and hence Mathematical.

(iii) Different formal statements in the descriptions, because of the propagation of the regularity of phenomena, can be closely connected with each other. Because these connections help the practical understanding of the regularities, they have been extensively examined and in some cases reduced to formal proofs, by rules of inference, from astutely chosen axioms.

(iv) In many cases, the results of these proofs fit the facts–not just the particular immediate facts from which the concepts were extracted, but other facts which, because of the propagation of regularity, also fit these deductions. By the choice of the successful cases of concept formation and deductions, various branches of Mathematics are selected and developed. From time to time the development is supported by additional formal explanations of the regularity of new phenomena.

A development of this analysis might account for the "unreasonable effectiveness of Mathematics in providing methods for science". For example the concept of a group, originally formulated to analyze certain symmetries in Galois theory and in geometry, turns out to be relevant to Mathematical physics, primarily because corresponding symmetry phenomena arise there.

To summarize: The world has many underlying regularities which, once extracted, can be analyzed and understood by Mathematical form. Because it is formal, the same Mathematical notions can apply to widely different phenomena.

Question IV. What is the boundary between Mathematics and (say) Physical Science?

Mathematics has been described here as a formal development of ideas suggested by phenomena. Some portions of this description apply to other sciences, notably to Theoretical Physics. These, too, make considerable use of models and forms–and hence of Mathematics. There, however, one develops only the forms that seem to be useful in fitting a specific type of phenomena. If it turns out that they don't fit, they are discarded. In Mathematics, one pursues the forms whichever they may lead.

10. Platonism

Our view of Mathematics provides both for formal concepts and for guiding ideas, and hence accounts for the role of intuition in Mathematics. It is, however, in sharp contrast to all variants of Platonism. Let us consider this contrast.

We are not concerned with the actual historical doctrines formulated by Plato but with the current views going under his name. These views typically hold that Mathematical concepts are about externally given objects which therefore impose their nature upon the results of Mathematics. That Euclidean geometry can lead to Platonism is clear. This geometry is not "about" the imperfect lines and figures which we may draw on an uneven blackboard or on crinkly paper. It must be "about" something, so it is about ideal straight lines, perfectly rounded circles and absolutely flat planes which have an objective existence. Once formulated so, this doctrine extends naturally to encompass ideal numbers and their properties, functions and their derivatives and finally the whole world of sets. Thus Platonism in Mathematics views Mathematics as dealing with a domain of abstract or ideal "objects" which have being independent of our thought about them and whose being therefore determines what can be truly thought about them.

There are various versions of this view; I will follow here the careful distinctions used by Michael Resnik on page 162 of his book *Frege and the Philosophy of Mathematics*. First a *methodological Platonist* is one whose activities fully endorse all the standard, infinite, and nonconstructive methods of Mathematics, as if Mathematics were "dealing with a mind-independent infinite domain of abstract entities". An *ontological* Platonist is one who "recognizes the existence of numbers, sets, . . . on a par with ordinary objects". An *epistemological Platonist* holds that "knowledge of Mathematics objects is . . . in part based upon a direct acquaintance with them". Finally a *realist* "believes that the objects of Mathematics . . . exist independently of us and our mental lives".

These several views are considerably different. First observe that the practise of Mathematical research does require deep concentration and complex understanding of what can follow from known axioms and theorems. This process often may depend on vivid intuitive imagination that the concepts at issue concern objects which are really "there". This description of Mathematical practise has been called "mythological Platonism"; it seems to be essentially the same as the methodological Platonism defined above. It is our view that it is not a philosophical doctrine about the nature of Mathematics, but a description of the process of Mathematical research. As such, it is an appropriate start on such a description. For example, if I wish to make (as I have done) extensive computations of the cohomology groups of the so-called Eilenberg–Mac Lane spaces $K(\pi,n)$, I know that such a space is defined to be one in which there is just one homotopy group π in dimension n, and I recall the theorem that this specification completely determines the cohomology and homology groups of the space. Therefore my computations are much assisted by an unquestioned belief that the spaces are there before me, firm and unyielding (even though I also know that there can be several different ways of realising one such space $K(\pi,n)$). When my calculations are done, I can return to the world of "ordinary" topological spaces. There I know that such a space X ordinarily has several homotopy groups–a fundamental group $\pi_1(X)$ and a whole string of higher dimensional homotopy group $\pi_n(X)$, together with certain connections between them, given by certain "obstruction cocycles". I can then recall that the Eilenberg–Mac Lane spaces are an abstract but convenient way of describing all the natural "operations" which one can apply to the cohomology of ordinary spaces. In their turn, many of these "ordinary" topological spaces may be pathological–they are accidental instances of a general definition of topology which was really devised in order to get at quite specific properties of continuous functions and topological manifolds. In their turn these manifolds in "reality" are not just topological; they actually have a complex structure or a Riemann metric or are embedded in Euclidean space–and we have for the time being forgotten about that structure in order to concentrate just on the concept "topological manifold".

In other words, Mathematics deals with a heaping pile of successive abstractions, each based on parts of the ones before, referring ultimately (but at many removes) to human activities or to questions about real phenomena. The advance of Mathematical understanding depends on the contemplation of each abstraction in itself. That is why the abstractions are there–to get rid of the "real" things which are extraneous to the particular questions at issue. In other words, the elaborate formal nature of Mathematics which we have observed requires us to deal with highly abstract concepts "as if" they are real. This is mythological Platonism. It is a description of the process of Mathematical understanding. It has nothing to do with ontology.

Ontological Platonism ascribes "reality" to the "objects" of Mathematics. This assumption first developed long ago when these objects were just numbers and geometrical figures (lines and circles). But now the objects of Mathematics come in such profuse abundance that they would hopelessly overpopulate reality. Moreover, all these new objects were considered not because they had been discovered "out there" but because they had proved to be handy means of understanding counting and moving. We have found no place in the practice of Mathematics and its applications where there is any use for the notion "real Mathematical object". The notion "Mathematical form" is completely sufficient.

Epistemological Platonism involves some sort of direct acquaintance with Mathematical objects, perhaps by some means not included in the five senses of sight, hearing, taste, smell, and touch. This doctrine seems to me to be speculation. Nowhere does such acquaintance, sensory or otherwise, enter into the abstract practice of Mathematics as we have described it. What is there is not acquaintance with objects, but the play of ideas.

The process of building all of Mathematics out of sets has led to a related doctrine which might be called *Set-theoretic Platonism*: There is a world of real and objective sets, pictured in the cumulative hierarchy, which we can apprehend. From this apprehension flows the axioms for sets and the deeper understanding of their properties–and thence, all the parts of Mathematics. This is a doctrine often espoused by set-theorists. For example, Kurt Gödel held firmly to the idea that sets are real and that their properties could be known; for example, he held that the axiom of choice is true, while the continuum hypotheses is false, as is the axiom $V=L$ of constructibility (every set is constructible, in the sense of §XI.8).

This realist view contrasts sharply with our views on set theory. It seems to us that realism in sets faces great difficulties. First, it is open to all the objections made above to ontological and to epistemological Platonism. Second, that "apprehension" of sets is pretty obscure, since abstract sets are not apprehended by our usual senses. Third, there are many viable variants of the Zermelo–Fraenkel axioms, to say nothing of intuitionistic set theory. Fourth, because of the artificial constructions required this doctrine does not account well for the actual "objects" studied by Mathematicians. For example, holomorphic functions of a complex variable are central objects of Mathematics–and they are certainly not to be understood well as "sets of ordered pairs".

In short, save for mythology, all the variants of Platonism shatter on the actual practice of Mathematics.

11. Preferred Directions for Research

A thorough description or analysis of the form and function of Mathematics should provide not only insights into the Philosophy of Mathematics but also some guidance in the effective pursuit of Mathematical research. The latter is a subtle task, on which we may make a few tentative comments, omitting the psychological aspects (see Hadamard [1954]).

Mathematics, in our description, rests on ideas and problems arising from human experience and scientific phenomena and consists in many successive and interconnected steps in formalizing and generalizing these inputs. Thus Mathematical research can be directed in a wide variety of overlapping ways:

(a) Extracting ideas and problems from the (scientific) environment,
(b) Formulating ideas;
(c) Solving externally posed problems;
(d) Establishing new connections between of Mathematical concepts;

(e) Rigorous formulation of concepts;
(f) Further development of concepts (e.g., new theorems);
(g) Solving (or partially solving) internal Mathematical problems;
(h) Formulating new conjectures and problems;
(i) Understanding aspects of all of the above.

This listing does put special emphasis upon extracting and formulating ideas and on understanding their import: this is in some contrast to the traditional view that research consists primarily of finding new theorems (item (f)). But all of these tasks are hard; the search for new ideas and new formalizations is inevitably adventurous and uncertain. Here are some examples.

(a) The idea of using a Fourier series to represent periodic functions was extracted from the study of heat and eventually led to a better understanding of the notion of a function and to an extensive and still continuing development of harmonic analysis (as under item (f)). More recently, the notions of game theory were extracted from practical concerns: they have had extensive applications in Mathematical economics, but somewhat less influence in Mathematics proper.

(b) The ideas of complex number, group, and set were first disentangled and formulated in the 19th century. They have each proved to be extraordinarily important; in particular the study of groups has led to all sorts of interconnections with other parts of Mathematics, for example in the harmonic analysis on a group.

The concept of a topological space was developed in its present general form only after the wide exploration of all sorts of particular examples—and it has served to establish many new connections with complex analysis, differential geometry, algebraic geometry, and Galois theory, for instance. Other formal concepts may be developed "before their time" to find use only much later. For example, the notions of lattice theory were found about 1900 by Dedekind and others, but did not at that time find any noticeable resonance. The same notions, when rediscovered by Garrett Birkhoff and Oystein Ore in the early 1930's, were immediately put to use in projective geometries, continuous geometries and in the analysis of subobjects of algebraic systems. It would seem that by 1930 there were at hand more uses for such abstract notions. Subsequently, after a lively decade of research, lattice theory became of less central interest to algebraic developments–it may be because the principal uses were already worked out and the remaining problems were artificial. There are many other examples of failure, success, and partial or temporary success in the introduction of new mathematical concepts. Exploring the unknown is bound to be an adventurous and chancy business!

(c) Solving problems arising in science is a major activity of applied Mathematics. There are many examples not covered in our text–but the development of the calculus to handle the problems of celestial and terrestrial mechanics is an outstanding example.

(d) We have seen all sorts of discoveries of interconnections. Linear algebra ties in with higher dimensional geometry, linear differential equations, functional analysis, and Galois theory. Riemannian geometry develops applications to relativity theory. The phase spaces of mechanics are tied to the tangent bundles of differential geometry. Each such tie leads to better understanding.

(e) The rigorous, $\epsilon-\delta$ formulation of the calculus in the 19th century was a development of great significance. It was necessary to the calculus itself, in the understanding of the role of functions; it was a necessary prerequisite to the development of complex analysis, for which clear concepts of convergence and differentiability were essential; it was a preliminary to much wider (unanticipated) subsequent abstract developments. There have been other analogous steps toward rigor. The first steps toward algebraic topology (Riemann, Poincaré, Schoenflies) were inexact; their correction (Brouwer, Alexander) made further development possible. Hilbert's precise treatment of the foundations of Euclidean geometry was the starting point of a lively extension of axiomatic methods. Algebraic geometry had a steady development until about 1930, but by that time its intuitions were so convoluted that they could not be communicated beyond a group of specialists. Extensive efforts (Krull, van der Waerden, Zariski, Weil, Chevalley, Serre, Grothendieck) provided a careful and more extensive foundation on which remarkable further progress has been possible. Currently, some parts of geometric topology may be in need of similar reformulation. Geometers are often prone to emphasize the importance of intuition (= ideas). These geometric ideas are important; however they cannot stand by themselves, but require rigorous formulation so that they can be clearly communicated and used.

(f) Solving problems *can* have wide effects. For centuries, Mathematicians tried to prove the parallel axiom from the other axioms of plane geometry; the negative solution represented by non-Euclidean geometry not only settled this problem but opened up an array of new possibilities for the notion of space. Similarly, the attempt to prove the continuum hypothesis led to a new method of forcing in set theory, with unexpected connections to geometry. But not all famous problems lead to such extensive effects.

In sum, Mathematical research is not just the proving of theorems.

But what about the choice of direction or speciality? Recently, the increasing number of active Mathematicians has led to the recognition of a large number of highly specialized fields of study. An individual choice of field might depend on many considerations, including for example:

(i) Habit, and talent,
(ii) Authority,
(iii) Fashion,
(iv) Opportunity,
(v) Insight.

The influence of habit is clear: it is easiest to work in the field that one learned first; moreover the field may have been chosen to fit the talent of the individual. Some Mathematicians are natural analysts (good at approximations), some are algebraically minded (manipulations of formulas), some are inspired by applications, and some have well developed geometrical intuitions. However, current specializations are much more specific than these varieties of talent.

Some specialities dry up, but not the mainstream of Mathematics. The voice of authority recommends work in the "mainstream"–a handy but inexact label to cover the essential portions of our subject. Our chapters have described some of the sources of these mainstreams: Number theory, geometry, calculus, algebra, mechanics, and complex variables (with logic on the side). Research in mainstream topics is likely to be relevant, but it is also likely to be difficult–much of the streambed has already been well raked over. The evidence suggests that it is also useful to keep an eye on possible new sources!

In the nineteenth century, synthetic projective geometry was in fashion, and the fashion persisted till at least 1935, when some graduate schools still required courses in both synthetic and analytic projective geometry. Today, no one is concerned about the difference in method, and there is little attempt to find new theorems by either method. Today, graph theory is in high fashion, perhaps because of uses for computers or purported applications to social science.

Some fashions are deeper, and depend on new concepts which open new opportunities. Thus the remarkable properties of holomorphic functions of one complex variable, as revealed in the work of Cauchy and Riemann, put the study of these functions in a central position in Mathematics for a considerable period (say 1854–1930). Toward the end of this period, most of the opportunities had been explored and the subject became the center of authority; a major (but narrow) objective appeared to be that of better understanding the big Picard theorem by finding more elementary proofs for it. By now, the emphasis has shifted, so that there is relatively much more attention to functions of several complex variables.

The calculus of variations was another central interest in analysis, because the precise methods of Weierstrass made possible a careful study of this calculus and the more general variants (such as the problem of Bolza) of its standard problems. This emphasis died down about 1930, but the field was rapidly reinvigorated by new ideas; first the use of topology (Morse theory) and then the applications to optimal control.

Finite group theory is a striking example of a field developed by a new opportunity. This field has been quite inactive, and seemed subordinate to related fields such as algebraic groups. Then powerful new techniques (Hall–Higman, Thompson and Feit–Thompson) appeared. These techniques suddenly made it possible to imagine a determination of all finite simple groups. This became the specific program of the whole field for the period 1962–1982. There are many other examples of such programs of special research–not always so successful.

This is also an example of insight: With these new ideas, such and such should be possible. There are many smaller and more specific examples of such insight by individual Mathematicians. There are also major examples. Riemann's introduction of Riemann surfaces; Hamilton's use of canonical coordinates; Galois' recognition of the use of groups; the emphasis upon power series by Weierstrass. These are examples of the important role of ideas.

Mathematics develops from ideas in a network of interlocking formal systems. This accounts for the inevitable specializations of research limited to one particular node of the network–but it also indicates that such specialization is by itself not enough. One needs also awareness of the relevance of other related parts of the network. Understanding Mathematics goes beyond specializations.

12. Summary

Now we return to the six questions raised in the introduction, pp. 1–4.

Origin. There are many origins for Mathematics: In the practices which develop in various human activities, leading to procedures, to ideas, and then to formal rules; in the questions raised by the scientific study of phenomena old and new; finally, in the human capacity to form ideas, to extract information, and to make generalizations and abstractions. These sources continue to supply new material for Mathematical thought.

Organization. The longstanding concern with number, space, time, and motion leads to corresponding branches of Mathematics: arithmetic, geometry, calculus and mechanics, respectively. But these branches turn out to be intimately connected with each other: The same real numbers arise from completing the rationals and from measuring the line or the circle. Geometry interacts with calculus in differential geometry. The concepts of calculus are precisely those needed by and forshadowed in mechanics and its Hamiltonian development. The four branches are supplemented by others. The combinations of numbers or of equations or of vectors or of differential operators demand the formal algebraic rules of their manipulation. Hence arises the branch, algebra. With its emphasis on the formal it becomes clear that Mathematics is no longer the science of number, space, time, and motion–it is the study of formal concepts and systems suggested by number, space, time, and motion, and also suggested by other activities: proving, computing, composing, and decomposing. Then geometry is no longer just the analysis of the "usual" space, but becomes the use of spatial intuitions and ideas to analyse surfaces extracted from space as well as analytic objects which can be understood by the appropriate geometric forms. The analysis of change, functions, and summation initiated by the calculus has remarkable extensions to holomorphic functions and to distributions and measures which are not functions at all. In short, once Mathematics is recognized for what it is, the study of ideas expressed by form, it acquires a wide variety of branches and subfields, some characterized by the objects exhibited. Since the same form can have many exemplars, there can be many close connections between different subfields. This is what probably accounts for the remarkably ramified uses of group theory: The forms presented by the composition of substitutions occurs whenever there are operations to be composed.

It is for these reasons that our study (especially §4 in subjects, specialties, and subdivisions) did not provide any single neat table of organization, with explicit lines of command, control, and communication. Some branches of Mathematics are defined by axioms (ring theory), some by concepts (holomorphic function) and some by intended application (mechanics), and they all need to communicate about common formal concerns.

Formalization. How are the forms in Mathematics derived? Some are suggested by the facts, but they become forms only when extracted from the facts, next considered vaguely as ideas and then finally pinned down by meticulous definitions or axioms. Other Mathematical forms are abstracted from more elaborate pieces of Mathematics, in order the better to understand those pieces. It may be for this reason that simple forms, like the forms of group theory, were discovered late in the development of Mathematics: They required first exemplification by more concrete forms in algebra and geometry. Mathematical forms are both discovered (within the examples) and invented (developed by thought and ideas).

It is the formal nature of Mathematics which makes it interpersonal, objective, and exact.

Dynamics. Mathematics involves repeated new discoveries, but there is no one simple description of the forces driving these discoveries: Curiosity about scientific questions, famous problems, or the wish to generalize. What is common is perhaps the desire to understand: To understand what that question from science really involves, or to see why that old problem, though it appears simple, is so subtle, or to understand that related features apparent in several different situations have a common explanation. In this sense the desire to understand is the most important dynamic for the advance of Mathematics.

Foundations. Mathematics has access to absolute rigor—because it is about form and not about fact. However, there is no single and absolute foundation for Mathematics. Any such fixed foundation would preclude the novelty which might result from the discovery of new form. A form is any development which proceeds by rule rather than by appeal to fact as to meaning. Among the many contemporary discussions of the philosophy of Mathematics we cite a number of authors: Bernays [1935], Curry [1951], Dummett [1977], Gödel [1947], Goodman [1979], Kitcher [1983], Lehman [1979], Mac Lane [1981], Quine [1963], Resnik [1980], Robinson [1965], Steiner [1975], Weyl [1949], Wilder [1981], and Wittgenstein [1964].

However, none of the usual systematic foundations or philosophies, as we have listed them in the introduction, seem to us satisfactory. They may be summarized (too briefly) as follows:

Logicism. Mathematics needs logic to provide firm canons of proof, but logic does not tell us which proofs to canonize; Mathematicians must search for the most illuminating proof. What matters most are proofs which are "deep"—but not in a logistic sense of depth. The logical rules of inference provide a careful syntax for each proof, but they do not serve to indicate the steps that are crucial—these points where one understands why the proof works. For this reason, Mathematics is not just logical deduction. Moreover, Mathematics requires not just the rules of logic but also some non-logical axioms; in particular (some version of) the axiom of infinity. This is a (generally recognized) technical reason why Mathematics is not a branch of logic, as was originally asserted by Frege and Bertrand Russell.

Set theory is strong enough to provide a formulation of most of Mathematics but this provision is often artificial and does not yield a "coordinate-free" or invariant foundation for Mathematics. Moreover, there is no unique notion of a "set".

Platonism is a useful mythology and a speculative ontology.

Formalism. Strict formalism can't explain *which* of many formulas matter and hasn't yet absorbed the profound consequences of Gödel's incompleteness theorem. The formal aspect is essential to Mathematics, but the choice of form is determined by ideas and experience.

Intuitionism can be dogmatic, when it endeavors to restrict Mathematical intuitions to the whole numbers. This simply fails to account for the many other sources of Mathematical ideas. On the other hand, intuitionism leads to important Mathematical structures which are related, through topos theory, to geometry and sheaf theory. Thus intuitionism has revealed important Mathematical structure.

Constructivism can be narrowly dogmatic (only constructive existence proofs are allowed), but can be more liberally interpreted as an emphasis on the algorithmic aspects of Mathematics.

Finitism recognizes the finite character of Mathematical argument, but does not account for the extraordinary success of arguments with infinite sets and limiting processes.

Empiricism cannot succeed in exhibiting an empirical exemplar for each piece of mathematical form; as we have repeatedly seen, the same Mathematical form can have many different exemplars, and that form can be many steps removed from the original facts. Mathematics originates not just in facts, but also in human activities and in scientific conundrums.

The opposite of empiricism is the view that Mathematics is in part a search for austere forms of beauty, and that Mathematical developments can be splendid even though they have no conceivable practical use. Such views are often expressed by the adjective "elegant". For example, Euclid's proof that there are infinitely many primes or the proof of the Cauchy–Goursat theorem (§X.6) are said to be elegant because each gets a "deep" result with a minimum of complexity. Elegance can be viewed as a desirable property of a well-constructed Mathematical form.

Each of these philosophies illuminates a relevant aspect of Mathematics, but none of them is remotely adequate as a description or foundation of the actual extensive network of Mathematics. Instead, our study has revealed Mathematics as an array of forms, codifying ideas extracted from human activities and scientific problems and deployed in a network of formal rules, formal definitions, formal axiom systems, explicit theorems with their careful proof and the manifold interconnections of these forms. More briefly, Mathematics aims to understand, to manipulate, to develop, and to apply those aspects of the universe which are formal. This view, as expounded in this book, might be called *formal functionalism*.

Bibliography

Ahlfors, Lars [1966]. *Complex Analysis: An introduction to the theory of analytic functions of one complex variable*, 2nd ed. 317 pp. New York: McGraw-Hill Book Co.

Artin, Emil [1959]. *Galois theory*, edited by Arthur N. Milgram. 2nd ed. 82 pp. South Bend, Indiana: University of Notre Dame Press.

Bachmann, Friedrich [1973]. *Aufbau der Geometrie aus dem Spiegelungsbegriff*. 2nd enl. ed. 374 pp. Heidelberg-New York: Springer-Verlag.

Barwise, Jon [1977]. Handbook of Mathematical Logic. *Studies in Logic and the foundations of Mathematics* **90**, 1165 pp. Edited by Jon Barwise. Amsterdam: North Holland Publishing Co.

Bernays, Paul [1935]. Sur la platonisme dans les mathématiques. *L'enseignement mathématique* **34**, 52–69.

Birkhoff, Garrett [1967]. *Lattice theory*, 3rd ed. 418 pp. The American Mathematical Society Colloquium Publications **25**.

Birkhoff, Garrett and Saunders Mac Lane [1977]. *A survey of modern algebra*, 4th ed. 500 pp. New York: Macmillan.

Bishop, Errett [1967]. *Foundations of constructive analysis*. 370 pp. New York: McGraw-Hill Book Co.

Bott, Raoul and Tu W. Loring [1982]. *Differential forms in Algebraic Topology*. 331 pp. Heidelberg: Springer-Verlag.

Bourbaki, N. [1946]. *Éléments de Mathématique*. Premier Partie, les structures fondamental de l'analyse. Livre III: Topologie générale, chap. 2: structures uniform. Actualités Sci. et industrielles, #838. Paris: Hermann & Cie.

Caratheodory, Constantin [1965]. *Calculus of variations and partial differential equations of the first order*. Part I [1965], 171 pp. Part II [1967], pp. 175–398. San Francisco: Holden-Day, Inc.

Curry, Haskell B. [1951]. Outlines of a formalist philosophy of mathematics. *Studies in logic and the foundations of mathematics*. 75 pp. Amsterdam: North Holland Publishing Company.

Davis, Philip J. and Reuben Hersh [1981]. *The mathematical experience*. 440 pp. Boston: Birkhauser.

Dieudonné, Jean [1977]. *Panorama des Mathématiques pures, le choix boubachique*. 302 pp. Paris: Gauthiers-Villars.

_____[1978]. The difficult birth of mathematical structure 1840–1940. *Scientific culture in the contemporary world*. pp. 7–23. Milan: Scientia.

Dodson, C. T. J. and Tim Poston [1978]. *Tensor geometry*. The geometric viewpoint and its uses. 598 pp. London and San Francisco: Putnam.

Dummett, Michael [1977]. *Elements of intuitionism.* Written with the assistance of Robert Minio. 467 pp. Oxford, England: Clarendon Press.

Eklof, Paul C. [1976]. Whitehead's problem is undecidable. *Amer. Math. Monthly* **83** 775–788.

Feynman, Richard, Robert Leighton, and Matthew Sands [1964]. *The Feynman lectures on physics.* 3 vols. 513 pp., 569 pp., and 365 pp. Reading, Mass.: Addison Wesley.

Frank, Philip and Richard V. Mises [1930]. *Die Differentialgleichungen und Integralgleichungen der Mechanik und Physik*, 2nd ed. Erster, mathematischer Teil. 916 pp. Braunschweig: F. Vieweg & Sohn.

Gärding, Lars [1977]. *Encounter with mathematics.* 270 pp. New York: Springer-Verlag.

Gleason, Andrew M. [1966]. *Fundamentals of abstract analysis.* 404 pp. Reading, Mass.: Addison-Wesley Pub. Co.

Gödel, Durt [1940]. The consistency of the continuum hypothesis,. *Annals of Math. Studies* No. 3. 66 pp. Princeton University Press.

——— [1947]. What is Cantor's continuum problem. *Amer. Math. Monthly* **54**, 515–525.

Goodman, Nicholas D. [1979]. Mathematics as an objective science. *Am. Math. Monthly* **86**, 540–551.

——— [1981]. The experiential foundations of mathematical knowledge. *Hist. Philos. Logic* **2**, 55–65.

Guggenheim, Heinrich W. [1967]. *Plane geometry and its groups.* 288 pp. San Francisco: Holden-Day, Inc.

Hadamard, Jacques [1954]. *An essay on the psychology of invention in the mathematical field.* 145 pp. New York: Dover Publications, Inc.

Halmos, Paul R. [1958]. *Finite dimensional vector spaces*, 2nd ed. D. van Nostrand; reprinted 1974. 200 pp. New York: Springer-Verlag. [1st ed., Princeton University Press, 1942].

Hamilton, Alan G. [1978]. *Logic for mathematics.* 224 pp. Cambridge-New York: Cambridge University Press.

Hardy, G. H. and E. M. Wright [1954]. *An introduction to the theory of numbers.* 3rd ed. 419 pp. Oxford, England: Clarendon Press.

Hausdorff, Felix [1914]. *Grundzúge du Magenlehre.* 476 pp. Leipzig: W. A. Gruyter & Co.; New York: Reprinted Chelsea Pub. Co. 1945.

Hilbert, David [1899; 1971]. *Foundations of Geometry*, 2nd ed. Translated from the tenth German edition by Leo Unger. 226 pp. LaSalle, Illinois: Open Court.

Hungerford, Thomas W. [1974]. *Algebra.* 502 pp. New York: Holt, Rinehart and Winston.

Hurwitz, A. and R. Courant [1964]. *Functionentheorie*, 4th ed. 706 pp. Heidelberg: Springer-Verlag.

Johnstone, Peter [1978]. *Topos theory.* 367 pp. New York: Academic Press.

——— [1982]. Stone spaces. *Cambridge studies in advanced mathematics*, Vol. 3. 370 pp. Cambridge, England: Cambridge University Press.

Keisler, H. J. [1976]. *Elementary calculus.* An approach using infinitesimals. 880 pp. Boston: Prindle, Weber & Schmidt.

Kleene, S. C. [1952]. *Introduction to metamathematics.* New York: Van Nostrand.

Kitcher, Philip [1983]. *The nature of mathematical knowledge.* 287 pp. New York and Oxford: Oxford University Press.

Klein, Felix and Arnold Sommerfeld [1965]. Uber die Theorie des Kreisels. *Bibliotheca Math Taubneriana*, Vol. 1. 966 pp. (Original published in four volumes 1897, 1898, 1903, 1910.) New York: Johnson Repr. Corp.

Kock, Anders [1981]. *Synthetic differential geometry*. London Math. Soc. Lecture Notes Series 51. 311 pp. Cambridge and New York: Cambridge University Press.

Landau, Edmund [1951]. Foundations of analysis. *The arithmetic of whole, rational, irrational, and complex numbers*. Translated by F. Steinhardt. 134 pp. New York: Chelsea Pub. Co.

Lang, Serge [1967]. *Introduction to differentiable manifolds*. 125 pp. New York: Interscience (John Wiley & Sons].

Lehman, Hugh [1979]. *Introduction to the philosophy of mathematics*. 169 pp. Totowa, New Jersey: Rownan and Littlefield.

Lightstone, A. H. and Abraham Robinson [1975]. *Non-archimedean fields and asymptotic expansions*. 204 pp. North Holland Mathematical Library. Vol. 13. Amsterdam-Oxford: North Holland Publishing Company.

Mackey, George W. [1978]. *Unitary group representation in physics, probability and number theory*. 402 pp. Math lecture notes series #55, Reading, Mass.

Mac Lane, Saunders [1963]. Homology. *Die Grundlehren der Math. Wissenschaften*, Vol. 114. 422 pp. Heidelberg: Springer-Verlag.

—— [1971]. *Categories for the working mathematician*, Graduate texts in mathematics, Vol. 5. 262 pp. Heidelberg: Springer-Verlag.

—— [1981]. Mathematical models: a sketch for the philosophy of mathematics. *Am. Math. Monthly* **88**: 462–472.

Mac Lane, Saunders and Garrett Birkhoff [1979]. *Algebra*, 2nd ed. 586 pp. (1st ed. 1967) New York: Macmillan Publishing Co.

Massey, William S. [1967]. *Algebraic topology: An introduction*. 261 pp. New York: Springer-Verlag.

Monna, A. F. [1975]. *Dirichlet's principle*. A mathematical comedy of errors and its influence on the development of analysis. 138 pp. Utrecht, The Netherlands: Oosthoek Scheltema & Holkema.

Myhill, John [1972]. What is a real number? *Am. Math. Monthly* **79**: 748–754.

Narasimhan, R. [1985]. *Complex analysis in one variable*. 216 pp. Boston: Burkhauser.

O'Neill, Barrett [1966]. *Elementary differential geometry*. 411 pp. New York and London: Academic Press.

Osgood, William Fogg [1937]. *Mechanics*. 495 pp. New York: The Macmillan Co.

Paige, Leigh [1928]. *Introduction to theoretical physics*. 587 pp. New York: D. van Nostrand.

Pars, L. A. [1965]. *A treatise on analytical dynamics*. 641 pp. New York: John Wiley & Sons.

Pontryagin, L. S. [1939]. *Topological groups*. Translated from the Russian by Emma Lehmer. Princeton math series vol. 2. 299 pp. Princeton, New Jersey: Princeton University Press.

Quine, W. V. O. [1963]. From a logical point of view. *Logico-philosophical essays*, 2nd ed., rev. 184 pp. New York: Harper and Row.

Resnik, Michael D. [1980]. *Frege and the philosophy of mathematics*. 243 pp. Ithaca, New York: Cornell University Press.

Robinson, Abraham [1965]. Formalism 64, pp. 228–246. In *Proc. Internat. Congress for Logic, Methodology, and Philosophy*. Jerusalem 1964. Amsterdam: North Holland Pub. Co.

Russell, Bertrand A. W. [1908]. Mathematical logic as based on the theory of types. *Amer. J. Math.* **30**: 222–262.

Shoenfield, J. R. [1975]. Martin's axiom. *Am. Math. Monthly* **82**: 610–619.

Sondheimer, Ernst and Alan Rogerson [1981]. *Numbers and infinity*. A historical account of mathematical concepts. 172 pp. London and New York: Cambridge University Press.

Spivak, Michael [1965]. *Calculus on manifolds*. 144 pp. New York and Amsterdam: W. A. Benjamin Inc.

Steiner, Mark [1975]. *Mathematical knowledge*. 164 pp. Ithaca, New York: Cornell University Press.

Titchmarsh, E. C. [1932]. *The theory of functions*. 454 pp. Oxford, England: The Clarendon Press.

Troelstra, A. S. [1972]. *Choice sequences*. A chapter of intuitionist mathematics. Oxford logic guides. 170 pp. Oxford Clarendon Press.

Weyl, Hermann [1949]. *Philosophy of mathematics and natural science*. 311 pp. Rev. English ed. Trans. by O. Helmer. Princeton, New Jersey: Princeton University Press.

——— [1923]. Raum, Zeit, Materie. *Vorlesungen uber allgemeine Relativitatstheorie*, 5th ed. 338 pp. Berlin: Springer-Verlag.

Whitehead, A. N. and Bertrand Russell [1910]. *Principla mathematica*. Vol. 1. 666 pp. 2nd ed. 1925. 674 pp. Cambridge, England: Cambridge University Press, 1925.

Wilder, Raymond L. [1981]. *Mathematics as a cultural system*. 182 pp. Oxford-New York: Pergamon Press.

Wilson, Edwin B. [1912]. *Advanced calculus*. 566 pp. Boston: Ginn & Co.

Wittgenstein, Ludwig [1964]. *Remarks on the foundation of mathematics*, 2nd ed. Edited by G. H. von Wright, R. Rhees, and G. E. M. Anscombe. Oxford, England: Basil Blackwell.

Zermelo, Ernst. 1908]. Untersuchungen über die Grundlagen der Mengenlehre I. *Mathematische Annalen* **85**: 261–281.

List of Symbols

Index